PRINCIPLES OF PHYSICS SERIES

MECHANICS, HEAT, AND SOUND

by

FRANCIS WESTON SEARS

Department of Physics
Massachusetts Institute of Technology

ADDISON-WESLEY PUBLISHING COMPANY, INC.

READING, MASSACHUSETTS, U.S.A.

SECOND EDITION

Seventh Printing — June 1958

PREFACE

This book is the first volume of a series of texts written for a two-year course in general physics. It is assumed that students using the book have completed a course in calculus or that they are studying calculus concurrently.

The title of the series, Principles of Physics, has been chosen deliberately to indicate that its emphasis is on physical principles. Historical background and practical applications have been given a place of secondary importance.

This volume opens with several chapters on statics in order that kinematics may be postponed until the student has acquired some familiarity with the concepts and notation of calculus. Beginning with Chapter 4, simple differentiation and integration are introduced to supplement and extend the algebraic development of the equations of linear motion with constant acceleration. From that point on, the calculus is used freely wherever its inclusion is warranted.

Three systems of units are used: the English gravitational because it is the one used in engineering work throughout this country, the cgs system because some familiarity with it is essential for any intelligent reading of the literature of physics, and the mks system because of its increasing use in electricity and magnetism as well as because it seems destined eventually to supplant the cgs system.

The author is particularly indebted to Professor M. Stanley Livingston for many stimulating and informative discussions and for his encouragement in the task of developing a set of lecture notes into this book.

The multiflash photographs were taken with the advice and assistance of Professor Harold E. Edgerton, to whom the author is duly grateful. Collective acknowledgment is made to numerous contributors to the *American Journal of Physics* (formerly the *American Physics Teacher*) since its inception.

F.W.S.

Cambridge, Mass.
March 1944.

CONTENTS

MECHANICS

HEAT

MECHANICS

CHAPTER 1

COMPOSITION AND RESOLUTION OF VECTORS

1–1 Force. Mechanics is the branch of physics and engineering which deals with the interrelations of force, matter, and motion. We shall begin with a study of forces. The term force, as used in mechanics, refers to what is known in everyday language as a *push* or a *pull*. We can exert a force on a body by muscular effort; a stretched spring exerts forces on the bodies to which its ends are attached; compressed air exerts a force on the walls of its container; a locomotive exerts a force on the train which it is drawing. In all of these instances the body exerting the force is in contact with the body on which the force is exerted, and forces of this sort are known as *contact* forces. There are also forces which act through empty space without contact, and are called *action-at-a-distance* forces. The force of gravitational attraction exerted on a body by the earth, and known as the *weight* of the body, is the most important of these for our present study. Electrical and magnetic forces are also action-at-a-distance forces, but we shall not be concerned with them for the present.

All forces fall into one or the other of these two classes, a fact that will be found useful later when deciding just what forces are acting on a given body. It is only necessary to observe what bodies are in contact with the one under consideration. The only forces on the body are then those exerted by the bodies in contact with it, together with the gravitational force or the weight of the body.

Those forces acting on a given body which are exerted by other bodies are referred to as *external* forces. Forces exerted on one part of a body by other parts of the same body are called *internal* forces.

1–2 Units and standards. The early Greek philosophers confined their activities largely to speculations about Nature, and to attempts to reconcile the observed behavior of bodies with theological doctrines. What has been called the *scientific method* began to appear in the time of Galileo Galilei (1564–1642). Galileo's studies of the laws of freely falling bodies were made not in an attempt to explain *why* bodies fell toward the earth, but rather to determine *how far* they fell in a given time, and *how fast* they moved. Physics as it exists today has been called the science of measurement, and the importance of quantitative knowledge and reasoning has

1

been expressed by Lord Kelvin (1824–1907) as follows: "I often say that when you can measure what you are speaking about, and express it in numbers, you know something about it; but when you cannot express it in numbers, your knowledge is of a meagre and unsatisfactory kind; it may be the beginning of knowledge, but you have scarcely, in your thoughts, advanced to the stage of *Science*, whatever the matter may be."

The first step in the measurement of a physical quantity consists in choosing a *unit* of that quantity. As the result of international collaboration over a long period, practically all of the units used in physics are now the same throughout the world. The second step is an experiment that determines the ratio of the magnitude of the quantity to the magnitude of the unit. Thus, when we say that the length of a rod is 10 centimeters, we state that its length is ten times as great as the unit of length, the centimeter.

It is possible to simplify many of the equations of physics by the proper choice of units of physical quantities. Any set of units which is chosen so that these simplified equations can be used is called a *system* of units. We shall use three such systems in this book. They are, first, the *English gravitational* system; second, the meter-kilogram-second or *mks* system; and third, the centimeter-gram-second or *cgs* system. The units of these systems will be defined as the need for them arises.

Most of the fundamental units of physics are embodied in a physical object called a *standard*. One of the functions of the National Bureau of Standards in Washington, D. C. is to maintain in its vaults standards of various quantities with which commercial and technical measuring instruments can be compared for accuracy.

1–3 The pound. The unit of force which we shall use for the present is the English gravitational unit, the *pound*. Other units will be discussed in Chapter 5. This unit is embodied in a cylinder of platinum-iridium called the *standard pound*. The unit of force is defined as the weight of the standard pound. That is, it is a force equal to the force of gravitational attraction which the earth exerts on the standard pound. Since the earth's gravitational attraction for a given body varies slightly from one point to another on the earth's surface it is further stipulated that the unit force shall equal the weight of the standard pound *at sea level and 45° latitude.**

In order that an unknown force can be compared with the force unit (and thereby measured) some measurable effect produced by a force must be used. One common effect of a force is to alter the dimensions or shape of a body on which the force is exerted; another is to alter the state of motion of the body. Both of these effects are used in the measurement of

---

*See Section 15–3 for a more precise definition.

forces. In this chapter we shall consider only the former; the latter will be discussed in Chapter 5.

The instrument used to measure forces is the spring balance, which consists of a coil spring enclosed in a case for protection and carrying at one end a pointer that moves over a scale. A force exerted on the balance increases the length of the spring. The balance can be calibrated as follows. The standard pound is first suspended from the balance and the position of the pointer marked 1 lb. Any number of duplicates of the standard can then be prepared by suspending each of them in turn from the balance and removing or adding material until the index stands at 1 lb. Then, when two, three, or more of these are suspended simultaneously from the balance, the force stretching it is 2 lb, 3 lb, etc., and the corresponding positions of the pointer can be labeled 2 lb, 3 lb, etc. This procedure makes no assumptions about the elastic properties of the spring, except that the force exerted on it is always the same when its index stands at the same point. The calibrated balance can then be used to measure any unknown force.

1–4 Graphical representation of forces. Vectors. Suppose we are to slide a box along the floor by pulling it with a string or pushing it with a stick, as in Fig. 1–1. That is, we are to slide it by exerting a force on it. The point of view which we now adopt is that the motion of the box is caused not by the *objects* which push or pull on it, but by the *forces* which these exert. For concreteness assume the magnitude of the push or pull to be 10 lb. It is clear that simply to write "10 lb" on the diagram would not completely describe the force, since it would not indicate

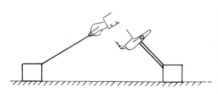

Fig. 1–1. The box is pulled by the string or pushed by the stick.

the direction in which the force was acting. One might write "10 lb, 30° above horizontal to the right," or "10 lb, 45° below horizontal to the right," but all of the above information may be conveyed more briefly if we adopt the convention of representing a force by an arrow. The length of the arrow, to some chosen scale, indicates the size or magnitude of the force, and the direction in which the arrow points indicates the direction of the force. Thus Fig. 1–2 (in which a scale of $\frac{1}{8}$ in = 1 lb has been chosen) is the force diagram corresponding to Fig. 1–1. (There are other forces acting on the box, but these are not shown in the figure.)

Force is not the only physical quantity which requires the specification of direction as well as magnitude. For example, the velocity of a plane is not completely specified by stating that it is 300 miles per hour; we need

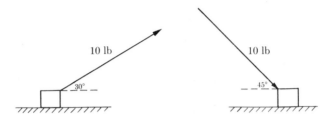

Fig. 1–2. The force diagram corresponding to Fig. 1–1.

to know the direction also. The concept of density, on the other hand, has no direction associated with it.

Quantities like force and velocity, which involve both magnitude and direction, are called *vector* quantities. Those like density, which involve magnitude only, are called *scalars*. Any vector quantity can be represented by an arrow, and this arrow is called a vector (or if a more specific statement is needed, a force vector or a velocity vector). We shall first consider force vectors only, but the ideas developed in dealing with them can be applied to any other vector quantity.

1–5 Components of a vector. When a box is pulled or pushed along the floor by an inclined force as in Fig. 1–1, it is clear that the effectiveness of the force in moving the box along the floor depends upon the direction in which the force acts. Everyone knows by experience that a given force is more effective for moving the box the more nearly the direction of the force approaches the horizontal. It is also clear that if the force is applied at an angle, as in Fig. 1–1, it is producing another effect in addition to moving the box ahead. That is, the pull of the string is in part tending to lift the box off the floor, and the push of the stick is in part forcing the box down against the floor. We are thus led to the idea of the *components* of a force, that is, the effective values of a force in directions other than that of the force itself.

The component of a force in any direction can be found by a simple graphical method. Suppose we wish to know how much force is available for sliding the box in Fig. 1–1 if the applied force is a pull of 10 lb directed 30° above the horizontal. Let the given force be represented by the vector OA in Fig. 1–3, in the proper direction and to some convenient scale. Line OX is the direction of the desired component. From point A drop a perpendicular to OX, intersecting it at B. The vector OB, to the same scale as that used for the given vector, represents the component of OA in the direction OX. Measurements of the diagram show that if OA represents a force of 10 lb, then OB is about 8.7 lb. That is, the 10-lb force at an angle of 30° above the horizontal has an effective value of only about 8.7 lb in producing forward motion.

The component OB may also be computed as follows. Since OAB is a right triangle, it follows that

$$\cos 30° = \frac{OB}{OA},$$

$$OB = OA \cos 30°.$$

The lengths OB and OA, however, are proportional to the magnitudes of the forces they represent. Therefore the desired component OB, in pounds, equals the given force OA, in pounds, multiplied by the cosine of the angle between OA and OB. The magnitude of OB is therefore

$$\begin{aligned}
OB \text{ (lb)} &= OA \text{ (lb)} \times \cos 30° \\
&= 10 \text{ lb} \quad \times 0.866 \\
&= 8.66 \text{ lb.}
\end{aligned}$$

This result agrees as well as could be expected with that obtained from measurements of the diagram. The superiority of the trigonometric method is evident, however, since it does not depend for accuracy on the careful construction and measurement of a scale diagram.

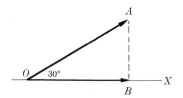

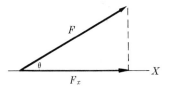

FIG. 1–3. Vector OB is the component of vector OA in the direction OX.

FIG. 1–4. $F_x = F \cos \theta$ is the X-component of F.

The line OX in Fig. 1–3 is called the X-axis, and the foregoing analysis may be generalized as follows. If a force F makes an angle θ with the X-axis (Fig. 1–4), its component F_x along the X-axis is

$$F_x = F \cos \theta. \tag{1-1}$$

It should be obvious that if the force F is at right angles to the X-axis, its component along that axis is zero (since $\cos 90° = 0$), and if the force lies along the axis, its component is equal to the force itself (since $\cos 0° = 1$).

The lifting component of an inclined force can be found as in Fig. 1–5. Line OY, called the Y-axis, is constructed in a vertical direction through O and a perpendicular dropped to this axis from the head of the arrow F. Evidently

$$F_y = F \cos \phi, \qquad (1\text{--}2)$$

where ϕ is the angle between F and the Y-axis.

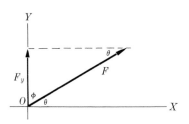

FIG. 1–5. $F_y = F \cos \phi = F \sin \theta$ is the Y-component of F.

FIG. 1–6. The force F may be replaced by its rectangular components F_x and F_y.

It is also evident from Fig. 1–5 that

$$F_y = F \sin \theta. \qquad (1\text{--}3)$$

If $F = 10$ lb and $\theta = 30°$, then $\phi = 60°$ and $\cos \phi = \sin \theta = 0.50$. Hence $F_y = 5$ lb.

Just as we may find the component of a given force in any direction, so may we find the component of any of its components, and so on. It will be seen from Fig. 1–6, however, that F_x has no component along the Y-axis and F_y has no component along the X-axis. No further resolution of the force into X- and Y-components is therefore possible. Physically this means that the two forces F_x and F_y, acting simultaneously, are equivalent in all respects to the original force F. Since the axes OX and OY are at right angles to one another, F_x and F_y are called the *rectangular components* of the force F. *Any force may be replaced by its rectangular components.* The fact that the force F has been replaced by its components F_x and F_y is indicated in Fig. 1–6 by crossing out lightly the vector F.

The process of finding the components of a vector is called the *resolution* of the vector, and one speaks of *resolving* a given vector into its rectangular components.

An experiment to show that a force may be replaced by its rectangular components is illustrated in Fig. 1–7. A small ring, to which are attached three cords, is placed on a pin set in a vertical board. Two of the cords pass over pulleys as shown. When weights of 8.66, 5, and 10 lb are sus-

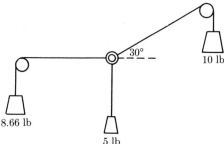

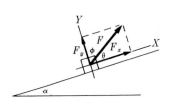

FIGURE 1–7

FIG. 1–8. F_x and F_y are the components of F, parallel and perpendicular to the surface of the plane.

pended from the cords, with the cord carrying the 10-lb weight making an angle of 30° with the horizontal, it will be found that the pin can be removed and that the ring will remain at rest under the combined action of the pulls in the three cords. This shows that the 10-lb force, at an angle of 30° above the horizontal, is equivalent to a horizontal force of 8.66 lb to the right and a vertical force of 5 lb upward, since the ring can be held at rest by the application of two forces equal to these but oppositely directed.

It is frequently necessary to find the components of a force in other than horizontal and vertical directions. Thus in Fig. 1–8, where a block is being drawn up an inclined plane by the force F, it is desired to find the components of this force parallel and perpendicular to the surface of the plane. The X- and Y-axes are now drawn parallel and perpendicular to this surface, and the same procedure followed as before.

1–6 Composition of forces. When a number of forces are simultaneously applied at a point, it is found that the same effect can always be produced by a single force having the proper magnitude and direction. We wish to find this force, called the *resultant*, when the separate forces are known. The process is known as the *composition* of forces, and is evidently the converse problem to that of resolving a given force into components. Let us begin by considering some simple cases.

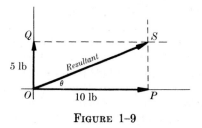

FIGURE 1–9

(1) Two forces at right angles. Suppose the two forces of 10 lb and 5 lb are applied simultaneously at the point O as in Fig. 1–9. To find the resultant force graphically, lay off the given forces OP and OQ to scale, and draw horizontal and vertical construction lines from P and Q,

intersecting at S. The arrow drawn from O to S represents the resultant of the given forces. Its length, to the same scale as that used for the original forces, gives the magnitude of the resultant, and the angle θ gives its direction.

Since the length PS or OQ represents 5 lb, and the length OP represents 10 lb, the magnitude of the resultant may be computed from the right triangle OPS. Thus

$$OS = \sqrt{\overline{OP}^2 + \overline{PS}^2} = \sqrt{10^2 + 5^2} = 11.2 \text{ lb.}$$

The angle θ may also be computed from either its sine, cosine, or tangent. Thus

$$\sin \theta = \frac{5}{11.2} = 0.447,$$

$$\cos \theta = \frac{10}{11.2} = 0.893,$$

$$\tan \theta = \frac{15}{10} = 0.500.$$

Using any one of these values we find from tables of natural functions

$$\theta = 26.5°.$$

We conclude, then, that a single force of 11.2 lb, at an angle of 26.5° above the horizontal, will produce the same effect as the two forces of 10 lb horizontally and 5 lb vertically. Notice that the resultant is not the arithmetic sum of 5 lb and 10 lb. That is, the two forces are not equivalent to a single force of 15 lb.

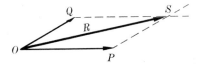

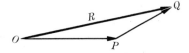

FIG. 1–10. Parallelogram method for finding the resultant of two vectors.

FIG. 1–11. Triangle method for finding the resultant of two vectors.

(2) Two forces not at right angles. (a) Parallelogram method. Let OP and OQ in Fig. 1–10 represent the forces whose resultant is desired. Draw construction lines from P parallel to OQ, and from Q parallel to OP, intersecting at S. The arrow OS represents the resultant R in magnitude and direction. Since $OPSQ$ is a parallelogram, this method is called the parallelogram method. The magnitude and direction of the resultant may

be found by measurement or may be computed from the triangle OPS with the help of the sine and cosine laws.

NOTE. The diagonal QP is *not* the resultant of the given forces.

(b) Triangle method. Draw one force vector with its tail at the head of the other as in Fig. 1–11 (the construction may be started with either vector), and complete the triangle. The closing side of the triangle, OQ, represents the resultant. A comparison of Figs. 1–11 and 1–10 will show that the same resultant is obtained by either method.

FIG. 1–12. Vector R is the resultant of vectors P and Q.

(3) Special case. Both forces in the same line. When both forces lie in the same straight line the triangle of Fig. 1–11 flattens out into a line also. To be able to see all of the force vectors, it is customary to displace them slightly as in Fig. 1–12. We then have Fig. 1–12(a) or 1–12(b), depending upon whether the two forces are in the same or opposite directions. Only in this case is the magnitude of the resultant equal to the sum (or difference) of the magnitudes of the components.

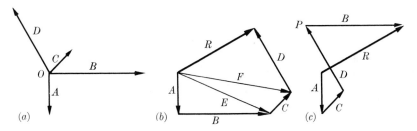

FIG. 1–13. Polygon method.

(4) More than two forces. Polygon method. When more than two forces are to be combined, one may first find the resultant of any two, then combine this resultant with a third, and so on. The process is illustrated in Fig. 1–13, which shows the four forces A, B, C, and D acting simultaneously at the point O. In Fig. 1–13(b), forces A and B are first combined by the triangle method giving a resultant E; force E is then combined by the same process with C giving a resultant F; finally F and D are combined to obtain the resultant R. Evidently the vectors E and F need not have been drawn—one need only draw the given vectors in succession with the tail of each at the head of the one preceding it, and complete the polygon by a vector R from the tail of the first to the head

of the last vector. The order in which the vectors are drawn makes no difference, as shown in Fig. 1–13(c).

It has been assumed in the preceding discussion that all of the forces lie in the same plane. Such forces are called *coplanar*, and, except in a few instances, we shall consider only situations involving coplanar forces.

1–7 Composition of forces by rectangular resolution. While the polygon method is a satisfactory graphical one for finding the resultant of a number of forces, it is awkward for computation because one must work, in general, with oblique triangles. Therefore the usual method for finding the resultant of a number of forces is first to resolve all of the forces into their rectangular components along any convenient pair of axes; second, to find the algebraic sum of all of the X- and all of the Y-components; and third, combine these sums to obtain the final resultant. This process makes it possible to work with right triangles only, and is called the *method of rectangular resolution*. As an example, let us compute the resultant of the four forces in Fig. 1–14, which are the same as those in Fig. 1–13.

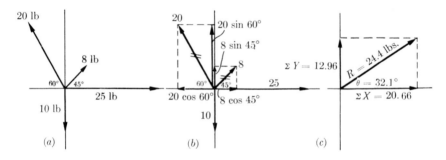

FIGURE 1–14

The forces are shown in Fig. 1–14(b) resolved into X- and Y-components. The 25-lb and the 10-lb forces are already along the axes and need not be resolved. It is customary to consider X-components which are directed toward the right as positive and those toward the left, negative. Similarly Y-components in an upward direction are considered positive and those downward, negative. This convention of signs is not always adhered to, however. In general one chooses positive and negative directions so as to avoid minus signs if possible.

The X-component of the 8-lb force is $+8 \cos 45° = +5.66$ lb, and its Y-component is $+8 \sin 45° = +5.66$ lb. The X-component of the 20-lb force is $-20 \cos 60° = -10$ lb, its Y-component is $+20 \sin 60° = +17.3$ lb. The algebraic sum of the X-components is a force of $25 + 5.66 - 10 = +20.66$ lb toward the right. The algebraic sum of the Y-components is a force of $17.3 + 5.66 - 10 = +12.96$ lb upward. The resultant is equal to the square root of the sums of the squares of the resultant X- and Y-components (Fig. 1–14c). The angle which it makes with the X-axis can be found from its tangent. Thus

$$R = \sqrt{20.66^2 + 12.96^2} = 24.4 \text{ lb,}$$

$$\tan \theta = \frac{12.96}{20.66} = 0.627,$$

$$\theta = 32.1°.$$

While three separate diagrams are shown in Fig. 1–14 for clarity, in practice one would carry out the entire construction in a single diagram.

The mathematical symbol for the algebraic sum of the X- or Y-components is ΣX or ΣY. (Σ is the Greek letter sigma or S, meaning "the sum of.") Hence one can write in general

$$R = \sqrt{(\Sigma X)^2 + (\Sigma Y)^2},$$

$$\tan \theta = \frac{\Sigma Y}{\Sigma X} \cdot$$

1–8 Resultant of a set of nonconcurrent forces. Figure 1–15 represents a rod upon which are exerted the three forces F_1, F_2, and F_3. These forces are not all applied at the same point, and even if their lines of action are extended as shown by the dotted lines, these do not intersect at a common point. Nevertheless the three forces have a resultant in the sense that it is possible to find a single force which will produce the same effect as is produced by the simultaneous action of the given forces. This resultant may be found graphically as follows.

Start with any two of the given forces, say F_1 and F_2, and extend their lines of action until they intersect (point x). Transfer F_1 and F_2 to point x, and find their resultant R_1 by any convenient method. The parallelogram method is used in the figure. Next, extend the lines of action of R_1 and F_3 until they intersect (point y) and combine R_1 and F_3 to find the resultant R. Finally, extend the line of action of R until it intersects the rod at point z. Then a single force having the magnitude and direction of R, and applied at point z of the rod, will produce the same effect as the given forces.

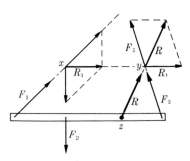

FIG. 1–15. Vector R is the resultant of F_1, F_2, and F_3.

1–9 Vector difference. The resultant of two vectors is also called their *vector sum*, and the process of finding the resultant is called *vector addition*. In many instances, as when computing accelerations or relative velocities, it is necessary to subtract one vector from another or to find their *vector difference*. This is done as follows. If A and B are the vectors, the vector difference $A - B$ can be written $A + (-B)$, that is, it is the vector *sum* of the vectors A and $-B$. The negative of a given vector has the same length as the given vector but points in the opposite direction.*

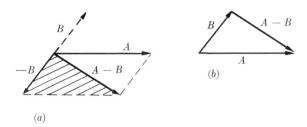

(a)   (b)

Fig. 1–16. Illustrates two methods of finding the vector difference $A - B$.

The method of finding a vector difference is illustrated in Fig. 1–16. A and B are the given vectors. The vector B is shown in Fig. 1–16(a) by a dashed line, the vector $-B$ by a full line. The vector sum of A and $-B$, or the vector difference $A - B$, is found by the parallelogram method.

The triangle method may also be used to find vector differences. Vectors A and B are laid off from a common origin as in Fig. 1–16(b). The vector from the end point of A to the end point of B then represents the vector difference $A - B$. That this is so may be seen by comparison of Fig. 1–16(b) with the shaded triangle in Fig. 1–16(a), or by noting that in Fig. 1–16(b) the vector A may be considered the vector sum of B and $(A - B)$, that is,

$$A = B + (A - B) \qquad \text{(vector sum)}.$$

Vector differences may also be found by the method of rectangular resolution. Each vector is resolved into its X- and Y-components. The difference between the X-components is the X-component of the desired vector difference; the difference between the Y-components is the Y-component of the vector difference.

---

*The "direction" of a vector is sometimes considered to mean the direction of a geometrical line of indefinite length along which the vector lies. Vectors B and $-B$ then have the same direction and are distinguished by saying that they are opposite in *sense*.

1-1. What is the basis for saying that force is a vector quantity?

1-2. A box is pushed along the floor as in Fig. 1-1 by a force of 40 lb making an angle of 30° with the horizontal. Using a scale of 1 in. = 10 lb, find the horizontal and vertical components of the force by the graphical method. Check your results by calculating the components.

1-3. (a) How large a force F must be exerted on a block as in Fig. 1-8 in order that the component parallel to the plane shall be 16 lb? (b) How large will be the component Fy? Let $\alpha = 20°$, $\theta = 30°$, $\phi = 60°$. Solve graphically, letting 1 in. = 8 lb.

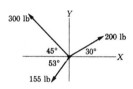

FIGURE 1-17

1-4. The three forces shown in Fig. 1-17 act on a body located at the origin. (a) Find the X- and Y-components of each of the three forces. Use the graphical method and any convenient scale. (b) Use the method of rectangular resolution to find the resultant of the forces. (c) Find the magnitude and direction of a fourth force which must be added to make the resultant force zero. Indicate the fourth force by a diagram.

FIGURE 1-18

1-5. Two men and a boy want to push a crate in the direction marked "X" in Fig. 1-18. The two men push with forces F_1 and F_2, whose magnitudes and directions are indicated in the figure. Find the magnitude and direction of the smallest force which the boy should exert.

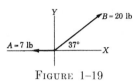

FIGURE 1-19

1-6. (a) Find graphically the vector sum $A + B$ and the vector difference $A - B$ in Fig. 1-19. (b) Use the method of rectangular resolution to find the magnitude and direction of the resultant of vectors A and B.

1-7. Two forces, F_1 and F_2, act at a point. The magnitude of F_1 is 8 lb and its direction is 60° above the X-axis in the first quadrant. The magnitude of F_2 is 5 lb and its direction is 53° below the X-axis in the fourth quadrant. (a) What are the horizontal and vertical components of the resultant force? (b) What is the magnitude of the resultant? (c) Show in a diagram a graphical method for finding the magnitude and direction of the resultant. (d) What is the magnitude of the vector difference $F_1 - F_2$?

1-8. Two forces, F_1 and F_2, act upon a body in such a manner that the resultant force, R, has a magnitude equal to F_1 and makes an angle of 90° with F_1. Let $F_1 = R = 10$ lb. Find the magnitude and direction (relative to F_1) of the second force, F_2.

1-9. Find the resultant of the following set of forces by the method of rectangular resolution: 80 lb, vertically down; 100 lb, 53° above horizontal to the right; 60 lb, horizontal to the left. Check by the polygon method.

CHAPTER 2

STATICS

2-1 Introduction. The science of mechanics is based on three natural laws which were clearly recognized for the first time by Sir Isaac Newton (1643–1727) and were published in 1686 in his *Philosophiae Naturalis Principia Mathematica* ("The Mathematical Principles of Natural Science"). It should not be inferred, however, that the science of mechanics began with Newton. Many men had preceded him in this field, perhaps the most outstanding being Galileo, who in his studies of accelerated motion had laid much of the groundwork for Newton's formulation of his three laws.

In this chapter we shall make use of only two of Newton's laws, the first and the third. Newton's second law will be discussed in Chapter 5.

2-2 Newton's first law. Newton's first law states that *when a body is at rest or moving with constant speed in a straight line, the* **resultant** *of all of the forces exerted* **on** *the body is zero.* The various girders, beams, columns, etc., which form the structure of a building or bridge, are bodies at rest. The forces exerted on them are their own weights, those exerted by other parts of the structure, and whatever loads the structure must carry. Since the resultant of all of the forces must be zero, if some of the forces are known the others may be computed. By successive application of Newton's first law to the various elements of a structure the engineer can compute how much force each part must withstand and therefore how strong each girder, beam, or column must be.

In most instances the forces on a structural element are so distributed that the turning effect, or the *moment* of each force, must also be taken into account. We shall postpone this complication until the next chapter, and consider for the present only structures in which all of the forces intersect at a common point. We shall also limit the discussion to coplanar forces.

Notice particularly that three words are emphasized in the preceding statement of Newton's first law—"the *resultant* of *all* of the forces exerted *on* the body." Most of the difficulties encountered in the application of this law to specific problems are due to a failure to use the *resultant* force, or to include *all* of the forces, or to use the forces exerted *on* the body. Furthermore, since Newton's second law also involves the resultant of all of the forces exerted on a body, it is extremely important that one should learn as early as possible how to recognize just what forces are exerted on a particular body.

When the resultant of all of the forces on a body is zero the body is said to be in *equilibrium*. This will be the case if it is at rest or moving with constant speed in a straight line. Both of these cases are grouped under the common heading of problems in *statics*. From what has been said in the preceding chapter, the forces on a body in equilibrium must satisfy the following conditions:

$$\Sigma X = 0, \qquad \Sigma Y = 0. \tag{2-1}$$

These equations are sometimes called the "first condition of equilibrium."

A carefully drawn diagram, in which each force exerted on a body is represented by an arrow, is essential in the solution of problems of this sort. The standard procedure is as follows. First, make a neat sketch of the apparatus or structure. Second, choose some one body which is in equilibrium and in a separate sketch show all of the forces exerted on it. This is called "isolating" the chosen body. Write on the diagram, which should be sufficiently large to avoid crowding, the numerical values of all given forces, angles, and dimensions, and assign letters to all unknown quantities. When a structure is composed of several members, a separate force diagram must be constructed for each. Third, construct a set of rectangular axes and indicate on each force diagram the rectangular components of any inclined forces. Cross out lightly those forces which have been resolved. Fourth, obtain the necessary algebraic or trigonometric equations from the conditions that the algebraic sum of the X- and Y-components of the forces must be zero.

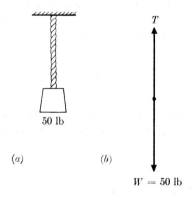

(a)   (b)

$W = 50$ lb

Fig. 2-1. Forces on a hanging block.

To begin with a simple example, consider the 50-lb block in Fig. 2-1(a), hanging by a vertical cord. One contact force, the upward pull of the cord, acts on the block, and also one action-at-a-distance force, the downward gravitational pull of the earth or the weight of the block. Fig. 2-1(b) is the force diagram of the block, which itself is represented by the black dot. The force exerted on the block by the cord is lettered T; the gravitational force exerted on the block by the earth, or the weight of the block, is lettered W.

Then
$$\Sigma Y = T - W,$$

and since the block is in equilibrium and $\Sigma Y = 0$,

$$T - W = 0, \quad \text{or} \quad T = W = 50 \text{ lb.}$$

That is, we find from Newton's first law that the cord pulls up on the block with a force equal to that with which the earth pulls down on the block.

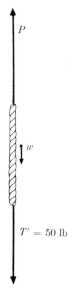

FIG. 2–2. Forces on the cord in Fig. 2–1. Force T' is the reaction to force T in Fig. 2–1.

2–3 Newton's third law. Newton's third law states that *whenever one body exerts a force on another, the second always exerts on the first a force which is equal in magnitude but oppositely directed*. These two forces are commonly referred to as an "action" and a "reaction," and the third law is often stated: "Action and reaction are equal and opposite."

We shall illustrate the third law by a further consideration of the forces in Fig. 2–1. The force T in Fig. 2–1 is an upward force exerted *on the block* by the cord. The reaction to this force is an equal downward force exerted *on the cord* by the block and is represented by T' in Fig. 2–2, which is the force diagram of the cord. The other forces on the cord are its own weight w and the upward force P exerted on it at its upper end by the ceiling. Since the cord is also in equilibrium,

$$\Sigma Y = P - w - T' = 0,$$
$$P = T' + w. \tag{2-2}$$

But since T' in Fig. 2–2 is the reaction to T in Fig. 2–1 its magnitude, by Newton's third law, is 50 lb. Let the weight of the cord be 1 lb. Then from Eq. (2–2)

$$P = 51 \text{ lb,}$$

and the ceiling pulls up on the cord with a force of 51 lb.

Finally, since the cord pulls down on the ceiling with a force equal to that with which the ceiling pulls up on the cord (third law), the downward force on the ceiling (P' in Fig. 2–3) is 51 lb.

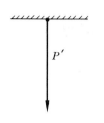

FIG. 2–3. Force P' is the reaction to P in Fig. 2–2.

If the weight of the cord is so small that it can be neglected, then from Eq. (2–2) $P = T' = 50$ lb, and since P and P' are equal, $P' = 50$ lb also. The downward pull on the ceiling (P') is then equal to the downward pull on the lower end of the cord (T'), and *a weightless cord may therefore be considered to transmit a force from one end to the other without change.*

It should be noted carefully that although the forces T and W in Fig. 2–1 are equal and oppositely directed, one is *not* the reaction to the other. The reaction to the force T is, as we have seen, the force T' which the block exerts on the cord. What is the reaction to W? The force W is the force exerted on the block by the earth. The reaction to W must then be an equal but opposite force exerted on the earth by the block. That is, if the earth attracts the block with a force of 50 lb, the block attracts the earth with a force of 50 lb also.

A body subjected to pulls at its ends, as is the cord in Fig. 2–2, is said to be in *tension*. The force exerted by (or on) the cord at either end is called the tension in the cord. For the present we shall consider only weightless cords in which the tension is the same at both ends. For example, if the weight of the cord in Fig. 2–2 is zero, then $P = T' = 50$ lb, and the tension in the cord is 50 lb (*not* 100 lb).

2–4 Simple structures. Figure 2–4(a) shows a 100-lb block hanging from a vertical cord which in turn is knotted to two cords making angles of 30° and 60° with the horizontal. The weights of the cords may be neglected. We wish to find the tension in each cord.

The tension in the vertical cord is evidently 100 lb. The inclined cords are not in contact with the block, and hence no information regarding the tensions in them can be obtained from a force diagram of the block. All three cords, however, exert forces on the knot which joins them. Hence the knot can be considered as a small body whose own weight is negligible and which is in equilibrium under the combined forces exerted by the three cords. In general, in any instance where a number of intersecting forces are in equilibrium, their *point of intersection* is treated as a small body in equilibrium.

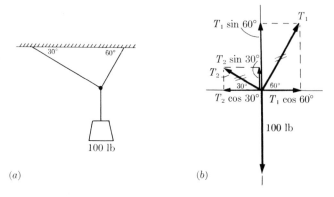

FIGURE 2–4

The force diagram of the knot is constructed in Fig. 2–4(b). Although the tensions T_1 and T_2 are not known in advance, an attempt should be made to draw them approximately to scale. We do know that their horizontal components are equal and opposite, so it is clear that T_1 must be larger than T_2. Do not make the mistake of drawing the vectors T_1 and T_2 equal in length to the cords which exert these forces. It can be seen that the larger force is exerted by the shorter cord.

From the conditions of equilibrium, we have

$$\Sigma X = T_1 \cos 60° - T_2 \cos 30° = 0,$$

$$\Sigma Y = T_1 \sin 60° + T_2 \sin 30° - 100 = 0,$$

whence

$$T_1 = 86.6 \text{ lb}, \qquad T_2 = 50 \text{ lb}.$$

Forces equal and opposite to T_1 and T_2 are exerted on the ceiling by the cords at their upper ends. These forces do not appear in the diagram since they are not forces exerted on the body for which the force diagram is drawn.

A common type of structure in which pushes, as well as pulls, are involved, is shown in Fig. 2–5(a). The hanging body may be a street lamp or a sign. We wish to compute the tension in the supporting cable and the compression in the strut when the weight of the suspended body is known.

Three forces intersect at the outer end of the strut, and accordingly a force diagram, Fig. 2–5(b), is constructed for this point. The three forces are the downward pull of the vertical cable, the outward push of the strut, and the tension in the slanting cable. (The weights of the strut and cable are neglected.) When the force exerted by a body is a push in the direction of its length, as is the case with the strut in Fig. 2–5, the body exerting it is said to be in *compression*. It is customary engineering practice to draw all force vectors with their tails at the origin. Hence vector C, representing the force exerted by the strut at its outer end, is shown as though it were a pull. For equilibrium, we must have

$$\Sigma X = C - T \cos 45° = 0,$$

$$\Sigma Y = T \sin 45° - 80 = 0,$$

whence,

$$T = 113 \text{ lb}, \qquad C = 80 \text{ lb}.$$

From Newton's third law, it follows that the strut pushes on the wall to the left with a force which is equal and opposite to C, and the slanting cable pulls down and to the right on the wall with a force which is equal and opposite to T.

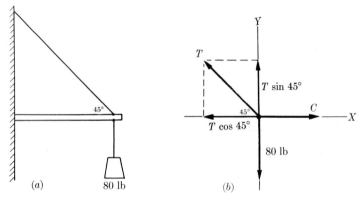

(a)   80 lb   (b)

FIGURE 2–5

2–5 Other examples of equilibrium. According to Newton's first law, the resultant force on a body is zero, both when the body is at rest and when it is moving in a straight line with constant speed. The examples in the preceding section dealt with bodies at rest. We now consider a few cases of linear motion with constant speed.

Before the time of Galileo and Newton, it was thought that in order to maintain such motion it was necessary that a force should continually be applied to the moving body, and indeed our experience seems to show that this is the case. For example, a steady push must be exerted on a book to move it at constant speed along a table top. However, if the surfaces of the book and table are made more and more smooth, that is, as friction is reduced, the force required to maintain the motion becomes smaller and smaller. The inference is that if friction could be eliminated no force at all would be necessary to keep the book in motion, *once it had been started*. We shall postpone for the present the problem of finding the forces while the body is being set in motion, as well as the consideration of friction forces, and take up a number of examples of linear motion with constant speed in the absence of friction.

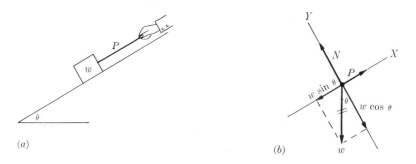

Fig. 2–6. P, N, and w are the external forces exerted on the block.

Figure 2–6 shows a block of weight w on a frictionless inclined plane. What force acting parallel to the plane is required to draw the block up the plane at constant speed?

The first step is to construct a force diagram as in Fig. 2–6(b) showing all of the forces exerted on the block. These forces are, first, the weight of the block w which acts vertically downward even though the block is on an inclined plane; second, the force P which we are to find; and third, the force N with which the plane pushes on the block. If there is no friction, the plane cannot exert any tangential force on the block, so the force N must be perpendicular or *normal* to the surface of the plane. Since P and N are at right angles, it will be simpler to choose X- and Y-axes parallel and perpendicular to the surface of the plane. Forces N and P are then along the axes and need not be resolved. The components of w are $w \sin \theta$ down the plane and $w \cos \theta$ perpendicular to the plane. Then for equilibrium,

$$\Sigma X = P - w \sin \theta = 0,$$

$$\Sigma Y = N - w \cos \theta = 0.$$

Thus if the weight and the slope angle of the plane are known, the desired force P can be found from the first equation and the push of the plane from the second. Notice that, since the sine of an angle is always less than unity, the force P will always be smaller than the weight of the body. The ratio of the weight, which is the force that would be needed

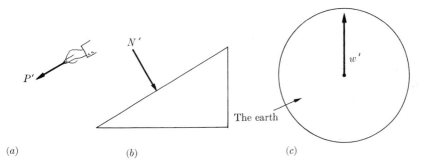

(a)   (b)   (c)

Fig. 2–7. Forces P', N', and w' are the reactions to the forces P, N, and w in Fig. 2–6.

to lift the body vertically, to the force P is called the *mechanical advantage* of the inclined plane.

What are the reactions to the forces P, N, and w in Fig. 2–6? P is a force exerted on the block by the hand; the reaction to P is an equal and opposite force exerted on the hand by the block. It is lettered P' in Fig. 2–7(a). The force N is exerted on the block by the plane. The reaction to N, lettered N' in Fig. 2–7(b), is equal and opposite to N, and is exerted on the plane by the block. The force w is the gravitational force exerted on the block by the earth. The reaction to w is an equal and opposite force exerted on the earth by the block. It is lettered w' in Fig. 2–7(c) (obviously not to scale).

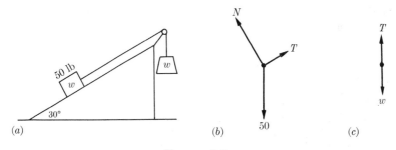

(a)   (b)   50   (c)

FIGURE 2–8

Figure 2–8 is a simple example of a case which is common in many types of moving machinery, the motion of two (or more) bodies connected together in some way. In this case the two are connected by a flexible weightless cord passing over a frictionless pulley. We wish to compute the weight of the hanging block which will just suffice to draw the 50-lb block up the plane with constant speed, once it has been set in motion.

In all problems such as this, which involve the motion of more than one body, it is essential to draw a force diagram for each body. The two force diagrams are shown in Figs. 2–8(b) and 2–8(c). An important point to be noted is that the cord exerts forces of equal magnitude on the bodies to which its ends are attached. It pulls vertically upward on the hanging block with a force T (as yet unknown), and pulls parallel to the plane on the other block with an equal force. From Fig. 2–8(b) we find

$$T = 25 \text{ lb},$$

and from Fig. 2–8(c)

$$w = 25 \text{ lb}.$$

The working out of the solution is left to the reader.

2–6 Friction. In the preceding section we discussed the motion of a body at constant speed when the friction forces were neglected. The next step will be to take friction into account, although the simple laws which will be given here can be expected to hold only approximately in an actual case.

Sliding friction means the force of opposition offered to the sliding of one surface over another. The force is due to the minute irregularities of one surface engaging in those of the other. In the case of metallic surfaces there may be an actual welding together of the "high spots" of the surfaces. The friction force between two surfaces depends on their nature, being smaller for hard smooth surfaces than for rough ones. It also depends upon the force with which the two surfaces are pressed together, but is nearly independent of the area of the surfaces in contact and of their relative speed.

Rolling friction, the opposition offered to the rolling of one body on another, is in general very much smaller than sliding friction. For this reason the sliding friction force exerted on a shaft rotating in bearings can be reduced by installing ball or roller bearings, which replace sliding friction by rolling friction. However, the sliding friction of a well-lubricated shaft is already so small that there is no great gain in efficiency on mounting it in roller bearings. The chief reason for the use of such bearings in machinery is to reduce wear and to simplify lubrication problems.

Viscous friction. The motion of a body through a liquid or gas is opposed by a type of friction called viscous friction, which does not follow the same laws as sliding friction. The friction force, instead of being independent of the speed, is directly proportional to it, provided the speed is not too high. At higher speeds the force may increase with the square or some higher power of the speed. Viscous friction is discussed further in Chapter 17.

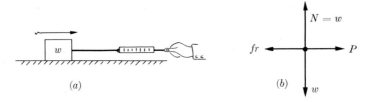

FIGURE 2–9

2–7 Coefficient of friction. Figure 2–9 represents a block of weight w being pulled toward the right along a rough horizontal surface at constant speed by a force P. The normal force exerted by the surface on the block is lettered N. The direction of the friction force fr, which always opposes the relative motion, is toward the left. (The forces N and fr are actually the rectangular components of the force exerted on the block by the plane, but it is convenient to consider them as separate forces.) The friction force may be measured, indirectly, by attaching a spring balance to the block, as shown, and measuring the force P. Since the block moves at constant speed, the force P and the friction force are equal in magnitude. While the line of action of the friction force is actually along the lower face of the block, it is assumed here for simplicity that all forces intersect at the center of the block. We shall see in Chapter 3 how to treat the problem, taking into account the actual lines of action of the various forces.

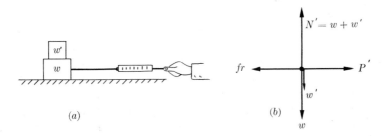

FIGURE 2–10

Suppose now that a second block of weight w' is placed on the first, as in Fig. 2–10(a), and the experiment repeated. The combined weight is now $w + w'$, and it is clear that the upward force N' will be larger than before, being now equal to $w + w'$. It is found by experiment that the friction force increases in the same proportion that the normal force increases. That is,

$$\frac{fr'}{fr} = \frac{N'}{N}, \quad \text{or} \quad \frac{fr}{N} = \frac{fr'}{N'} = \text{constant.}$$

If we represent the constant by the Greek letter μ (mu), we can write

$$\frac{fr}{N} = \mu, \qquad \text{or} \qquad fr = \mu N. \qquad (2\text{-}3)$$

The proportionality constant μ is called the *coefficient of sliding friction* or the *coefficient of kinetic friction*. Since fr and N are both expressed in the same unit, μ is a pure number whose value is characteristic of the surfaces. Some typical coefficients of sliding friction are given in Table I.

TABLE I

COEFFICIENTS OF SLIDING FRICTION

Materials	μ
Wood on wood, dry	0.25–0.50
Metals on metals, dry	0.15–1.4
Smooth surfaces, greased	0.05–0.08

The direct proportion between the force of sliding friction and the normal force was first discovered by Charles Augustin Coulomb (1736–1806), who is more widely known for his experiments on the forces between electric charges.

If the block in Fig. 2–9 is initially at rest and the pull exerted by the spring balance is gradually increased from zero, it is found that a somewhat larger force is required to start the block in motion than is needed to maintain the motion at constant speed once it has been started. One of the surfaces can be felt to "stick," momentarily, to the other. In other words, the coefficient of friction just before motion begins is larger than the coefficient of friction when there is actual sliding of one surface over the other. The former is referred to as the coefficient of static friction, and the latter as the coefficient of sliding friction. The product of the static coefficient and the normal force equals the minimum force required to start the motion; the product of the sliding coefficient and the normal force is the minimum force needed to maintain the motion once it has been started.

The retarding force which stops an automobile when its brakes are applied is the friction force between the tires and the road, and the larger the coefficient of friction, the greater is the available force. Hence an automobile can be stopped in the shortest distance if the pressure on the brake pedal is such that the wheels are just on the point of "locking," without actually doing so. As long as the wheels do not actually slide, the coefficient of friction is the static coefficient. As soon as sliding starts, the coefficient drops to the sliding coefficient, and the braking effect decreases.

As a numerical illustration, suppose that the block in Fig. 2–9 weighs 50 lb, the coefficient of static friction is 0.30, and the coefficient of sliding friction is 0.20. The normal force N is equal in magnitude to the weight of the block and is therefore 50 lb. If the block is initially at rest, the minimum force required to start it moving is equal and opposite to the force of static friction and is therefore $0.30 \times 50 = 15$ lb. As soon as the block starts to slide, the coefficient of friction decreases to 0.20 and the force needed to maintain the block in motion at constant speed is $0.20 \times 15 = 10$ lb.

Let us now ask, how great is the friction force if the block is initially at rest and a horizontal force of 5 lb is exerted on it? We have shown that a force of at least 15 lb is required to start the block moving, so it is clear that when the applied force is only 5 lb the block remains at rest. Since the block is in equilibrium the friction force must be 5 lb also, in a direction opposite to that of the external force. That is, if the block is at rest and any horizontal force less than 15 lb is exerted on it, it remains at rest and the friction force automatically adjusts itself to a value equal and opposite to the applied force.

It follows that the force of friction between two surfaces can be computed by the relation $fr = \mu N$ *only* when one surface is actually sliding over the other or is just on the point of sliding. In the former case the sliding coefficient applies, in the latter the static coefficient. If neither of these conditions holds, the friction force adjusts itself to balance the applied force and may have any value from zero up to that given by the product of the static coefficient and the normal force.

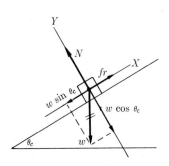

FIG. 2–11. The critical angle θ_c is that at which the block slides down the plane at constant speed.

EXAMPLE. One end of a board upon which rests a block of weight w is raised until the angle of inclination is such that the block slides down the board at constant speed when it has once been set in motion (Fig. 2–11). For a given coefficient of sliding friction what will be the angle of inclination?

The direction of the friction force is up the plane, opposite to the direction of motion. Let θ_c represent the critical angle at which the block just slides with constant speed. Take axes parallel and perpendicular to the surface of the plane and resolve the weight w into components. Then

$$\Sigma X = fr - w \sin \theta_c = 0,$$

$$\Sigma Y = N - w \cos \theta_c = 0,$$

$$fr = \mu N.$$

From the simultaneous solution of these equations we find

$$\mu = \tan \theta_c.$$

This result provides a simple experimental method of measuring coefficients of sliding friction. The static coefficient may be measured in a similar way, by slowly increasing the angle of the plane until the block starts to slide. This angle is always steeper than that at which the block slides at constant speed.

Problems

2–1. A block rests on a horizontal surface. (a) What two forces act on it? (b) By what bodies are each of these forces exerted? (c) What are the reactions to these forces? (d) On what body is each reaction exerted, and by what body is each exerted?

2–2. A block is given a push along a table top, and slides off the edge of the table. (a) What force or forces are exerted on it while it is falling from the table to the floor? (b) What is the reaction to each force, that is, on what body and by what body is the reaction exerted? Neglect air resistance.

2–3. A block is at rest on an inclined plane. (a) Show in a diagram all of the forces acting on the block. (b) What is the reaction to each force?

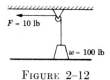

FIGURE 2–12

2–4. A 10-lb force is applied horizontally to a string which runs over a pul-

ley and is fastened to a 100-lb block resting on the floor, as shown in Fig. 2–12. (a) What is the direction and magnitude of the resultant force exerted by the string on the pulley? (b) What is the magnitude of the force which the 100–lb block exerts on the floor? (c) What is the magnitude of the force which the floor exerts on the 100-lb block? (d) What is the gravitational force on the 100-lb block?

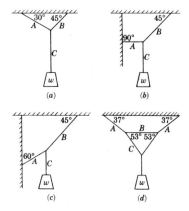

FIGURE 2–13

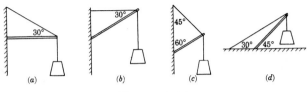

FIGURE 2-14

2-5. Find the tension in each cord in Fig. 2-13. The suspended weight is 200 lb.

2-6. Find the tension in the cable and the compression in the strut in Fig. 2-14. Let the weight of the suspended object in each case be 1000 lb. Neglect the weight of the strut.

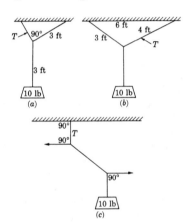

FIGURE 2-15

2-7. (a) In which of the parts of Fig. 2-15 can the tension T be computed if the only quantities known are those explicitly given? (b) For each case in which insufficient information is given, state one additional quantity, a knowledge of which would permit solution.

2-8. A horizontal boom 8 ft long is hinged to a vertical wall at one end, and a 500-lb body hangs from its outer end. The boom is supported by a guy wire from its outer end to a point on the wall directly above the boom. (a) If the tension in this wire is not to exceed 1000 lb, what is the minimum height above the boom at which it may be fastened to the wall? (b) By how many pounds would the tension be increased if the wire were fastened 1 ft below this point, the boom remaining horizontal? Neglect the weight of the boom.

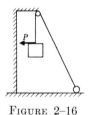

FIGURE 2-16

2-9. A load of building material weighing 300 lb is hoisted from the ground as in Fig. 2-16, where it hangs 20 ft below the pulley. What horizontal force P is needed to pull it a horizontal distance of (a) 6 inches toward the building? (b) 6 ft? (c) What will be the tension in the supporting cord in part (b)? The length of the cable is kept fixed.

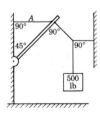

FIGURE 2-17

2-10. Find the tension in cord A in Fig. 2-17. Neglect the weight of the strut.

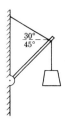

FIGURE 2–18

2–11. Find the largest weight which can be supported by the structure in Fig. 2–18 if the maximum tension the upper rope can withstand is 1000 lb and the maximum compression the strut can withstand is 2000 lb. The vertical rope is strong enough to carry any load required.

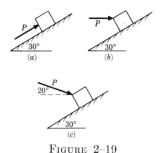

FIGURE 2–19

2–12. (a) What force P parallel to the sloping surface of the plane will push a 20-lb block up a frictionless 30° plane at constant speed? (b) What horizontal force will push it up the plane at constant speed? (c) What force at an angle of 20° with the horizontal? (See Fig. 2–19.) (d) What is the normal force exerted on the block by the plane in each instance?

2–13. A block weighing 20 lb rests on a horizontal surface. The coefficient of static friction between block and surface is 0.4 and the coefficient of sliding friction is 0.2. (a) How large is the friction force exerted on the block? (b) How great will the friction force be if a horizontal force of 5 lb is exerted on the block? (c) What is the minimum force which will start the block in motion? (d) What is the minimum force which will keep the block in motion once it has been started? (e) If the horizontal force is 10 lb, how great is the friction force?

2–14. Coefficients of sliding friction are sometimes expressed in "pounds per ton," that is, the number of pounds of horizontal force needed to maintain a load of 1 ton in steady motion on a level surface. If the coefficient of sliding friction between two surfaces is 0.2, express this in "pounds per ton."

2–15. What force P at an angle ϕ above the horizontal is needed to drag a box weighing w pounds at constant speed along a level floor if the coefficient of sliding friction between box and floor is μ?

2–16. A safe weighing 600 lb is to be lowered at constant speed down skids 8 ft long, from a truck 4 ft high. (a) If the coefficient of sliding friction between safe and skids is 0.3, will the safe need to be pulled down or held back? (b) How great a force parallel to the plane is needed?

2–17. A block weighing 14 lb is placed on an inclined plane and connected to a 10-lb block by a cord passing over a small frictionless pulley as in Fig. 2–8. The coefficient of sliding friction between the block and the plane is $1/7$. For what two values of θ will the system move with constant velocity? Hint: $\cos \theta = \sqrt{1 - \sin^2 \theta}$.

FIGURE 2–20

2–18. Block A, of weight w, slides down an inclined plane S of slope angle 37° at constant velocity while the plank B, also of weight w, rests on top of A. The plank is attached by a cord to the top of the plane (Fig. 2–20). (a) Draw a diagram of all the forces acting on block *A*. (b) If the coefficient of friction is the same between the surfaces *A* and *B* and between *S* and *A*, determine the coefficient of friction.

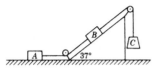

FIGURE 2–21

2–19. Two blocks, *A* and *B*, are placed as in Fig. 2–21 and connected by ropes to block *C*. Both *A* and *B* weigh 20 lb and the coefficient of sliding friction between each block and the surface is 0.5. Block *C* descends with constant velocity. (a) Draw two separate force diagrams (not necessarily to scale) showing the forces acting on *A* and *B*. (b) Find the tension in the rope connecting blocks *A* and *B*. (c) What is the weight of block *C*?

2–20. (a) If a force of 86 lb parallel to the surface of a 20° inclined plane will push a 120-lb block up the plane at constant speed, what force parallel to the plane will push it down at constant speed? (b) What is the coefficient of sliding friction?

FIGURE 2–22

2–21. How great a force *P*, at an angle *α* with the surface of the plane, is required to pull a block of weight *w* pounds up the surface of a plane inclined at an angle *β* with the horizontal as in Fig. 2–22, if the coefficient of sliding friction is *μ*?

2–22. (a) Show that the general expression for the horizontal force *P* required to push a block of weight *w* at constant velocity up the sloping surface of an inclined plane of slope angle *θ*, when the coefficient of sliding friction is *μ*, is

$$P = w \frac{\sin\theta + \mu\cos\theta}{\cos\theta - \mu\sin\theta}.$$

(See Figure 2–19(b) for a diagram.) (b) Let $w = 100$ lb, $\mu = 0.75$. Compute and show in a graph the values of the force *P* for the following angles *θ*: 0, 37°, 45°, 53°, 60°. (c) From the general equation above, find the general expression for the angle *θ* for which an infinite force is required. Does the result agree with the graph in part (b)? (d) Discuss the situation when *θ* is greater than this angle.

FIGURE 2–23

2–23. A flexible chain of weight *w* hangs between two hooks at the same height, as shown in Fig. 2–23. At each end the chain makes an angle *θ* with the horizontal. (a) What is the magnitude and direction of the force exerted by the chain on the hook at the left? (b) What is the tension in the chain at its lowest point?

CHAPTER 3

MOMENTS—CENTER OF GRAVITY

3–1 Introduction. Units and standards of length. It was pointed out in Chapter 2 that the forces on a structural element are often distributed in such a way that the turning effect of the force must be considered. For example, the weight w in Fig. 3–1, if acting alone, would cause the rod to rotate in a clockwise direction about the pivot at O, but its turning effect is counteracted by that of the tension in the cord AB. The turning effect of a force about a pivot is found to depend on the distance of the force from the pivot. Hence it is necessary at this point to define some of the common units of distance or length.

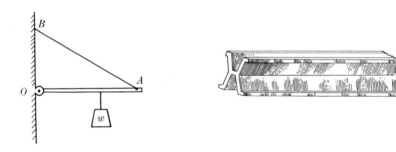

Fig. 3–1. The turning effect of the weight is counteracted by that of the cord AB.

Fig. 3–2. Portion of the standard meter bar.

The unit of length in the English gravitational system is the *foot*. This is the unit used in engineering in the United States and Great Britain. In scientific work throughout the world the unit of length is the *meter*, in the mks system, or the *centimeter* ($= 1/100$ meter) in the cgs system.

The *standard meter* is a bar of platinum-iridium of X-shaped cross section. The meter is defined as the distance between two fine transverse lines engraved on this bar. For many years the foot was defined as one-third of the distance between two lines on a similar one-yard standard, but to avoid the necessity of maintaining two standards of length when one is sufficient, the United States yard is now defined by the relation

$$1 \text{ yard} = \frac{3600}{3937} \text{ meter (exactly)}.$$

From this it follows that

$$1 \text{ foot } (= 1/3 \text{ yard}) = 0.3048006 \text{ meter}$$

$$= 30.48006 \text{ centimeters}.$$

$$1 \text{ inch } (= 1/12 \text{ foot}) = 2.5400 \text{ centimeters}.$$

A useful approximation, good to within 1%, is

$$1 \text{ foot } = 30 \text{ centimeters}.$$

It may be mentioned here that for the purpose of securing a standard of length which is as nearly permanent as anything can be, the length of the standard meter has been carefully compared with the wave length of one particular color of light emitted by cadmium vapor in an electric discharge. In one sense, then, the wave length of this light is the ultimate standard of length. The precise value found is

$$1 \text{ meter } = 1{,}553{,}164.13 \text{ wave lengths}.$$

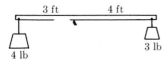

FIGURE 3–3

3–2 Moment of a force. Torque. Suppose that a rigid uniform rod is supported at its center on a frictionless knife-edge as in Fig. 3–3, with a 4-lb weight suspended from a point 3 ft to the left of the knife-edge. It is clear that this weight alone would cause the rod to rotate in a counterclockwise direction about the knife-edge. Suppose we wish to balance the rotary effect of the 4-lb weight by hanging a 3-lb weight at some point to the right of the pivot. Everyone realizes that the 3-lb weight must be suspended at a greater distance from the pivot than the 4-lb weight, and trial would show that if it were hung at a distance of exactly 4 ft from the pivot, the rod would be in equilibrium. The rotary effect of a force about a pivot must then depend on something besides the magnitude of the force. Experiments such as the one described show that the effectiveness of a force in producing rotation about an axis is determined by the product of the force and the perpendicular distance from the axis to the line of action of the force. This perpendicular distance is called the *force arm* or the *moment arm* of the force. Thus the moment arm of the 3-lb weight in Fig. 3–3 is 4 ft, and of the 4-lb weight it is 3 ft.

The product of a force and its force arm is called the *moment* of the force, or the *torque* produced by the force. We shall represent torque by the Greek letter τ (tau). If forces are expressed in pounds and distances in feet, the unit of torque is one pound-foot. One pound-foot is the torque produced by a force of one pound at a perpendicular distance of one foot from an axis. Thus the torque due to the 3-lb force in Fig. 3–3 is $3 \times 4 = 12$ lb·ft clockwise, and that of the 4-lb force is 12 lb·ft counterclockwise. Other units occasionally used are the pound-inch, ounce-inch, etc.

While the units of force in the cgs and mks systems have not yet been defined, it may be stated here for completeness that the cgs unit of torque is one centimeter-dyne and the mks unit is one meter-newton.

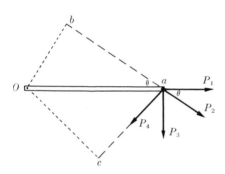

FIG. 3–4. The moment of each force about an axis through O is the product of the force and its force arm.

Figure 3–4 shows a rod pivoted about an axis through O perpendicular to the plane of the diagram. Vectors P_1, P_2, etc., indicate various directions in which a force P might be exerted on the rod at point a. It is evident from experience that the force will be most effective in producing rotation about the axis if its direction is that of P_3. It will be less effective in the directions of P_2 and P_4, while if its direction is that of P_1, no rotation whatever will result, although the distance Oa is the same in all instances. Oa, however, is not the moment arm of all of the forces. The moment arm is defined as "the perpendicular distance from the axis to the line of action of the force." The line of action of a force is a line of indefinite length formed by extending the force vector in either direction, and the moment arm is the perpendicular distance from the axis to this line.

Thus in Fig. 3–4 the line of action of P_4 is shown extended to c, and the moment arm of P_4 is the distance Oc. The line of action of P_2 is extended to point b, and Ob is the moment arm of P_2. Since P_3 is at right angles to Oa, Oa is the moment arm of P_3. The line of action of P_1 passes through O. Hence the distance *from* O *to* the line of action is zero, and the moment of P_1 about O is zero.

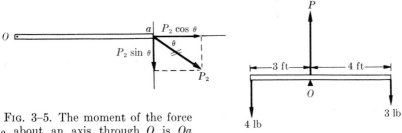

FIG. 3–5. The moment of the force P_2 about an axis through O is $Oa \times P \sin \theta$.

FIGURE 3–6

Another point of view which may be adopted, and which is sometimes more convenient, is illustrated in Fig. 3–5. Force P_2 may be resolved into its rectangular components $P_2 \sin \theta$ and $P_2 \cos \theta$. Since the line of action of $P_2 \cos \theta$ passes through O, this component has no moment about O. The moment arm of the component $P \sin \theta$ is the distance Oa. Hence the moment of the force P_2 about an axis through O is $P_2 \sin \theta \times Oa$. Comparison of Fig. 3–5 with the force P_2 of Fig. 3–4 will show that the same result is obtained whichever method may be used, since the distance Ob in Fig. 3–4 is equal to $Oa \sin \theta$.

3–3 Rotational equilibrium. When a number of coplanar forces act on a pivoted body, the resultant torque on the body is the algebraic sum of the torques due to the individual forces. In forming this algebraic sum, moments in one direction (say clockwise) are considered positive; those in the opposite direction, negative. If the resultant torque is zero we have the rotational analogue of Newton's first law; that is, if the body is already in rotation it continues to rotate uniformly, if at rest it remains at rest. In either case the body is in *rotational equilibrium*. Hence for complete equilibrium, including rotation, the following conditions must be satisfied

$$\Sigma X = 0, \qquad \Sigma Y = 0, \qquad \Sigma \tau = 0. \qquad (3\text{–}1)$$

These equations are called the three conditions of equilibrium.

The second condition may be used to compute the upward force exerted on the rod in Fig. 3–3 by the pivot. If this upward force is represented by P in Fig. 3–6, and if the weight of the rod is negligible, then

$$\Sigma Y = 0,$$
$$P - 4 - 3 = 0,$$
$$P = 7 \text{ lb.}$$

In the particular arrangement shown in Figs. 3–3 and 3–6 it seemed most natural to compute moments about an axis through the pivot O.

But since the rod does not rotate, we could equally well have stated that it does not rotate about an axis through the left end, or the right end, or, in fact, through any point whatsoever. It might be expected, then, that the clockwise and the counterclockwise moments of the forces acting on the rod would be equal no matter what point was considered to be the pivot.

To show that this is the case, let us compute moments about an axis through the point of attachment of the 4-lb weight. The moment of the 4-lb force is now zero since its force arm about the new axis is zero. The moment of the 7-lb force is $7 \times 3 = 21$ lb·ft counterclockwise, and the moment of the 3-lb force is $3 \times 7 = 21$ lb·ft clockwise. The moments are thus equal and opposite if the axis is considered to pass through this point, and it can be shown that the same result will be obtained for any point whatever. As an exercise, calculate the moments of the forces about an axis 4 ft to the left of point O, and show that they are in equilibrium.

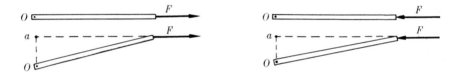

Fig. 3–7. (a) Stable and (b) unstable equilibrium.

3–4 Stable and unstable equilibrium. The upper diagram in Fig. 3-7(a) is a top view of a rod resting on a smooth horizontal surface and pivoted at one end O. A force F is exerted on the other end of the rod as shown. Figure 3–7(b) is the same except that the force F is opposite in sense. Whether the force acts toward the right or left, its moment about the pivot is zero and the rod is in equilibrium. The two cases differ, however, in the following respect. If the rod is given a small angular displacement as in the lower part of the figures, and the force F remains parallel to its original direction, a torque equal to $F \times Oa$ acts about the pivot. It is evident from the diagrams that if the force is directed toward the right as in (a) the torque tends to return the rod to its initial position, while if the force is toward the left as in (b) the effect of the torque is to increase the displacement still further. In the former case the equilibrium is said to be *stable*, in the latter case it is *unstable*, and we can say in general:

When a body is in rotational equilibrium, the equilibrium is stable if a small displacement gives rise to a restoring torque and unstable if a small displacement gives rise to a torque which tends to increase the displacement.

If the torque remains zero when the body is displaced, the equilibrium is *neutral*.

A circular cone resting on its base is in stable equilibrium; balanced on its apex it is in unstable equilibrium; resting on its side on a level surface it is in neutral equilibrium.

3–5 The resultant of a set of parallel forces. The resultant of two (or more) forces is defined as the single force which, if acting alone, will produce the same effect as the simultaneous action of its components. The method of finding the resultant of a number of *intersecting* forces has been explained in Chapter 1. However, if the forces are parallel their lines of action do not intersect and these methods cannot be used.

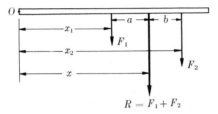

$$R = F_1 + F_2$$

FIG. 3–8. Force R is the resultant of the parallel forces F_1 and F_2.

Let Fig. 3–8 represent a rod pivoted at O and acted upon by the parallel forces F_1 and F_2, at distances x_1 and x_2 from an axis through O. We wish to find the resultant of F_1 and F_2. Since both forces are in the same direction, the *magnitude* of their resultant R must equal their algebraic sum.

$$R = F_1 + F_2.$$

The effect of F_1 and F_2, in this instance, is to produce rotation about O, and since the resultant must produce the same effect as its components, the *line of action* of the resultant must be in such a position that the moment of the resultant about O is equal to the algebraic sum of the moments of the components. That is, the line of action of the resultant R must be at such a distance x from the axis that

$$Rx = F_1x_1 + F_2x_2$$

or, since

$$R = F_1 + F_2,$$

$$(F_1 + F_2)x = F_1x_1 + F_2x_2,$$

$$x = \frac{F_1x_1 + F_2x_2}{F_1 + F_2}. \tag{3-2}$$

The expression for the position of the line of action of R can be put in a somewhat different form. Referring to Fig. 3–8 we have

$$R(x_1 + a) = (F_1 + F_2)(x_1 + a) = F_1 x_1 + F_2(x_1 + a + b).$$

Expanding and cancelling, we find that

$$F_1 a = F_2 b,$$

or

$$\frac{a}{b} = \frac{F_2}{F_1}. \tag{3–3}$$

That is, the line of action of the resultant of two parallel forces in the same sense, divides the distance between the forces into two parts which are inversely proportional to the magnitudes of the forces.

Since the distance from the axis, x_1, does not appear in Eq. (3–3), the line of action of the resultant is the same regardless of the point at which the rod is pivoted, or even if it is not pivoted at all. That is, if the pivot were removed from the rod in Fig. 3–8, the combined effect of F_1 and F_2 would be to cause the rod as a whole to move down, and at the same time to rotate in a clockwise direction. The single force R at the position x would be found to produce exactly the same effect.

This process can evidently be extended to include any number of parallel forces. The general expressions for the magnitude of the resultant and the position of its line of action become

$$R = \Sigma F, \tag{3–4}$$

$$x = \frac{\Sigma F x}{\Sigma F}, \tag{3–5}$$

where ΣF is the algebraic sum of the forces, and $\Sigma F x$ is the algebraic sum of the moments. The position of the line of action of the resultant does not depend on the position of the axis, which may be taken through any convenient point. When applying these equations, whatever convention of signs is adopted for the F's and x's must be used consistently. The algebraic sign of each Fx product is determined by the signs of its factors.

EXAMPLE. Find the magnitude and line of action of the resultant of the three forces in Fig. 3–9.

Let us first take the axis through point O, and consider forces positive if upward and distances positive if to the right of O. Then

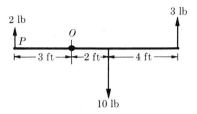

FIGURE 3–9

$$R = \Sigma F = +2 - 10 + 3 = -5 \text{ lb,}$$

$$x = \frac{\Sigma Fx}{\Sigma F} = \frac{(+2)(-3) + (-10)(+2) + (+3)(+6)}{+2 - 10 + 3}$$

$$= \frac{-8}{-5} = +1.6 \text{ ft,}$$

and the three forces are equivalent to a downward force of 5 lb whose line of action is 1.6 ft to the right of O.

If the axis is taken through point P,

$$x = \frac{(+2)(0) + (-10)(+5) + (+3)(+9)}{+2 - 10 + 3}$$

$$= +4.6 \text{ ft,}$$

and the line of action is 4.6 ft to the right of P, which is the same as 1.6 ft to the right of O.

3–6 Center of gravity. The weight of a body is defined as the force of gravitational attraction exerted on the body by the earth. This gravitational pull, however, is not merely one force exerted on the body as a whole. Each small element of the body is attracted by the earth, and the force which is called the weight of the body is in reality the resultant of all of these small parallel forces. Its magnitude, and the position of its line of action, may be calculated by the methods explained in the preceding section.

The direction of the gravitational force on each element of a body is vertically down, so the direction of the resultant is vertically down also, regardless of the orientation of the body. The line of action of the resultant will, however, occupy a different position relative to the body as the orientation of the body is altered. (See Fig. 3–10.)

It is found that however a body may be oriented there is always a common point through which all of these lines of action pass. This point is called the *center of gravity* of the body, and its position is shown in Fig. 3–10(d), in which the lines of action of the weight in the three previous orientations are indicated. The body's weight may therefore be treated

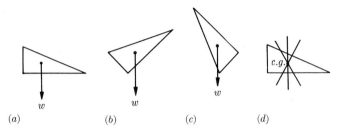

FIG. 3–10. The line of action of the weight passes through the center of gravity.

as a single force whose point of application is the center of gravity, although actually the "point of application" has no significance. All one can say is that the line of action of the weight passes through the center of gravity.

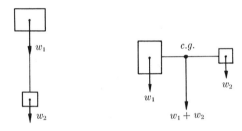

FIGURE 3–11

The center of gravity of any number of bodies whose own centers of gravity are known is located as follows. The center of gravity of any *two* bodies must lie on the line joining their centers of gravity. This follows at once from the fact that if the bodies are placed with their centers in a vertical line as in Fig. 3–11(a), both of their weights lie along this line and hence the resultant lies on this line also. The exact position of the common center of gravity on this line may be found by placing the bodies with the line of centers horizontal as in Fig. 3–11(b), when the line of action of their combined weights is simply that of the resultant of two parallel forces. From Eq. (3–3) we may then say that the center of gravity of two bodies lies in the line joining their centers of gravity, and divides the distance between the centers into two parts which are inversely proportional to the two weights. The two bodies may be replaced by a single body whose center of gravity is at this point and whose weight equals the sum of the weights of the two bodies. This fictitious body may then be combined with a third, and so on.

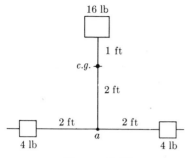

FIGURE 3–12

EXAMPLE. Find the position of the center of gravity of the three weights in Fig. 3–12.

The center of gravity of the two 4-lb weights is at point a, halfway between them. That is, they may be replaced by a single 8–lb weight at point a. When this is combined with the 16-lb weight, the center of gravity is found to be at a point 1 ft below the 16-lb weight, and the group is equivalent to a single 24-lb weight at this point.

A more general method is to construct a pair of X- and Y-axes, specify the position of the center of gravity of each weight by its X- and Y-coordinates, and imagine the pull of gravity to be first parallel to the Y-axis and then parallel to the X-axis. In the former case the X-coordinate of the line of action of the combined weight, and therefore the X-coordinate of the center of gravity, is

$$\bar{x} = \frac{w_1 x_1 + w_2 x_2 + \cdots}{w_1 + w_2 + \cdots} = \frac{\Sigma wx}{\Sigma w}, \tag{3–6}$$

where $\bar{x}$ is the X-coordinate of the center of gravity, and x_1, x_2, etc., are the X-coordinates of the separate weights. Similarly the Y-coordinate of the center of gravity is

$$\bar{y} = \frac{\Sigma wy}{\Sigma w}. \tag{3–7}$$

EXAMPLE. Find the position of the center of gravity of the three weights in Fig. 3–12 by the general method explained above.

Construct a convenient set of rectangular axes as in Fig. 3–13.

$$w_1 = 4 \text{ lb}, \qquad x_1 = 0, \qquad y_1 = 0,$$

$$w_2 = 4 \text{ lb}, \qquad x_2 = 4 \text{ ft}, \qquad y_2 = 0,$$

$$w_3 = 16 \text{ lb}, \qquad x_3 = 2 \text{ ft}, \qquad y_3 = 3 \text{ ft},$$

$$\bar{x} = \frac{\Sigma wx}{\Sigma w} = \frac{4 \times 0 + 4 \times 4 + 16 \times 2}{4 + 4 + 16} = 2 \text{ ft},$$

$$\bar{y} = \frac{\Sigma wy}{\Sigma w} = \frac{4 \times 0 + 4 \times 0 + 16 \times 3}{4 + 4 + 16} = 2 \text{ ft},$$

which checks the previous answer.

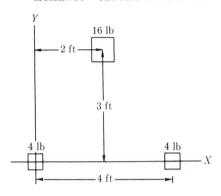

FIGURE 3–13

To find the center of gravity of a flat body of any shape, it may in imagination be subdivided into infinitesimal elements of weight dw, and the sums in Eqs. (3–6) and (3–7) replaced by integrals. That is,

$$\bar{x} = \frac{\int x \, dw}{\int dw}, \qquad \bar{y} = \frac{\int y \, dw}{\int dw}. \qquad (3\text{–}8)$$

The weight of a volume element can be expressed as the product of its weight per unit volume, or weight-density, times its volume dV. Representing the weight-density by ρ, Eq. (3–8) becomes

$$\bar{x} = \frac{\int x\rho \, dV}{\int \rho \, dV} = \frac{\int x \, dV}{\int dV}, \qquad \bar{y} = \frac{\int y \, dV}{\int dV}. \qquad (3\text{–}9)$$

If the thickness of the body is uniform, the volume of an element is the product of its area, dA, times the thickness t. Then

$$\bar{x} = \frac{\int xt \, dA}{\int t \, dA} = \frac{\int x \, dA}{\int dA}, \qquad \bar{y} = \frac{\int y \, dA}{\int dA}. \qquad (3\text{–}10)$$

The calculation of the positions of centers of gravity affords an excellent opportunity to practice integration. Since such calculations form the basis of many problems in courses in integral calculus,* they will not be discussed further here. From the physical standpoint the important thing to remember is that the weight of a body is the resultant of an infinite number of infinitesimal forces, and that its line of action passes through the center of gravity in all orientations.

While the expressions for the coordinates of the center of gravity are simple, the evaluation of the integrals may be difficult if not impossible

---

*See Thomas, *Calculus and Analytic Geometry*, Chapter 5.

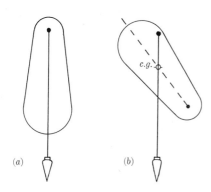

except when the shape of the body is relatively simple. The position of the center of gravity of a body of complex shape is best found by experiment, based on the fact that a pivoted body is in stable equilibrium only when its center of gravity is vertically below the pivot. By pivoting a body successively from two points, two lines may be located at whose point of intersection the center of gravity must lie. Figure 3–14 illustrates this method as applied to an irregularly shaped flat plate.

Fig. 3–14. Locating the center of gravity of a flat object.

EXAMPLES. (1) Compute the tension in the supporting cable in Fig. 3–15, if the strut weighs 40 lb and its center of gravity is at its center. Compute also the force exerted on the strut at its point of attachment to the wall.

NOTE. The arrangement is the same as that of Fig. 2–5 on page 19 except that the weight of the strut is now to be taken into account.

Isolate the strut and show all of the forces exerted on it, as in Fig. 3–15(b). The force at the wall is unknown in magnitude and direction. Instead of working with an unknown force at an unknown angle it is simpler to treat the horizontal and vertical *components* as unknowns, and combine them later to find their resultant. These components are lettered H and V in Fig. 3–15(b). Resolve the tension T into components as shown. Then from the three conditions of equilibrium,

$$\Sigma Y = T \sin 45° + V - 40 - 80 = 0,$$
$$\Sigma X = H - T \cos 45° = 0,$$
$$\Sigma \tau = (80 \times 8) + (40 \times 4) - (T \sin 45° \times 8) = 0.$$

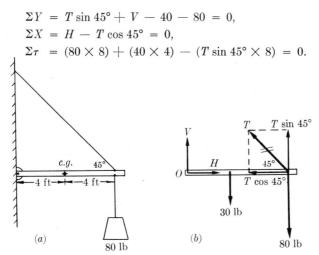

FIGURE 3–15

where the moments are computed about an axis through point O. Solution of these simultaneous equations gives $T = 141$ lb, $V = 20$ lb, $H = 100$ lb.

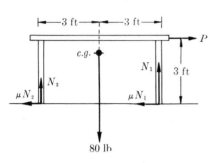

(2) In Chapter 2 a number of examples were discussed in which a body was dragged along a rough surface where for simplicity all of the forces exerted on the body were assumed to intersect at its center. We are now in a position to take into account the actual lines of action of the forces in such cases. Figure 3–16 shows a table which is being pulled to the right at constant speed by a horizontal force P. The center of gravity of the table is midway between front and rear legs. N_1 and N_2 are the upward forces at the front and rear legs, and μN_1 and μN_2 are the friction forces. The coefficient of friction is 0.40. We wish to find the forces P, N_1, and N_2.

80 lb

FIGURE 3–16

From the three conditions of equilibrium,

$$\Sigma Y = P - \mu N_1 - \mu N_2 = 0,$$
$$\Sigma Y = N_1 + N_2 - 80 = 0,$$
$$\Sigma \tau = (P \times 3) + (N_2 \times 6) - (80 \times 3) = 0,$$

where the moments are computed about an axis through the point of contact between the front legs and the floor. From these simultaneous equations, we find

$$N_1 = 56 \text{ lb}, \qquad N_2 = 24 \text{ lb}, \qquad P = 32 \text{ lb}.$$

(3) Figure 3–17 represents a common type of structure known as an A-frame or truss. Each arm of the truss weighs 30 lb, and the center of gravity of each is at its center. A 20-lb weight hangs from the center of one arm. We wish to find the upward force at the lower end of each arm, the tension in the connecting cable, and the force exerted by each arm on the other at the top pin. The frame rests on a frictionless surface, so the forces at the bottom of the truss arms are vertical.

This problem is similar to those discussed in Chapter 2 in which more than one body was involved, and it was desired to find the forces exerted by one part of the system on another. Just as before, a complete solution can be obtained only by *isolating each part of the system*.

The first step in the solution of problems such as this is usually to consider the system as a whole before isolating its individual parts. The force diagram for the whole system is shown in Fig. 3–17(b). The tension in the cord and the forces at the top pin do not appear in this figure, as they are *internal* forces when the truss is considered as a whole. Let us take moments about an axis through A. Then

$$\Sigma \tau_A = 0,$$
$$(50 \times 3) + (30 \times 9) - (V_2 \times 12) = 0,$$
$$V_2 = 35 \text{ lb}.$$

The force V_1 can next be found, either from the condition that $\Sigma Y = 0$, or by taking moments about an axis through B.

From $\Sigma Y = 0$,
$$V_1 + V_2 - 50 - 30 = 0,$$
$$V_1 = 45 \text{ lb}.$$

From $\Sigma \tau_B = 0$,
$$(V_1 \times 12) - (50 \times 9) - (30 \times 3) = 0,$$
$$V_1 = 45 \text{ lb}.$$

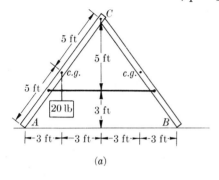

(a)

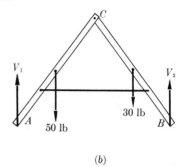

(b)

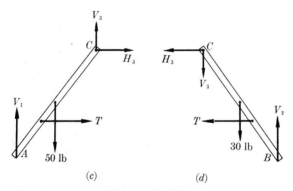

(c)   (d)

FIGURE 3–17

Next isolate the left-hand member of the truss as in Fig. 3–17(c). Assume that the right-hand member exerts a force on it, at the point C, whose components are H_3 and V_3.

One important fact should be noted here. It is not obvious whether the horizontal component H_3 is to the right or the left, and whether the vertical component V_3 is up or down. However, it is not necessary that these directions be known in advance. The components may be assumed in *either* direction, and the problem worked through on the basis of this assumption. The correct *magnitude* of the component will be obtained in any case. If the direction assumed was the correct one, the algebraic sign of the answer will be positive; if the asssumed direction was incorrect the algebraic sign will be negative.

The conditions of equilibrium lead to the equations:

$$\Sigma X = H_3 + T = 0,$$
$$\Sigma Y = V_1 + V_3 - 50 = 0, \tag{3-11}$$
$$\Sigma \tau_C = (V_1 \times 6) - (T \times 5) - (50 \times 3) = 0,$$

where the moments are computed about point C.

The forces on the right-hand member of the truss are shown in Fig. 3–17(d). Notice that the forces exerted on this member at C are the *reactions* to the forces at point C in Fig. 3–17(c), and are therefore equal and opposite to those forces. From the equilibrium conditions,

$$\Sigma X = -H_3 - T_3 = 0,$$
$$\Sigma Y = V_2 - V_3 - 30 = 0, \tag{3-12}$$
$$\Sigma \tau_C = (30 \times 3) + (T \times 5) - (V_2 \times 6) = 0.$$

From Eqs. (3–11), we find

$$T = 24 \text{ lb}, \qquad V_3 = 5 \text{ lb}, \qquad H_3 = -24 \text{ lb}.$$

That is, forces T and V_3 were assumed in the correct direction, but the force H_3 is one of 24 lb, directed toward the *left* in Fig. 3–17(c) and toward the *right* in Fig. 3–17(d). All of the forces are now known and Eqs. (3–12) need not be solved. It is customary to do so, however, as a check on the computations. We have

$$\Sigma X = +24 - 24 = 0,$$
$$\Sigma Y = 35 - 5 - 30 = 0,$$
$$\Sigma \tau_C = (30 \times 3) + (24 \times 5) - (35 \times 6) = 0.$$

(4) A uniform ladder 20 ft long and weighing 80 lb leans against a vertical frictionless wall as in Fig. 3–17A, with its lower end 12 ft from the foot of the wall. Find the magnitudes and directions of the forces exerted on the ladder at each of its ends.

The vertical wall is frictionless, so the force exerted by it on the ladder can have no component parallel to the wall. Hence the force at the upper end is

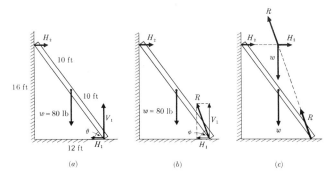

(a)   (b)   (c)

FIGURE 3–17A

horizontal and is represented by H_2 in the figure. The force at the lower end is unknown in magnitude and direction. Let H_1 and V_1 represent its horizontal and vertical components. Then

$$\Sigma X = H_2 - H_1 = 0,$$
$$\Sigma Y = V_1 - 80 = 0,$$
$$\Sigma \tau_0 = (H_2 \times 16) - (80 \times 6) = 0.$$

From the second equation,

$$V_1 = 80 \text{ lb.}$$

From the third equation,

$$H_2 = \frac{80 \times 6}{16} = 30 \text{ lb.}$$

From the first equation,

$$H_1 = H_2 = 30 \text{ lb.}$$

The resultant force R exerted on the foot of the ladder is shown in Fig. 3–17A(b). Evidently,

$$R = \sqrt{80^2 + 30^2} = 85.5 \text{ lb,}$$

$$\tan \varphi = \frac{80}{30}, \quad \varphi = 69.5°.$$

Notice carefully that at neither end of the ladder does the direction of the force coincide with the direction of the ladder. At the upper end the force is horizontal; at the lower end the resultant force is at an angle φ of 69.5° above the horizontal, while the ladder itself makes an angle θ of only 53° with the horizontal.

In Fig. 3–17A(c) the forces acting on the ladder are extended as in Fig. 1–15 to find their resultant. The reader can readily verify that the lines of action of all three forces intersect at a common point and that their resultant is zero, as it must be for equilibrium.

3–7 Couples. An interesting and important case involving parallel forces occurs when a body is acted on by two forces which are equal in magnitude but opposite in direction (or sense) and whose lines of action do not coincide. Such a pair of forces is called a *couple*. (Fig. 3–18.) If we set out to find the resultant of these parallel forces by the methods used in Section 3–3, we find

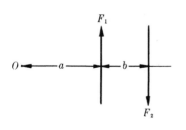

FIG. 3–18. The equal and oppositely directed forces F_1 and F_2 constitute a couple.

$$R = F_1 - F_2 = 0.$$

Since the resultant is of zero magnitude, there is no point at which it can be applied to produce the same effect as the given forces. Looked at in another way, it is impossible for a single force to produce the same effect as a couple, and conversely, there is no single force which will balance a couple. The sole effect of a couple is to produce rotation, and a couple can be balanced only by another couple which is equal and opposite.

The moment of a couple is found as follows. If moments are taken about an axis through O and perpendicular to the plane of the diagram (Fig. 3–18), we find

$$\Sigma\tau_0 = F_2 \times (a + b) - F_1 \times a$$

$$= F_2a + F_2b - F_1a.$$

But since F_1 and F_2 are equal in magnitude,

$$\Sigma\tau_0 = F_2b = F_1b.$$

That is, the resultant torque produced by a couple is equal to the product of *either* of the forces constituting the couple times the perpendicular distance between their lines of action. This product is called the moment of the couple. Furthermore, since the distance a does not appear in the expression for the moment, it follows that the moment of a couple is the same about *all* axes perpendicular to the plane of the forces constituting the couple.

A common example of a couple is afforded by the forces acting on a compass needle in the earth's magnetic field. The north and south poles are urged in opposite directions with equal forces. Since the resultant *force* on the needle is zero there is no tendency for it to move north or south as a whole. If it is originally in position (a) in Fig. 3–19, the effect of the couple is to rotate it clockwise to position (b). In this position the moment of the couple becomes zero and the north-south direction is therefore the position of stable equilibrium.

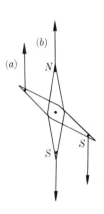

FIGURE 3–19

PROBLEMS

3–1. Give an example to show that the following statement is false: Any two forces acting on a body can be combined into a single resultant force that would have the same effect.

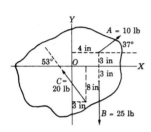

FIGURE 3–20

3–2. (a) Find the X- and Y-components of forces A, B, and C in Fig. 3–20. (b) Find the sum of the X-components and the sum of the Y-components of all the forces. (c) What is the resultant force on the body (give both magnitude and direction)? (d) What is the resultant torque about an axis through point O, perpendicular to the plane of the diagram? Is the torque clockwise or counterclockwise?

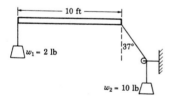

FIGURE 3–21

3–3. A single force is to be applied to the bar in Fig. 3–21, to maintain it in equilibrium in the position shown. The weight of the bar can be neglected. (a) What are the X- and Y-components of the required force? (b) What is the tangent of the angle which the force must make with the bar? (c) What is

the magnitude of the required force? (d) Where should the force be applied?

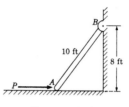

FIGURE 3–22

3–4. End A of the bar AB in Fig. 3–22 rests on a frictionless horizontal surface, while end B in hinged. A horizontal force P of 12 lb is exerted on end A. Neglect the weight of the bar. (a) What are the horizontal and vertical components of the force exerted by the bar on the hinge at B? (b) Show in a sketch all of the forces acting on the bar.

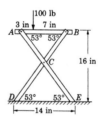

FIGURE 3–23

3–5. A weightless camp stool pivoted at points A, B, and C rests on a frictionless floor, as in Fig. 3–23. A vertical force of 100 lb is exerted on the member AB. (a) Calculate the forces exerted by the floor on the stool at D and E. (b) Calculate the vertical components of the forces exerted on member AB by the legs. (c) Calculate and show in a diagram the horizontal and vertical components of all the forces acting on leg AE.

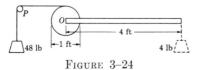

FIGURE 3–24

3–6. A circular disk 1 ft in diameter, pivoted about a horizontal axis through its center O, has a cord wrapped around its rim. The cord passes over a frictionless pulley P and is attached to a body of weight 48 lb. A uniform rod 4 ft long is fastened to the disk with one end at the center of the disk. The apparatus is in equilibrium, with the rod horizontal as shown in Fig. 3–24. (a) What is the weight of the rod? (b) What is the new equilibrium direction of the rod when a second body weighing 4 lb is suspended from the outer end of the rod as shown by the dotted line?

3–7. Find the tension in cord A in Fig. 2–17, assuming that the boom is uniform and weighs 400 lb.

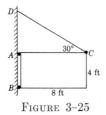

FIGURE 3–25

3–8. A gate 8 ft long and 4 ft high weighs 80 lb. Its center of gravity is at its center, and it is hinged at A and B. To relieve the strain on the top hinge a wire CD is connected, as shown in Fig. 3–25. The tension in CD is increased until the horizontal force at hinge A is zero. (a) What is the tension in the wire CD? (b) What is the magnitude of the horizontal component of force at

hinge B? (c) What is the combined vertical force exerted by hinges A and B?

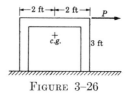

FIGURE 3–26

3–9. The bench in Fig. 3–26 weighs 80 lb and its center of gravity is 2 ft above the floor. The coefficient of sliding friction between bench and floor is 0.2. (a) What horizontal force P is required to drag the bench toward the right with constant velocity? (b) What are the upward forces exerted on the bench at its front and rear legs?

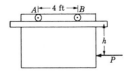

FIGURE 3–27

3–10. A garage door is mounted on an overhead rail as in Fig. 3–27. The wheels at A and B have rusted so that they do not roll, but slide along the track. The coefficient of sliding friction is 0.5. The distance between the wheels is 4 ft, and each is 1 ft in from the vertical sides of the door. The door is symmetrical and weighs 160 lb. It is pushed to the left at constant velocity by a horizontal force P. (a) If the distance h is 3 ft, what is the vertical component of the force exerted on each wheel by the track? (b) Find the maximum value h can have without causing one wheel to leave the track.

3–11. A freight elevator weighing 2000 lb and having dimensions of 8 × 8 × 12 ft hangs from a cable with a

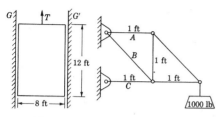

FIGURE 3–28   FIGURE 3–29

small clearance between vertical frictionless guides, G and G', as shown in Fig. 3–28. A load of 1200 lb is placed in the elevator with its center of gravity 2 ft to the left of the center of the floor. The elevator is then moved upward at constant velocity. (a) Show in a diagram the location and direction of the forces exerted by the guides on the elevator. (b) Compute the magnitude of these forces.

3–12. (a) Neglecting the weights of the bars in Fig. 3–29, find the forces exerted on the hinges by bars $A, B,$ and C. (b) Are the bars in tension or compression?

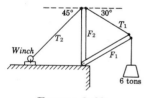

FIGURE 3–30

3–13. The derrick in Fig. 3–30 has two booms of equal length pivoted together at the base and with a fixed cable of the same length joining the ends, to form an equilateral triangle. When one boom is vertical the cable to the winch makes an angle of 45° with the horizontal. A load of 6 tons hangs from the end of the second boom. Assume the cables and booms to have negligible weight. Find the tensions in the two cables, T_1

and T_2, and the compressional forces F_1 and F_2 in the two booms.

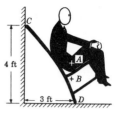

FIGURE 3–31

3–14. A man who weighs 180 lb sits on the seat of a chair weighing 20 lb. The center of gravity of the man is at point A, 2.5 ft out from the wall and 3.5 ft above the floor. The center of gravity of the chair is at point B, 2.5 ft out from the wall and 1.5 ft above the floor (Fig. 3–31). The coefficient of friction between chair and floor is 0.2; between chair and wall is zero. (a) Find the forces on the chair at C and D. (b) Will the chair slip? Explain why or why not.

3–15. A roller 2 ft in diameter weighs 1000 lb. What is the horizontal force necessary to pull the roller over a brick 2 inches high?

3–16. Find the magnitude and line of action of the resultant of the four forces in Fig. 3–32.

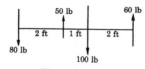

FIGURE 3–32

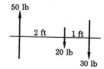

FIGURE 3–33

3–17. Find the resultant of the three forces in Fig. 3–33.

3-18. Weights of 3, 6, 9, and 12 lb are fastened to the corners of a light wire frame 2 ft square. Find the position of the center of gravity of the weights.

3-19. Prove by the methods of calculus that the center of gravity of a uniform rectangular plate is at its center.

FIGURE 3-34

3-20. Find the position of the center of gravity of the T-shaped plate in Fig. 3-34.

3-21. The diameter of a tapered pole of uniform density increases uniformly from 6 inches at one end to 12 inches at the other. If the pole is 12 ft long, find the position of its center of gravity.

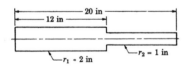

FIGURE 3-35

3-22. A machine part, shown in cross section in Fig. 3-35, consists of two homogeneous, solid, coaxial cylinders. Where is its center of gravity?

3-23. A cube having sides of length a rests on an inclined plane making an angle θ with the horizontal. The angle θ is slowly increased until the block begins to slide. (Assume static and sliding coefficients to be equal.) (a) What is the maximum coefficient of friction μ for which the cube will slide rather than tip? (b) What is the angle

θ at which it will begin to slide if μ has this maximum value?

3-24. A uniform ladder leans against a vertical frictionless wall with its lower end on a horizontal surface. The coefficient of static friction between the bottom of the ladder and the horizontal surface is μ. (a) Find the minimum angle θ between the ladder and the horizontal surface for which the ladder will remain in equilibrium. (b) Find the angle ϕ between the horizontal surface and the resultant force exerted on the bottom of the ladder when θ has the value computed in part (a).

3-25. A uniform ladder 20 ft long, weighing 80 lb, leans against a vertical frictionless wall with its lower end 12 ft from the wall. (a) Compute the magnitude and direction of the resultant force exerted on the lower end of the ladder. (b) In a carefully drawn diagram to scale show all of the forces acting on the ladder.

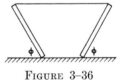

FIGURE 3-36

3-26. Two uniform boards, tied together by a string, are balanced on a surface, as shown in Fig. 3-36. The coefficient of static friction between boards and floor is 0.5. (a) What is the minimum angle ϕ at which the boards can be balanced without slipping? (b) If each board weighs 20 lb, what is the maximum tension in the string when the boards are balanced?

3-27. The chair in Fig. 3-37 is to be dragged to the right at constant speed along a horizontal surface, the coefficient of sliding friction being 0.3. The chair weighs 50 lb. (a) What horizon-

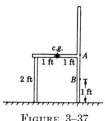

FIGURE 3–37

by the cable without causing the truck to tip. (c) How great is the friction force at the rear wheels when the truck is just on the point of tipping?

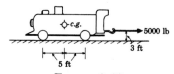

FIGURE 3–39

tal force is needed? (b) What is the upward force at each leg if the force dragging the chair is applied at point A? (c) What is the upward force at each leg if the force is applied at point B? (d) What is the maximum height at which the dragging force can be applied without causing the chair to tip?

3–29. The locomotive in Fig. 3–39 weighs 25 tons. Find the vertical forces at the front and rear wheels when the draw-bar pull is 5000 lb.

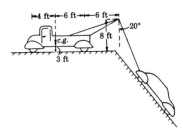

FIGURE 3–38

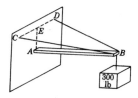

FIGURE 3–40

3–28. The wrecking truck in Fig. 3–38 weighs 3000 lb and has brakes on rear wheels only. (a) Compute the horizontal and vertical components of the forces exerted by the ground at the bottom of each wheel when the tension in the cable is 1000 lb. (b) Compute the maximum force that can be exerted

3–30. A horizontal rigid bar AB, 16 feet long and of negligible weight, rests against a vertical wall at A and carries a 300-lb block at its outer end. Two guy wires run from B to points C and D in the wall. Points C and D are in the same horizontal line. The guy wires are each 25 ft long. The distance CE = ED = 15 ft, AE = 12 ft (Fig. 3–40). Find (a) the tension in each wire, and (b) the force with which the wall presses against the bar.

CHAPTER 4

LINEAR MOTION

4–1 Motion. At the beginning of Chapter 1 it was stated that mechanics is concerned with the relations of force, matter, and motion. The preceding chapters have dealt with forces and we are now ready to discuss the mathematical and graphical methods of describing motion. This branch of mechanics is called *kinematics*.

Motion may be defined as a continuous change in position. We shall restrict the discussion in this chapter to motion along a straight line, or *linear* motion. In order to specify the position of a body moving along a line, some fixed reference point on the line is chosen as the *origin*. The distance from the origin to the body is called the *coordinate* of the body. The coordinate is usually considered positive if the body is at the right of

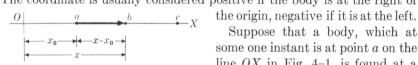

Fig. 4–1. The vector from a to b is the displacement.

the origin, negative if it is at the left.

Suppose that a body, which at some one instant is at point a on the line OX in Fig. 4–1, is found at a later instant at point b. The origin is at O, the coordinate of point a is x_0, and the coordinate of point b is x. The *displacement* of the body is defined as the vector drawn from a to b; its magnitude is evidently $x - x_0$. The displacement is the same whatever motion the body may have performed. For example, if the body moves from a to c and back to b, its displacement is still defined as the vector from a to b. That is, the displacement is always the vector from the initial point to the end point.

The total space moved over by the body, or the sum of the segments ac and cb, is called the *length of path*. Length of path is considered a scalar, not a vector.

4–2 Average velocity and average speed. The *average velocity* of a moving body is defined as the ratio of its displacement to the length of the time interval in which the displacement occurred.

$$\text{Average velocity (a vector)} = \frac{\text{displacement (a vector)}}{\text{elapsed time (a scalar)}}.$$

Average velocity is a vector, since the ratio of a vector to a scalar is itself a vector, and its direction is the same as that of the displacement.

Let t_0 be the time at which the body is at point a, Fig. 4–1, and let t be the time at which it passes point b. The elapsed time is $t - t_0$ and the average velocity is therefore

$$\bar{v} = \frac{x - x_0}{t - t_0}. \tag{4-1}$$

(A bar over the symbol for a quantity signifies an average value.)

If the final position of the body is at the right of its initial position, the displacement $x - x_0$ is positive. If the final position is at the left of the initial position the displacement is negative. The elapsed time $t - t_0$ is positive always. Hence the algebraic sign of the average velocity is the same as that of the displacement, and a positive average velocity indicates a displacement toward the right and vice versa.

The *average speed* of a moving body is defined as the ratio of length of path to elapsed time.

$$\text{Average speed (a scalar)} = \frac{\text{length of path (a scalar)}}{\text{elapsed time (a scalar)}}.$$

Average speed is the ratio of a scalar to a scalar and is itself a scalar. Since the length of path cannot be expressed as the difference between initial and final coordinates no expression like Eq. (4–1) can be written for average speed. Except in special cases the displacement and length of path of a moving body are not numerically equal. Therefore the average speed and average velocity of a moving body will in general differ numerically. However, velocity and speed are both the ratio of a length to a time and both are therefore expressed in terms of the same unit.

All systems of units employ the same unit of time, the *second*. One second (strictly speaking, one "mean solar second") is defined as 1/86,400 of a mean solar day. A mean solar day is the mean or average time for the earth to make one rotation on its axis relative to the sun. The figure 86,400 comes from dividing the day into 24 hours and the hour into 3600 seconds. (24 × 3600 = 86,400.) There is no physical standard of time corresponding to the standards of length or force, except insofar as the earth and sun may be considered to constitute the standard.

The unit of velocity (or speed) in the English system is one foot per second, abbreviated ft/sec. The corresponding unit in the mks system is one meter per second (m/sec), and in the cgs system it is one centimeter per second (cm/sec). Many other units such as the mile per hour are in common use.

Equation (4–1) may be cleared of fractions and written

$$(x - x_0) = \bar{v}(t - t_0), \qquad (4\text{–}2)$$

or in words, displacement equals the product of average velocity and elapsed time.

It is often useful to solve Eq. (4–2) for the coordinate x.

$$x = x_0 + \bar{v}(t - t_0). \qquad (4\text{–}3)$$

If time is counted from the instant when the body is at point a, then $t_0 = 0$ and Eq. (4–3) becomes

$$x = x_0 + \bar{v}t. \qquad (4\text{–}4)$$

If point a is at the origin then $x_0 = 0$ and Eq. (4–4) simplifies further to

$$x = \bar{v}t. \qquad (4\text{–}5)$$

EXAMPLE. A runner on a straight track passes a point 50 ft from the starting line at the instant when a stop-watch reads $12\frac{3}{5}$ sec, and passes a second point 158 ft from the starting line when the same watch reads $16\frac{1}{5}$ sec. What was his average velocity in ft/sec?

Take the origin at the starting line. Then $x_0 = 50$ ft, $x = 158$ ft, $t_0 = 12.6$ sec, $t = 16.2$ sec. Hence

$$\bar{v} = \frac{158 - 50}{16.2 - 12.6} = \frac{108}{3.6} = 30 \text{ ft/sec.}$$

4–3 Instantaneous velocity. The velocity of a moving body at some one instant of time, or at some one point of its path, is called its *instantaneous velocity*. Instantaneous velocity is a concept that requires careful definition. Velocity is the ratio of a displacement to a time interval. An instant of time, however, has no duration and consequently a body can undergo no displacement precisely at an instant. This logical difficulty is avoided as follows.

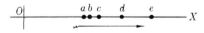

FIG. 4–2. Instantaneous velocity is the limiting ratio of displacement to elapsed time.

The lettered points in Fig. 4–2 represent successive positions of a body moving toward the right along the X-axis. Consider the average velocity of the body first over the displacement ae, then over the successively shorter displacements *ad*, *ac*, and *ab*. The shorter the displacement the more nearly will the average velocity over this displacement equal the instantaneous velocity at point *a*. *The instantaneous velocity at a point* may therefore be defined as *the average velocity over an extremely small displacement which includes the point.*

Notice that although the displacement is extremely small the time interval by which it must be divided to obtain the instantaneous velocity is small also. The quotient is not necessarily a small quantity.

In the notation of calculus the displacement *ab* is written Δx and the corresponding time interval Δt. The average velocity is then

$$\bar{v} = \frac{\Delta x}{\Delta t}.$$

The limiting value of the average velocity, when Δx and Δt are infinitesimally small, is the instantaneous velocity v, and the limiting value of $\Delta x/\Delta t$ is dx/dt. Hence

$$v = \lim_{\Delta t \to 0} \frac{\Delta x}{\Delta t} = \frac{dx}{dt}. \tag{4–6}$$

Equation (4–6) may be considered the definition of instantaneous velocity.

When the change in a quantity is divided by the time interval during which the change occurred, the quotient is called the *time rate of change*, or simply the *rate of change of the quantity*. Average velocity is therefore the *average rate of change of position* and instantaneous velocity is the *instantaneous rate of change of position*.

Since Δt (or dt) is necessarily a positive quantity, it follows that v has the same algebraic sign as Δx (or dx). Hence a positive velocity indicates motion toward the right and vice versa, if we use the usual convention of signs.

EXAMPLE. The coordinate of a body moving along the *X*-axis is given by

$$x = 10t^2,$$

where x is in cm and t in sec. Compute the average velocity of the body over the time interval from: (a) 2 to 2.1 sec; (b) 2 to 2.001 sec; (c) 2 to 2.00001 sec; (d) What is its instantaneous velocity exactly at 2 sec?

(a) The time at the start of the interval under consideration, or t_0, is 2 sec. The corresponding coordinate, x_0, is

$$x_0 = 10 \times 2^2 = 40 \text{ cm.}$$

At the end of the first interval $t = 2.1$ sec and

$$x = 10 \times (2.1)^2 = 44.1 \text{ cm.}$$

Hence

$$\bar{v} = \frac{44.1 - 40}{2.1 - 2} = \frac{4.1}{0.1} = 41 \text{ cm/sec.}$$

(b) When $t = 2.001$ sec,

$$x = 10 \times (2.001)^2 = 40.04001 \text{ cm,}$$

$$\bar{v} = \frac{40.04001 - 40}{2.001 - 2} = \frac{.04001}{0.001} = 40.01 \text{ cm/sec.}$$

(c) When $t = 2.00001$ sec,

$$x = 10 \times (2.00001)^2 = 40.000400001 \text{ cm,}$$

$$\bar{v} = \frac{40.000400001 - 40}{2.00001 - 2} = \frac{.000400001}{0.00001} = 40.0001 \text{ cm/sec.}$$

(d) From Eq. (4-6)

$$v = \frac{dx}{dt} = \frac{d}{dt}(10t^2) = 20t,$$

and when $t = 2$ sec

$$v = 20 \times 2 = 40 \text{ cm/sec.}$$

This example shows how the average velocity becomes more and more nearly equal to the instantaneous velocity as the time interval is made smaller and smaller.

4–4 Average acceleration. Except in certain special cases the velocity of a moving body changes continuously as the motion proceeds. When this is the case the body is said to move with *accelerated motion*, or to have an *acceleration*.

Figure 4–3 again represents a body moving to the right along the X-axis. Suppose that by the methods explained in the preceding section we have found its instantaneous velocity at point a to have the value v_0, represented by the vector v_0 in Fig. 4–3. Similarly the

Fig. 4–3. Average acceleration is the ratio of change in velocity to elapsed time.

instantaneous velocity· at point b has been found to be v. The *average acceleration* during the interval while the body moves from a to b is defined as the *ratio of the change in velocity to the elapsed time.*

$$\text{Average acceleration (a vector)} = \frac{\text{change in velocity (a vector)}}{\text{elapsed time (a scalar)}},$$

$$\bar{a} = \frac{v - v_0}{t - t_0}, \tag{4-7}$$

where t_0 and t are the times corresponding to the velocities v_0 and v. Since v and v_0 are vectors, the quantity $(v - v_0)$ is a *vector difference* and must be found by the methods explained in Section 1–8. However, since in linear motion both vectors lie in the same straight line, the magnitude of the vector difference in this special case equals the difference between the magnitudes of the vectors. The more general case, in which v and v_0 are not in the same direction, will be considered in Chapter 10.

In the English system of units, where the unit of velocity is the ft/sec and the unit of time the second, the unit of acceleration is one foot per second, per second, abbreviated ft/sec^2. In the mks and cgs systems the units of acceleration are respectively one meter per second, per second (m/sec^2) and one centimeter per second, per second (cm/sec^2).

In accordance with the usual convention of sign, if $v - v_0$ is a positive quantity, the acceleration is positive also and directed toward the right. See Problem 4-6 at the end of the Chapter. When the absolute magnitude of the velocity of a body decreases, that is, when the body is slowing down, it is said to have a *deceleration* or to be *decelerated*.

EXAMPLE. The instantaneous velocity of an automobile is 10 ft/sec, 3 sec after it starts, and increases to 40 ft/sec at 6 sec after the start. Find the average acceleration. $t_0 = 3$ sec, $v_0 = 10$ ft/sec, $t = 6$ sec, $v = 40$ ft/sec. The change in velocity is $40 - 10 = 30$ ft/sec and the elapsed time is $6 - 3 = 3$ sec. Hence

$$\bar{a} = \frac{30}{3} = 10 \text{ ft/sec}^2.$$

4-5 Instantaneous acceleration. The instantaneous acceleration of a body, that is, its acceleration at some one instant of time or at some one point of its path, is defined in the same way as instantaneous velocity. Let the points a and b in Fig. 4-3 be taken closer and closer together. The smaller the distance between them, the more nearly will the average acceleration over this distance equal the instantaneous acceleration at point a. We accordingly define the *instantaneous acceleration* at a point as the *average acceleration over an extremely small displacement which includes the point*.

Let Δv represent the change in velocity during a time interval Δt. The average acceleration during this time is then

$$\bar{a} = \frac{\Delta v}{\Delta t}.$$

The limiting value of the average acceleration, when Δt and Δv are infinitesimally small, is the instantaneous acceleration a, and the limiting value of $\Delta v/\Delta t$ is dv/dt. Hence

$$a = \lim_{\Delta t \to 0} \frac{\Delta v}{\Delta t} = \frac{dv}{dt}, \tag{4-8}$$

and since $v = dx/dt$,

$$a = \frac{d}{dt}\left(\frac{dx}{dt}\right) = \frac{d^2x}{dt^2}. \tag{4-9}$$

Either Eq. (4–8) or (4–9) may be considered the definition of instantaneous acceleration.

Since acceleration is a change in velocity divided by the time interval during which the change takes place, average acceleration is the *average rate of change of velocity* and instantaneous acceleration is the *instantaneous rate of change of velocity*. Instantaneous acceleration plays an important part in the laws of mechanics. Average acceleration is less frequently used. Hence from now on when the term "acceleration" is used we shall understand it to mean "instantaneous acceleration" unless otherwise specified.

The definition of acceleration just given applies to motion along a path of any shape, straight or curved. When a body moves in a curved path the *direction* of its velocity changes, and this change in direction also gives rise to an acceleration as will be explained in Chapter 10.

4–6 Linear motion with constant acceleration. The simplest kind of accelerated motion is one in which the acceleration is constant, that is, the velocity changes at the same rate throughout the motion. The *velocity*, of course, is not constant in accelerated motion, and to state that the acceleration is constant simply means that the velocity increases (or decreases) by the same amount in each unit of time.

Now the average value of a quantity that does not change is simply the constant value of the quantity. Hence in linear motion with constant acceleration the average acceleration $\bar{a}$ can be replaced by the constant acceleration a and Eq. (4–7) becomes

$$a = \frac{v - v_0}{t - t_0}. \tag{4–10}$$

If Eq. (4–10) is solved for v we obtain

$$v = v_0 + a(t - t_0). \tag{4–11}$$

This equation can be interpreted as follows. The quantity a is the rate of change of velocity or the change per unit time. The quantity $(t - t_0)$ is the duration of the time interval in which we are interested. The product of (change in velocity per unit time) and (duration of time interval), or the product $a(t - t_0)$, is simply the total change in velocity. When this is added to the initial velocity, v_0, the sum is the velocity at the end of the interval.

If we start counting time from the instant when the velocity is v_0, then $t_0 = 0$ and

$$\boxed{v = v_0 + at.} \tag{4–12}$$

Having found an expression for the velocity of the body at any time, we next derive an expression for its coordinate at any time. We have shown in Eq. (4–2) that the displacement of a body moving along the X-axis is

$$x - x_0 = \bar{v}(t - t_0), \tag{4–2}$$

where $\bar{v}$ is the average velocity.

If the velocity of a body increases at a constant rate, i.e. if its acceleration is constant, its average velocity during any time interval equals one-half the sum of the velocities at the beginning and at the end of the interval. That is,

$$\bar{v} = \frac{v_0 + v}{2}. \tag{4–13}$$

Note carefully that Eq. (4–13) is *not* true, in general, except when the acceleration is constant.

When the expression for v from Eq. (4–11) is substituted in Eq. (4–13) we obtain

$$\bar{v} = \frac{v_0 + [v_0 + a(t - t_0)]}{2}$$

$$= v_0 + \tfrac{1}{2}a(t - t_0). \tag{4–14}$$

Insertion of this expression for $\bar{v}$ in Eq. (4–2) gives

$$x - x_0 = v_0(t - t_0) + \tfrac{1}{2}a(t - t_0)^2. \tag{4–15}$$

If time is measured from the instant when the velocity is v_0, then $t_0 = 0$ and

$$x - x_0 = v_0 t + \tfrac{1}{2}at^2. \tag{4–16}$$

Finally, if the initial position of the body is at the origin, then $x_0 = 0$ and

$$\boxed{x = v_0 t + \tfrac{1}{2}at^2.} \tag{4–17}$$

When the initial velocity v_0 and the constant acceleration a are known, Eq. (4–12) gives the velocity at any time and Eq. (4–17) the coordinate at any time. It is also useful to have an expression for the velocity at any coordinate. This can readily be obtained by solving Eq. (4–12) for t and substituting in Eq. (4–17). The result is

$$v^2 = v_0^2 + 2a(x - x_0). \tag{4–18}$$

If $x_0 = 0$, this reduces to

$$v^2 = v_0^2 + 2ax. \qquad (4\text{--}19)$$

Equations (4–12), (4–17), and (4–19), which have been boxed in to show their importance, are the usual forms of the equations of motion with constant acceleration. They are special cases where x_0 and t_0 are both zero. The general forms are Eqs. (4–11), (4–15), and (4–18).

The equations of linear motion with constant acceleration can be derived in a simple and elegant way from the calculus definitions of velocity and acceleration. We have, from Eq. (4–8),

$$a = \frac{dv}{dt}, \qquad (a = \text{constant})$$

$$\int dv = \int a\,dt,$$

$$v = at + C_1,$$

where C_1 is a constant of integration. If $v = v_0$ when $t = 0$, then $v_0 = 0 + C_1$ and

$$v = v_0 + at,$$

which is Eq. (4–12).

But

$$v = \frac{dx}{dt}.$$

Hence

$$\frac{dx}{dt} = v_0 + at,$$

$$\int dx = \int v_0\,dt + \int at\,dt,$$

$$x = v_0 t + \tfrac{1}{2}at^2 + C_2.$$

If $x = 0$ when $t = 0$, then $C_2 = 0$ and

$$x = v_0 t + \tfrac{1}{2}at^2,$$

which is Eq. (4–17).

To obtain Eq. (4–19), write

$$a = \frac{dv}{dt} = \frac{dv}{dt}\frac{dx}{dx} = \frac{dx}{dt}\frac{dv}{dx} = v\frac{dv}{dx}. \qquad (4\text{--}20)$$

Then from the first and last terms,

$$\int v\,dv = \int a\,dx,$$

$$\frac{v^2}{2} = ax + C_3.$$

If $v = v_0$ when $x = 0$, then $C_3 = v_0^2/2$ and

$$v^2 = v_0^2 + 2ax.$$

4–7 Uniform motion. One further special case is of interest, namely, when the acceleration is zero. Since acceleration is rate of change of velocity, if the acceleration is zero the velocity does not change but remains constant. Hence *constant velocity* and *zero acceleration* are synonymous.

Since the equations of motion with constant acceleration are true whatever the value of the acceleration, they must apply when the acceleration is always zero. From Eq. (4–12), we find when $a = 0$,

$$v = v_0.$$

That is, the velocity is constant and equal to the initial velocity. From Eq. (4–16) when $a = 0$,

$$x = x_0 + vt,$$

and if $x_0 = 0$,

$$x = vt.$$

4–8 Freely falling bodies. The most common example of motion with (nearly) constant acceleration is that of a body falling toward the earth. In the absence of air resistance it is found that all bodies, regardless of their size or weight, fall with the same acceleration at the same point on the earth's surface, and if the distance covered is not too great the acceleration remains constant throughout the fall. The effect of air resistance is discussed in Chapter 17, and the decrease in acceleration with altitude in Chapter 15. For the present we shall neglect both of these factors. This idealized motion is spoken of as "free fall," although the term includes rising as well as falling.

The acceleration of a freely falling body is called the acceleration due to gravity, or the acceleration of gravity, and is denoted by the letter g. At or near the earth's surface it is approximately 32 ft/sec^2, 9.8 m/sec^2, or 980 cm/sec^2. More precise values, and small variations with latitude and elevation, will be considered later.

NOTE. The quantity g is sometimes referred to simply as "gravity," or as "the force of gravity," both of which are incorrect. "Gravity" is a phenomenon, and the "force of gravity" means the force with which the earth attracts a body, otherwise known as the weight of the body. The letter g represents the *acceleration* caused by the force resulting from the phenomenon of gravity.

It is customary when using Eqs. (4–12), (4–17), and (4–19) in an analysis of the motion of a freely falling body to replace a by g and to consider the motion as taking place along the Y-axis. Thus these equations become

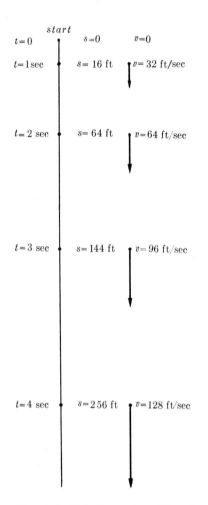

$$v = v_0 + gt,$$

$$y = v_0 t + \tfrac{1}{2}gt^2,$$

$$v^2 = v_0^2 + 2gy.$$

Figure 4–4 shows the velocities and positions of a freely falling body released from rest ($v_0 = 0$) for the first few seconds of its fall. The positive direction has been chosen downward to avoid negative signs.

EXAMPLE. To illustrate the application of the equations of motion with constant acceleration, the following example will be analyzed in detail. A ball is thrown (nearly) vertically upward from the cornice of a building, leaving the thrower's hand with a velocity of 48 ft/sec and just missing the cornice on the way down. (Fig. 4–5.) Find the maximum height reached, the time to reach the highest point, and the position and velocity of the ball 2 sec and 5 sec after leaving the thrower's hand. Neglect air resistance.

In elementary courses in physics, a problem such as this is usually solved by first finding the maximum height reached and then assuming the ball to be dropped from that point. This procedure is necessary because the equations of motion taught in such courses do not include the initial velocity but are written simply

FIG. 4–4. Velocity and position of a freely falling body.

$$v = gt,$$
$$h = \tfrac{1}{2}gt^2,$$
$$v^2 = 2gh.$$

Our equations, which include the initial velocity, are more general and avoid the necessity of breaking up the problem into two parts. However, it is important that proper attention be paid to the algebraic signs of distances, velocities, and accelerations. Let us take the origin at the point where the ball leaves the thrower's hand, and the upward direction as positive. Then the initial velocity, being upward, is positive, and

$$v_0 = +48 \text{ ft/sec.}$$

The acceleration, however, is downward, even though the velocity at the start is in an upward direction. Hence

$$g = -32 \text{ ft/sec}^2.$$

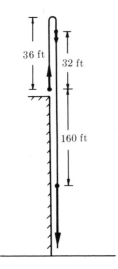

FIGURE 4–5

To find the maximum height reached, we can make use of the fact that the velocity at that point is zero. Then from $v = v_0 + gt$ one may find the time to reach the highest point, and from $v^2 = v_0^2 + 2gy$ one may find its position. Substituting the given data, we obtain

$$0 = 48 + (-32)t,$$
$$t = 1.5 \text{ sec},$$
$$0^2 = (48)^2 + 2(-32)y,$$
$$y = +36 \text{ ft}.$$

So the ball rises 36 ft above the origin and reaches the highest point in 1.5 sec.

The height can also be found from $y = v_0 t + \frac{1}{2}gt^2$, using for t the computed value of 1.5 sec.

$$y = 48 \times 1.5 + \frac{1}{2}(-32)(1.5)^2$$
$$= +36 \text{ ft}.$$

We next compute the position and velocity 2 sec after the ball is thrown.

$$y = 48 \times 2 + \frac{1}{2}(-32)(2)^2$$
$$= +32 \text{ ft};$$
$$v = 48 + (-32)(2)$$
$$= -16 \text{ ft/sec}.$$

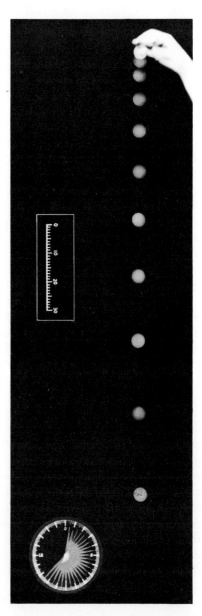

FIG. 4–6. Multiflash photograph (retouched) of a freely falling golf ball.

In other words, the ball is 32 ft above its starting point, and is moving down (v is negative) with a velocity of 16 ft/sec.

At 5 sec after the start,

$$y = (48 \times 5) + \tfrac{1}{2}(-32)(5)^2$$
$$= -160 \text{ ft};$$
$$v = 48 + (-32)(5)$$
$$= -112 \text{ ft/sec}.$$

That is, the ball is now 160 ft below the starting point (y is negative) and is moving down with a velocity of 112 ft/sec. Notice that y does not represent the total space moved over by the ball, or its length of path, but only its distance from the origin or its displacement.

Figure 4–6 is a "multiflash" photograph of a freely falling golf ball. This photograph was taken with the aid of the ultra-high-speed stroboscopic light source developed by Dr. Harold E. Edgerton of the Massachusetts Institute of Technology. By means of this source a series of intense flashes of light can be produced. The interval between successive flashes is controllable at will, and the duration of each flash is so short (a few millionths of a second) that there is no blur in the image of even a rapidly moving body. The camera shutter is left open during the entire motion, and as each flash occurs the position of the ball at that instant is recorded on the photographic film.

Included in the photograph are a clock and a scale. The clock hand rotates continuously, requiring two seconds to make one complete revolution. The small divisions around its circumference correspond to 1/100 sec each. Since the position of the clock hand is photographed at each flash, the time interval between flashes is automatically recorded. The divisions on the scale are 1 cm apart.

The equally spaced light flashes subdivide the motion into equal time intervals Δt. The corresponding displacements, Δx, can be read from the photograph, using the centimeter scale. The average velocity between each pair of flashes, $\Delta x/\Delta t$, can therefore be calculated, and since the time interval Δt can be made very small (of the order of a few hundredths of a second) these average velocities are a good approximation to instantaneous velocities. Since the time intervals are all equal, the velocity of the ball between any two flashes is directly proportional to the separation of its corresponding images in the photograph. If the velocity were constant the images would be equally spaced. The increasing separation of the images during the fall shows that the velocity is continually increasing, or, the motion is accelerated. By comparing two successive displacements of the ball the *change* in velocity in the corresponding time interval can be found. Careful measurements, preferably on an enlarged print, show that this change in velocity is the same in each time interval. In order words, the motion is one of *constant* acceleration.

4–9 Motion with variable acceleration.   If the acceleration is not constant, then Eqs. (4–10) to (4–19) do not apply. We shall consider two cases, (a) when the acceleration is known as a function of *time*, and (b) when the acceleration is known as a function of *position*.

(a) Given that

$$a = \frac{dv}{dt} = f(t).$$

(The symbol $f(t)$ stands for any function of t such as $2t^2$ or $\log t$.)

Then

$$dv = f(t)\, dt,$$

$$v = \int f(t)\, dt + C_1. \tag{4–21}$$

Provided the integral can be evaluated, the velocity can be found from Eq. (4–21) as a function of time. The constant C_1 is determined if the velocity is known at any time. Let us call the expression for the velocity obtained from Eq. (4–21) $g(t)$. Then

$$v = \frac{dx}{dt} = g(t),$$

$$dx = g(t)\, dt,$$

$$x = \int g(t)\, dt + C_2. \tag{4–22}$$

The constant C_2 is determined if the position is known at any time.

(b) Given that

$$a = \frac{dv}{dt} = f(x).$$

It does no good to write

$$dv = f(x)\, dt,$$

$$v = \int f(x)\, dt + C,$$

since $\int f(x)\, dt$ can not be evaluated as it stands. A solution can be obtained, however, from the substitution of Eq. (4–20)

$$a = v\frac{dv}{dx}.$$

Then, if $a = f(x)$,

$$a = v\frac{dv}{dx} = f(x),$$

$$\int v\, dv = \int f(x)\, dx.$$

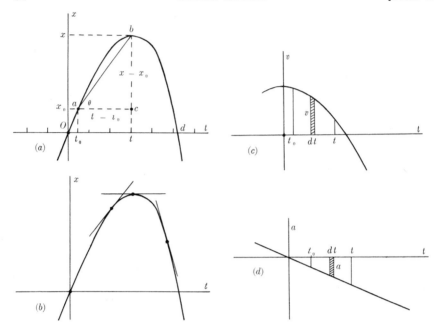

FIG. 4–7. Graphs of the coordinate, velocity, and acceleration of a body moving on the X-axis.

The left side integrates to $v^2/2$. Therefore

$$\frac{v^2}{2} = \int f(x)\, dx + C_1.$$

4–10 Graphical methods. It is often convenient to represent by a graph the position, velocity, and acceleration of a body in linear motion. Such graphs are also helpful in understanding the interrelations of these quantities. The curve in Fig. 4–7(a) is a coordinate-time graph of the motion of a body moving along the X-axis. The ordinate of the curve at any time represents the X-coordinate of the body. In this example the body is at the origin when $t = 0$, moves away from the origin toward the right until point b on the graph is reached, returns to the origin at point d, and then moves off to the left of the origin, since x is negative beyond this point.

At point a on the curve the coordinate of the body is x_0 and the corresponding time is t_0. At point b the coordinate is x and the time is t. The displacement in the interval between t_0 and t is represented by the line bc, whose length is proportional to $x - x_0 = \Delta x$. The corresponding time interval is represented by the line ac, whose length is proportional to $t - t_0 = \Delta t$. The slope of the chord ab, or the tangent of the angle θ, is

$\tan \theta = bc/ac$, and from the proportions just cited this slope is proportional to

$$\frac{x - x_0}{t - t_0} = \frac{\Delta x}{\Delta t} = \bar{v}.$$

That is, *the slope of the chord between two points on a coordinate-time graph is proportional to the average velocity in the interval between the points.* If both x and t are plotted to the same scale, the average velocity is *equal* to the tangent of the angle θ. In general, different scales are used for x and t.

The instantaneous velocity at any point on the curve is proportional to the limiting value of the slope of the chord drawn from the point to a second point at an infinitesimal distance from the first. But by definition the slope of the chord between two points at an infinitesimal separation is the slope of the curve at either point or the slope of the tangent to the curve, dx/dt. Hence *the slope of the tangent line at any point on a coordinate-time graph is proportional to the instantaneous velocity at that point.* The steeper the curve, the greater the velocity. At a point such as b, where the tangent is horizontal, the velocity is zero. At points to the right of b the slope, and therefore the velocity, is negative, that is, the body is moving toward the left.

By drawing tangents to the curve at a number of points and measuring their slopes, as in Fig. 4–7(b), the velocity of the body at these points may be found. These velocities may then be plotted on a velocity-time graph as in Fig. 4–7(c). The *ordinate* of this graph at any point is proportional to the *slope* of the first graph at the same point. It should be evident from the preceding discussion that *the slope of the chord between any two points on the velocity-time graph is proportional to the average acceleration between the points,* and that *the slope of the tangent at any point is proportional to the instantaneous acceleration at that point.*

Another feature of the velocity-time graph is shown in Fig. 4–7(c). The height of the shaded strip is v and its width is dt. The area of the strip is therefore $v\,dt$ which, from the definition $v = dx/dt$, is equal to dx. In other words, the shaded area represents the infinitesimal displacement in the interval dt. The total area between the curve and the X-axis and bounded by the vertical lines at t_0 and t is the sum of the areas of all the strips, and therefore represents the total displacement in the interval from t_0 to t. Mathematically this is equivalent to

$$x = \int_{t_0}^{t} v\,dt.$$

Hence *the area under a velocity-time graph represents the displacement.*

Finally, the instantaneous acceleration may be plotted as a function of time. The slope of this graph, which is $da/dt = d^3x/dt^3$, is not of physical significance. Evidently, however, the area between the curve and the X-axis and bounded by vertical lines at t_0 and t represents the change in velocity in the interval $t - t_0$, since

$$v = \int_0^t a\, dt.$$

4–11 Velocity components. Relative velocity. Velocity is a vector quantity involving both magnitude and direction. A velocity may therefore be resolved into components, or a number of velocity components combined into a resultant. As an example of the former process, suppose that a ship is steaming 30° E of N at 20 mi/hr in still water. Its velocity may be represented by the arrow in Fig. 4–8, and one finds by the usual method that its velocity component toward the east is 10 mi/hr, while toward the north it is 17.3 mi/hr.

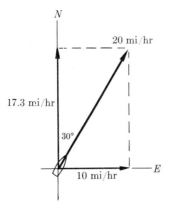

FIG. 4–8. Resolution of a velocity vector into components.

Velocity, like position, can only be specified relative to some reference frame or set of axes; the axes themselves may or may not be in motion. Ordinarily, velocities are specified relative to axes fixed with respect to the earth and considered to be "at rest," although of course they partake of the motion of the earth through space. In what follows, the expression "the velocity of a body" is understood to mean its velocity relative to the earth. *The velocity of one body relative to another* when the second is in motion (relative to the earth) *is the vector difference between the velocities of the bodies* (relative to the earth). Specifically, if the bodies are designated by A and B, and their velocities (relative to the earth) by v_A and v_B, the velocity of A relative to B is

$$v_{AB} = v_A - v_B \text{ (vector difference)}, \tag{4–23}$$

and the velocity of B relative to A is

$$v_{BA} = v_B - v_A \text{ (vector difference)}.$$

EXAMPLE. Automobile A, traveling at 30 mi/hr on a straight level road, is ahead of automobile B traveling in the same direction at 20 mi/hr. What is the velocity of A relative to B and the velocity of B relative to A?

Since both vectors are in the same straight line the magnitude of their vector difference equals their arithmetic difference. The velocity of A relative to B is

$$v_{AB} = v_A - v_B = 30 - 20 = +10 \text{ mi/hr},$$

and the operator of car B sees car A pulling away from him at the rate of 10 mi/hr. The velocity of B relative to A is

$$v_{BA} = v_B - v_A = 20 - 30 = -10 \text{ mi/hr},$$

and the operator of car A (if he looks back) sees car B dropping behind him (v_{BA} is negative) at 10 mi/hr.

Equation (4–23) may be written

$$v_A = v_B + v_{AB} \text{ (vector sum)}.$$

That is, the velocity of body A (relative to the earth) is the vector sum of the velocity of B (relative to the earth) and the velocity of A relative to B. In general, then, when one body is in motion relative to a second, *the velocity of the first is the vector sum of the velocity of the second and the velocity of the first relative to the second.*

EXAMPLES. (1) The compass of a ship indicates that it is headed due north, and its log shows that it is moving through the water at 20 mi/hr. If there is a tidal current of 5 mi/hr due east, what is the velocity of the ship relative to the earth?

The velocity of the water is 5 mi/hr, due east. The velocity of the ship *relative to the water* is 20 mi/hr, due north. The velocity of the ship is the vector sum of these velocities, and from the construction in Fig. 4–9 it is 20.6 mi/hr, 14° E of N. The two velocities of 20 mi/hr and 5 mi/hr can be considered the components of the actual velocity of the ship.

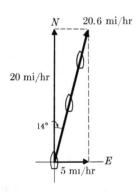

FIGURE 4–9

(2) In what direction should the pilot of the ship set his course in order to travel due north? What will then be his velocity relative to the earth?

The course set by the pilot is the direction in which the ship would actually travel in still water. It is thus in the direction of the velocity of the ship relative to the water. The resultant velocity must be due north. These velocities are related as in Fig. 4–10, from which we find the angle θ to be 14.5° W of N, and the resultant velocity to be 19.4 mi/hr, due north.

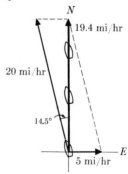

FIGURE 4–10

PROBLEMS

4–1. The 2-mile record on an indoor track is 8 min, 51 sec. To what average velocity does this correspond in (a) mi/sec? (b) mi/hr? (c) cm/sec? (d) ft/sec?

4–2. The timer in the example in Sec. 4–2 may have made an error of $\frac{1}{5}$ sec in each reading of his stop watch. (a) What are the longest and shortest values the time interval might have had? (b) What are the corresponding velocities? (c) Is it justifiable to keep more than two figures in the answer? (d) Is the last figure certain? (e) How much may it be in error?

4–3. A body moves along a straight line, its distance from the origin at any instant being given by the equation $x = 8t - 3t^2$, where x is in centimeters and t is in seconds. (a) Find the average velocity of the body in the interval from $t = 0$ to $t = 1$ sec, and in the interval from $t = 0$ to $t = 4$ sec. (b) Find the expression for the average velocity in the interval from t to $t + \Delta t$. (c) What is the limiting value of this expression as Δt approaches zero? (d) Find the time or times at which the body is at rest. (e) Find the expression for the acceleration at any time.

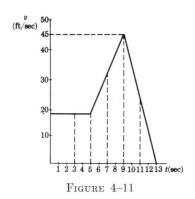

FIGURE 4–11

4–4. The graph in Fig. 4–11 shows the velocity of a body plotted as a function of time.

(a) What is the instantaneous acceleration at $t = 3$ sec?
(b) What is the instantaneous acceleration at $t = 7$ sec?
(c) What is the instantaneous acceleration at $t = 11$ sec?
(d) How far does the body go in the first 5 sec?
(e) How far does the body go in the first 9 sec?
(f) How far does the body go in the first 13 sec?

4–5. Each of the following changes in velocity takes place in a 10-sec interval. What is the magnitude, the algebraic sign, and the direction of the average acceleration in each interval?

(a) At the beginning of the interval a body is moving toward the right along the X-axis at 5 ft/sec, and at the end of the interval it is moving toward the right at 20 ft/sec.

(b) At the beginning it is moving toward the right at 20 ft/sec, and at the end it is moving toward the right at 5 ft/sec.

(c) At the beginning it is moving toward the left at 5 ft/sec, and at the end it is moving toward the left at 20 ft/sec.

(d) At the beginning it is moving toward the left at 20 ft/sec, and at the end it is moving toward the left at 5 ft/sec.

(e) At the beginning it is moving toward the right at 20 ft/sec, and at the end it is moving toward the left at 20 ft/sec.

(f) At the beginning it is moving toward the left at 20 ft/sec, and at the end it is moving toward the right at 20 ft/sec.

(g) In which of the above instances is the body decelerated?

4–6. The makers of a certain automobile advertise that it will accelerate from 15 to 50 mi/hr in high in 13 sec. Compute the acceleration in ft/sec^2, and the distance the car travels in this time, assuming the acceleration to be constant.

4–7. An airplane taking off from a landing field has a run of 1200 ft. If it starts from rest, moves with constant acceleration, and makes the run in 30 sec, with what velocity in ft/sec did it take off?

4–8. A subway train starts from rest at a station and accelerates at a rate of 4 ft/sec^2 for 10 sec. It then runs at constant speed for 30 sec, and decelerates at 8 ft/sec^2 until it stops at the next station. Find the *total* distance covered.

4–9. The "reaction time" of the average automobile driver is about 0.7 sec. (The reaction time is the interval between the perception of a signal to stop and the application of the brakes.) If an automobile can decelerate at 16 ft/sec^2, compute the total distance covered in coming to a stop after a signal is observed: (a) from an initial velocity of 30 mi/hr, (b) from an initial velocity of 60 mi/hr.

4–10. An automobile and a truck start from rest at the same instant, with the automobile initially at some distance behind the truck. The truck has a constant acceleration of 4 ft/sec^2, and the automobile an acceleration of 6 ft/sec^2. The automobile overtakes the truck after the truck has moved 150 ft. (a) How long does it take the auto to overtake the truck? (b) How far was the auto behind the truck initially? (c) What is the velocity of each when they are abreast?

4–11. A juggler performs in a room whose ceiling is 9 ft above the level of his hands. He throws a ball vertically upward so that it just reaches the ceiling. (a) With what initial velocity does he throw the ball? (b) How many seconds are required for the ball to reach the ceiling?

He throws a second ball upward with the same initial velocity, at the instant that the first ball is at the ceiling. (c) How long after the second ball is thrown do the two balls pass each other? (d) When the balls pass each other, how far are they above the juggler's hands?

4–12. An object is thrown vertically upward. It has a speed of 32 ft/sec when it has reached one-half its maximum height. (a) How high does it rise? (b) What is its velocity and acceleration 1 sec after it is thrown? (c) 3 sec after? (d) What is the average velocity during the first half-second?

4–13. A ball is thrown vertically downward from the top of a building, leaving the thrower's hand with a velocity of 30 ft/sec.

(a) What will be its velocity after falling for 2 sec?

(b) How far will it fall in 2 sec?

(c) What will be its velocity after falling 30 ft?

(d) If it moved a distance of 3 ft while in the thrower's hand, find its acceleration while in his hand.

(e) If the ball was released at a point 120 ft above the ground, in how many seconds will it strike the ground?

(f) What will be its velocity when it strikes?

4–14. A student determined to test the law of gravity for himself walks off a skyscraper 900 ft high, stop watch in hand, and starts his free fall (zero initial velocity). Five seconds later, superman arrives at the scene and dives off the roof to save the student. (a) What must superman's initial velocity be in order that he catch the student just before the ground is reached? (b) What must be the height of the skyscraper so that even superman can't save him? (Assume that superman's acceleration is that of any freely falling body.)

4–15. A ball is thrown nearly vertically upward from a point near the cornice of a tall building. It just misses the cornice on the way down, and passes a point 160 ft below its starting point 5 sec after it leaves the thrower's hand. (a) What was the initial velocity of the ball? (b) How high did it rise above its starting point? (c) What was the magnitude of its velocity as it passed a point 64 ft below the starting point?

4–16. A ball is thrown vertically upward from the ground and a student gazing out of the window sees it moving upward past him at 16 ft/sec. The window is 32 ft above the ground. (a) How high does the ball go above the ground? (b) How long does it take to go from a height of 32 ft to its highest point? (c) Find its velocity and acceleration $\frac{1}{2}$ sec after it left the ground, and 2 sec after it left the ground.

4–17. A ball rolling on an inclined plane moves with a constant acceleration. One ball is released from rest at the top of an inclined plane 18 m long and reaches the bottom 3 sec later. At the same instant that the first ball is released a second ball is projected upward along the plane from its bottom with a certain initial velocity. The second ball is to travel part way up the plane, stop, and return to the bottom so that it arrives simultaneously with the first ball. (a) Find the acceleration. (b) What must be the initial velocity of the second ball? (c) How far up the plane will it travel?

4–18. "Near the earth's surface a body falls with uniformly increasing acceleration." Is this statement correct? If not, correct it.

4–19. The acceleration of a body is constant and equal to 8 cm/sec². Find by integration the expressions for its velocity and displacement in terms of t. What further information is needed to determine the motion completely?

4–20. The position of a body moving along the X-axis is given by $x = 10t^2 - 5t$, where x is in cm and t in sec. (a) Find by differentiation the expressions for its velocity and acceleration. (b) What is its initial velocity? (c) What is the significance of the negative sign? (d) Find the maximum distance the body moves to the left of the origin. (e) At what two instants is $x = 0$? Explain. (f) Find the velocity at each of these instants.

4–21. A body moves along a straight line, its position at any instant being given by the equation $x = 32t - 8t^3/3$,

where x is in feet and t is in seconds. (a) Find the average velocity in the time interval from $t = 0$ to $t = 3$ sec. (b) Find the time or times at which the body is at rest. (c) Find the acceleration at $t = 2$ sec.

4–22. The coordinate of a body moving along the X-axis is given by $x = A \cos 2\pi f t$, where A and f are constants. Find the expressions for the velocity and acceleration of the body at any time.

4–23. The distance of a body from a point A is given by the equation $x = 20t + 8t^2 - t^3$, where x is in feet and t in seconds. (a) Write the equation in modified form so as to give the distance x_1 between the body and a point B, 10 ft in the negative direction from A. (b) Write another equation which gives the distance x_2 from the body to a moving point C. At $t = 0$, the point C is at the same position as point A. Point C has a velocity of $+10$ ft/sec. (c) Write the expressions for the velocity and acceleration of the body relative to point A.

4–24. The velocity of a body moving along the X-axis is given by the equation $v = 3t^2$, where v is in meters/sec and t is in seconds. The body is 36 meters to the right of the origin when $t = 0$. (a) Find its coordinate x, its velocity v, and its acceleration a at the times $t = 0$ and $t = 4$ sec. (b) What is the average velocity of the body in the time interval between $t = 0$ and $t = 4$ sec?

4–25. A body moves along a straight line, its velocity at any time being given by the equation $v = 16 - 8t$, where v is in ft/sec and t is in seconds. (a) What will be the displacement between $t = 1$ sec and $t = 2$ sec? (b) Find the expression for the acceleration at any time. What is the acceleration at $t = 2$ sec? (c) Find the time or times at which the body is at rest. (d) Describe the motion of the body in words during the time interval between $t = 0$ and $t = 4$ sec.

4–26. The acceleration of a body moving on the X-axis is given by $a =$

$4t^2 - 2t + 8$, where a is in cm/sec^2 and t is in seconds. Find its average velocity during the first three seconds.

4–27. The acceleration of a body moving on the X-axis is given by $a = 2 + 6x$, where a is in cm/sec^2 and x is in cm. The velocity of the body is 10 cm/sec at the point $x = 0$. Find the velocity of the body at any position.

4–28. The equation of motion of the body in Fig. 4–7 is $x = 27t - t^3$ where x is in cm and t in sec.

(a) Plot the coordinate-time graph for the interval between $t = 0$ and $t = 6$ sec. Let 1 in $= 1$ sec horizontally and 1 in $= 10$ cm vertically.

(b) Construct tangents to this curve at time $t = 1$ sec, $t = 2$ sec, and $t = 4$ sec. Measure their slopes and find the instantaneous velocities at these times.

(c) By differentiation, find the equation for the velocity at any time. From this equation compute the velocities at $t = 1$ sec, 2 sec, and 4 sec, and compare with the values found in (b).

(d) Construct a velocity-time graph of the motion. Let 1 in $= 1$ sec and 1 in $= 20$ cm/sec.

(e) Construct tangents to this curve at times $t = 1$ sec, 2 sec, and 4 sec. Measure their slopes and find the instantaneous accelerations at these times.

(f) By differentiation, find the equation for the acceleration at any time and compute the accelerations at $t = 1$ sec, 2 sec, and 4 sec.

(g) Construct the acceleration-time graph. Use any convenient scale.

4–29. Two piers A and B are located on a river, one mile apart. Two men must make round trips from pier A to pier B and return. One man is to row a boat at a velocity of 4 mi/hr relative to the water, and the other man is to walk on the shore at a velocity of 4 mi/hr. The velocity of the river is 2 mi/hr in the direction from A to B. How long does it take each man to make the round trip?

4–30. A river flows due north with a velocity of 3 mi/hr. A man rows a boat across the river, his velocity relative to the water being 4 mi/hr due east. (a) What is his velocity relative to the earth? (b) If the river is 1 mile wide, how far north of his starting point will he reach the opposite bank? (c) How long a time is required to cross the river?

4–31. When a train has a speed of 10 mi per hour eastward, raindrops which are falling vertically with respect to the earth make traces on the windows of the train which are inclined 30° to the vertical. (a) What is the horizontal component of a drop's velocity with respect to the earth? with respect to the train? (b) What is the velocity of the raindrop with respect to the earth? with respect to the train?

4–32. An airplane pilot sets a compass course due west and maintains an air speed of 120 mi per hour. After flying for one-half hour he finds himself over a town which is 75 mi west and 20 mi south of his starting point. (a) Find the wind velocity, in magnitude and direction. (b) If the wind velocity were 60 mi per hour due south, in what direction should the pilot set his course in order to travel due west? Take the same air speed of 120 mi per hour.

CHAPTER 5

NEWTON'S SECOND LAW

5-1 Introduction. In the preceding chapters we have discussed separately the concepts of force and acceleration. We have made use, in statics, of Newton's first law, which states that when the resultant force on a body is zero, the acceleration of the body is also zero. The next logical step is to ask how a body behaves when the resultant force on it is *not* zero. The answer to this question is contained in Newton's second law, which states, in part, that when the resultant force is not zero, the body moves with accelerated motion. The acceleration, with a given force, depends on a property of the body known as its mass, and before proceeding with the discussion of the second law, we devote the next section to the concept of mass.

This part of mechanics, which includes both the study of motion and the forces that bring about the motion, is called *dynamics*. In its broadest sense, dynamics includes nearly the whole of mechanics. Statics treats of special cases in which the acceleration is zero, and kinematics deals with motion only.

5-2 Mass. The term mass, as used in mechanics, refers to that property of matter which in everyday language is described by the word *inertia*. We know from experience that an object at rest will never start to move of itself—a push or pull must be exerted on it by some other body. In more technical language, an external force is required to accelerate the body, and we say the force is needed because the body has inertia.

It is also a familiar fact that a force is required to slow down or stop a body which is already in motion, and that a sidewise force must be exerted on a moving body to deviate it from a straight line. In these instances also, we say the force is necessary because the body possesses inertia.

It will be seen that the processes above (i.e., speeding up, slowing down, or changing direction) involve a change in either the magnitude or the direction of the velocity of the body. In other words, in every case the body is accelerated. We may therefore say that inertia is that property of matter because of which a force must be exerted on a body in order to accelerate it.

To assign a numerical value to the inertia of any given body, we choose as a standard some one body whose inertia is arbitrarily taken as unity, and state the inertia of all other bodies in terms of this standard. The inertia of a body, when stated in this quantitative way, is called its *mass*. *Mass is a quantitative measure of inertia.*

FIG. 5–1. Kilogram No. 20, the national standard of mass.

The mass of a body is an invariant property of the body, independent of its velocity,* acceleration, position on the earth's surface or height above the earth's surface. In the latter two respects it differs from the weight of the body, which varies with position and elevation. See Section 15–3.

The standard of mass in both the mks and cgs systems is a platinum-iridium cylinder called the *standard kilogram*. The original standard is, presumably, kept in Sèvres, France, and one or more accurate duplicates are possessed by most other countries. They are not all identical in mass with the original standard, but this is not of importance since their masses relative to the standard are accurately known.

The unit of mass in the mks system is the mass of the standard kilogram. The unit of mass in the cgs system is 1/1000 as great as the mass of the standard kilogram and is called one gram.

There is no mass standard in the English gravitational system of units. That is, government laboratories do not preserve in their vaults a certain piece of matter whose mass is equal to the unit mass. The English system is based on standards of *force*, length, and time, and the unit of mass is defined in terms of these standards, as will be explained shortly.

The pound of force was defined on page 2 as the force of the earth's gravitational attraction at sea level and 45° latitude on a specified body called the standard pound. To avoid the unnecessary duplication in maintaining two such standard bodies, the standard kilogram and the standard pound, the latter is now *defined* in terms of the standard kilogram by the relation that its mass shall equal 0.4535924277 kgm.

---

*Except that at very high velocities, approaching the velocity of light, relativity effects result in an appreciable increase in mass.

5–3 Newton's second law. The observations described at the beginning of the preceding section point to a connection between force, mass, and acceleration. To obtain the quantitative relation between them, consider the following series of (idealized) experiments.

(1) A block of any arbitrary mass is placed on a level, frictionless surface and accelerated along the surface by a horizontal force exerted on it by a spring balance. For concreteness, suppose the balance has been calibrated in pounds as described on page 3. With the calibrated balance we can exert forces of 1, 2, 3, etc., lb on the block, and measure with a scale and stop watch the corresponding accelerations. The results of this series of experiments will show that with a constant mass, the acceleration is directly proportional to the accelerating force and is in the same direction as the force.

$$a \propto F \text{ (when } m \text{ is constant)}. \tag{5–1}$$

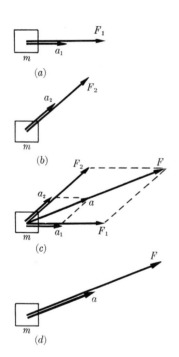

(2) For the second series of experiments, we may start with our standard kilogram and prepare a number of duplicates of it, testing them for equality of mass by observing that all accelerate at the same rate when acted on by the same force. Combinations of these will give us masses of 2 kgm, 3 kgm, etc.

Now let us apply in successive experiments the same force (any force will do) to masses of 1 kgm, 2 kgm, 3 kgm, etc., and measure the accelerations. This series of experiments leads to the result that with a constant force, the acceleration is inversely proportional to the mass.

$$a \propto \frac{1}{m} \text{ (when } F \text{ is constant)}. \tag{5–2}$$

The results of both series of experiments may now be expressed by the single relation

Fig. 5–2. The acceleration of a body is proportional to the resultant force exerted on the body and is in the direction of the resultant force.

$$a \propto \frac{F}{m}. \tag{5–3}$$

This obviously reduces to Eq. (5–1) or Eq. (5–2) when m or F is constant.

(3) Let us next experiment with more than one force acting on the body. Suppose we have found that the force F_1, acting alone, produces an acceleration a_1 in the same direction as F_1 as in Fig. 5–2(a). Similarly, force F_2 produces an acceleration a_2 as in Fig. 5–2(b). If we now apply forces F_1 and F_2 simultaneously as in Fig. 5–2(c), we find that the observed acceleration a is the same as the vector sum of the accelerations a_1 and a_2, and furthermore that the same acceleration results if instead of applying F_1 and F_2 simultaneously, we apply a single force F equal to the vector sum of F_1 and F_2 as in Fig. 5–2(d).

These experiments show, first, that when a number of forces are exerted on a body at the same time, each force acts independently of the others and produces the same acceleration as if it alone were present, and second, that the resultant acceleration is proportional to the resultant force and is in the same direction as this resultant. Hence, the proportion, Eq. (5–3), holds both for the resultant force and for each of its components. That is,

$$a_1 \propto \frac{F_1}{m}, \qquad a_2 \propto \frac{F_2}{m}, \qquad a \propto \frac{F}{m}.$$

The component accelerations produced by the components of a force will be considered in the next chapter. For the moment we shall discuss resultant forces and accelerations only.

Newton's second law is simply a formal statement of the results of experiments such as those just described. If we restrict it for the present to resultant forces and accelerations, it may be stated: *The acceleration of a body is proportional to the resultant force exerted on the body, is inversely proportional to the mass of the body, and is in the same direction as the resultant force.* That is,

$$a \propto \frac{F}{m}, \qquad \text{or} \qquad F \propto ma.$$

The second form is equivalent to

$$F = kma, \tag{5–4}$$

where k is a constant of proportionality. The magnitude of the proportionality constant will depend on the units in which force, mass, and acceleration are expressed. For example, it is found by experiment that a resultant force of one pound imparts an acceleration of 14.6 ft/sec^2 to a mass of one kilogram. Then if these units are used,

$$k = \frac{F}{ma} = \frac{1 \text{ lb}}{1 \text{ kgm} \times 14.6 \text{ ft/sec}^2} = 0.0685 \frac{\text{lb·sec}^2}{\text{kgm·ft}},$$

and $F = 0.0685\, ma$ (F in pounds, m in kgm, a in ft/sec²).

It is obviously inconvenient to have to remember all of the values of k that would be required to take care of all possible combinations of units, but since the value of k is determined solely by the units chosen for F, m, and a, why not use a combination that will give k some simple, easily remembered value? The simplest choice, of course, is to make $k = 1$, and the so-called *systems* of mechanical units are all set up with this end in view. It may well happen, as indeed it does, that some of the required units are unfamiliar ones but the advantage of making $k = 1$ outweighs the disadvantage of defining new units.

If we use a system of units in which $k = 1$, Eq. (5–4) reduces to

$$\boxed{F = ma.} \tag{5–5}$$

This equation is usually considered to be the mathematical formulation of Newton's second law. It is probably the most important equation in mechanics. Note carefully that it is a *vector* equation; that is, the resultant acceleration a is in the same direction as the resultant force F. The *algebraic* relation $F = ma$ alone is not a complete statement of the law.

We can see from Eq. (5–5) the physical conditions that must be fulfilled for motion with constant acceleration; namely, if a is constant, then F must be constant also. In other words, motion with constant acceleration is motion under the action of a constant force. If the force varies, the acceleration varies in direct proportion, since the mass m is constant.

It is also evident from Eq. (5–5) that if the resultant force on a body is zero, the acceleration of the body is zero and its velocity is constant. Hence, if the body is in motion, it continues to move with no change in the magnitude or direction of its velocity; if at rest, it remains at rest (its velocity is then constant and equal to zero). But these are evidently the conditions to which Newton's *first* law applies, and we see that the first law is merely a special case of the second when F and a are both zero. There are thus only two independent laws of Newton, the second and the third.

In calculus notation, Newton's second law becomes, for motion along the X-axis,

$$F = ma = m\frac{dv}{dt} = m\frac{d^2x}{dt^2} = mv\frac{dv}{dx}. \tag{5–6}$$

5–4 Systems of units. If we adopt $F = ma$ as the expression of Newton's second law, it follows that when $m = 1$ unit of mass and $a = 1$ unit of acceleration, then $F = 1$ unit of force. In other words, the units of force, mass, and acceleration must be so chosen that *unit force imparts unit acceleration to unit mass.* Obviously we are not at liberty to choose all three units arbitrarily. We may, however, choose any two of them and use Eq. (5–5) to fix the magnitude of the third.

In the meter-kilogram-second system of units, the kilogram fixes the unit of mass, and the meter and second together fix the unit of acceleration. The unit of force in this system must then be of such magnitude that it imparts an acceleration of one meter per second, per second, to a mass of one kilogram. This force is called *one newton.*

A newton is that force which imparts to a mass of one kilogram an acceleration of one meter per second, per second.

The newton is equal to a force of 0.224 lb.

In the centimeter-gram-second system, the unit of mass is the gram, and the unit of acceleration is the centimeter per second, per second. The unit force in this system must be of such magnitude that it imparts an acceleration of one centimeter per second, per second, to a mass of one gram. This force is called *one dyne.*

A dyne is that force which imparts to a mass of one gram an acceleration of one centimeter per second, per second.

Since 1 kgm = 1000 gm and 1 meter = 100 cm, it follows that 1 newton = 100,000 dynes = 10^5 dynes. One dyne is a force of 2.24×10^{-6} lb.

We have already defined the English gravitational unit of force, the pound; and of acceleration, the foot per second, per second. As in the other systems, we wish to have unit force impart unit acceleration to unit mass. The unit mass in this system must then be of such magnitude that when acted on by a force of one pound, its acceleration is one foot per second, per second. This mass is called *one slug.*

A slug is that mass to which a force of one pound imparts an acceleration of one foot per second, per second.

The slug is equal to 14.6 kilograms.

The newton, dyne, and slug are called *derived* units, as distinguished from the *arbitrary* units, the pound force and the kilogram mass.

As a summary, then, when Newton's second law is written in the form $F = ma$ (with $k = 1$), the following combinations of units may be used:

$$F \text{ (in newtons)} = m \text{ (in kgm)} \times a \text{ (in m/sec}^2).$$

$$F \text{ (in dynes)} = m \text{ (in grams)} \times a \text{ (in cm/sec}^2).$$

$$F \text{ (in pounds)} = m \text{ (in slugs)} \times a \text{ (in ft/sec}^2).$$

There is no general agreement on the precise roles played by experiment and definition in Newton's second law. The law was stated on page 77 as being derived from experimental observations. If we use the restricted form $F = ma$, the statement that the law is an experimental one implies that a number of experiments were performed in which a force F was exerted on a mass m, causing an acceleration a, and when F, m, and a were measured (and expressed in the appropriate units of any one system), the numerical value of F was found in every case to equal the product of the numerical values of m and a.

When we look more closely at the precise method used for measuring these quantities, we discover that our definitions of the newton, dyne, and slug were based on the assumption that F *is* equal to the product of m and a. It is not surprising, then, that we find the equality to hold in some later experiment. That is, if we use the equation $F = ma$ to *define* F, as in the mks and cgs systems, or the equivalent relation $m = F/a$ to *define* m, the law becomes merely a definition of F or m and is not susceptible of experimental verification.

A book of this nature is not the place for an extended philosophical discussion of this question. The interested reader is referred to some of the books listed immediately following Chapter 17.

5–5 Weight and mass. Every body in the universe exerts a force of gravitational attraction on every other body. The earth attracts the book and pencil lying on your desk; each of these attracts the other; the earth attracts the moon; the sun attracts the earth and the other planets of the solar system as well as the most distant stars; and each of these bodies pulls back on each of the others which attract it, with an equal and opposite force. This phenomenon of universal gravitational attraction will be considered in more detail in Chapter 15. At present we are concerned with one aspect of it only, namely, the force of gravitational attraction between the earth and bodies on or near its surface. *The force of gravitational attraction which the earth exerts on a body is called the* **weight** *of the body.*

Thus the statement that a man weighs 160 lb is equivalent to stating that he is attracted by the earth with a force of 160 lb. Since the weight of a body is a force, it must be expressed in force units, that is, in *pounds* in the English system, and in *newtons* or *dynes* in the mks or cgs systems.

The mass of a body, although it is not the same thing as the body's weight, is directly proportional to the weight as will be shown shortly. Hence the weight of any body of known mass can be found by direct proportion if one measures once and for all the weight of each unit of mass. That is, one must measure the force of the earth's attraction, in pounds, for a mass of one slug; the force of attraction, in newtons, for a mass of one kilogram; and the force of attraction, in dynes, for a mass of one gram.

The experimental method consists simply of allowing a unit mass to fall freely. While it is falling, the only force acting on it is its weight, which we wish to know, and its acceleration is that of a freely falling body.

Consider a one-slug mass falling freely. Its acceleration is the acceleration of gravity, g, at the point where the experiment is performed. In round numbers, this is 32 ft/sec^2. By definition, a unit force (one pound) imparts to a one-slug mass an acceleration of only 1 ft/sec^2. Since the freely falling slug has an acceleration of g ft/sec^2, the force accelerating it must be g times as great as the unit force, or g pounds. In other words, *one slug weighs g pounds* where g is the local acceleration of gravity expressed in ft/sec^2. In round numbers, *one slug weighs about 32 lb.*

If the body is one kilogram, falling with an acceleration of g m/sec^2, or about 9.8 m/sec^2, the accelerating force must be g newtons, since by definition a force of one newton imparts to a mass of one kilogram an acceleration of only 1 m/sec^2. Hence *one kilogram weighs g newtons*, where g is the local acceleration of gravity expressed in m/sec^2. In round numbers, *one kilogram weighs about 9.8 newtons.*

Similar reasoning shows that *one gram weighs g dynes*, where g is the local acceleration of gravity expressed in cm/sec^2. In round numbers, *one gram weighs about 980 dynes.*

All bodies, whatever their mass, fall with the same acceleration at the same point on the earth's surface. It follows that the accelerating force or the weight of a body is directly proportional to its mass. If this were not the case—if, for example, the weight of a 2-slug mass were slightly more or less than twice the weight of a 1-slug mass—then the acceleration of a 2-slug mass in free fall would not equal that of a 1-slug mass. Since weight and mass are proportional, and since the weight of each mass unit is known, the weight of any body of known mass can be found, and vice versa.

The preceding analyses can be carried out more briefly by simply applying Newton's second law to any freely falling body of mass m. The resultant force on the body is its weight w, its acceleration is g, and the equation $F = ma$ becomes $w = mg$. In other words, the weight of a body, when expressed in terms of the force unit of any system, is numerically equal to the mass of the body, in the mass unit of that system, multiplied by the corresponding value of the acceleration of gravity.

$$w = mg, \qquad m = \frac{w}{g}. \qquad (5\text{–}7)$$

The reader undoubtedly knows that the force of gravitational attraction between two bodies decreases as the distance between them increases. Therefore, the weight of a body, or the force of gravitational attraction

between the body and the earth, is not an invariant property of the body
but diminishes as the elevation of the body is increased, because of the
increased distance to the earth's center. Since the mass of a body is an
invariant property of the body, entirely independent of its position, it
follows from Eq. (5–7) that the acceleration of gravity varies in direct
proportion to the variation in a body's weight. That is, the reason g is
smaller at high altitudes than at low is that a body weighs less at high
altitudes and therefore accelerates more slowly in free fall.

Much of the confusion existing between the concepts of weight and mass arises
from the fact that the terms pound, gram, and kilogram are often used with
meanings other than those they have in the English gravitational, the mks, and
the cgs systems. In the English absolute system of units, the unit of mass is the
mass of the standard pound, the same body whose weight is the force unit in the
English gravitational system. The name "pound" is given to this unit of mass,
so we have the same name for a unit of force in one system and a unit of mass in
another. The force unit in the English absolute system is the *poundal*, defined
as the force which imparts to a "one-pound mass" an acceleration of one ft/sec².
Since the "pound mass" is $\frac{1}{32}$ as large as the slug, the poundal is $\frac{1}{32}$ as large as a
one-pound force, or about one half an ounce. The English absolute system is not
used at all in this country (except in textbooks) and we shall make no use of it.*
When the word pound is used, it will refer to a force only.

There are also two other (incomplete) systems of units in which as in the
English gravitational system, the unit of force rather than mass is arbitrarily
defined. These systems take as their units of force the weight of the standard
kilogram and the weight of a gram. The former force is called one kilogram of
force and the latter one gram of force. One kilogram of force is about 2.2 lb of
force or about 9.8 newtons; one gram of force is about 0.0022 lb or 980 dynes.
The gram of force is commonly used as a force unit in elementary physics texts,
where the reader has probably met it. The kilogram force is used as a force
unit in engineering work in those countries which use the metric system exclu-
sively. We shall use neither of these units in this book, and the words gram and
kilogram will refer to mass only.

EXAMPLES. (1) The acceleration of gravity at St. Michael, Alaska, is 32.221
ft/sec². At Panama, in the Canal Zone, it is 32.094 ft/sec². What is the weight
in pounds, at each of these points, of a body whose mass is exactly 3 slugs?
Ans.: 96.663 lb; 96.282 lb.

(2) What is the mass, in slugs, of a man whose weight is 160 lb at a point
where $g = 32.0$ ft/sec²? What would be his weight at a point where $g = 32.2$ ft/sec²?
Ans.: 5 slugs, 161 lb.

(3) Compute your own mass in slugs. Take $g = 32.2$ ft/sec².

*Except that in Chapters 18–25 on the subject of Heat, to conform to current
practice, we shall use the mass of the standard pound as a unit of mass.

(4) What is the mass of a body, hanging at rest from a cord, which produces a tension of 10^6 dynes in the cord? What is the weight of the body, in cgs and mks units? Let $g = 980$ cm/sec^2.

Ans.: 1020 gm, 10^6 dynes, 10 newtons.

(5) 454 gm weigh 1 lb. Compute your own mass in kilograms and your own weight in newtons.

(6) What is the mass, in grams, of a body which weighs exactly one dyne at a point where $g = 980$ cm/sec^2? What is the mass, in kilograms, of a body which weighs exactly one newton at this point? What is the mass, in slugs, of a body whose weight is 1 lb at a point where $g = 32$ ft/sec^2?

Ans.: 1/980 gm (about 1 milligram); 1/9.8 kgm (about one-tenth of a kilogram or 100 gm); 1/32 slug.

(Unless otherwise stated, the value of g in the following examples will be taken as 32 ft/sec^2, 9.8 m/sec^2 or 980 cm/sec^2.)

(7) What resultant force is necessary to accelerate a block weighing 48 lb at the rate of 6 ft/sec^2?

It is first necessary to find the mass of the block in slugs. Since one slug weighs 32 lb, a body weighing 48 lb has a mass of 1.5 slugs. Then from Newton's second law

$$F = ma = 1.5 \times 6 = 9 \text{ lb.}$$

(8) What resultant force is necessary to accelerate a block whose mass is 48 gm at the rate of 6 cm/sec^2?

Since the mass of the block is given directly,

$$F = ma = 48 \times 6 = 288 \text{ dynes.}$$

(9) A 10-kgm block rests on a horizontal surface. What constant horizontal force is required to give it a velocity of 4 m/sec in 2 sec, starting from rest, if the friction force between block and surface is constant and equal to 5 newtons?

Since the forces are constant, the block moves with constant acceleration, and since the velocity increases from zero to 4 m/sec in 2 sec, the acceleration is

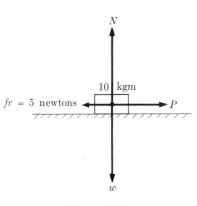

$$a = \frac{v - v_0}{t} = \frac{4}{2} = 2 \text{ m/sec}^2.$$

Let P, Fig. 5–3, represent the required horizontal force. The *resultant* force F exerted on the block is then

$$F = P - fr = P - 5.$$

(F is not shown in the figure.)

Finally, since resultant force equals mass times acceleration,

$$F = ma,$$
$$P - 5 = 10 \times 2 = 20 \text{ newtons,}$$
$$P = 20 + 5 = 25 \text{ newtons.}$$

FIGURE 5–3

FIG. 5-4. The resultant force is $T - w$.

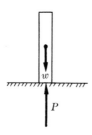

FIG. 5-5. The resultant force is $P - w$.

(10) An elevator weighing 8 tons is given an upward acceleration of 4 ft/sec². Find the tension in the supporting cable.

Let T, Fig. 5-4, represent the tension in pounds. The resultant force F acting on the elevator is $T - w = T - 16,000$ lb. The mass of the elevator is 500 slugs. Hence,

$$F = ma$$
$$T - 16,000 = 500 \times 4$$
$$T = 18,000 \text{ lb or 9 tons.}$$

(11) With what force will the elevator push upward on a 160-lb passenger, while the elevator is accelerating at the above rate?

The passenger is represented schematically in Fig. 5-5. The forces on him are the upward push P of the elevator floor, and his weight of 160 lb acting down. The resultant force is therefore $P - w = P - 160$ lb. Hence,

$$F = ma$$
$$P - 160 = 5 \times 4$$
$$P = 180 \text{ lb.}$$

According to Newton's third law the passenger exerts an equal and opposite force on the elevator floor. Hence, while the elevator is accelerating upward at 4 ft/sec², a 160-lb passenger presses down on the floor with a force of 180 lb.

(12) The driver of an automobile, traveling at 30 mi/hr on a level road, applies the brakes and comes to rest in a distance of 100 ft. If the weight of car and load is 1600 lb, and the acceleration is constant, find the friction force between tires and road.

The mass of the automobile is 50 slugs. Its acceleration may be found from

$$v^2 = v_0{}^2 + 2ax,$$
$$0 = (44)^2 + 2a(100),$$
$$a = -9.68 \text{ ft/sec}^2.$$

The magnitude of the braking force P is therefore

$$P = ma = 50 \times (-9.68) = 484 \text{ lb.}$$

The minus sign means that the force is toward the left if the car was originally moving toward the right.

(13) With what acceleration will a block slide down a frictionless plane, inclined at an angle θ with the horizontal?

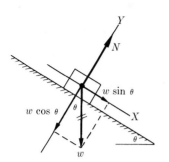

FIG. 5–6. N and w are the forces exerted on the block. The resultant force is $w \sin \theta$.

The forces acting on the block are its weight and the normal force exerted by the plane. (Fig. 5–6.) Neither the weight nor the mass of the block is given as part of the data. Hence a letter must be used to represent one or the other. Let the weight be called w. Take axes parallel and perpendicular to the surface of the plane and resolve w into its components. Since the block remains on the plane, the Y-components are in equilibrium and $N = w \cos \theta$. The only remaining force is then $w \sin \theta$, which is therefore the resultant force exerted on the block. In terms of its weight w, the mass of the block is $m = w/g$. Hence,

$$F = ma,$$

$$w \sin \theta = \frac{w}{g} a,$$

$$a = g \sin \theta.$$

Since the weight does not appear in the final result, it follows that any block, regardless of its weight, will slide down a frictionless inclined plane of slope angle θ with an acceleration $g \sin \theta$.

The following examples illustrate situations in which more than one body is involved. A similar example has been considered earlier (see Section 2–5). It will be emphasized again that in such instances it is necessary to consider each part of the system separately, and to show in separate force diagrams *all* the forces exerted *on* that part of the system under consideration. This procedure is spoken of as *isolating* one part of the system at a time. The complete group of forces acting *on* the isolated part of the system is called a *set of forces*. It is extremely important to understand this process of isolating a part of a system and to be able to recognize the set of forces acting on it.

(14) A 16-lb and an 8-lb block (Fig. 5–7) on a horizontal frictionless surface are connected by cord A and are pulled along the surface with a uniform acceleration of 4 ft/sec^2 by a second cord B. Show in a diagram the set of forces acting on each body and find the tension in each cord.

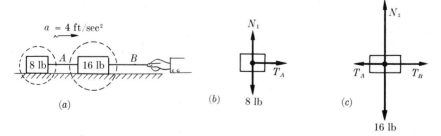

FIGURE 5–7

Each body is to be isolated as indicated by the dotted lines, and a force diagram drawn for each. Let T_A and T_B represent the tensions in cords A and B. The set of forces acting on the 8-lb block consist of (a) its weight of 8 lb, vertically down; (b) the normal push of the surface on it, N_1; (c) the tension T_A toward the right. The set of forces on the 16-lb block are (a) its weight of 16 lb; (b) the normal force N_2; (c) the tension T_A toward the left; (d) the tension T_B toward the right.

Cord A simply serves to transmit a force from one block to the other, so the forces labelled T_A constitute an action-and-reaction pair and are numerically equal. Since the vertical forces on each block are in equilibrium, $N_1 = 8$ lb and $N_2 = 16$ lb. Hence the resultant force on the 8-lb block is T_A, and the resultant force on the 16-lb block is $T_B - T_A$. The acceleration of each block is 4 ft/sec^2 (given). Applying the second law to *each* body, we have

$$T_A = \frac{8}{32} \times 4, \tag{5-8}$$

$$T_B - T_A = \frac{16}{32} \times 4. \tag{5-9}$$

Hence

$$T_A = 1 \text{ lb}, \qquad T_B = 3 \text{ lb}.$$

Notice carefully that although the hand exerts a pull of 3 lb on the system through cord B, this pull is *not* transmitted as a 3-lb force to the 8-lb block. It is cord A which pulls on the 8-lb block, and the tension in cord A is only one pound.

If the tension in cord B only is desired, the two blocks may be considered together. Their combined mass is $\frac{3}{4}$ slug, and the resultant force exerted *on the combination* is simply the tension in cord B. Therefore

$$T_B = \tfrac{3}{4} \times 4 = 3 \text{ lb}. \tag{5-10}$$

If Eqs. (5–8) and (5–9) are added, one obtains

$$T_B = \left(\frac{8}{32} + \frac{16}{32} \right) \times 4 = 3 \text{ lb},$$

which is the same as Eq. (5–10). That is, the algebraic equation obtained by adding the equations for the isolated parts of a system will always be the same as that secured by considering the system as a whole.

(15) Suppose the 8-lb block hangs vertically as in Fig. 5–8. Neglect all friction forces and the inertia of the pulley. Find the acceleration and the tension in the cord.

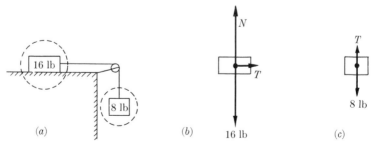

FIGURE 5–8

Isolate as indicated by the dotted lines. The only effect of the pulley is to change the direction of the cord, which pulls with the same force, T, on each block. The resultant force on the block on the table is T lb. The resultant force on the hanging block is $8 - T$ lb. Hence,

$$T = \frac{16}{32} \times a, \qquad 8 - T = \frac{8}{32} \times a,$$

and

$$a = 10\tfrac{2}{3} \text{ ft/sec}^2, \qquad T = 5\tfrac{1}{3} \text{ lb}.$$

Notice carefully that although the earth pulls on the hanging block with a force of 8 lb, this force is not transmitted to the 16-lb block. The force on the latter is the tension in the connecting cord, and this must be less than 8 lb; otherwise the 8-lb block would not accelerate downward.

5–6 D'Alembert's principle. Newton's second law,

$$F = ma,$$

can be written

$$F - ma = 0. \tag{5–11}$$

It was pointed out by D'Alembert that this form of the equation could be interpreted as follows. Suppose that in addition to the actual forces exerted on a body there was also exerted a *fictitious* force, equal in magnitude but opposite in sense, to the product ma; in other words, a fictitious force $-ma$. This force is sometimes called an "inertial force" or an "inertial reaction." Then, since F represents the resultant of the *actual* external forces, $F - ma$ represents the resultant of *all* the forces including the fictitious force $-ma$. Equation (5–11) then states that the resultant force on the body is zero. Hence the problem reduces to an equilibrium problem and can be treated by the methods of statics. That is, every body, whether accelerated or not, can be considered in equilibrium under the combined effect of the actual forces exerted on it, together with a fictitious force equal in magnitude to ma but oppositely directed. This is D'Alembert's principle.

(a)   (b)

Fig. 5–9. (a) The Newtonian viewpoint. (b) The D'Alembert viewpoint.

The Newtonian and the D'Alembert points of view are illustrated in Fig. 5–9, which represents a body of mass m, pulled to the right on a level, frictionless surface by an external force F. The Newtonian viewpoint, in

Fig. 5–9(a), is that the resultant force is the force F, and this resultant force equals the product of mass and acceleration. The D'Alembert viewpoint, in Fig. 5–9(b), is that the resultant force is $F - ma$ and is equal to zero.

Note that if D'Alembert's principle is used, one must abandon or modify the usual statement of Newton's second law. That is, it is not true that the resultant force (in the D'Alembert sense) equals the product of mass and acceleration, since the resultant force (in the D'Alembert sense) is always zero, even when a body is accelerated.

Considered merely as a technique for solving problems, there is little to choose between the viewpoints of Newton and D'Alembert, since both lead to the same algebraic equations, but for the purpose of understanding the principles of dynamics, the Newtonian method is much to be preferred and we shall make no use of fictitious D'Alembert forces in this book.

It should be stated for completeness that there is, in fact, more to D'Alembert's principle than the mere addition of a fictitious force $-ma$ to the actual set of forces on a body. If the observer's reference system moves with the same acceleration as the body, the body has no acceleration relative to the observer. The observer (who knows about Newton's second law) therefore reasons that since the acceleration of the body (as far as he can tell) is zero, the resultant force on the body must be zero. He accordingly concludes that in addition to the "real" forces whose resultant is F, another force $-ma$ is acting on the body to preserve equilibrium. For further analysis of the equations of motion in accelerated systems, the reader is referred to a more advanced text on Mechanics.

5–7 Density. The density of a homogeneous material is defined as its mass per unit volume. Densities are therefore expressed in grams per cubic centimeter, kilograms per cubic meter, or slugs per cubic foot. We shall represent density by the Greek letter ρ (rho).

$$\rho = \frac{m}{V}, \qquad m = \rho V.$$

The definition above gives the *average* density of a body. If the density varies from point to point, the density at a point is defined by constructing a small volume element dV at the point, and taking the ratio of the mass of the element, dm, to its volume dV.

$$\rho = \frac{dm}{dV}, \qquad dm = \rho \, dV.$$

The total mass of the body may then be written

$$m = \int dm = \int \rho \, dV,$$

where the limits of integration must be chosen so as to include the entire volume of the body, and ρ may be a function of the coordinates of dV.

TABLE II

DENSITIES

Material	Density (gm/cm^3)
Aluminum	2.7
Brass	8.6
Copper	8.9
Gold	19.3
Ice	0.92
Iron	7.8
Lead	11.3
Platinum	21.4
Silver	10.5
Steel	7.8
Mercury	13.6
Ethyl alcohol	0.81
Benzene	0.90
Glycerin	1.26
Water	1.00

In engineering work in this country, and in everyday life as well, the term density is used for the *weight* per unit volume, the common unit being the pound per cubic foot. This quantity may be distinguished from that defined above by calling it the "weight-density." For example, the weight-density of water is 62.5 lb per cubic foot; its density is $62.5/32.2 = 1.94$ slugs per cubic foot.

The *specific gravity* of a material is the ratio of its density to that of water and is therefore a pure number. The specific gravity of lead, for example, is 11.3 in any system of units. The density of lead, in English units, is 21.9 slugs/ft^3 and its "weight-density" is 706 lb/ft^3. In cgs units the density of water* is 1 gm/cm^3 and the density of lead is 11.3 gm/cm^3. In mks units the density of water is 1000 kgm/m^3 and the density of lead is 11,300 kgm/m^3.

"Specific gravity" is an exceedingly poor term since it has nothing to do with gravity. "Relative density" would describe the concept more precisely.

5–8 The equal-arm analytical balance. The equal-arm analytical balance is a common laboratory instrument for measuring masses with a

---

*Small variations with temperature are neglected here. See Section 18–6.

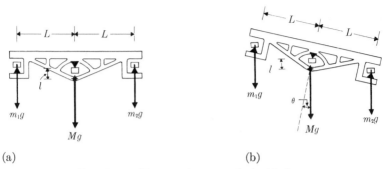

(a)   (b)

FIG. 5–10. The equal-arm analytical balance.

high degree of precision. Although the process of using a balance is spoken of as "weighing," and the standard masses used with the balance are called a "set of weights," it is actually mass and not weight which the balance measures.

The essential feature of the equal-arm analytical balance is a light, rigid beam on which are firmly mounted three equally spaced agate knife-edges, parallel to one another and perpendicular to the length of the beam. The central knife-edge rests on a polished plane agate plate, supported from the floor of the balance case. The scale pans are hung from two similar plates resting on the knife-edges at the ends of the beam. A vertical pointer, fastened to the beam, swings in front of a scale.

The knife-edges act as practically frictionless pivots. Since the scale pans can swing freely about their supporting knife-edges, the center of gravity of pan and weights will always be directly below the knife-edges. The center of gravity of the beam is directly below the central knife-edge when the beam is horizontal. The beam is, therefore, a body in equilibrium under the action of a number of parallel forces.

In using the balance, a body of unknown mass, say m_1, is placed in the left pan, and known masses m_2 are placed in the right. Suppose m_2 is slightly larger than m_1. The forces acting on the balance beam are shown in Fig. 5–10(a). Mg is the weight of the beam. Since this force has no moment about the central knife-edge, the resultant torque on the beam is

$$(m_2 g)L - (m_1 g)L = (m_2 - m_1)gL$$

in a clockwise direction. This unbalanced torque causes the beam to deflect clockwise as in Fig. 5–10(b). As it does so, the deflecting torque decreases, becoming

$$(m_2 - m_1)gL \cos \theta,$$

while at the same time the restoring torque $Mgl \sin \theta$ comes into play. A position of equilibrium will eventually be reached, in which these two torques become equal. If θ is the equilibrium angle, then from $\Sigma\tau = 0$,

$$Mgl \sin \theta = (m_2 - m_1)gL \cos \theta,$$

$$\frac{\sin \theta}{\cos \theta} = \tan \theta = \frac{m_2 - m_1}{M} \frac{L}{l}. \tag{5–11}$$

Hence, if the standard mass m_2 is adjusted until $\theta = 0$, it follows that $m_2 = m_1$, and the unknown mass is equal to the standard mass.

The most sensitive balance is one which deflects through the largest angle θ for a given difference between m_1 and m_2. But if θ is to be large, then from Eq. (5–11), L should be large and M and l should be small. In other words, by using a long light beam, with its center of gravity only a short distance below the central pivot, the sensitivity of the balance may be made as large as desired. Unfortunately, the same conditions which make for great sensitivity also make the balance extremely slow in taking up its final position. Hence a compromise must be struck between sensitivity and time of swing. A good balance will measure to a tenth of a milligram, and a really excellent instrument to a hundredth or a thousandth of a milligram.

If the lengths of the balance arms are not exactly equal, the mass needed to balance an unknown mass will depend upon which scale pan is used for the known and which for the unknown. Corrections for this error may be made by balancing with the unknown first on one side and then on the other. The true mass is then the geometric mean, or the square root of the product, of the standard masses required for balance.

In work of the highest precision, correction must be made for the buoyancy of the air if the density of the unknown mass differs from that of the standard masses. This correction will be explained in Chapter 16.

PROBLEMS

(For problem work, use the approximate values of $g = 32$ ft/sec^2 = 9.8 m/sec^2 = 980 cm/sec^2. A force diagram should be constructed for each problem.)

5–1. The Springfield rifle bullet weighs 150 grains (7000 grains = 1 lb), its muzzle velocity is 2700 ft/sec, and the length of the rifle barrel is 30 in. Compute the resultant force accelerating the bullet, assuming it to be constant.

5–2. A body of mass 15 kgm, initially at rest on a frictionless horizontal plane, is acted on by a horizontal force of 30 newtons. (a) What acceleration is produced? (b) How far will the body travel in 10 sec? (c) What will be its velocity at the end of 10 sec?

5–3. A body of mass 50 gm is at rest at the origin, $x = 0$, on a horizontal frictionless surface. At time $t = 0$ a force of 10 dynes is applied to the body parallel to the X-axis, and 5 sec later this force is removed. (a) What are the position and velocity of the body at $t = 5$ sec? (b) If the same force (10 dynes) is again applied at $t = 15$ sec, what are the position and velocity of the body at $t = 20$ sec?

5–4. A .22 rifle bullet, traveling at 36,000 cm/sec, strikes a tree which it penetrates to a depth of 10 cm. The mass of the bullet is 1.8 gm. Assume a constant retarding force. (a) How long a time was required for the bullet to stop? (b) What was the decelerating force, in dynes? in lbs?

5–5. An electron (mass = 9×10^{-28} gm) leaves the cathode of a radio tube with zero initial velocity and travels in a straight line to the anode, which is 1 cm away. It reaches the anode with a velocity of 6×10^8 cm/sec. If the accelerating force was constant, compute (a) the accelerating force, in dynes, (b) the time to reach the anode, (c) the acceleration. The gravitational force on the electron may be neglected.

5–6. If action and reaction are always equal and opposite, why don't they always cancel one another and leave no net force for accelerating a body?

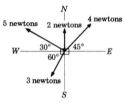

FIGURE 5–11

5–7. Fig. 5–11 is a top view of a block whose mass is 5 kgm, resting on a horizontal frictionless surface and acted on by four horizontal forces. Find the magnitude and direction of the acceleration of the block.

5–8. The mass of a certain object is 10 gm. (a) What would its mass be if taken to the planet Mars? (b) Is the expression $F = ma$ valid on Mars? (c) Newton's second law is sometimes written in the form $F = Wa/g$ instead of $F = ma$. Would this expression be valid on Mars? (d) If a Martian scientist hangs this 10-gm mass on a spring balance calibrated correctly on the earth, would the spring balance read 10 gm? (e) How do you reconcile this with your answer to (a)?

5–9. A block weighing 10 lb is held up by a string which can be moved up or down. What conclusions can you draw regarding the magnitude and direction of the acceleration and velocity of the upper end of the string, when the tension in the string is (a) 5 lb, (b) 10 lb, (c) 15 lb?

5–10. A body hangs from a spring balance supported from the roof of an elevator. (a) If the elevator has an upward acceleration of 4 ft/sec^2 and the balance reads 45 lb, what is the true weight of the body? (b) Under what circumstances will the balance read 35 lb? (c) What will the balance read if the elevator cable breaks?

5–11. A 5-kgm block is supported by a cord and pulled upward with an acceleration of 2 m/sec^2. (a) What is the tension in the cord? (b) After the block has been set in motion the tension in the cord is reduced to 49 newtons. What sort of motion will the block perform? (c) If the cord is now slackened completely, the block is observed to move up 2 meters farther before coming to rest. With what velocity was it traveling?

5–12. A balloon is descending with a constant acceleration a, less than the acceleration of gravity g. The weight of the balloon, with its basket and contents, is w. What weight, W, of ballast should be released so that the balloon will begin to be accelerated upward with constant acceleration a? Neglect air resistance.

5–13. A body starting from rest is pulled along a horizontal frictionless surface by a constant force for 10 sec. At the end of this time the velocity is 10 cm/sec. For the next 10-second interval the force is zero. At the end of this interval a force of one-half the original force, and in the opposite direction, is applied until the body comes to rest. (a) Make a graph of velocity vs time. (b) Give the time when the body comes to rest. (c) How far did the body move in the first 10-second interval? (d) What is the total distance traveled by the body? (*Note:* All these questions can be answered by reference to the graph.)

5–14. A particle of mass 2 gm is acted upon by two forces, a constant force of 540 dynes to the left, and a variable force of 120t dynes to the right (where t is in seconds). The particle starts from rest at the origin. (a) Give a brief verbal description of the motion. (b) What is the greatest distance that the particle moves to the left of the origin?

5–15. A body of mass 10 kgm moves along the X-axis. Its velocity at any

time is given by the equation $v = 2 + 9t^2$, where v is in meters/sec and t is in seconds. The body is 5 meters to the right of the origin at time $t = 0$. (a) Find the force acting on the body at the instant when $t = 4$ sec. (b) Find the displacement of the body during the interval between $t = 1$ sec and $t = 3$ sec. (c) Find the average velocity of the body during the same time interval as in part (b).

5–16. A body of mass m moves along the X-axis, its velocity being given by $v = A \cos \omega t$ where A and ω are constants. What is the resultant force acting on the body at any time t? What is the force when the coordinate of the body is x?

5–17. If the coefficient of friction between tires and road is 0.5, what is the shortest distance in which an automobile can be stopped when traveling at 60 mi/hr?

5–18. A hockey puck leaves a player's stick with a velocity of 30 ft/sec and slides 120 ft before coming to rest. Find the coefficient of friction between the puck and the ice.

5–19. A transport plane is to take off from a level landing field with two gliders in tow, one behind the other. Each glider weighs 2400 lb, and the friction force or drag on each may be assumed constant and equal to 400 lb. The tension in the towrope between the transport plane and the first glider is not to exceed 2000 lb. (a) If a velocity of 100 ft/sec is required for the take-off, how long a runway is needed? (b) What is the tension in the towrope between the two gliders, while the planes are accelerating for the take-off?

5–20. A locomotive weighing 100 tons is to pull a train of n cars on a level track. Each car weighs 10 tons. The coefficient of friction between wheels and the track is 0.2. If the train is to be capable of an acceleration of 0.64 ft/sec^2 without slipping, how many cars can the locomotive haul? Neglect all friction except that between the wheels and the track.

5–21. An 80-lb packing case is on the floor of a truck. The coefficient of static friction is 0.25. Calculate all the forces acting on the packing case and show them in a force diagram for each of the following cases: (a) The truck is at rest. (b) The truck is traveling in a straight line with a constant velocity. (c) The truck is accelerating at 6 ft/sec².

5–22. A block of mass 2.5 slugs rests on a horizontal surface. The coefficient of static friction between the block and the surface is 0.3 and the coefficient of sliding friction is 0.25. The block is acted upon by a variable horizontal force P. This force is initially zero and increases with time at the constant rate of 2 lb/sec. (a) When will the block start to move? (b) What is its acceleration 8 sec after it starts to move? (c) What is its velocity 8 sec after it starts to move?

5–23. A body weighing 64 lb slides down an inclined plane of angle 37° and length 16 ft in 2 sec. If the body started from rest at the top of the plane, what was the coefficient of sliding friction?

5–24. A body of mass 5 kgm starts from rest at the foot of a smooth inclined plane of angle 30° and length 4.9 meters, and reaches the top of the plane in 10 sec. What external force parallel to the plane was exerted on the body?

5–25. A block slides with constant velocity down an inclined plane of slope angle α. With what acceleration will it slide down the same plane when the slope angle is increased to a larger value, θ?

5–26. A 64-lb block is pushed up a 37° inclined plane by a horizontal force of 100 lb. The coefficient of sliding friction is 0.25. Find (a) the acceleration, (b) the velocity of the block after it has moved a distance of 20 ft along the plane, (c) the normal force exerted by the plane.

5–27. A block rests on an inclined plane which makes an angle θ with the horizontal. The coefficient of sliding friction is 0.5 and the coefficient of static friction is 0.75. (a) As the angle θ is increased, find the minimum angle at which the block starts to slip. (b) At this angle, find the acceleration once the block has begun to move. (c) How long a time is required for the block to slip 20 ft along the inclined plane?

5–28. A block slides with constant velocity down an inclined plane of slope angle α. If it is projected up the same plane with an initial velocity v_0, how far up the plane will it move before coming to rest?

5–29. A block whose mass is 1 slug rests on a horizontal frictionless surface and is connected by a cord passing over a pulley to a hanging weight. (a) What must be the tension in the cord to accelerate the block at 1 ft/sec²? (b) What hanging weight will produce this acceleration?

5–30. Block A rests on a horizontal frictionless surface and is connected by a cord passing over a pulley to a hanging block B. The inertia of cord and pulley can be neglected. The mass of block B is 10 kgm. The system is released from rest and block B is observed to descend 80 cm in 4 sec. (a) Show in a diagram all of the forces acting on block B, and compute the tension in the cord. (b) Show in a second diagram all of the forces exerted on block A, and compute its mass.

5–31. A man who weighs 160 lb stands on a platform which weighs 80 lb. He pulls a rope which is fastened to the platform and runs over a pulley on the ceiling. With what force does he have to pull in order to give himself and the platform an upward acceleration of 2 ft/sec²?

5–32. A plumb bob hangs from the roof of a passenger car. (a) Find the angle of inclination of the cord with the vertical when the train has a forward acceleration of 4 ft/sec². (b) What is the general expression for the angle when the acceleration is a? (This problem illustrates the principle of one form of accelerometer.)

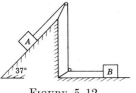

FIGURE 5–12

5–33. Two blocks, each having mass 20 kgm, rest on frictionless surfaces as shown in Fig. 5–12. Assuming the pulleys to be light and frictionless, compute: (a) the time required for block A to move 1 meter down the plane, starting from rest, (b) the tension in the cord connecting the blocks.

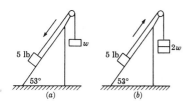

FIGURE 5–13

5–34. A block weighing 5 lb, connected to an unknown weight w as in Fig. 5–13(a), is observed to slide down a 53° inclined plane with an acceleration of 4 ft/sec². When the weight w is doubled, the block slides up the plane with an acceleration of 4 ft/sec². (a) What is the weight w? (b) What is the coefficient of sliding friction between the 5-lb block and the plane?

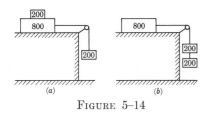

FIGURE 5–14

5–35. A block of mass 200 gm rests on the top of a block of mass 800 gm. The combination is dragged along a level surface at constant velocity by a hanging block of mass 200 gm as in Fig. 5–14(a). (a) The first 200-gm block is removed from the 800-gm block and attached to the hanging block, as in Fig. 5–14(b). What is now the acceleration of the system? (b) What is the tension in the cord attached to the 800-gm block in part (b) of the figure?

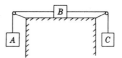

FIGURE 5–15

5–36. Block A in Fig. 5–15 weighs 3 lb and block B weighs 30 lb. The coefficient of friction between B and the horizontal surface is 0.1. (a) What is the weight of block C if the acceleration of B is 6 ft/sec² toward the right? (b) What is the tension in each cord when B has the acceleration stated above?

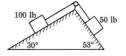

FIGURE 5–16

5–37. Two blocks connected by a cord passing over a small frictionless pulley rest on frictionless planes as shown in Fig. 5–16. (a) Which way will the system move? (b) What is the acceleration of the blocks? (c) What is the tension in the cord?

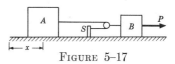

FIGURE 5–17

5–38. Two blocks are connected as shown in Fig. 5–17. A string is fastened to block A, passes over the weightless pulley on block B, and is fastened to a support S. Block A weighs 16 lb, block

B weighs 4 lb. The blocks, pulled by a force P, slide along a frictionless surface. The displacement of block A is found to be $x = 2t^3$, where x is in feet and t in seconds. (a) What is the acceleration of block A when $t = 5$ sec? (b) What is the tension T in the string at this moment? (c) What is the force P at this moment?

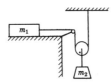

FIGURE 5–18

5–39. In terms of m_1, m_2, and g, find the accelerations of both blocks in Fig. 5–18. Neglect all friction and the masses of the pulleys.

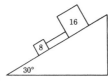

FIGURE 5–19

5–40. Two blocks, weighing 8 and 16 lb respectively, are connected by a string and slide down a 30° inclined plane, as in Fig. 5–19. The coefficient of friction between the 8-lb block and the plane is 0.25 and between the 16-lb block and the plane it is 0.5. (a) Calculate the acceleration of each block. (b) Calculate the tension in the string.

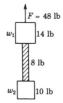

FIGURE 5–20

5–41. The two blocks in Fig. 5–20 are connected by a heavy uniform rope which weighs 8 lb. An upward force of 48 lb is applied as shown. (a) What is the acceleration of the system? (b) What is the tension at the top of the 8-lb rope? (c) What is the tension at the mid-point of the rope?

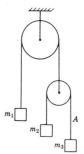

FIGURE 5–21

5–42. Find the tension in cord A, Fig. 5–21, in terms of the masses m_1, m_2, m_3, and the acceleration of gravity, g. (Neglect the masses of the pulleys.)

5–43. The weights w_1 and w_2 in Fig. 5–22 are initially at rest on the floor. They are connected by a weightless string passing over a weightless and frictionless pulley. An upward force F is applied to the pulley. Find the accelerations a_1 of w_1 and a_2 of w_2 when F is (a) 24 lb, (b) 40 lb, (c) 72 lb, (d) 90 lb, (e) 120 lb.

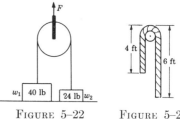

FIGURE 5–22   FIGURE 5–23

5–44. A flexible rope 10 ft long, weighing $\frac{1}{2}$ lb per foot, passes over a small frictionless pulley. It is released from rest with 4 ft of the rope hanging from one side and 6 ft from the other side of the pulley, as in Fig. 5–23. (a) What is the initial acceleration of the rope? (b) What is the acceleration of the rope when a length x hangs from the right-hand side of the pulley?

CHAPTER 6

MOTION OF A PROJECTILE

6–1 Projectiles. In this chapter we shall discuss the motion of a projectile, such as a baseball or golf ball, a bomb released from a plane, a rifle bullet, or the shell of a gun. The path followed by a projectile is called its *trajectory*. The trajectory is affected to a large extent by air resistance, which makes an exact analysis of the motion extremely complex. In fact, the subject of exterior ballistics, which is the term applied to the calculation of the trajectories of bullets or shells, is a science in itself. We shall, however, neglect the (important) effects of air resistance and assume that the motion takes place in empty space.

The motion of a projectile is most readily analyzed with the aid of Newton's second law expressed in component form. As we saw in the preceding chapter, each component of the force exerted on a body can be considered to produce its own component of acceleration. Then if F_x and F_y are the X- and Y-components of a force F exerted on a body of mass m, the X-component of the force equals the product of the mass and the X-component of acceleration, and the Y-component of the force equals the product of the mass and the Y-component of acceleration.

$$F_x = ma_x, \qquad F_y = ma_y. \tag{6–1}$$

The force F is often the resultant of a number of applied forces. Then F_x and F_y have the same meaning as ΣX and ΣY in Chapter 2. Hence Eq. (6–1) can be written

$$\Sigma X = ma_x, \qquad \Sigma Y = ma_y. \tag{6–2}$$

If the mass is in equilibrium, a_x and a_y are both zero. Hence, for equilibrium,

$$\Sigma X = 0, \qquad \Sigma Y = 0.$$

That is, Eq. (6–2) includes the so-called conditions of equilibrium as a special case.

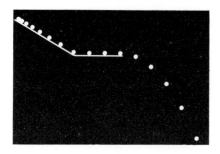

FIG. 6–1. Constant acceleration down the incline, constant velocity on the level track, and a combination of constant velocity and constant acceleration after leaving the track.

6–2 Motion of a body projected horizontally. Figure 6–1 is a multiflash photograph of a ball which rolls down an inclined track, then along a horizontal track, and finally leaves the track and moves as a projectile. After leaving the track, the only force on the ball is its weight. Hence

$$F_x = 0 = ma_x, \qquad (6\text{--}3)$$

$$F_y = mg = ma_y. \qquad (6\text{--}4)$$

Since the horizontal force component is zero, there is no horizontal acceleration, and the horizontal velocity component remains constant and equal to the velocity on the level portion of the track. This is proved by the fact that the *horizontal* spacing of the images remains the same throughout the trajectory. On the other hand, since there is a resultant vertical force, there will be a vertical acceleration in the direction of this force. The *vertical* spacing of the images therefore increases along the trajectory.

The vertical acceleration is found from Eq. (6–4); namely, $a_y = g$. That is, the vertical acceleration is the same as that of a body falling in a vertical line and is quite unaffected by the fact that the body has at the same time a horizontal velocity component. The forward velocity of the body does not "support" it in flight.

An interesting demonstration of this fact is afforded by the experiment shown in Fig. 6–2. One ball is projected horizontally from a spring gun at the upper left of the picture. As it leaves the muzzle of the gun, it operates a small tripswitch which opens the circuit of an electromagnet and releases a second ball at the upper right. It will be seen that both balls fall at precisely the same rate, and that when the first reaches the line of motion of the second, a collision takes place in mid-air.

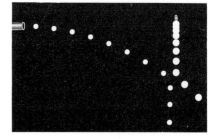

FIG. 6–2. The vertical acceleration is the same for both bodies.

Figure 6–3 is a drawing corresponding to a portion of the trajectory of Fig. 6–1. X- and Y-axes have been constructed with origin at the point where the ball leaves the track and begins its flight as a projectile. Let

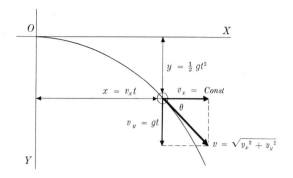

Fɪɢ. 6–3. Trajectory of a body projected horizontally.

$t_0 = 0$ at the origin, and t the time when the ball is in the position shown. The velocity of the ball can be found by computing separately its horizontal and vertical components and combining them by the usual methods for vector addition. The horizontal velocity component is denoted in Fig. 6–3 by v_x. We have seen that the horizontal acceleration is zero and that the horizontal velocity component, v_x, remains constant throughout the motion. Since the vertical acceleration is g (taking the downward direction as positive), the vertical velocity component at time t is

$$v_y = gt$$

(the *initial vertical* velocity is zero).

The magnitude of the velocity is therefore

$$v = \sqrt{v_x^2 + v_y^2},$$

and its direction can be found from

$$\tan \theta = \frac{v_y}{v_x}.$$

The velocity vector, v, is tangent to the trajectory, and its direction at any instant is the direction in which the projectile is moving at that instant.

The horizontal *displacement* at time t is

$$x = v_x t,$$

and the vertical displacement is

$$y = \tfrac{1}{2} g t^2.$$

The equation of the trajectory may be found by eliminating t from the two preceding equations. From the first we have $t^2 = x^2/v_x^2$, and introducing this in the second gives

$$y = \left(\frac{1}{2}\frac{g}{v_x^2}\right)x^2.$$

Since g and v_x are constants, the expression in parentheses is a constant, say k. Hence the equation of the trajectory has the form

$$y = kx^2,$$

which will be recognized as the equation of a parabola.

EXAMPLE. The ball in Fig. 6–3 leaves the track with a velocity v_x of 8 ft/sec. Find its position and velocity after $\frac{1}{4}$ sec.

The horizontal displacement is

$$x = v_x t = 8 \times \tfrac{1}{4} = 2 \text{ ft,}$$

and the vertical displacement is

$$y = \tfrac{1}{2}gt^2 = \tfrac{1}{2} \times 32 \times (\tfrac{1}{4})^2 = 1 \text{ ft.}$$

The ball is therefore two feet out from its starting point and one foot down.

The horizontal velocity component is

$$v_x = \text{constant} = 8 \text{ ft/sec,}$$

and the vertical component at this instant is

$$v_y = gt = 32 \times \tfrac{1}{4} = 8 \text{ ft/sec.}$$

The resultant velocity is therefore

$$v = \sqrt{v_x^2 + v_y^2} = \sqrt{(8)^2 + (8)^2} = 8\sqrt{2} \text{ ft/sec,}$$

and since

$$\tan \theta = \frac{v_y}{v_x} = \frac{8}{8} = 1,$$

the direction of the velocity is 45° below the horizontal.

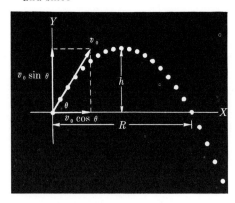

FIG. 6–4. Trajectory of a body projected at an angle with the horizontal.

6–3 Body projected at an angle. In the most general case of projectile motion the body is given an initial velocity at some angle θ above (or below) the horizontal. Such a trajectory is shown in the multiflash photograph of Fig. 6–4, to which have been added X- and Y-axes and velocity vectors. Let v_0 represent the initial velocity (called the muzzle velocity if the projectile is a bullet or shell). Its horizontal and vertical components are

$$v_{0x} = v_0 \cos \theta, \qquad v_{0y} = v_0 \sin \theta.$$

(The upward direction is considered positive.)

As in Fig. 6–2, the horizontal velocity component remains constant throughout the motion. The vertical part of the motion is one of constant downward acceleration, and is the same as that of a body projected vertically upward with an initial velocity $v_0 \sin \theta$. At a time t after the start, the horizontal velocity is

$$v_x = v_{0x} = v_0 \cos \theta = \text{constant}, \tag{6–5}$$

and the vertical velocity

$$v_y = v_{0y} - gt = v_0 \sin \theta - gt. \tag{6–6}$$

The horizontal displacement is

$$x = v_{0x}t = (v_0 \cos \theta)t, \tag{6–7}$$

and the vertical displacement

$$y = v_{0y}t - \tfrac{1}{2}gt^2 = (v_0 \sin \theta)t - \tfrac{1}{2}gt^2. \tag{6–8}$$

The maximum height, h, is reached at a time when the vertical velocity component has decreased to zero. Setting $v_y = 0$ in Eq. (6–6), we find for this time

$$t = \frac{v_0 \sin \theta}{g}.$$

Hence from Eq. (6–8) the maximum height is

$$h = \frac{v_0^2 \sin^2 \theta}{2g}. \tag{6–9}$$

The time for the body to return to its initial elevation is found from Eq. (6–8) by setting $y = 0$. This gives

$$t = \frac{2v_0 \sin \theta}{g}. \tag{6–10}$$

Notice that this is just twice the time to reach the highest point.

The horizontal displacement when the ball returns to its initial elevation is called the *horizontal range*. Introducing the time to reach this point in Eq. (6–7), we find

$$R = \frac{2v_0^2 \sin \theta \cos \theta}{g}. \tag{6–11}$$

Since $2 \sin \theta \cos \theta = \sin 2\theta$, Eq. (6–11) may be written

$$R = \frac{v_0^2 \sin 2\theta}{g}.\qquad(6\text{–}12)$$

The horizontal range is thus proportional to the square of the initial velocity for a given angle of elevation. Since the maximum value of $\sin 2\theta$ is unity, the maximum horizontal range $R_{\max}$ is v_0^2/g. But if $\sin 2\theta = 1$, $2\theta = 90°$, and $\theta = 45°$. Hence the maximum horizontal range, in the absence of air resistance, is attained with an angle of elevation of $45°$.

From the standpoint of gunnery, what one usually wishes to know is what the angle of elevation should be for a given muzzle velocity v_0 in order to hit a traget whose position is known. If target and gun are at the same elevation and the target is at a distance R, Eq. (6–12) may be solved for θ.

$$\theta = \frac{1}{2}\sin^{-1}\left(\frac{Rg}{v_0^2}\right) = \frac{1}{2}\sin^{-1}\left(\frac{R}{R_{\max}}\right).$$

Provided R is less than the maximum range, this equation has two solutions for values of θ between $0°$ and $90°$. Thus if $R = 800$ ft, $g = 32$ ft/sec^2, and $v_0 = 200$ ft/sec,

$$2\theta = \sin^{-1}\left(\frac{800 \times 32}{200^2}\right) = \sin^{-1}(0.64)$$

$$= 40°, \quad\text{or}\quad 180° - 40° = 140°.$$

$$\theta = 20° \quad\text{or}\quad 70°.$$

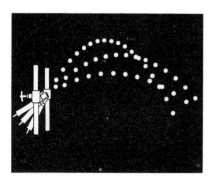

FIG. 6–5. An angle of elevation of $45°$ gives the maximum horizontal range.

Either of these angles gives the same range. Of course the time of flight and the maximum height reached are both greater for the high angle trajectory.

Figure 6–5 is a photograph of three trajectories, taken on the same film, of a ball projected from a spring gun with angles of elevation of $30°$, $45°$ and $60°$. It will be seen that the horizontal ranges are (very nearly) the same for the $30°$ and $60°$ elevations, and that both are less than the range when the angle is $45°$.

(The spring gun does not impart exactly the same initial velocity to the ball as the angle of elevation is altered.)

If the angle of elevation is below the horizontal, as for instance in the motion of a ball after rolling off a sloping roof, or the trajectory of a bomb released by a dive bomber, exactly the same principles apply. The horizontal velocity component remains constant and equal to $v_0 \cos \theta$. The vertical motion is the same as that of a body projected *downward* with an initial velocity $v_0 \sin \theta$. Minus signs can be avoided by taking the downward direction as positive.

EXAMPLE. A projectile is fired with a muzzle velocity of 1200 ft/sec at an angle of elevation of 15° above the horizontal as in Fig. 6–6. At what height will it strike a vertical cliff distant 15,000 ft horizontally from the gun? Find the magnitude and direction of its velocity when it strikes. Neglect air resistance.

The horizontal component of velocity is

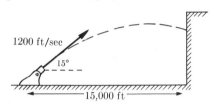

FIGURE 6–6

$$1200 \times \cos 15° = 1200 \times 0.966 = 1160 \text{ ft/sec.}$$

The time to travel a horizontal distance of 15,000 ft is

$$\frac{15,000}{1160} = 12.9 \text{ sec.}$$

The initial vertical component of velocity is

$$1200 \times \sin 15° = 1200 \times 0.259 = 311 \text{ ft/sec.}$$

The vertical height 12.9 sec after firing is

$$311 \times 12.9 - \tfrac{1}{2} \times 32 \times (12.9)^2 = 1360 \text{ ft}$$

and hence this is the height at which it strikes the cliff.

The vertical component of velocity is

$$311 - 32 \times 12.9 = -102 \text{ ft/sec.}$$

The resultant velocity is

$$\sqrt{1160^2 + 102^2} = 1160 \text{ ft/sec.}$$

The angle below the horizontal is

$$\tan^{-1} \frac{102}{1160} = \tan^{-1} 0.088 = 5 \text{ degrees.}$$

PROBLEMS

6–1. A golf ball is driven horizontally from a point which is 96 ft above a level fairway. If it strikes the ground at a point distant 150 yd *horizontally* from the start, what was its initial velocity?

6–2. A block weighing 8 lb rests on a frictionless horizontal table top 4 ft high. The block is initially 6 ft from the edge of the table. A horizontal force P pushes it from rest to the edge of the table top, at which point P is removed. If the block leaves the table with a velocity of 12 ft/sec, find (a) the force P, (b) the horizontal distance from the table at which the block strikes the floor, (c) the horizontal and vertical components of its velocity when it reaches the floor.

6–3. A .22 rifle bullet is fired in a horizontal direction with a muzzle velocity of 900 ft/sec. In the absence of air resistance, how far will it have dropped in traveling a horizontal distance of (a) 50 yd? (b) 100 yd? (c) 150 yd? (d) How far will it drop in one second?

6–4. A block passes a point 10 ft from the edge of a table with a velocity of 12 ft/sec. It slides off the edge of the table, which is 4 ft high, and strikes the floor 4 ft from the edge of the table. What was the coefficient of sliding friction between block and table?

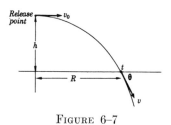

FIGURE 6–7

6–5. The following table charts the results of experimental studies of the ballistics of aircraft bombs (100–2000 lb) released from a plane in level flight traveling at 200 mi/hr. See Fig. 6–7 for the meaning of the symbols.

h(ft)	t(sec)	R(ft)	v(ft/sec)	θ(deg)
1000	7.97	2270	377	40.0
5000	18.1	4950	590	64.5
10000	26.0	6990	754	72.3
25000	42.7	10450	980	80.5

Calculate a few values of t, R, v, and θ, neglecting air resistance, and compare with the observed values.

6–6. A bombing plane in level flight releases three bombs at intervals of 2 sec. Compute the vertical distance between the first and second, and between the second and third, (a) at the instant the third is released, (b) after the first bomb has fallen 900 ft. Neglect air resistance.

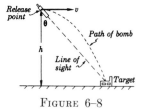

FIGURE 6–8

6–7. Fig. 6–8 illustrates the principle of the bomb sight used in level-flight bombing. The height of the plane h and its velocity v over the ground are known to the bomber. The bombs are released when the line of sight to the target makes an angle θ with the vertical. Find an expression for θ in terms of h and v.

6–8. A train is traveling at a constant velocity $v_1 = 22$ ft/sec across a

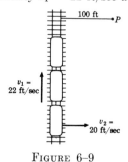

FIGURE 6–9

bridge over a river. A man inside one of the cars throws a stone horizontally out of a window in a direction perpendicular to the direction of motion of the train. The initial velocity of stone in this direction is $v_2 = 20$ ft/sec. The stone hits the water at the point P, 100 ft in a horizontal direction from the bridge. (See Fig. 6–9.) (a) Where is the car window from which the stone was thrown when the stone hits the water? (Neglect air friction.) (b) How far below the car window is the surface of the water?

6–9. A bomber is making a horizontal bombing run on a destroyer from an altitude of 25,600 ft. The magnitude of the velocity of the bomber is 300 mi per hour. (a) How much time is available for the destroyer to change its course after the bombs are released? Neglect air resistance. (b) If the bomber is to be shot down before its bombs can reach the ship, what is the maximum angle that the line of sight from ship to bomber can make with the horizontal? Draw a diagram, showing distances approximately to scale.

6–10. A trench mortar fires a projectile at an angle of 53° above the horizontal with a muzzle velocity of 200 ft/sec. A tank is advancing directly toward the mortar on level ground at a speed of 10 ft/sec. What should be the distance from mortar to tank at the instant the mortar is fired in order to score a hit? Neglect air resistance.

6–11. A player kicks a ball at an angle of 37° with the horizontal and with an initial velocity of 48 ft/sec. A second player standing at a distance of 100 ft from the first in the direction of the kick starts running to meet the ball at the instant it is kicked. How fast must he run in order to catch the ball before it hits the ground?

6–12. A ball is projected with an initial upward velocity component of 80 ft/sec and a horizontal velocity component of 100 ft/sec. (a) Find the position and velocity of the ball after 2 sec; 3 sec; 6 sec. (b) How long a time

is required to reach the highest point of the trajectory? (c) How high is this point? (d) How long a time is required for the ball to return to its original level? (e) How far has it traveled horizontally during this time? Show your results in a neat sketch, large enough to show all features clearly.

6–13. A golf ball is driven with a velocity of 200 ft/sec at an angle of 37° above the horizontal. It strikes a green at a horizontal distance of 800 ft from the tee. (a) What was the elevation of the green above the tee? (b) What was the velocity of the ball when it struck the green?

6–14. During the first World War, the Germans bombarded Paris with a specially constructed long-range gun. A newspaper report listed the following statistics on this gun:

Length of barrel . . . . 118 ft
Diameter of bore . . . 8.26 in.
Weight of projectile . 264 lb
Muzzle energy . . . . . . 46,700 ft-tons
Muzzle velocity . . . . 4760 ft/sec
Angle of elevation . . 55°
Range . . . . 132,000 yd
Weight of gun . . . . . 318,000 lb

Use the given values of muzzle velocity and angle of elevation to compute the range, in the absence of air resistance, and compare with the actual range. Compute also, neglecting air resistance, the maximum height reached, in miles, and the time of flight. (Note: the rotation of the earth and its curvature, as well as air resistance, are important factors influencing the trajectories of long-range projectiles.)

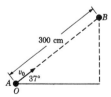

FIGURE 6–10

6–15. A ball A is projected from O with an initial velocity $v_0 = 700$ cm/sec in a direction 37° above the horizontal. A ball B 300 cm from O on a line 37° above the horizontal is released from rest at the instant A starts. (a) How far will B have fallen when it is hit by A? (b) In what direction is A moving when it hits B?

6–16. A baseball leaves the bat at a height of 4 ft above the ground, traveling at an angle of 45° with the horizontal, and with a velocity such that the horizontal range would be 400 ft. At a distance of 360 ft from home plate is a fence 30 ft high. Will the ball be a home run?

6–17. The projectile of a trench mortar has a muzzle velocity of 300 ft/sec. (a) Find the two angles of elevation to hit a target at the same level as the mortar and 300 yd distant. (b) Compute the maximum height of each trajectory, and the time of flight of each. Make a neat sketch of the trajectories, approximately to scale.

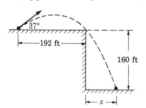

FIGURE 6–11

6–18. A ball is thrown, as shown in Fig. 6–11, with an initial velocity v_0 at an angle of 37° above the horizontal, from a point 192 ft from the edge of a vertical cliff 160 ft high. The ball just misses the edge of the cliff. (a) Find the initial velocity v_0. (b) Find the distance x beyond the foot of the cliff where the ball strikes the ground.

FIGURE 6–12

6–19. A particle is projected with an initial velocity of 160 ft/sec at an angle of 53° above the horizontal (Fig. 6–12). The particle passes into a tube pointing 45° to the vertical, and so placed that as the particle enters the tube its direction of motion coincides with the tube axis. (a) How long is the particle in flight? (b) What are the X- and Y-coordinates of the tube opening?

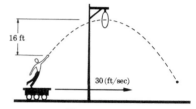

FIGURE 6–13

6–20. A man is riding on a flat car traveling with a constant velocity of 30 ft/sec (Fig. 6–13). He wishes to throw a ball through a stationary hoop 16 ft above the height of his hands in such a manner that the ball will move horizontally as it passes through the hoop. He throws the ball with a velocity of 40 ft/sec with respect to himself. (a) What must be the vertical component of the initial velocity of the ball? (b) How many seconds after he releases the ball will it pass through the hoop? (c) At what horizontal distance in front of the hoop must he release the ball?

6–21. A bomber, diving at an angle of 53° with the vertical, releases a bomb at an altitude of 2400 ft. The bomb is observed to strike the ground 5 sec after its release. (a) What was the velocity of the bomber, in ft/sec? (b) How far did the bomb travel horizontally during its flight? (c) What were the horizontal and vertical components of its velocity just before striking? Air resistance may be neglected.

6–22. A 15-lb stone is dropped from a cliff in a high wind. The wind exerts a steady horizontal 10-lb force on the stone as it falls. Is the path of the stone a straight line, a parabola, or some more complicated path? Explain.

CHAPTER 7

CENTER OF MASS

7–1 Center of mass. The series of multiflash photographs in Fig. 7–1 illustrates an extremely useful and important concept, namely, that of the *center of mass* of a body or of a system of bodies. The body in Fig. 7–1 consists of a light wooden rod with balls of different masses attached to its ends. In each photograph the body is initially at rest in a horizontal position and is struck a vertical blow by a small spring-driven hammer.

In the first photograph the body rotates in a clockwise direction as it flies upward after the blow is struck. In the second photograph, where the blow was struck further to the right, the rotation is counterclockwise. But when the blow is struck at the point marked with a black band, as in the third photograph, the body does not rotate at all. We say that the acceleration imparted to it by the hammer is one of *pure translation,* and the point at which an external force must be applied to produce pure translational acceleration is called the *center of mass.*

More generally, when any number of external forces act on a body, *the body moves with pure translational acceleration when the line of action of the resultant force passes through the center of mass.*

We have in fact already made use of the center of mass concept in Chapter 5, where in all of the examples of the acceleration of a body under the action of a number of external forces it was tacitly assumed that the

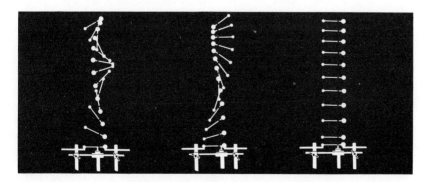

FIG. 7–1. The body is struck (a) at the left of its center of mass; (b) at the right of its center of mass; (c) at its center of mass.

forces intersected at a common point. We now see that this point must have been the center of mass of the body.

7–2 Coordinates of the center of mass. It is easy to see why the rod in Fig. 7–1 starts to rotate unless the blow is struck at the proper point. Each of the masses at the ends of the rod is accelerated because of the force exerted on it by the rod. By Newton's third law, each mass exerts on the rod a force which is equal and opposite to the force the rod exerts on it. If we consider the point of application of the external force as a pivot, one of these reactions exerts a torque one way on the rod, the other, a torque the other way. The center of mass is the particular point of application of the external force for which these torques are equal and opposite.

We shall first derive an expression for the coordinate of the center of mass of a system consisting of two point masses m_1 and m_2 attached to the

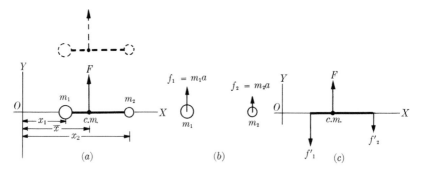

FIGURE 7–2

ends of a rigid rod whose own mass is negligible. The system is shown in Fig. 7–2(a). Gravitational and friction forces are neglected. The external force F acts at the center of mass whose coordinate x we wish to find. The coordinates of m_1 and m_2 are x_1 and x_2 respectively. The force diagrams of the masses m_1 and m_2 are shown in Fig. 7–2(b), where f_1 and f_2 represent the forces exerted on m_1 and m_2 by the rod. By hypothesis, the force F acts at the center of mass so the system moves with pure translational acceleration and both masses have the same acceleration a. Then from Newton's second law,

$$f_1 = m_1 a, \qquad f_2 = m_2 a.$$

The force diagram of the rod is given in Fig. 7–2(c) where the forces f'_1 and f'_2 are the reactions to the forces f_1 and f_2 in Fig. 7–2(b). From the second law,

$$F - f'_1 - f'_2 = 0.$$

The right side of this equation is zero since the rod is assumed of zero mass.

When the three preceding equations are added we get

$$F + (f_1 - f_1') + (f_2 - f_2') = (m_1 + m_2)a.$$

But since the forces f_1' and f_2' are the reactions to the forces f_1 and f_2, it follows that $f_1 - f_1' = 0$ and $f_2 - f_2' = 0$. Hence

$$F = (m_1 + m_2)a,$$

or

$$a = \frac{F}{m_1 + m_2},$$

and the acceleration of the system is the same as would be that of a point mass equal to the sum of the masses m_1 and m_2. The fact that the masses are separated in space does not affect the acceleration imparted to them by the external force F.

Consider next the consequences of the requirement that the rod shall not rotate. The resultant torque on the rod must be zero, and if we take moments about an axis through O this gives

$$F\bar{x} = f_1' x_1 + f_2' x_2,$$

or, since

$$f_1' = f_1 = m_1 a \qquad \text{and} \qquad f_2' = f_2 = m_2 a,$$

$$F\bar{x} = m_1 x_1 a + m_2 x_2 a = (m_1 x_1 + m_2 x_2)a.$$

Finally, since we have shown that $a = F/(m_1 + m_2)$,

$$F\bar{x} = (m_1 x_1 + m_2 x_2)\frac{F}{m_1 + m_2},$$

and

$$\bar{x} = \frac{m_1 x_1 + m_2 x_2}{m_1 + m_2}. \tag{7-1}$$

This is the desired expression for the X-coordinate of the center of mass of the system.

An equivalent method of procedure is to find the line of action of the parallel forces f_1' and f_2'. The line of action of F must lie in this line also if the rod is to be in rotational equilibrium, and hence the center of mass lies on the line of action of the resultant of the reaction forces f_1' and f_2'.

For simplicity, the preceding derivation was carried out for the special case of two point masses lying on the X-axis. It is not difficult to show that for any number of point masses m_1, m_2, etc., whose coordinates are x_1 and y_1, x_2 and y_2, etc., the coordinates $\bar{x}$ and $\bar{y}$ of the center of mass are

$$\bar{x} = \frac{\Sigma mx}{\Sigma m}, \qquad \bar{y} = \frac{\Sigma my}{\Sigma m}. \qquad (7\text{-}2)$$

A further generalization of Eq. (7-2) to include bodies of finite size is obvious. The body is divided, in imagination, into infinitesimal elements of mass dm. Then if x and y are the coordinates of any dm,

$$\bar{x} = \frac{\int x\, dm}{\int dm}, \qquad \bar{y} = \frac{\int y\, dm}{\int dm}. \qquad (7\text{-}3)$$

The limits of integration must be taken so as to include the entire body.

It will be recognized that Eq. (7-3) is of the same form as Eq. (3-8) on page 40 for the coordinates of the center of gravity. In fact, if we write $dm = dw/g$, the equations become identical and hence the center of mass coincides with the center of gravity.* However, it should be realized that essentially different concepts are involved in the definitions of the two centers. The center of gravity of a body is the common point through which pass the lines of action of the resultant gravitational force acting on the body, as the orientation of the body is altered. The center of mass is the point through which passes the resultant of the reaction forces (such as f_1' and f_2' in Fig. 7-2) when a body is accelerated. If the acceleration of gravity were not the same in magnitude and direction at all points of a body, or if gravitation were to be eliminated from our world, center of gravity would cease to have a meaning although the concept of center of mass would remain (that is, if we preserved inertia while removing gravitation).

The position of the center of mass of a body or of a system of bodies can easily be shown to be independent of the origin to which $\bar{x}$ and $\bar{y}$ are referred, and of the orientation of the X- and Y- axes. An example of this will be given later.

If a body is symmetrical, it can also be readily shown that its center of mass coincides with its center of symmetry. This follows from the fact that if the origin is placed at the center of symmetry, then for every dm at a positive value of x there is an equal dm at an equal negative value of x. Hence

$$\int x\, dm = 0 \qquad \text{and} \qquad \bar{x} = 0.$$

By the same reasoning, $\bar{y} = 0$ and the center of mass coincides with the center of symmetry.

---

*Provided the acceleration of gravity has the same magnitude and direction at all points of the body, which is true in any case of practical interest.

EXAMPLES. (1) Find the coordinates of the center of mass of the four point masses in Fig. 7–3. The side of each small square represents 1 cm.

$$\bar{x} = \frac{\Sigma mx}{\Sigma m} = \frac{m_1x_1 + m_2x_2 + m_3x_3 + m_4x_4}{m_1 + m_2 + m_3 + m_4}$$

$$= \frac{10(-1) + 10(1) + 20(3) + 30(2)}{10 + 10 + 20 + 30}$$

$$= +1.7 \text{ cm.}$$

$$\bar{y} = \frac{\Sigma my}{\Sigma m} = \frac{10(1) + 10(3) + 20(2) + 20(0)}{10 + 10 + 20 + 30}$$

$$= +1.15 \text{ cm.}$$

and the center of mass lies 1.7 cm to the right of the Y-axis and 1.15 cm above the X-axis.

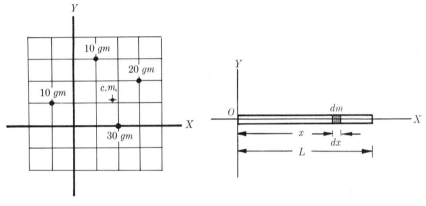

FIGURE 7–3   FIGURE 7–4

(2) Prove that the center of mass of a homogeneous rod of constant cross section lies at the center of the rod.

Since the rod is symmetrical about its center, the center of mass is at the center. However, we shall prove this to be the case as an illustration of the application of Eq. (7–3). Take the origin of coordinates at one end of the rod (Fig. 7–4), with the X-axis along the rod. Let the length of the rod be L, its cross section A, and its density ρ. Consider an element of the rod of length dx at a distance x from the origin. The mass of the element, dm, is

$$dm = \rho A\, dx.$$

Hence

$$\bar{x} = \frac{\int x\, dm}{\int dm} = \frac{\int_0^L x\rho A\, dx}{\int_0^L \rho A\, dx} = \frac{\dfrac{x^2}{2}\bigg]_0^L}{x\bigg]_0^L}$$

$$= \frac{L}{2}.$$

(3) As an illustration (not a general proof) of the fact that the center of mass is found at the same point regardless of the choice of coordinate system, let us work out the example above, locating the origin at a distance b to the left of the rod. The reader may construct his own diagram. We now have

$$\bar{x} = \frac{\displaystyle\int_b^{b+L} \rho A x\, dx}{\displaystyle\int_b^{b+L} \rho A\, dx} = \frac{\left. \dfrac{x^2}{2} \right]_b^{b+L}}{\left. x \right]_b^{b+L}}$$

$$= \frac{1}{2} \frac{(b+L)^2 - b^2}{(b+L) - b} = b + \frac{L}{2}.$$

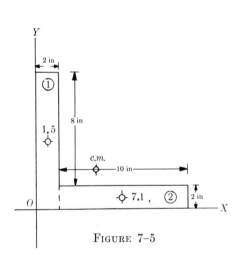

FIGURE 7–5

(4) Find the center of mass of the L-shaped section in Fig. 7–5.

Similar problems have been discussed in connection with centers of gravity. (See page 49.) The section may be subdivided into two rectangles as shown. The center of mass of each is at its center, so rectangle (1) may be replaced by a point mass whose coordinates are $x_1 = 1$, $y_1 = 5$. Similarly, rectangle (2) may be replaced by a point mass whose coordinates are $x_2 = 7$, $y_2 = 1$. The masses of the two sections are proportional to their areas, if the slab is of uniform thickness and density. Hence, assuming thickness and density each equal to unity, $m_1 = m_2 = 20$. The coordinates of the center of mass of the section are now found by

$$\bar{x} = \frac{m_1 x_1 + m_2 x_2}{m_1 + m_2} = \frac{20(1) + 20(7)}{20 + 20} = 4 \text{ in.}$$

$$\bar{y} = \frac{m_1 y_1 + m_2 y_2}{m_1 + m_2} = \frac{20(5) + 20(1)}{20 + 20} = 3 \text{ in.}$$

The center of mass is labeled c.m., and is seen to lie outside the material of the slab.

7–3 Acceleration of the center of mass. When the resultant force on a body does not pass through its center of mass, the motion is a combination of translation and rotation. We have not yet laid the groundwork for a complete analysis of the motion, but there is one aspect of it which can readily be derived; namely, in every case the acceleration of the center of mass itself is the same as if all of the mass of the body were concentrated at the center of mass, and the line of action of the resultant force passed through this point. In other words, as far as the motion of its center of mass is concerned a body of any shape, acted on by any number of forces, can be replaced by a point mass at its center of mass, and all of the forces on it can be considered to act at this point. We shall illustrate by a simple example.

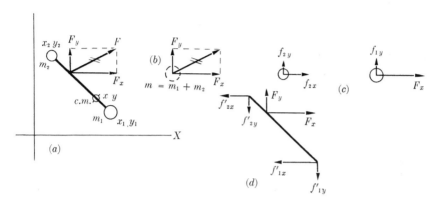

FIGURE 7–6

Figure 7–6(a) shows the same body as in Fig. 7–2 acted on this time by a force F which does not pass through the center of mass. Figure 7–6(b) shows a concentrated mass $m = m_1 + m_2$ acted on by the resultant external force F. Its acceleration a is

$$a = \frac{F}{m_1 + m_2} = \frac{F}{m}$$

or in component form

$$a_x = \frac{F_x}{m}, \qquad a_y = \frac{F_y}{m}. \tag{7-4}$$

We shall now show that these accelerations are equal to the X- and Y-components of the acceleration of the center of mass. Let f_{1x} and f_{1y} in Fig. 7–6(c) represent the components of the force exerted on m_1 by the rod. Similarly, f_{2x} and f_{2y} are the components of the force exerted on m_2. The accelerations of m_1 and m_2 in this case are not equal but are given by the equations

$$f_{1x} = m_1 a_{1x}, \qquad f_{1y} = m_1 a_{1y}.$$
$$f_{2x} = m_2 a_{2x}, \qquad f_{2y} = m_2 a_{2y}.$$

The forces on the rod are shown in Fig. 7–6(d), and since the mass of the rod is assumed zero

$$F_x - f'_{1x} - f'_{2x} = 0, \qquad F_y - f'_{1y} - f'_{2y} = 0.$$

When each set of the equations above is added we get

$$F_x = m_1a_{1x} + m_2a_{2x}, \qquad F_y = m_1a_{1y} + m_2a_{2y}. \qquad (7\text{--}5)$$

We have shown that the coordinates $\bar{x}$ and $\bar{y}$ of the center of mass are

$$\bar{x} = \frac{m_1x_1 + m_2x_2}{m_1 + m_2}, \qquad \bar{y} = \frac{m_1y_1 + m_2y_2}{m_1 + m_2}.$$

When both sides of these equations are differentiated twice with respect to the time t we find as the components of the acceleration of the center of mass

$$\frac{d^2\bar{x}}{dt^2} = \bar{a}_x = \frac{m_1a_{1x} + m_2a_{2x}}{m_1 + m_2}, \qquad \frac{d^2\bar{y}}{dt^2} = \bar{a}_y = \frac{m_1a_{1y} + m_2a_{2y}}{m_1 + m_2}.$$

The numerators of the right sides of the equations are, from Eq. (7–5), simply F_x and F_y. Hence,

$$\bar{a}_x = \frac{F_x}{m_1 + m_2}, \qquad \bar{a}_y = \frac{F_y}{m_1 + m_2},$$

or

$$\bar{a}_x = \frac{F_x}{m}, \qquad\qquad \bar{a}_y = \frac{F_y}{m}. \qquad (7\text{--}6)$$

Comparison with Eq. (7–4) shows that the acceleration of the center of mass is the same as that of a point mass $m = m_1 + m_2$, acted on by the force F. Since F may represent the resultant of any number of external forces, Eqs. (7–6) are equivalent to

$$\boxed{\Sigma X = m\bar{a}_x, \qquad \Sigma Y = m\bar{a}_y.} \qquad (7\text{--}7)$$

FIG. 7–7. The trajectory of the center of mass is a parabola in every case.

These are obvious generalizations of Eq. (6–2) on page 97, and their significance is illustrated by the photographs in Fig. 7–7. The trajectory of the center of mass of the body (marked by the black band) is seen in every case to be a smooth parabola. That is, the motion of the center of mass is the same as would be that of a body of small dimensions when projected with an initial horizontal velocity.

Notice that the *internal* forces f_1 and f_2 in Fig. 7–6 do not appear in Eq. (7–7). The motion of the center of mass is not affected by these internal forces, although, of course, the motions of the individual masses of a system are. A consequence of this fact is that if the external force acting on a system of bodies is zero, the acceleration of the center of mass of the system is zero also, and hence (generalization of Newton's first law) *the center of mass* remains at rest or moves with constant speed in a straight line. Thus in the solar system, although the sun and planets exert forces on one another and each moves in a complex curve, the center of mass of the entire solar system moves through space with constant speed in a straight line (if one neglects the extremely small external forces exerted on the system by the stars, the nearest of which is 4.3 light-years from the sun).

EXAMPLES. (1) Suppose the section whose center of mass was located in example (4) on page 112 is placed on a level frictionless surface and forces of 3 lb and 4 lb, constant in magnitude and direction, are exerted on it as shown in Fig. 7–8. Let the section be 1 in thick and weigh $\frac{1}{2}$ lb/cu in. Find the position of its center of mass after 4 sec.

The volume of the section is 40 in^3, its weight 20 lb, and its mass $20/32 = 0.625$ slugs. Then, since $\Sigma X = 3$ lb and $\Sigma Y = 4$ lb,

$$\bar{a}_x = \frac{\Sigma X}{m} = \frac{3}{.625} = 4.8 \text{ ft/sec}^2$$

$$\bar{a}_y = \frac{\Sigma Y}{m} = \frac{4}{.625} = 6.4 \text{ ft/sec}^2$$

$$\bar{a} = \sqrt{\bar{a}_x^2 + \bar{a}_y^2} = 8 \text{ ft/sec}^2$$

at 53° above the X-axis.

An alternate procedure is to first combine the 3- and 4-lb forces into a resultant of 5 lb at an angle of 53° above the X-axis. One may then write

$$\bar{a} = \frac{F}{m} = \frac{5}{.625} = 8 \text{ ft/sec}^2.$$

Hence, after 4 sec, the center of mass will be displaced a distance

$$\bar{s} = \frac{1}{2}\bar{a}t^2 = \frac{1}{2} \times 8 \times (4)^2 = 64 \text{ ft}$$

along a line 53° above the X-axis.

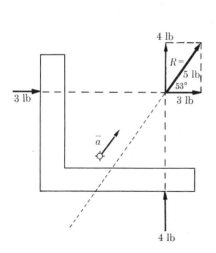

FIGURE 7–8

Since the line of action of the resultant force does not pass through the center of mass, the section will not move with pure translation. Nevertheless, the center of mass moves with constant acceleration in a straight line, since the forces are constant in magnitude and direction.

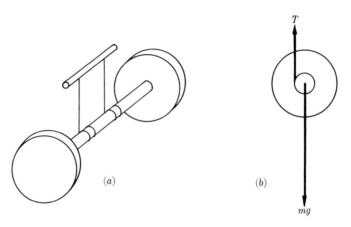

(a) (b)

FIGURE 7–9

(2) A spool, shown in Fig. 7–9, has strings wrapped around its axle and fastened to a fixed horizontal bar. When released, the spool falls and also rotates. If the downward acceleration of the spool is observed to be 16 ft/sec², compute the tension in each string.

The center of mass of the spool is at the center of its axis. The forces acting on it are its weight mg and the tension T. Since there are no X-forces, $\bar{a}_x = 0$, and the spool will fall vertically when released. The resultant Y-force is

$$\Sigma Y = mg - T.$$

Hence

$$mg - T = m\bar{a}_y = m \times 16$$

$$T = \tfrac{1}{2}mg,$$

and the tension equals half the weight of the spool. (This is the combined tension in the two strings.)

(3) What must be the tension in the supporting strings of the preceding example in order that the center of mass of the spool shall remain at rest?

Since $\bar{a}_y = 0$,

$$\Sigma Y = mg - T = 0, \qquad T = mg.$$

In order to exert a tension of this magnitude, the ends of the strings would have to be pulled upward with accelerated motion. The acceleration depends on the rotational aspects of the motion and will be discussed later.

7–4 Pure translational acceleration. The general equations that must hold when a body moves with pure translational acceleration can now be written down. We have shown that under these circumstances the line of action of the resultant external force passes through the center of mass. It follows that the resultant force has no moment about an axis through the center of mass. More generally, if a line is constructed through the center of mass in the direction of the acceleration, the resultant force has no moment about any axis passing through this line.

For example, Fig. 7–10 shows a body moving with pure translational acceleration a. The resultant force on the body is represented by F, and $A-A$ is a line through the center of mass in the direction of the acceleration. It is evident that the moment of the resultant force is zero about *any* axis through the line $A-A$.

The complete set of equations determining the motion of the body are therefore

$$\Sigma X = m\bar{a}_x, \qquad \Sigma Y = m\bar{a}_y, \qquad \Sigma \tau = 0. \qquad (7\text{–}8)$$

Note carefully that the torque may not be set equal to zero about any arbitrary axis, as may be done in statics, but only about an axis through a line in the direction of the acceleration and passing through the center of mass.

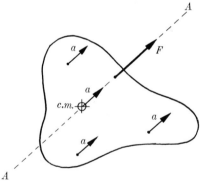

Fig. 7–10. When a body moves with pure translational acceleration, the line of action of the resultant force passes through its center of mass.

EXAMPLES. (1) With the help of the preceding equations we can analyze in further detail the set of forces acting on a body undergoing pure translational acceleration. In our earlier work, as in Example 9 on page 83, it was assumed for simplicity that the lines of action of all of the forces on such a body passed through its center of mass. Actually, the friction force and the normal force, in a case such as that of Fig. 5–2, are distributed over the lower face of the block. We shall still simplify the problem to some extent by considering that the block has projections on its lower edges as in Fig. 7–11, so that the normal and friction forces act at these edges only.

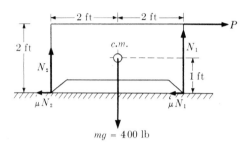

$$mg = 400 \text{ lb}$$

FIGURE 7–11

The block in Fig. 7–11 is 4 ft long and 2 ft high, with its center of mass at its center. It weighs 400 lb and the coefficient of sliding friction between block and surface is 0.20. What force P is needed to accelerate the block at 4 ft/sec², and what are the normal forces N_1 and N_2?

$$\Sigma X = P - \mu N_1 - \mu N_2 = m\bar{a}_x$$
$$\Sigma Y = N_1 + N_2 - mg = m\bar{a}_y.$$

The acceleration is parallel to the X-axis. The resultant torque is therefore zero about any axis perpendicular to the plane of the diagram and passing through any point on a horizontal line through the center of mass. Let us take the axis through the center of mass itself. Then

$$\Sigma\tau = 2 \times N_2 + 1 \times P + 1 \times \mu N_1 + 1 \times \mu N_2 - 2 \times N_1 = 0$$
$$\bar{a}_x = 4 \text{ ft/sec}^2, \qquad \bar{a}_y = 0, \qquad \mu = 0.2, \qquad m = 12.5 \text{ slugs}.$$

We find from these equations that

$$P = 130 \text{ lb}, \qquad N_1 = 252 \text{ lb}, \qquad N_2 = 148 \text{ lb}.$$

(2) An automobile weighs 2400 lb, its wheelbase is 10 ft, and its center of mass is half-way between front and rear axles and 3 ft above the ground. It is brought to rest in 6 sec from a speed of 60 mi/hr. Find the upward forces at front and rear wheels during braking.

The forces on the automobile are shown in Fig. 7–12. If the car has 4-wheel brakes, friction forces act as shown at the bottom of all four wheels. Note, however, that unless the car slides to a stop with wheels locked, these forces are not equal to the product of the coefficient of sliding friction and the normal force.

$$\Sigma Y = -fr_1 - fr_2 = m\bar{a}_x$$
$$\Sigma Y = N_1 + N_2 - mg = m\bar{a}_y = 0$$
$$\Sigma\tau = 3 \times (fr_1 + fr_2) + 5N_2 - 5N_1 = 0.$$

(The torque axis is taken through the center of mass.)

From the given data the acceleration of the car, $\bar{a}_x$, is $-14\frac{2}{3}$ ft/sec².

Hence

$$N_1 = 1530 \text{ lb}, \qquad N_2 = 870 \text{ lb}.$$

That is, the upward force at the front wheels is, in this particular instance, about twice as great as that at the rear wheels, although by symmetry these forces would be equal if the car were at rest or moving with constant velocity. This effect is one reason for the introduction of 4-wheel brakes. Since the maximum friction, or braking force, available at a wheel is proportional to the normal force, the rear wheels are obviously much less suitable for applying a braking force than are the front wheels, since the normal force at the rear wheels is lessened as soon as the brakes are applied. By using brakes on all wheels, the maximum braking force available is made independent of the distribution of load, since $N_1 + N_2 = w$, regardless of how the load is distributed.

The effect of the increased upward force at the front of the car, and the decreased force at the rear, is to compress the front springs and extend the rear springs, causing the car to nose down when its brakes are applied. Everyone has probably observed this effect.

When the car is accelerating, the force diagram appears as in Fig. 7–13(a). The rear wheels (if the car has a rear-wheel drive) press backward against the surface of the road. The reaction is a forward force P which accelerates the car. Let us assume that the acceleration is 8 ft/sec^2. Then

$$\Sigma X = P = m\bar{a}_x$$
$$\Sigma Y = N_1 + N_2 - mg = m\bar{a}_y$$
$$\Sigma\tau_{\text{c.m.}} = 5 \times N_2 - 5 \times N_1 - 3 \times P.$$

Simultaneous solution of these equations gives

$$P = 600 \text{ lb},$$
$$N_1 = 1020 \text{ lb},$$
$$N_2 = 1380 \text{ lb}.$$

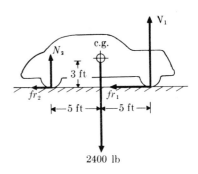

Fig. 7–12. Forces on an automobile while braking.

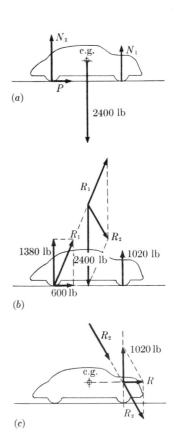

Fig. 7–13. Forces on an accelerating automobile.

Thus the force N_2 is increased above its value when the car is in equilibrium, while N_1 is decreased. As a result, the force on the front springs is lessened and the car noses up when it accelerates, another familiar effect.

To show that the line of action of the resultant force passes through the center of mass, the resultant of the four forces acting on the car is found graphically in Figs. 7–13(b) and (c). In part (b) the forces at the rear wheels are first combined to find their resultant R_1. The lines of action of this force and the weight of the car are then extended to their point of intersection, the vectors transferred to this point, and their resultant R_2 determined. In part (c), this resultant is combined with the force N_1 to find the resultant R of the entire set of forces, and it will be seen that the line of action of R does, in fact, pass through the center of mass.

PROBLEMS

7–1. "If a plane passing through a body contains the center of mass, as much mass lies on one side of the plane as on the other side of the plane." Is the above statement always true? Explain.

7–2. Three masses of 60 gm each are placed at the corners of an equilateral triangle 20 cm on a side. Find the position of their center of mass.

7–3. Masses of 10, 20, 30, and 40 gm are placed at the corners of a square 20 cm on a side. Find the X- and Y-coordinates of their center of mass.

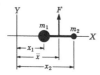

FIGURE 7–14

7–4. Two masses are fastened to a rod of negligible mass, and a force is exerted perpendicular to the rod so that the system will move in pure translation (Fig. 7–14). Neglect gravity. (a) Draw force diagrams for both masses and the rod. (b) Derive an expression for the acceleration of the system in terms of F, m_1, and m_2. (c) Derive an expression for the distance $\bar{x}$ in terms of m_1, m_2, x_1, and x_2.

7–5. A uniform rod 1 meter long, of mass 100 gm, is loaded with a 20-gm

mass 20 cm from one end and a 40-gm mass 40 cm from the same end. Find the position of the center of mass of the entire system.

7–6. Attached to one end of a rod is a body whose mass is twice that of the rod. At what fraction of its length from the loaded end should the rod be struck if it is to move with pure translation as in Fig. 7–1(c)?

7–7. The mass of the moon is 1/80 that of the earth. The distance of the moon from the center of the earth is 240,000 mi and the radius of the earth is (about) 4000 mi. How far from the center of the earth is the center of mass about which the earth and the moon revolve?

7–8. A slender rod is bent into a semicircle of radius R. Find the position of its center of mass.

7–9. The density ρ of a slender rod of length L and uniform cross section, increases uniformly from one end to the other according to the relation $\rho = \rho_0 + ax$. Find the position of the center of mass of the rod.

7–10. Two small bodies of masses 100 gm and 400 gm respectively rest on a horizontal frictionless surface and attract one another with a constant force of 100 dynes. Initially they are 100 cm apart. Where and when do they collide?

7–11. A 160-lb man is standing at the end of a 12-ft plank weighing 32 lb.

The plank is resting on ice; there is no friction between the plank and the ice. The man walks to the other end of the plank. How far does he move relative to the ice?

7–12. A block of mass m slides with velocity v along a frictionless level surface toward a block of mass $4m$ initially at rest. The blocks collide and stick together. (a) What was the velocity of the center of mass of the system before the impact? (b) What was the velocity of the center of mass after the impact?

7–13. (a) A bomb weighing 8 lb is thrown out in a horizontal direction with a velocity of 8 ft/sec from the top of a building 400 ft high. The ground around the building is level. How far out from the building will the bomb hit the ground? (b) An identical bomb is thrown out in exactly the same way. This one, however, explodes into two pieces before it hits the ground. The two pieces are ejected horizontally, so that both hit the ground at the same time. One piece weighs 3 lb and lands just at the edge of the building, directly below the starting point at the top of the building. How far out from the building does the 5–lb piece land?

FIGURE 7–15

7–14. A monkey is at rest on a weightless rope which goes over a pulley and is tied to a bunch of bananas at the other end (Fig. 7–15). The bananas weigh exactly the same as the monkey. The pulley is frictionless and of negligible mass. The monkey starts to climb up the rope to reach the bananas. As he climbs, does the distance between him and the bananas increase, decrease, or stay the same? Why?

7–15. A load of baseballs is being shipped from Earth to Mars. The rocket engines break down. How can you change the direction of the ship? What principle are you applying?

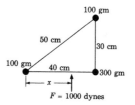

FIGURE 7–16

7–16. Masses of 100 gm, 100 gm, and 300 gm are connected by a light rigid frame and rest on a level frictionless surface, as shown in Fig. 7–16. (a) At what distance x should the constant force F of 1000 dynes be applied in order that the system move with pure translational acceleration? (b) How far will the 300-gm mass move in the first 5 sec?

FIGURE 7–17

7–17. A rectangular piece of wood rests on a smooth level surface. A force F is applied to one corner as in Fig. 7–17. At what point should the parallel force $2F$ be applied in order that the body shall move with pure translation?

7–18. Fig. 7–18 shows a top view of a triangular block of mass 10 kgm which is pulled along a horizontal frictionless

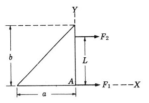

FIGURE 7–18

surface by two forces. The block moves with pure translational acceleration. The force F_1 is fixed at corner A and is equal to 20 newtons. (a) Find by integration the Y-coordinate of the center of mass of the triangle. (b) If the forces F_1 and F_2 are equal, find the distance L. (c) If the distance L were equal to b, what would be the acceleration of the center of mass?

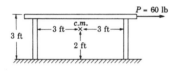

FIGURE 7–19

7–19. A table weighing 64 lb is pulled along the floor by a constant horizontal force of 60 lb. The coefficient of sliding friction is 0.5 (Fig. 7–19). (a) Find the acceleration of the table. (b) Find the normal force on each leg.

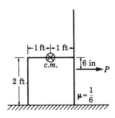

FIGURE 7–20

7–20. A chair weighing 48 lb is pulled to the right by a horizontal force P applied 6 inches below the center of mass of the chair (Fig. 7–20). The coefficient of sliding friction is 1/6. Find the upward forces exerted on the chair legs by the floor if the force P is great enough to give the chair an acceleration of 16 ft/sec².

7–21. Refer to the automobile in Fig. 7–12. Find the shortest distance in which it may be brought to rest from a velocity of 60 mi/hr without causing the rear wheels to leave the ground. How great must the coefficient of friction be?

7–22. Refer to Fig. 7–13. (a) Compute the upward forces at the front and rear wheels when the acceleration of the automobile is 4 ft/sec². (b) What is the maximum acceleration the automobile can have without causing the front wheels to leave the ground?

FIGURE 7–21

7–23. An elevator is 8 ft high and 6 ft wide. It moves without friction between vertical guides whose distance of separation is slightly more than 6 ft. The cable is attached at the center of the top of the elevator. The elevator and load together weigh 1600 lb. The center of mass of elevator and load is 2 ft to the left of the center line and 5 ft above the elevator floor (Fig. 7–21). Find the tension in the cable, and the forces exerted on the elevator by the guides, when the upward acceleration of the elevator is 4 ft/sec². Show the forces on the elevator in a diagram.

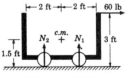

FIGURE 7–22

7–24. A cart weighing 96 lb is hitched to a horse which exerts a horizontal force of 60 lb. Neglect friction.

(a) What are the normal forces N_1 and N_2? (b) What is the acceleration?

7–25. (a) Find the location of the center of mass of a right circular cone 16 cm high and of radius 3 cm at the base. (b) If the cone has a mass of 400 gm, what horizontal force applied to the tip of the cone is required to tip it over? Assume that the base of the cone rests on a horizontal frictionless surface.

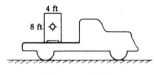

FIGURE 7–23

7–26. A packing case 4 ft square and 8 ft high stands as shown in Fig. 7–23 on a truck traveling at 50 ft/sec. The coefficient of friction between the truck tires and the road is 0.5; the coefficient of friction between the packing case and the truck is 0.3. The center of mass of the case is at its center. The truck driver suddenly applies his brakes so as to produce the maximum possible deceleration of the truck. Will the packing case tip over, slide without tipping, or remain at rest relative to the truck?

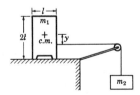

FIGURE 7–24

7–27. A box of mass m_1, width l, and height $2l$, has its center of mass at its center. It is pulled, without tipping, along a frictionless surface by a horizontal rope which passes over a pulley and down to the mass m_2 (Fig. 7–24). (a) What is the tension in the rope? (b) In terms of the given quantities, what is the largest value which y, the distance of the rope below the center of mass, can have and still not tip the box?

7–28. A rectangular block of height h and width w slides down an inclined plane of slope angle 37°, as in Fig. 7–25. The coefficient of friction between surfaces is 0.2. (a) Find the acceleration of the block. (b) Find the maximum ratio of height to width, h/w, for which the block will slide without tipping.

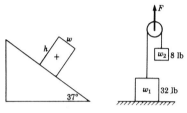

FIGURE 7–25   FIGURE 7–26

7–29. Weights w_1 and w_2 are connected by a weightless string passing over a weightless and frictionless pulley, as shown in Fig. 7–26. (a) What is the maximum upward force F which can be applied to the pulley without causing w_1 to leave the floor? (b) What is the acceleration of w_2 when this maximum force F is applied to the pulley? (c) What is the acceleration of the center of mass of the system when a force of 80 lb is applied?

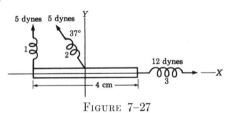

FIGURE 7–27

7–30. A bar of mass 2 gm and length 4 cm lies on a smooth, horizontal table, along the X-axis, with its center at the origin. Attached to it are three springs, exerting forces as shown in Fig. 7–27. (a) What is the magnitude, direction, and point of application of the force F which will hold the bar in equilibrium? (b) What is the initial direction and magnitude of the acceleration of the center of mass of the bar if the force F is removed?

CHAPTER 8

WORK AND ENERGY

8–1 Conservation of energy. One of the general principles that underlie all natural processes, whether they be physical, chemical, or biological, is the principle of the conservation of energy. This principle, as it relates to purely mechanical problems, is a necessary consequence of Newton's laws of motion, although it was never stated by Newton himself in the form in which we use it today. Its generalization to include heat as a form of energy was not fully appreciated until the work of Sir James Prescott Joule (1818–1889), and others, on the measurement of the mechanical equivalent of heat, which took place near the middle of the last century, about 150 years after Newton's death.

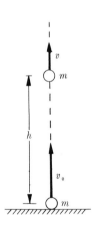

FIGURE 8–1

To illustrate the way in which the energy concept makes its appearance in mechanical problems, let us consider the simple example of a body of mass m projected vertically upward with an initial velocity v_0 (Fig. 8–1). When the body has risen to a height h above its initial position, its velocity has decreased from v_0 to v, where

$$v^2 = v_0^2 - 2gh. \qquad (8\text{–}1)$$

When both sides of this equation are multiplied by $m/2$ and the terms rearranged we get

$$\tfrac{1}{2}mv^2 + mgh = \tfrac{1}{2}mv_0^2. \qquad (8\text{–}2)$$

As the body rises, v decreases and h increases. Hence both terms on the left of Eq. (8–2) change during the motion, but their sum remains constant and equal to $\tfrac{1}{2}mv_0^2$. When a quantity remains constant in a process, it is said to be *conserved*, and we may therefore say that the sum of $\tfrac{1}{2}mv^2$ and mgh is conserved in this motion. (Of course, from Eq. (8–1), the sum $v^2 + 2gh$ is also conserved, but other considerations make it more useful to multiply by the constant $m/2$.)

Names are given to the combinations of quantities appearing in Eq. (8–2). The product of one-half the mass of a body and the square of its velocity is called its *kinetic energy*. Note that one does not prove that kinetic energy $= \frac{1}{2}mv^2$; it is a matter of definition. The product of the weight of a body (mg) and its height h above some reference level is called its *gravitational potential energy* relative to that reference level. Again, the fact that gravitational potential energy $= mgh$ is a matter of definition, not proof.

The sum of $\frac{1}{2}mv^2 + mgh$ (in this particular example) is the *total mechanical energy* of the body. The term $\frac{1}{2}mv_0^2$ is its initial kinetic energy. In terms of these definitions, Eq. (8–2) may be stated: *the total mechanical energy of the body, at all points of its path, is constant and equal to the initial kinetic energy.* Hence, although energy is in no sense a material substance, the body behaves *as if* it were given a supply of energy at the start in the form of kinetic energy, and during the motion it distributes this supply between the two forms of kinetic and potential energy, but always in such a way that the total amount remains constant.

Kinetic energy is often represented by the single letter K and potential energy by the letter V.

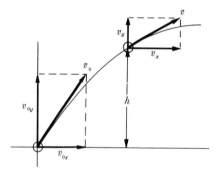

FIGURE 8–2

Conservation of energy is not restricted to vertical motion. Consider a projectile (Fig. 8–2) to which is given an initial velocity v_0 whose components are v_{0x} and v_{0y}. At some point on its trajectory at a height h

$$v_x^2 = v_{0x}^2$$
$$v_y^2 = v_{0y}^2 - 2gh.$$

Multiply both equations by $m/2$, add, and rearrange terms. This gives

$$\tfrac{1}{2}m(v_x^2 + v_y^2) + mgh = \tfrac{1}{2}m(v_{0x}^2 + v_{0y}^2).$$

But

$$v_x^2 + v_y^2 = v^2, \qquad v_{0x}^2 + v_{0y}^2 = v_0^2,$$

and therefore

$$\tfrac{1}{2}mv^2 + mgh = \tfrac{1}{2}mv_0^2.$$

Here also the sum of the kinetic and potential energies remains constant and equal to the original kinetic energy.

We shall return to the subject of kinetic and potential energies in a later paragraph. For the moment, however, we leave them and consider a related concept called *work*.

8–2 Work. In everyday life, the word "work" is applied to any form of activity which requires the exertion of muscular or mental effort. In physics, however, the term is used in a very restricted sense. Figure 8–3 represents a body moving in a horizontal direction which we shall take as the X-axis. A force F, at an angle θ with the direction of motion, is exerted on the body. *The work dW done by the force F, while the body undergoes a displacement dx, is defined as the product of the displacement and the component of the force in the direction of the displacement.*

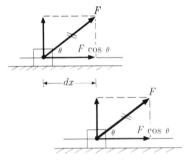

$$dW = F \cos \theta \, dx. \quad (8\text{–}3)$$

The work W done in a finite displacement from the coordinate x_1 to the coordinate x_2 is

FIG. 8–3. The work done by the force F in a displacement dx is $F \cos \theta \, dx$.

$$W = \int dW = \int_{x_1}^{x_2} F \cos \theta \, dx. \quad (8\text{–}4)$$

In the most general case both the magnitude and direction of the force may vary during the displacement, and in order to evaluate the integral both F and θ must be known in terms of x. In the special case when the force is constant in magnitude and direction

$$W = F \cos \theta \int_{x_1}^{x_2} dx = F \cos \theta (x_2 - x_1). \quad (8\text{–}5)$$

If the force is constant and in the same direction as the displacement, then $\theta = 0$, $\cos \theta = 1$, and

$$W = F(x_2 - x_1). \quad (8\text{–}6)$$

That is, in this very special case, the work done by the force is the product of the force and the displacement. In elementary texts work is usually defined as the product of force and distance, which is seen to be equivalent to Eq. (8–6). It is important to remember, however, that Eq. (8–4) is the general definition of the work done by a force; only if the force is constant and in the same direction as the displacement, is it true that "work equals force times distance."

The concept of work is so important that it will bear further discussion. Work is done only when a force is exerted on a body while the body at the same time moves in such a way that the force has a component in the direction of motion. Thus work is done when a weight is lifted, or a spring is stretched, or a gas is compressed in a cylinder. On the other hand, although it would be considered "hard work" to hold a heavy weight stationary at arm's length, no work would be done in the technical sense, since there is no motion. Even if one were to walk along a horizontal floor carrying the weight, no work would be done, since the (vertical) force has no component in the direction of the (horizontal) motion.

A locomotive does work while pulling a moving train, but if the brakes of the train should become locked so as to prevent motion, then no work would be done no matter how great a force the locomotive were to exert. The expanding gas in the cylinders of an automobile engine does work in pushing against the moving pistons, but if the motion of the pistons were to be prevented in some way, no work would be done by the gas in the cylinders no matter how great its pressure.

In the English system, the unit of force is the pound and the unit of distance is the foot. The unit of work in this system is therefore *one foot-pound*. (The order of terms is interchanged to distinguish this unit from the unit of torque, the pound-foot.) *One foot-pound* may be defined as *the work done by a constant force of one pound when the body on which the force is exerted moves a distance of one foot in the same direction as the force.**

In the mks system, where forces are expressed in newtons and distances in meters, the unit of work is the *newton-meter*. The reader can supply the definition of a newton-meter from the definition above of a foot-pound. In the cgs system the unit of work is the *dyne-centimeter*. One dyne-centimeter is called *one erg;* one newton-meter is called *one joule*. There is no corresponding single term for one foot-pound.

Since 1 meter = 100 cm and 1 newton = 10^5 dynes, it follows that 1 newton-meter = 10^7 dyne-centimeters, or

$$1 \text{ joule} = 10^7 \text{ ergs.}$$

Also, from the relations between the newton and pound, and the meter and foot,

$$1 \text{ joule} = 0.7376 \text{ foot-pound}$$
$$1 \text{ foot-pound} = 1.356 \text{ joules.}$$

---

*More generally, one foot-pound of work is done under any circumstances such that $\int_{x_1}^{x_2} F \cos \theta \, dx = 1$, when F is in pounds and dx in feet.

EXAMPLES. (1) A box weighing 100 lb is pushed 20 ft along a level floor at constant velocity by a horizontal force. The coefficient of sliding friction between box and floor is 0.30. How many foot-pounds of work are done by the force?

The force required to maintain the motion of the box is 30 lb. Since the distance moved in the direction of the force is 20 ft, the work done is 600 foot-pounds.

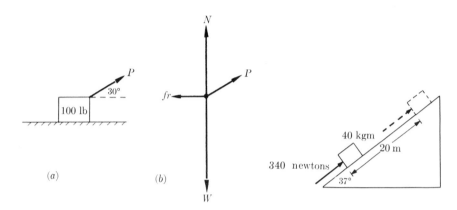

FIGURE 8–4   FIGURE 8–5

(2) How much work would be required to drag the same box 20 ft along the floor by a rope attached to the box, making an angle of 30° with the horizontal?

The force is not in the same direction as the motion. The first step is to find from a force diagram as in Fig. 8–4(b) how large a force P is required. This turns out to be 29.4 lb. Hence

$$W = P \cos \theta (x_2 - x_1)$$
$$= 29.4 \times 0.866 \times 20$$
$$= 512 \text{ foot-pounds.}$$

(3) A constant force of 340 newtons parallel to the sloping surface of a 37° inclined plane pushes a 40 kg block a distance of 20 meters up the sloping surface. (Fig. 8–5.) How much work is done by the force?

The force is constant and in the same direction as the displacement. Hence,

$$W = 340 \times 20 = 6800 \text{ newton-meters}$$
$$= 6800 \text{ joules.}$$

8–3 Energy and work. Figure 8–6 represents a body of mass m which is being drawn up a frictionless plane of slope angle ϕ by a constant force F parallel to the plane. The weight of the body and the normal force exerted on it by the plane are omitted from the diagram to avoid confusion. The body passes a point at elevation h_1 with velocity v_1, and passes a second point at elevation h_2 with velocity v_2. Let x_1 and x_2 be the coordinates of the body measured parallel to the plane.

Previous analyses of similar examples have shown that the resultant force accelerating the body up the plane is $F - mg \sin \phi$, and hence from Newton's second law

or
$$F - mg \sin \phi = ma,$$

$$a = \frac{F}{m} - g \sin \phi.$$

Furthermore, since the acceleration is constant

or
$$v_2^2 = v_1^2 + 2a(x_2 - x_1),$$

$$v_2^2 = v_1^2 + 2\left(\frac{F}{m} - g \sin \phi\right)(x_2 - x_1).$$

This equation can be rearranged in the form

$$(\tfrac{1}{2}mv_2^2 + mgx_2 \sin \phi) - (\tfrac{1}{2}mv_1^2 + mgx_1 \sin \phi) = F(x_2 - x_1),$$

and since $x_2 \sin \phi = h_2$ and $x_1 \sin \phi = h_1$ we have finally

$$(\tfrac{1}{2}mv_2^2 + mgh_2) - (\tfrac{1}{2}mv_1^2 + mgh_1) = F(x_2 - x_1). \qquad (8\text{–}7)$$

The terms in parentheses on the left side of the equation are, respectively, the final and initial energies of the body. The left side of the equation therefore represents the *increase in energy* of the body. The right side of the equation is the *work done by the force F.* Hence the work done by the force F is equal (in this case) to the increase in energy of the body on which the force is exerted. We say that work is done *on* the body *by* the force F.

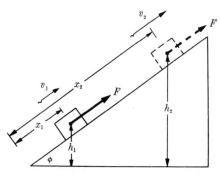

FIG. 8–6. The work done by the force F equals the sum of the increases in kinetic and potential energy of the mass m.

Although the energy of the body is increased in this process, it should not be inferred that energy has been created. The body (not shown in the figure) that exerted the force will in every instance be found to suffer a decrease in energy equal to the work done by it and hence equal to the increase in energy of the body on which the force was exerted. Thus there is conservation of energy in this process also when *all* of the bodies taking part are considered, and any instance in which work is done merely consists of a transfer of energy from one body to another.

Equation (8–7) can also be arranged as follows:

$$(\tfrac{1}{2}mv_2^2 - \tfrac{1}{2}mv_1^2) + (mgh_2 - mgh_1) = F(x_2 - x_1). \qquad (8\text{–}8)$$

The first term is the increase in kinetic energy, the second is the increase in potential energy, and the sum of these equals the work done on the body.

As a special case, if the force F is of such magnitude that it just balances the force $mg \sin \phi$ down the plane, the acceleration is zero, $v_2 = v_1$, there is no increase in kinetic energy, and the work done is accounted for by an increase in potential energy alone. The same result is found to be true if a body is raised from one elevation to another along any frictionless path. If its velocity has the same magnitude at the initial and final elevations, the work done in lifting it is equal to the increase in its potential energy.

If a body is lifted in such a way that not all parts of it rise through the same vertical height, the quantity to be used in computing the increase in potential energy is the vertical height through which the center of gravity is lifted.

Another special case of Eq. (8–8) is that in which the plane is horizontal and $h_2 = h_1$. Then there is no increase in potential energy and all of the work done on the body goes into increasing its kinetic energy.

$$\tfrac{1}{2}mv_2^2 - \tfrac{1}{2}mv_1^2 = F(x_2 - x_1). \qquad (8\text{–}9)$$

In the general case where the force makes an angle θ with the surface of the plane, and the force F or the angle θ or both may vary during the motion, the accelerating force at any point is $F \cos \theta - mg \sin \phi$ (the reader may supply his own diagram) and

$$F \cos \theta - mg \sin \phi = ma = m \frac{dv}{dt} = mv \frac{dv}{dx},$$

$$mv\,dv + mg \sin \phi\,dx = F \cos \theta\,dx,$$

and since $\sin \phi\,dx = dh$,

$$mv\,dv + mg\,dh = F \cos \theta\,dx,$$

$$\int_{v_1}^{v_2} mv\,dv + \int_{h_1}^{h_2} mg\,dh = \int_{x_1}^{x_2} F \cos \theta\,dx,$$

$$(\tfrac{1}{2}mv_2^2 - \tfrac{1}{2}mv_1^2) + (mgh_2 - mgh_1) = \int_{x_1}^{x_2} F \cos \theta\,dx. \qquad (8\text{–}10)$$

The right side is the work done by the force F, the left side is the increase in energy, and the two are equal even if the force and its angle with the plane are not constant. Note that the slope angle of the plane, ϕ, does not appear in Eq. (8–10). The result therefore holds for a frictionless plane of any angle whatever. Furthermore, since a path of any arbitrary curved shape is merely a succession of planes of different slopes, the path need not be restricted to a plane. Hence, in the special case where $v_2 = v_1$, the work done in lifting a body along any arbitrary path equals the increase in potential energy, which proves the statement above to that effect.

If $h_2 = h_1$, then

$$\tfrac{1}{2}mv_2^2 - \tfrac{1}{2}mv_1^2 = \int_{x_1}^{x_2} F \cos \theta \, dx, \tag{8-11}$$

which is an obvious generalization of Eq. (8-9).

8-4 Units of energy. Dimensions. The units of kinetic and potential energy were not discussed in Section 8-1, where these quantities were first defined. We can now see, however, that since work equals increase in energy, the units of both kinetic and potential energy must be the same as the units of work, that is, foot-pounds, joules, or ergs.

The unit of any "compound" quantity such as mgh or $\tfrac{1}{2}mv^2$ is obtained by combining the units of its terms. Thus, in the English system, where m is in slugs, g in ft/sec^2, h in ft and v in ft/sec, the unit of potential energy is

$$1 \text{ slug} \times 1 \, \frac{\text{ft}}{\text{sec}^2} \times 1 \text{ ft} = 1 \, \frac{\text{slug·ft}^2}{\text{sec}^2},$$

and the unit of kinetic energy is

$$1 \text{ slug} \times \left(1 \, \frac{\text{ft}}{\text{sec}}\right)^2 = 1 \, \frac{\text{slug·ft}^2}{\text{sec}^2}.$$

These units are the same, and to show that they are equivalent to one foot-pound, we may return to Newton's second law, $F = ma$, which in terms of its units becomes

$$1 \text{ lb} = 1 \, \frac{\text{slug·ft}}{\text{sec}^2},$$

or

$$1 \text{ slug} = 1 \, \frac{\text{lb·sec}^2}{\text{ft}}.$$

We may therefore replace the slug, in the expressions for kinetic and potential energy, by the lb·sec^2/ft, which gives

$$1 \, \frac{\text{slug·ft}^2}{\text{sec}^2} = 1 \, \frac{\text{lb·sec}^2}{\text{ft}} \times \frac{\text{ft}^2}{\text{sec}^2}.$$

After canceling and rearranging terms, this reduces to one foot-pound.

Similar reasoning shows that the unit of energy in the mks system is the newton-meter or joule, and in the cgs system it is the dyne-cm or erg.

Note carefully that if it is desired to compute the kinetic energy of a body in foot-pounds from the expression $\frac{1}{2}mv^2$, the mass *must* be expressed in slugs; to compute the kinetic energy in newton-meters or joules, the mass *must* be expressed in kilograms, etc.

A somewhat more formal method of handling units in physical equations is known as "dimensional analysis." The fundamental "dimensions" of all mechanical quantities are usually taken as mass, length, and time. Mass has the dimension M, length the dimension L, and time the dimension T. Velocity, the ratio of length to time, has the dimensions L/T or LT^{-1}. Acceleration, the ratio of velocity to time, has the dimensions LT^{-2}. Force, equal to the product of mass and acceleration, has the dimensions MLT^{-2}. Work, the product of force and length, has the dimensions ML^2T^{-2} and these are evidently the dimensions of kinetic and potential energy also.

EXAMPLES. (1) What is the kinetic energy of a 2400-lb automobile traveling at 100 ft/sec?
The mass of the automobile is 75 slugs and its kinetic energy is

$$KE = \tfrac{1}{2}mv^2 = \tfrac{1}{2} \times 75 \times (100)^2 = 375,000 \text{ ft·lb.}$$

(2) An electron of mass 9×10^{-28} gm strikes the target of an X-ray tube with a velocity of 6×10^9 cm/sec. What is its kinetic energy?
In cgs units,
$$KE = \tfrac{1}{2} \times 9 \times 10^{-28} \times (6 \times 10^9)^2$$
$$= 1.62 \times 10^{-8} \text{ erg.}$$
In mks units,
$$KE = \tfrac{1}{2} \times 9 \times 10^{-31} \times (6 \times 10^7)^2$$
$$= 1.62 \times 10^{-15} \text{ joule.}$$

Since 1 joule $= 10^7$ ergs, the answers are in agreement.
(3) Refer to example 3 on page 128. If the plane is frictionless and the body starts from rest, find from energy considerations its velocity after moving 20 m up the slope.
The work done on the block equals the sum of the increases in its kinetic energy and potential energy. The work done has been shown (see page 128) to be 6800 joules. The block ascends 20 m along the slope or 12 m vertically. The increase in potential energy is therefore $40 \times 9.8 \times 12 = 4710$ joules, and the increase in kinetic energy is $6800 - 4710 = 2090$ joules. Since the initial kinetic energy is zero, this equals the final kinetic energy and

$$\tfrac{1}{2} \times 40 \times v^2 = 2090,$$
$$v^2 = 104.5,$$
$$v = 10.2 \text{ m/sec.}$$

8–5 Absolute values of potential and kinetic energy. The definition of gravitational potential energy, mgh, implies that the potential energy is zero when $h = 0$, that is, when the body is at the reference level. To be specific, suppose this level to be the top of the laboratory table. If the reference level had been taken at some lower elevation such as the floor, the potential energy would not be zero at the table top. If it had been taken at the ceiling, the potential energy at the table top would be negative. It will be seen that the gravitational potential energy of a body at any point depends on the arbitrary choice of a reference level and hence is indeterminate to that extent. The indeterminacy is not of importance, however, since in any practical case one is concerned only with differences in potential energy, which are independent of the reference level. It is usually most convenient to choose this level at or below the lowest point in any specific problem, which avoids the appearance of negative energies.

Another aspect of potential energy should be pointed out here. When a man picks up a weight from the floor and raises it above his head, he has in effect inserted his body between the weight and the earth and pushed the weight one way with his hands, and the earth the other way with his feet. The general principle involved is obscured in this common example by the disparity in mass between the earth and the body lifted—the displacement of the earth is so much smaller than that of the body. Imagine an Atlas standing on one small planetoid and "lifting" another of the same size as the first. We may then ask, "To which body has potential energy been given?" If the bodies are of equal size, each will move an equal distance from its original position, and the process is better described as "separating" the bodies rather than "lifting" one of them. It is evident in this imaginary example that the potential energy should not be assigned to either body singly but rather that it represents a joint property of the *system*. The potential energy of the two planetoids is greater when they are separated than when they are close together.

The same is evidently true whatever the ratio of masses of the bodies, and therefore it applies to our original example of a man lifting a weight. The potential energy is not a property of the weight alone but a joint property of the system weight + earth. Although this aspect of potential energy must be kept in mind, we shall nevertheless, for convenience, continue to speak of "the potential energy of a raised weight" as if it were associated with the weight alone.

Considerations similar to those above apply to kinetic energy also. We say that an object at rest in the laboratory has no kinetic energy, since its velocity is zero. But although it has no velocity relative to the earth, it partakes in the motion of the earth about its axis, the revolution of the earth about the sun, and the motion of the whole solar system through space.

8–6 Potential energy of a stretched spring. Gravitational potential energy is only one of many forms of potential energy. Another type commonly encountered is the potential energy of an elastic body which has been distorted in some way. The subject of elasticity will be discussed more fully in Chapter 13. For the present it will suffice to state that when most solid bodies are distorted, the force needed to produce the distortion increases in direct proportion to the distortion, provided the latter is not too great. This property of matter was one of the first to be studied quantitatively and the statement above was published by Robert Hooke in 1678. It is known as Hooke's law.

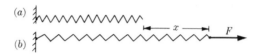

FIGURE 8-7

To be specific, let us consider a coil spring subjected to a stretching force. Figure 8–7(a) represents an unstretched spring. In Fig. 8–7(b) the spring has been stretched a distance x above its normal or no-load length. The force required to hold it in this position is F. Hooke's law, in mathematical form, states that

$$F \propto x \qquad \text{or} \qquad F = kx, \qquad (8\text{–}12)$$

where k is a constant of proportionality called the *force constant* or the *stiffness coefficient* of the spring. The force constant may be defined as the ratio of the force to the extension it produces above the no-load length, or the force per unit elongation. It is expressed in lb/ft, newtons/meter, or dynes/cm.

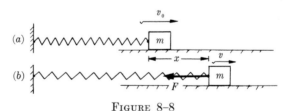

FIGURE 8-8

In Fig. 8–8(a), a block of mass m on a horizontal frictionless surface is connected to a fixed point by a spring. The spring is assumed to have its natural unstretched length and the force exerted by it on the block is zero. Let the block be suddenly given a velocity v_0 toward the right. As soon as it starts to move, the spring becomes stretched and when the coordinate of the block is x, the force exerted by the block on the spring, from Hooke's law, equals kx. The reaction to this force, the force exerted on the block by the spring, is $-kx$. Hence from Newton's second law

$$F = -kx = ma = mv\frac{dv}{dx},$$

$$mv\,dv = -kx\,dx,$$

$$\int_{v_0}^{v} mv\,dv = -\int_{0}^{x} kx\,dx,$$

$$\tfrac{1}{2}mv^2 - \tfrac{1}{2}mv_0^2 = -\tfrac{1}{2}kx^2,$$

$$\tfrac{1}{2}mv^2 + \tfrac{1}{2}kx^2 = \tfrac{1}{2}mv_0^2. \tag{8-13}$$

This equation should be compared with Eq. (8-2) on page 124. We see that in this motion the sum $\tfrac{1}{2}mv^2 + \tfrac{1}{2}kx^2$ is conserved, and is equal to the original kinetic energy $\tfrac{1}{2}mv_0^2$. We therefore conclude that $\tfrac{1}{2}kx^2$ represents the *elastic* potential energy of the stretched spring. It is easy to show from the definition of k as the ratio of a force to a length that the unit of elastic potential energy is the same as the unit of work in any system.

It will be left as an exercise for the reader to show that the work done in stretching a spring from an elongation x_1 to an elongation x_2 is equal to the increase in potential energy of the spring. This is analogous to the fact that the work done in raising a weight equals the increase in potential energy of the weight.

EXAMPLE. A force of 5 lb is found to stretch a screen door spring 6 in. What is the potential energy of the spring when opening the door stretches it 18 in.?

Since a 5-lb force stretches the spring 6 in., the force constant k, or the ratio of force to extension, is

$$k = \frac{5}{\tfrac{1}{2}} = 10\ \text{lb/ft}.$$

Hence

$$\text{PE} = \tfrac{1}{2} \times 10 \times (1.5)^2 = 11.3\ \text{ft·lb}.$$

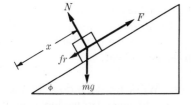

FIGURE 8-9

8-7 Work against friction. We shall consider next how friction forces, which have been neglected up to this point, affect the mechanical energy of a moving body. Again for illustration we make use of a block being drawn up an inclined plane. (Fig. 8-9.) The external force F and the friction force fr are both assumed constant. The resultant force up the plane is $F - mg \sin \phi - fr$, and Newton's second law leads to the equation

$$F - mg \sin \phi - fr = mv \frac{dv}{dx}.$$

By the usual procedure we obtain

$$(\tfrac{1}{2}mv_2^2 + mgh_2) - (\tfrac{1}{2}mv_1^2 + mgh_1) = F(x_2 - x_1) - fr(x_2 - x_1). \quad (8\text{–}14)$$

The term $fr(x_2 - x_1)$ is the work done against the friction force, and we see that the increase in the energy of the block is not equal to the work done by the force F, but is smaller by an amount equal to the work done against friction. Another point of view is to call the right side of the equation the *net* work done on the block, and this of course equals the increase in its energy.

If the force F were zero, and the body had simply been given an initial velocity v_1 up the plane and then left to itself, Eq. (8–14) would become

$$(\tfrac{1}{2}mv_2^2 + mgh_2) = (\tfrac{1}{2}mv_1^2 + mgh_1) - fr(x_2 - x_1).$$

The term $\tfrac{1}{2}mv_1^2 + mgh_1$ is the initial energy of the block; the term $\tfrac{1}{2}mv_2^2 + mgh_2$ is its energy at any higher point. We see that the mechanical energy of the block is *not* conserved but decreases continually as x_2 increases, because of the term $-fr(x_2 - x_1)$. Again, this does not mean that energy has been destroyed, since it is found that heat is developed whenever friction forces are present, and it can be shown by measuring this heat that it is exactly equivalent to the decrease in mechanical energy. "Work done against friction" is therefore the same as "energy converted to heat."

Let us write Eq. (8–14) once more as

$$F(x_2 - x_1) = (\tfrac{1}{2}mv_2^2 - \tfrac{1}{2}mv_1^2) + (mgh_2 - mgh_1) + fr(x_2 - x_1).$$

This equation sums up in a single expression all of the special cases considered thus far in this chapter. That is, whenever work is done by a force, this work can always be accounted for in one or more of three different ways: an increase in kinetic energy, an increase in potential energy, or as work done against friction and converted to heat.

8–8 Conservative and dissipative forces. Work must be done by some outside force to lift a body vertically at constant velocity; we have shown that this work is equal to the increase in gravitational potential energy of the body. Work must also be done by an outside force to slide a body at constant velocity along a rough horizontal surface; in this case the potential energy of the body does not change and the work done is converted to heat. Why is it that although external work must be done in both cases, we have an increase in potential energy in the first example and not in the second? The distinction becomes evident when we consider the process of returning the body to its initial position.

We may say that in the first instance work was done against the gravitational pull of the earth, while in the second instance work was done against the force of friction. If the weight is lowered at constant velocity to its original position, the gravitational force remains constant in magnitude and direction. The descending weight can therefore be made to do work (it could, for example, raise a second equal weight connected to it by a string passing over a pulley) and since the force and distance are the same in both ascent and descent, the work obtainable equals that originally expended. In other words, the work is *recoverable*, or in still other words, the net work done in a round trip is zero.

Contrast this with the behavior of the friction force. When we slide the body on the rough surface back to its original position, the friction force reverses, and instead of recovering the work done in the first displacement we must again do work on the return trip. The net work done in a round trip is not zero.

This difference between gravitational forces and friction forces is the criterion that determines whether or not there is an increase in potential energy when work is done. If the work can be recovered there is an increase in potential energy; if the work can not be recovered there is no increase. Forces such as those of gravity or the force exerted by a spring, where the work is recoverable, are called *conservative* forces. Forces like those of sliding friction are called *nonconservative* or *dissipative* forces. Only when all the forces are conservative is the mechanical energy of a system conserved, and only when work is done against a conservative force is there an increase in potential energy.

It may be objected that since the heat developed when friction is present is equivalent to the energy dissipated, this heat might be used to operate a heat engine whose output could be used to raise a weight. This question will be answered more fully in Chapter 24 in connection with the second law of thermodynamics. For the present we shall simply state that while part of the heat can be converted back to mechanical work, it is never possible to recover all of it.

In the light of the preceding discussion a general expression can be written for the increase in potential energy. Let a body be taken from point a to point b along a path of any shape as in Fig. 8–10. Let F represent a conservative force making an angle θ with an element ds of the path. The work done *against* the force F, in the distance ds, is $-F \cos \theta \, ds$ and this equals the increase in potential energy, dV. That is

$$dV = -F \cos \theta \, ds, \tag{8-15}$$

and

$$\int_{V_a}^{V_b} dV = V_b - V_a = - \int_a^b F \cos \theta \, ds. \tag{8-16}$$

The term $\int_a^b F \cos \theta \, ds$ is called the *line integral* of F between the points a and b. (See Thomas, *Calculus and Analytic Geometry*, page 482.)

The two special cases of gravitational potential energy and the potential energy of a stretched spring can be deduced at once from Eq. (8–16). For instance, when a body of mass m is lifted vertically from an elevation y_1 to an elevation y_2, $F = mg$, $\theta = 180°$, $\cos \theta = -1$, $ds = dy$, and

$$V_b - V_a = - \int_{y_1}^{y_2} mg \, (-1) \, dy = mg(y_2 - y_1).$$

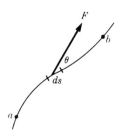

FIGURE 8–10

We shall illustrate still another useful feature of work-energy relations by a simple example. Consider a special case in which the force and displacement are both along the X-axis. Then Eq. (8–15) becomes

$$dV = -F \, dx,$$

$$F = - \frac{dV}{dx}.$$

Hence the force is equal to the negative of the derivative of the potential energy with respect to the X-coordinate. If we construct a graph of V as a function of x, the term dV/dx equals the slope of the graph, or, as it is called, the *potential energy gradient*. Hence we may say that *the force equals the negative of the potential energy gradient*. This relation between force and potential energy turns out to be very useful later on when we study gravitational, electrical, and magnetic forces.

In particular, if there is some point on the axis where the force is zero, a body at that point is in equilibrium. But if the force is zero then $dV/dx = 0$, which is the condition for the existence of a maximum or a minimum in the potential energy. It is not difficult to show that for the equilibrium to be stable, the potential energy must be a minimum. The equilibrium is unstable if the potential energy is a maximum. Hence when a body is acted on by conservative forces, it is in stable equilibrium at any point where its potential energy is a minimum and in unstable equilibrium at any point where its potential energy is a maximum. A familiar example, which is left to the reader to analyze, is that of a rod suspended from a pivot at one end.

There is one point in connection with work and potential energy that is sometimes confusing. Suppose a body is lifted vertically by a force P, greater in magnitude than the weight of the body. If v_1 and v_2 are the upward velocities at the heights h_1 and h_2, and if friction is neglected,

$$\text{Work} = \Delta \text{KE} + \Delta \text{PE},$$

$$P(h_2 - h_1) = \tfrac{1}{2}mv_2^2 - \tfrac{1}{2}mv_1^2 + mgh_2 - mgh_1.$$

If the potential energy terms are transferred to the left side of the equation we get

$$P(h_2 - h_1) - mg(h_2 - h_1) = \tfrac{1}{2}mv_2^2 - \tfrac{1}{2}mv_1^2.$$

The term $P(h_2 - h_1)$ is the work done by the force P. The term $-mg(h_2 - h_1)$ is the product of the gravitational force $-mg$, which is opposite to the direction of motion, and the distance moved, $h_2 - h_1$. It may therefore be described as *the work done against the gravitational force* and the third equation could be stated verbally: "The *net* work done on the body is equal to the increase in its kinetic energy alone."

The same algebraic process of transferring terms from one side of the work-energy equation to the other can be performed with any conservative force. That is, we may put the terms involving the force on the right side of the equation and call their difference an increase in potential energy, or we may transfer them to the left side of the equation with signs reversed and say that they represent work done against the conservative force. One cannot have it both ways, however. If, when a body is lifted vertically, we put the *mgh* terms on the left side of the equation and consider that work has been done against the force of gravity, then it is not legitimate to assign a gravitational potential energy to the body. On the other hand if the terms are written on the right and are considered to represent an increase in potential energy we must not include in the "work" term the work done against gravitational forces. It is a matter of indifference which point of view is adopted. The more common one, however, is to associate a potential energy with a conservative force.

Summary

If we let K_1 and V_1 represent the initial kinetic and potential energies of a body, K_2 and V_2 the energies at the end of some process, and H the energy converted to heat, then we may write

$$\text{Work done} = (K_2 - K_1) + (V_2 - V_1) + H.$$

Special cases. (1) If the system is *frictionless* then $H = 0$ and all of the work done is accounted for by increases in kinetic and potential energy.

(2) If no work is done on the system, we have

$$0 = (K_2 - K_1) + (V_2 - V_1) + H,$$

or

$$K_1 + V_1 = K_2 + V_2 + H.$$

That is, the total mechanical energy of an isolated system at the start of any process is equal to the mechanical energy at the end of the process plus any energy that has been converted to heat.

This equation can also be written

$$(V_1 - V_2) = (K_2 - K_1) + H.$$

That is, in any process taking place in an isolated system, the loss in potential energy equals the gain in kinetic energy plus the heat developed.

Still another arrangement of terms gives

$$(K_2 + V_2) = (K_1 + V_1) - H.$$

That is, the mechanical energy at the end of a process is less than that at the beginning by the amount of energy converted to heat.

(3) If the system is frictionless and no work is done on it, then we have

$$(K_1 + V_1) = (K_2 + V_2).$$

That is, the total mechanical energy of an isolated frictionless system is constant.

The latter equation can also be written

$$(V_1 - V_2) = (K_2 - K_1).$$

That is, in any process taking place in an isolated frictionless system the loss in potential energy equals the gain in kinetic energy.

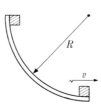

FIGURE 8–11

EXAMPLE. A body slides from rest without friction down a track consisting of one quadrant of a circle of radius R. (Fig. 8–11.) Show that the magnitude of its velocity at the lowest point is equal to that which would be acquired in free fall through a vertical height R.

The normal force exerted on the body by the track is always at right angles to the motion, so no work is done on the body as it descends and the total energy remains constant. At the starting point the kinetic energy is zero and the potential energy is mgR, if the reference level is that of the lowest point. At the end the potential energy is zero and the kinetic energy is $\frac{1}{2}mv^2$. Hence

$$0 + mgR = \tfrac{1}{2}mv^2 + 0,$$
$$v^2 = 2gR.$$

8–9 The principle of virtual work. The principle of *virtual work* provides an extremely powerful tool for the solution of problems in statics. In effect, it is equivalent to the three conditions of equilibrium.

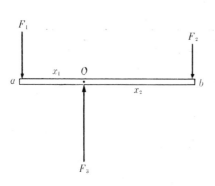

Figure 8–12 represents a rigid rod (whose own weight can be neglected) in equilibrium under the action of the forces F_1, F_2, and F_3. If we imagine the rod to be rotated through a small angle about an axis through O, point a moves down a small distance y_1 and point b moves up a small distance y_2. Work equal to $F_1 y_1$ is done *by* the force F_1, and work $F_2 y_2$ is done *against* the force F_2. No work is done by the force F_3. But from the conditions of equilibrium we know that $F_1 x_1 = F_2 x_2$, and from similar triangles $y_1/y_2 = x_1/x_2$. It follows from the last two equations that $F_1 y_1 = F_2 y_2$. Hence the work done by the force F_1 equals the work done against the force F_2 and *the net work is zero*.

FIGURE 8–12

The displacement of the system will of course not occur of itself; we simply imagine it to take place and compute the net work that *would* be done if it *did* take place. It is called a *virtual displacement*. The principle of virtual work is a generalization of this example. It states that *when a system is in equilibrium, the net work done in a virtual displacement is zero.*

In the example chosen for illustration the forces were known in advance. When this method is used as a technique for problem solution the forces are not known but are computed with the help of the principle.

8–10 Power. The time element is not involved in the definition of work. The same amount of work is done in raising a given weight through a given height whether the work is done in one second, or one hour, or one year. In many instances, however, it is necessary to consider the *rate* at which work is done as well as the total amount of work accomplished. The rate at which work is done by a working agent is called the *power* developed by that agent.

If a quantity of work W is done in a time interval $t_2 - t_1$, the average power $\overline{P}$ is defined as

$$\text{Average power} = \frac{\text{work done}}{\text{time interval}}.$$

$$\overline{P} = \frac{W}{t_2 - t_1}.$$

If the rate of doing work is not uniform, the power of any instant is the ratio of the work done to the time interval, when both are extremely small.

$$\text{Instantaneous power } P = dW/dt.$$

In the English system, where work is expressed in foot-pounds and time in seconds, the unit of power is one foot-pound per second. Since this unit is inconveniently small, a larger unit called the horsepower (hp) is in common use. 1 hp = 550 ft·lb/sec = 33,000 ft·lb/min. That is, a 1 hp motor, running at full load, is doing 33,000 ft·lb of work every minute it runs.

The mks unit of power is one joule per second, which is called one *watt*. This is also an inconveniently small unit, and power is more commonly expressed in *kilowatts* or kw (1 kw = 1000 watts = 1000 joules/sec), or *megawatts* (1 megawatt = 1000 kw = 1,000,000 watts).

The cgs power unit is one erg per second. No single term is assigned to this unit.

A common misconception is that there is something inherently *electrical* about a watt or a kilowatt. This is not the case. It is true that electrical power is usually expressed in watts or kilowatts, but the power consumption of an incandescent lamp could equally well be expressed in horsepower, or an automobile engine rated in kilowatts.

From the relations between the newton, pound, meter, and foot, it is easy to show that 1 hp = 746 watts = 0.746 kw, or about $\frac{3}{4}$ of a kilowatt. This is a useful figure to remember.

Having defined two units of power, the horsepower and the kilowatt, we may use these in turn to define two new units of work, the *horsepower-hour* and the *kilowatt-hour* (kwh).

One horsepower-hour is the work done in one hour by an agent working at the constant rate of one horsepower.

Since such an agent does 33,000 ft·lb of work each minute, the work done in one hour is 60 × 33,000 = 1,980,000 ft·lb.

$$1 \text{ horsepower-hour} = 1.98 \times 10^6 \text{ ft·lb.}$$

One kilowatt-hour is the work done in one hour by an agent working at the constant rate of one kilowatt.

Since such an agent does 1000 joules of work each second, the work done in one hour is 3600 × 1000 = 3,600,000 joules.

$$1 \text{ kwh} = 3.6 \times 10^6 \text{ joules.}$$

Notice carefully that the horsepower-hour and the kilowatt-hour are units of *work*, not power.

One aspect of work or energy which may be pointed out here is that although it is an abstract physical quantity, it nevertheless has a monetary value. A pound of force or a foot-per-second of velocity are not things which are bought and sold as such, but a foot-pound or a kilowatt-hour of energy are quantities offered for sale at a definite market rate. In the form of electrical energy, a kilowatt-hour can be purchased at a price varying from a few tenths of a cent to a few cents, depending on the locality and the quantity purchased. In the form of heat, 778 ft·lb (one Btu) costs about a thousandth of a cent.

8–11. Power and velocity. Suppose a constant force F is exerted on a body, while the body undergoes a displacement $x_2 - x_1$ in the direction of the force. The work done is

$$W = F(x_2 - x_1),$$

and the average power developed is

$$\overline{P} = \frac{W}{t_2 - t_1} = F\frac{(x_2 - x_1)}{(t_2 - t_1)}. \tag{8–17}$$

But $x_2 - x_1/t_2 - t_1$ is the average velocity, $\bar{v}$. Hence

$$\overline{P} = F\bar{v}.$$

If the time interval is made extremely short Eq. (8–17) reduces to

$$P = F\frac{dx}{dt},$$

or

$$P = Fv \tag{8–18}$$

where P and v are instantaneous values.

EXAMPLE. A locomotive traveling at 50 ft/sec exerts a draw-bar pull of 20,000 lb. What horsepower does it develop?

$$P = Fv$$

$$= 20,000 \times 50 = 1,000,000 \text{ ft·lb/sec}$$

$$= \frac{1,000,000}{550} = 1820 \text{ hp}.$$

PROBLEMS

8-1. The locomotive of a freight train exerts a constant force of 6 tons on the train while drawing it at 50 mi/hr on a level track. How many foot-pounds of work are done by the locomotive in a distance of 1 mi?

8-2. A horse is towing a canal boat, the towrope making an angle of 10° with the towpath. If the tension in the rope is 100 lb, how many foot-pounds of work are done by the horse while moving 100 ft along the towpath?

FIGURE 8-13

8-3. An 80-lb block is pushed a distance of 20 ft along a level floor at constant speed by a force at an angle of 30° with the horizontal, as in Fig. 8-13. The coefficient of friction between block and floor is 0.25. How many foot-pounds of work are done?

8-4. The force in pounds required to stretch a certain spring a distance of x ft beyond its unstretched length is given by $F = 10x$. What force will stretch the spring 6 in? 1 ft? 2 ft? How much work is required to stretch the spring 6 in? 1 ft? 2 ft?

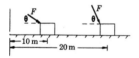

FIGURE 8-14

8-5. A 20-kgm block is pushed along a horizontal surface by a force F making an angle θ with the surface (Fig. 8-14). The force is increased during the motion according to the relation $F = 6x$, where F is in newtons and x is in meters. The angle is also increased according to the relation $\cos \theta = 0.70 - .02x$. How much work is done by the force as the body moves from $x = 10$ m to $x = 20$ m?

8-6. The force of gravitational attraction between two point masses m and m' is given by

$$F = \frac{Gmm'}{x^2},$$

where x is the distance between the masses and G is a constant. Compute the work done when the separation of the masses is increased from a to b.

8-7. A body is attracted toward the origin with a force given by $F = -6x^3$, where F is in lb and x in ft. What force is required to hold the body at point a, 1 ft from the origin? At point b, 2 ft from the origin? How much work must be done to move the body from point a to point b?

8-8. A block is pushed 4 ft along a horizontal surface by a force of 10 lb. The opposing force of friction is 2 lb. How much work is done by the 10-lb force? What is the *net* work done on the block?

8-9. A bullet of mass 2 gm is fired from a gun whose barrel is 100 cm long. The force accelerating the bullet, while it is in the gun barrel is given by the equation $F = 12 \times 10^6 - (6 \times 10^4)x$, where F is in dynes and x in cm. Find the muzzle velocity of the bullet.

8-10. Compute the kinetic energy of an 1800-lb automobile traveling at 30 mi/hr. How many times as great is the kinetic energy if the velocity is doubled?

8-11. An electron strikes the screen of a cathode-ray tube with a velocity of 10^9 cm/sec. Compute its kinetic energy in ergs and in joules. The mass of an electron is 9×10^{-28} gm.

FIGURE 8-15

8-12. A meter stick whose mass is 300 gm is pivoted at one end as in Fig.

8–15 and displaced through an angle of 60°. What is the increase in its potential energy?

8–13. The scale of a certain spring balance reads from zero to 400 lb and is 8 inches long. What is the potential energy of the spring when it is stretched 8 in? 4 in? When a 50-lb weight hangs from the spring?

8–14. A block weighing 16 lb is pushed 20 ft along a horizontal frictionless surface by a horizontal force of 8 lb. The block starts from rest. (a) How much work is done by the force? What becomes of the work? (b) Check your answer by computing the acceleration of the block, its final velocity, and its kinetic energy.

8–15. In the preceding problem, suppose the block had an initial velocity of 10 ft/sec, other quantities remaining the same. (a) How much work is done by the force? (b) Check by computing the final velocity and the increase in kinetic energy.

8–16. A 16-lb block is lifted vertically at a constant velocity of 10 ft/sec through a height of 20 ft. How great a force is required? How much work is done by the force? What becomes of this work?

8–17. A 25-lb block is pushed 100 ft up the sloping surface of a plane inclined at an angle of 37° to the horizontal by a constant force F of 32.5 lb acting parallel to the plane. The coefficient of friction between the block and plane is 0.25. (a) How much work is done by the force F? (b) Compute the increase in kinetic energy of the block. (c) Compute the increase in potential energy of the block. (d) Compute the work done against friction. What becomes of this work? (e) What can you say about the sum of b, c, and d?

8–18. A block of mass m slides a distance s down the slope of a plane inclined at an angle θ with the horizontal. The coefficient of sliding friction between block and plane is μ. The velocity of the block at the top of the plane is v_1. What is the expression for its velocity v_2 at the bottom? (Use energy considerations to obtain the answer.) From this expression, what is the minimum angle of inclination at which the block will slide down the plane at constant velocity?

8–19. The hammer of a pile driver weighs 1 ton. It drops 10 ft onto a pile which it drives in 3 inches. Compute the force exerted on the pile, from energy considerations. (Assume the force to be constant.)

8–20. The spring of a spring-gun has a stiffness coefficient of 3 lb per inch. It is compressed 2 inches and a ball weighing 0.02 lb is placed in the barrel against the compressed spring. (a) Compute the maximum velocity with which the ball leaves the gun when released. (b) Determine the maximum velocity if a resisting force of 2.25 lb acts on the ball.

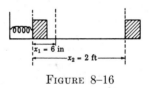

FIGURE 8–16

8–21. A block weighing 2 lb is forced against a horizontal spring of negligible mass, compressing the spring an amount $x_1 = 6$ inches. Upon releasing the block, it moves on a horizontal table top a distance $x_2 = 2$ ft before coming to rest. The force constant k is 8 lb/ft. (Fig. 8–16.) What is the coefficient of friction, μ, between the block and the table?

8–22. A block of mass 10 gm slides with velocity v along a frictionless plane until it strikes a perfectly elastic spring of negligible mass. The block comes to rest after compressing the spring 5 cm. The force constant of the spring is 1000 dynes/cm. (a) What is the potential energy of the spring in its compressed position? (b) What was the velocity v of the block before it struck the spring?

(c) What was the work done by the block on the spring?

8–23. A 2-kgm block is dropped from a height of 40 cm onto a spring whose force constant, k, is 1960 newtons/meter. Find the maximum distance the spring will be compressed.

FIGURE 8–18

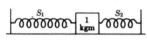

FIGURE 8–17

8–24. A block of mass 1 kgm is resting on a table as in Fig. 8–17. It is connected to two upright supports by springs S_1 and S_2. The coefficient of friction between the block and the table is 0.2. The force constant of S_1 can be determined by the fact that it takes 10 newtons to stretch the spring 0.25 meter. It takes 2.5 newtons to stretch S_2 0.25 meter.

Initially the block rests at its position of equilibrium between unstretched springs. It is then slowly displaced 0.5 meter by a horizontal force F, compressing one spring and stretching the other. (a) Compute the total work done by the horizontal force F in this process. (b) If the block is subsequently released, what is its velocity as it passes through the equilibrium position?

8–25. A 16-lb projectile is fired from a gun with a muzzle velocity of 800 ft/sec at an angle of elevation of 45°. The angle is then increased to 90° and a similar projectile is fired with the same muzzle velocity. (a) Find the maximum heights attained by the projectiles. (b) Show that the total energy at the top of the trajectory is the same in the two cases. (c) Using the energy principle, find the height attained by a similar projectile if fired at an elevation angle of 30°.

8–26. A pogo stick is essentially a platform on a stiff coil spring. Suppose that this spring is 1 ft long when uncompressed, and that it has a force constant of 1260 lb/ft. Fig. 8–18 shows a

200-lb man who has just landed on the spring and compressed it 8 inches (so that he is about to be propelled into the air). (a) What is the gravitational potential energy of the man at this instant? (Assume it is zero when the man's feet are on the ground, and that he doesn't bend his knees.) (b) How much potential energy is stored in the spring at this instant? (c) How high above the ground does the man go? (Note that the man is not fastened to the spring.)

FIGURE 8–19

8–27. Fig. 8–19 shows a frictionless track in the form of a quarter-circle of 4 ft radius, smoothly joined at the bottom to a flat horizontal surface. On the surface is a spring with its free end at the bottom of the circular track. A force of 1200 lb will compress this spring $\frac{3}{4}$ ft. An object of weight 12.5 lb is released, with zero velocity, at the top of the track, and is brought to rest on the level surface by the spring. (a) What is the velocity of the object just before it hits the spring? (b) How far has the spring been compressed when the object comes to rest? (c) Let the potential energy be zero just before the object hits the spring. What is the total mechanical energy of the system when the object has compressed the spring 0.1 ft?

8–28. A 10-lb block is projected up a 37° inclined plane, with a velocity at the foot of the plane of 32 ft/sec. It is observed to ascend a distance of 20 ft along the sloping surface of the plane, stop, and slide back to the bottom. (a) Compute the initial kinetic energy of the block, and its potential energy at the highest point. (b) Find the friction force acting on the block. (c) Find the velocity of the block when it returns to the foot of the plane.

FIGURE 8–20

8–29. The track shown in Fig. 8–20 consists of a frictionless quarter-circle of radius 4 ft, smoothly joined to a rough inclined plane of angle 37°. A block is released from rest at the top of the quarter-circle. The coefficient of friction between the block and the inclined plane is 0.3. What fraction of the kinetic energy which the block possesses at the bottom of the track is dissipated by friction as the block slides up the plane?

8–30. A meter stick whose mass is 100 gm is pivoted at one end, raised to a horizontal position, and released. Find its kinetic energy as it swings through a vertical position.

8–31. A small sphere of mass m is fastened to a weightless string of length 2 ft to form a pendulum. The pendulum is swinging so as to make a maximum angle of 60° with the vertical. (a) What is the velocity of the sphere when it passes through the vertical position? (b) What is the instantaneous acceleration when the pendulum is at its maximum deflection?

8–32. An elevator, with its load, weighs 2400 lb. The elevator starts from rest at the first floor, and 5 sec later it passes the 5th floor, 60 ft above the first, with a velocity of 30 ft/sec. Find

the total work done on the elevator during the 5-sec interval, and the average horsepower developed. Neglect friction.

8–33. Compute the horsepower developed by the locomotive in Problem 8–1.

8–34. The hammer of a pile driver weighs 1000 lb and must be lifted a vertical distance of 6 ft in 3 sec. What horsepower engine is required?

FIGURE 8–21

8–35. A gun fires a projectile at an angle of 37° above the horizontal, as shown in Fig. 8–21. The projectile comes down at A, just missing the edge of a cliff, which is at the same elevation as the gun, and 30,000 ft away. (a) Calculate the muzzle velocity of the gun. Neglect air resistance. (b) If the projectile weighs 64 lb, what is its kinetic energy as it passes point A? (c) The target, B, is 1000 ft below A. What is the kinetic energy of the projectile as it strikes the target?

8–36. A man stands at the top of a 37° incline and throws a ball horizontally. The ball strikes at a point 240 ft down the incline. (a) If the ball had been thrown with the same initial speed but at an angle of 37° above a horizontal surface, how far would it have traveled before striking? (Neglect air resistance.) (b) If the ball weighs 0.5 lb and is in the thrower's hand for $\frac{1}{8}$ sec, what is the average horsepower developed while the ball is being thrown?

8–37. A ski tow is to be operated on a 37° slope 800 ft long. The rope is to move at 8 mi/hr and power must be provided for 80 riders at one time, each weighing, on an average, 150 lb. Estimate the horsepower required to operate the tow.

8–38. If energy costs 5 cents per kwh, how much is one horsepower-hour worth? How many ft-lb can be purchased for one cent?

8–39. At 5 cents per kwh, what does it cost to operate a 10-hp motor for 8 hr?

8–40. A horizontal force of 16 lb acts on a body of mass 2 slugs, initially at rest on a horizontal frictionless surface. Find the instantaneous power developed by the force at the end of 1 sec; at the end of 5 sec. Find also the average power developed during the first second and during the first 5 sec. Explain why the power is not constant.

8–41. An automobile weighing 2200 lb has an engine which develops a constant power of 100 horsepower. The car starts from rest and all the work done by the motor is used to increase the kinetic energy of the car. (Neglect friction.) (a) At the end of 10 sec how much work has been done by the engine? (b) What is the velocity at the end of 10 sec?

8–42. The engine of a motorboat delivers 40 horsepower to the propeller while the boat is making 20 mi/hr. What would be the tension in the towline if the boat were being towed at the same speed?

8–43. An automobile weighing 2000 lb has a maximum speed of 100 ft/sec on a horizontal road when the engine is developing full power of 50 horsepower. What is its maximum speed if the road rises 1 ft in 20 ft? Assume all friction forces to be constant.

8–44. The resisting force against the hull of a motorboat is approximately proportional to the velocity of the boat. If a 10-horsepower motor will drive a boat at 8 mi/hr, what horsepower is required for a velocity of 16 mi/hr?

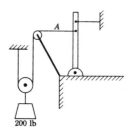

FIGURE 8–22

8–45. Use the principle of virtual work to compute the tension in cord A, Fig. 8–22. (Imagine the upper end of the lever to be pulled a short distance to the right.)

CHAPTER 9

IMPULSE AND MOMENTUM

9–1 Impulse and momentum. In the preceding chapter it was shown how the concepts of work and energy are developed from Newton's laws of motion. We shall next see how two similar concepts, those of *impulse* and *momentum*, also arise from Newton's laws. The most common use of these concepts is in connection with problems in impact.

Figure 9–1 represents two bodies that approach one another on a smooth horizontal surface, collide, and then recede from one another. Quantities relating to the first body are unprimed, those relating to the second are primed. The subscripts 1 and 2 refer to values before and after the collision respectively.

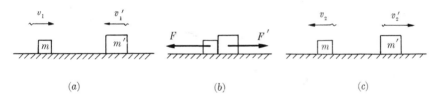

(a) (b) (c)

FIGURE 9–1

Since the plane is horizontal and frictionless, the only forces of interest are those which either body exerts on the other during the time the two are in contact. These forces are designated by F and F' in Fig. 9–1(b). From Newton's third law, F and F' are equal in magnitude and oppositely directed. That is, $F = -F'$. The forces F and F' will both vary during the collision. Of course, both are zero before contact. Both are small at the first instant of contact, then both increase to a maximum, and both decrease and become zero when the bodies separate.

At any instant while the bodies are in contact (Fig. 9–1b) we have from Newton's second law

$$F = m\frac{dv}{dt}, \qquad F' = m'\frac{dv'}{dt}.$$

Let t_1 be the instant at which the bodies first make contact and t_2 the instant at which they separate. Then

149

$$\int_{v_1}^{v_2} m\, dv = \int_{t_1}^{t_2} F\, dt, \qquad \int_{v_1}^{v_2} m'\, dv' = \int_{t_1}^{t_2} F'\, dt,$$

or

$$mv_2 - mv_1 = \int_{t_1}^{t_2} F\, dt, \qquad m'v_2' - m'v_1' = \int_{t_1}^{t_2} F'\, dt. \qquad (9\text{-}1)$$

It is useful to give names to the terms mv and $\int F\, dt$. The product of the mass and velocity of a body is called its *momentum*.

$$\text{Momentum} = mv.$$

The integral of a force over the time interval the force acts is called the *impulse* of the force and is represented by J.

$$\text{Impulse of a force} = J = \int_{t_1}^{t_2} F\, dt.$$

Equation (9–1) may therefore be stated verbally as follows: *The change in momentum of either body equals the impulse of the force exerted on the body.* The similarity of Eqs. (9–1) and (8–11) should be carefully noted. Change in momentum is related to the *time* integral of a force in the same way that change in kinetic energy is related to the *space* integral of the force.

If in a special case the force is constant, it may be taken outside the integral sign and

$$\text{Impulse of a constant force} = J = F \int_{t_1}^{t_2} dt = F(t_2 - t_1). \qquad (9\text{-}2)$$

The unit of impulse is one pound-second in the English system, one newton-second in the mks system, and one dyne-second in the cgs system. The units of momentum in the three systems are one slug·ft/sec, one kgm·m/sec and one gm·cm/sec. By the methods explained on page 131 it is easy to show that the unit of impulse in any system is equivalent to the corresponding unit of momentum. That is, one slug·ft/sec = one lb·sec, etc.

Momentum and impulse, unlike energy and work, are vector quantities. The momentum vector of a moving body is in the same direction as its velocity; the direction of an impulse vector is the same as that of the force producing the impulse.

EXAMPLES. (1) A golf ball weighs $1\frac{2}{3}$ oz. If its velocity immediately after being driven is 225 ft/sec, what was the impulse of the blow?

Since the ball is initially at rest, its change in momentum is equal to its final momentum, or

$$mv_2 - mv_1 = \frac{1.67}{16 \times 32} \times 225 = 0.734 \text{ slug·ft/sec.}$$

Hence, since impulse and change in momentum are numerically equal, the impulse of the blow is $J = 0.734$ lb·sec.

Note that the *force* of the blow cannot be computed from the data above. Any force such that $\int_{t_1}^{t_2} F\, dt = 0.734$ lb·sec, or any constant force acting for a time interval of such length that the product $F(t_2 - t_1) = 0.734$ lb·sec, would result in the same velocity. Only if the time of duration of the blow is known, can the force, assuming it to be constant, be computed. An analysis of golf strokes carried out by Dr. Edgerton of M.I.T. with the aid of the high speed stroboscope, shows that a golf ball remains in contact with the club face for about half a thousandth of a second (0.0005 sec). Let us assume the force constant during this time. Then

$$J = F(t_2 - t_1),$$

$$F = \frac{J}{t_2 - t_1} = \frac{0.734}{0.0005} = 1470 \text{ lb.}$$

The figure above gives the *time average* force whether it is constant or not.

(2) A ball weighing $\frac{1}{4}$ lb is thrown horizontally against a vertical wall. Its velocity before striking is 64 ft/sec and it rebounds with a velocity of 48 ft/sec. The time of contact with the wall is 0.05 sec. Compute the momentum of the ball before and after the collision and the force exerted on it by the wall. Assume the force constant.

Momentum before collision $= mv_1 = \dfrac{\frac{1}{4}}{32} \times 64 = 0.5$ slug·ft/sec.

Momentum after collision $= mv_2 = \dfrac{\frac{1}{4}}{32} \times (-48) = -0.375$ slug·ft/sec.

(Consider the initial direction of motion positive.)

The change in momentum is $mv_2 - mv_1 = -0.375 - 0.500 = -0.875$ slug·ft/sec. Hence the impulse of the force on the ball is $J = -0.875$ lb·sec, and if the time is 0.05 sec and the force is constant,

$$F = \frac{J}{t_2 - t_1} = \frac{-0.875}{0.05} = -17.5 \text{ lb.}$$

The minus sign means that the direction of the force on the ball is opposite to its initial velocity.

9–2 Conservation of momentum. Let us return to a consideration of Eq. (9–1),

$$mv_2 - mv_1 = \int_{t_1}^{t_2} F\, dt, \qquad m'v_2' - m'v_1' = \int_{t_1}^{t_2} F'\, dt.$$

We know from Newton's third law that at every instant $F = -F'$. Hence

$$\int_{t_1}^{t_2} F\, dt = -\int_{t_1}^{t_2} F'\, dt.$$

and therefore

$$(mv_2 - mv_1) = -(m'v_2' - m'v_1'),$$

or

$$mv_1 + m'v_1' = mv_2 + m'v_2'. \tag{9–3}$$

The left side of Eq. (9–3) is the total momentum of the system before the collision, the right side is the total momentum after the collision. We have therefore derived the extremely important result that *the total momentum of the colliding bodies is unaltered by the collision*. This fact is called the *principle of the conservation of momentum*. It is one of the most important principles in mechanics.

Notice that detailed knowledge of how the forces F and F' vary is unnecessary. The impulses of the forces are necessarily equal in magnitude and opposite in direction and hence they produce equal and opposite changes in momentum. The net change in momentum is therefore zero.

A more general statement of the principle of the conservation of momentum, which does not restrict it to a collision between two bodies, is as follows:

The total momentum of a system can only be changed by external forces acting on the system. The internal forces, being equal and opposite and acting for equal times, produce equal and opposite changes in momentum which cancel one another. Hence, *the total momentum of an isolated system is constant in magnitude and direction*.

EXAMPLES. (1) A billiard ball A (Fig. 9–2) moving with velocity v_0 of 36 cm/sec strikes a similar ball B initially at rest. After the collision, ball A rebounds with a velocity v_A of 15 cm/sec at an angle of 37° with its initial direction. Find the magnitude and direction of B's velocity.

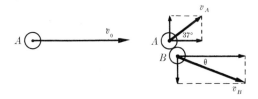

FIGURE 9–2

Momentum is a vector quantity and since it is conserved in a collision, its components must be conserved also. Let the X-axis be taken along the direction of v_0 and let m be the mass of either ball. The initial X-momentum of the system is $m \times 36$ gm·cm/sec and the initial Y-momentum is zero. After the collision the X-momentum of ball A is $m \times v_A \times \cos 37° = m \times 12$ gm·cm/sec. Hence the X-momentum of ball B is $m \times 36 - m \times 12 = m \times 24 = mv_B \cos \theta$ gm·cm/sec. The Y-momentum of ball A after the collision is $m \times v_A \times \sin 37° = m \times 9$ gm·cm/sec and the Y-momentum of ball B is therefore $-m \times 9 = mv_B \sin \theta$. Hence

$$\frac{mv_B \sin \theta}{mv_B \cos \theta} = \frac{-m \times 9}{m \times 24}, \ \tan \theta = -0.375, \ \theta = -20.5°, \text{ and } v_B = 25.6 \text{ cm/sec.}$$

(2) The Springfield rifle weighs 9.69 lb and fires a bullet weighing 150 grains (1 lb = 7000 grains) at a muzzle velocity of 2700 ft/sec. Compute the recoil velocity of the rifle if freely suspended.

The momentum of rifle and bullet before firing is zero. Hence, after firing, the forward momentum of the bullet is numerically equal to the backward momentum of the rifle. The mass of the rifle is 9.69/32 slugs and that of the bullet is 150/7000 × 32 slugs. Then

$$\frac{150}{7000 \times 32} \times 2700 = \frac{9.69}{32} \times v,$$

$$v = 5.9 \text{ ft/sec.}$$

(The forward momentum of the burnt gases, which is quite appreciable, has been neglected.)

It is important to note that the kinetic energies of the bullet and rifle are *not* equal. The explanation is evident when one considers that a body acquires kinetic energy when work is done on it, the work being the integral of the force over the distance moved. While the gases are propelling the bullet forward and the rifle backward, although the force on each is the same, the distance moved by the bullet is relatively large (the length of the barrel) while the distance moved by the slowly recoiling rifle is much less. Hence the work done on the bullet is much greater than the work done on the rifle and its kinetic energy is correspondingly greater. Momentum, however, being equal to the integral of force and *time*, is the same for both bullet and rifle.

Referring to the example of the Springfield, we find

$$KE_{\text{bullet}} = \tfrac{1}{2}mv^2 = \tfrac{1}{2}\left(\frac{150}{7000 \times 32}\right)(2700)^2 = 2440 \text{ ft·lb.}$$

$$KE_{\text{rifle}} = \tfrac{1}{2}MV^2 = \tfrac{1}{2}\left(\frac{9.69}{32}\right)(5.9)^2 = 5.25 \text{ ft·lb.}$$

9–3 Newton's third law. The point of view adopted in this discussion has been to accept Newton's third law as a law of nature and from it derive the principle of the conservation of momentum. However, if we wish to consider all of Newton's laws as derived from experiments, the argument may be reversed. The momenta of the bodies taking part in a collision can be measured both before and after the collision, and one finds by experiment that the momentum of the system remains constant. Then with the help of Newton's second law one can show that in order for this to be true the forces F and F', exerted on either body by the other, must at every instant have been equal in magnitude and oppositely directed. Hence the experimentally observed fact that momentum is conserved is the experimental proof of Newton's third law.

9–4 Elastic and inelastic collisions. Coefficient of restitution. Although the momentum remains constant when two (or more) bodies collide, the same is not necessarily true of the kinetic energy. If the kinetic energy *does* remain constant the collision is called *perfectly elastic*. At the opposite extreme from a perfectly elastic collision is one in which the colliding bodies stick together, as would two lumps of putty, and both move with the same velocity after the collision. In this case the collision is *perfectly inelastic*. Depending on the properties of the colliding bodies, all intermediate cases between perfectly elastic and perfectly inelastic collisions are possible. Collisions between bodies of finite size such as two billiard balls are never completely elastic and the only instances of perfectly elastic collisions known are those between atoms, molecules, and electrons. Even these may not be perfectly elastic if the kinetic energies of the particles are sufficiently great.

If a collision between two bodies is perfectly elastic, the equations

$$(\tfrac{1}{2}mv_1^2 + \tfrac{1}{2}m'v_1'^2) = (\tfrac{1}{2}mv_2^2 + \tfrac{1}{2}m'v_2'^2) \text{ (Conservation of energy)}$$

and

$$(mv_1 + m'v_1') = (mv_2 + m'v_2') \text{ (Conservation of momentum)}$$

must both be satisfied. (The primes and subscripts have the same significance as in Fig. 8–1.) These may be written

$$m(v_1^2 - v_2^2) = m'(v_2'^2 - v_1'^2),$$

$$m(v_1 - v_2) = m'(v_2' - v_1').$$

When the first is divided by the second we obtain

$$v_1 + v_2 = v_2' + v_1',$$

or finally,

$$v_1 - v_1' = -(v_2 - v_2').$$

But $v_1 - v_1'$ is the relative velocity before the collision and $v_2 - v_2'$ is the relative velocity after the collision. Hence, in a perfectly elastic collision the relative velocity is reversed in direction but unaltered in magnitude.

The degree to which a pair of colliding bodies approach perfect elasticity is expressed by their *coefficient of restitution, e,* which is defined as the negative ratio of the relative velocity after collision to the relative velocity before collision.

$$e = -\frac{v_2 - v_2'}{v_1 - v_1'}.$$

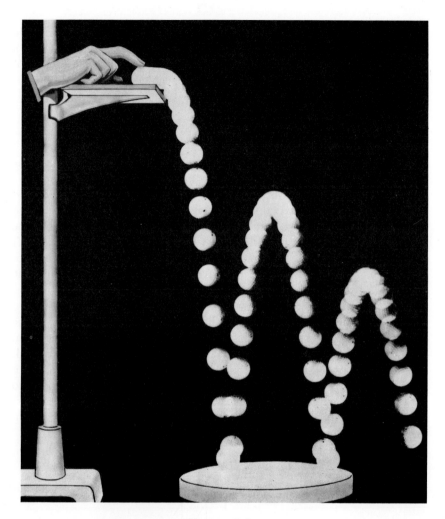

FIGURE 9–3

From what has just been shown, the coefficient of restitution is unity if the colliding bodies are perfectly elastic, and is zero if the bodies are perfectly inelastic. These are the two extremes and, in general, the coefficient of restitution has some value between zero and unity.

When a ball is dropped onto and rebounds from a fixed plate it has, in effect, collided with the earth. The mass of the earth is so large its velocity is practically unaltered by the collision. Hence in this special case,

$$e = - \frac{v_2}{v_1}.$$

The relative velocity before the collision is simply the velocity acquired in falling from a height h_1 or $\sqrt{2gh_1}$. If, after colliding, the ball rises to a height h_2, the relative velocity after the collision is $-\sqrt{2gh_2}$ (the downward direction is considered positive). Hence the coefficient of restitution is

$$e = - \frac{-\sqrt{2gh_2}}{\sqrt{2gh_1}} = \sqrt{\frac{h_2}{h_1}}$$

and a simple way of measuring it is to measure these two heights. The value obtained represents a joint property of the ball and the surface.

Figure 9–3 is a multiflash photograph of a golf ball dropping onto and rebounding from an iron plate. The heights h_1 and h_2 can be measured from the photograph and the velocities before and after colliding can be found from the spacing of the images before and after impact. It is left as an exercise to verify from the photograph that the ratio of the velocities equals the square root of the ratio of the heights, and to compute the coefficient of restitution.

9–5 The ballistic pendulum. Another example of the principle of the conservation of momentum is afforded by the ballistic pendulum, used to measure the velocity of a rifle bullet and illustrated in Fig. 9–4. A wooden block of mass M hangs vertically by a cord. A bullet of mass m, whose velocity v is to be measured, is fired horizontally into the block and remains embedded in it. After the bullet has come to rest in the block, both block and bullet have the common velocity V. From the principle of the conservation of momentum

$$mv = (M + m)V.$$

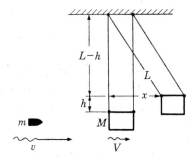

Fig. 9–4. The ballistic pendulum.

The block now swings until its center of gravity has risen through a vertical height h such that the potential energy at the top of the swing is equal to the kinetic energy at the bottom. That is,

or
$$(M + m)gh = \tfrac{1}{2}(M + m)V^2,$$
$$V^2 = 2gh.$$

The vertical rise h is usually small and is best obtained indirectly by measuring the horizontal displacement x. It will be seen from Fig. 9–4 that if L is the length of the pendulum,

or
$$(L - h)^2 + x^2 = L^2,$$
$$h = \frac{h^2 + x^2}{2L}.$$

If h is small compared with x, then h^2 may be neglected and $h = x^2/2L$.

In practice, the mass of the bullet is usually negligible compared to that of the pendulum. Then

$$mv = MV, \qquad V = \sqrt{2gh} = x\sqrt{g/L},$$

$$v = \frac{Mx}{m}\sqrt{g/L}. \tag{9–4}$$

EXAMPLE. A bullet of mass 20 gm is fired into a ballistic pendulum of mass 5 kgm. The center of gravity of the pendulum rises 10 cm after being struck. Find the initial velocity of the bullet.

The potential energy of the pendulum at the top of its swing was

$$Mgh = 5 \times 9.8 \times .10 = 4.9 \text{ joules.}$$

This is equal to the kinetic energy at the bottom of the swing and the velocity at this point was therefore

$$V = 1.4 \text{ m/sec.}$$

Hence the momentum of the pendulum at the start of its swing was

$$MV = 5 \times 1.4 = 7.0 \text{ kgm·m/sec.}$$

This is equal to the original momentum of the bullet, mv. Therefore

$$mv = 7.0 = 0.02v,$$
$$v = 350 \text{ m/sec} = 1150 \text{ ft/sec.}$$

It must be emphasized that this is an inelastic collision and the kinetic energy of the bullet before the collision is not equal to the kinetic energy of the pendulum after the collision. The latter, computed above, is 4.9 joules. The kinetic energy of the bullet was

$$\tfrac{1}{2}mv^2 = \tfrac{1}{2} \times 0.02 \times (350)^2$$
$$= 1220 \text{ joules.}$$

Hence only about one-half of one per cent of the kinetic energy of the bullet is transferred to the pendulum.

9–6 Newton's second law.   Newton himself did not state his second law in the form in which we have used it.   A free translation (Newton's *Principia* was written in Latin) is as follows:

"Change of motion is proportional to the applied force, and takes place in the direction of the force. . . . Quantity of motion is proportional to mass and velocity conjointly."

From Newton's definition of "motion," or "quantity of motion," it is evident that he used this term for the concept we now call momentum. It is also clear from his writings that the term "change" meant "rate of change" and that the "applied force" referred to the resultant force. Hence in current terminology Newton's statement is:

"Rate of change of momentum is proportional to the resultant force and is in the direction of this force."

In mathematical language this becomes

$$\frac{d}{dt}(mv) \propto F,$$

or

$$F = k\frac{d}{dt}(mv).$$

If the mass m is constant, this reduces to

$$F = km\frac{dv}{dt} = kma,$$

which is the form we have used, with k made equal to unity by proper choice of units.

Now while the mass of a body can ordinarily be considered constant, there is ample experimental evidence that actually it is a function of the velocity of the body, increasing with increasing velocity according to the relation

$$m = \frac{m_0}{\sqrt{1 - v^2/c^2}},$$

where m_0 is the "rest mass" of the body, c is the velocity of light, and v the velocity of the body.

This equation was predicted by Lorentz and Einstein on theoretical grounds based on relativity considerations and it has been directly verified by experiments on rapidly moving electrons and ions. The increase in mass is not appreciable until the velocity approaches that of light and therefore it ordinarily escapes detection. However, if the mass cannot be considered constant, we cannot set $F = ma$ and the original form of Newton's law must be used.

It is a striking example of Newton's genius that, although he could scarcely have previsioned the theory of relativity, he appreciated the fact that momentum is an entity more fundamental than mass.

9–7 Mass and energy. Let us derive the relativistic expression for the kinetic energy of a moving body, taking into account the increase of mass with increasing velocity. We shall neglect any changes in potential energy and any friction forces, so that the kinetic energy of the body is equal to the work done on it, starting from rest. The general form of Newton's second law is

$$F = \frac{d}{dt}(mv).$$

When the right side is multiplied by ds/ds, we get

$$F = \frac{ds}{ds}\frac{d}{dt}(mv) = \frac{ds}{dt}\frac{d}{ds}(mv)$$

$$= v\frac{d}{ds}(mv),$$

and hence

$$F\,ds = v\,d(mv) = v(m\,dv + v\,dm)$$

$$= mv\,dv + v^2\,dm.$$

Since the work $F\,ds$ equals the increase in kinetic energy dK, we can write

$$dK = mv\,dv + v^2\,dm. \tag{9–6}$$

Notice that if the mass is constant, the second term on the right is zero and the equation above integrates at once to give $K = \frac{1}{2}mv^2$.

For brevity, we now define a new variable β by the relation

$$\beta = \frac{v}{c}. \tag{9–7}$$

Then

$$v = c\beta, \qquad dv = c \, d\beta,$$

and Eq. (9–6) becomes

$$dK = mc^2\beta \, d\beta + c^2\beta^2 \, dm. \tag{9–8}$$

We next express m and dm in terms of β and $d\beta$, with the help of Eqs. (9–5) and (9–7).

$$m = m_0(1 - \beta^2)^{-1/2},$$
$$dm = m_0\beta(1 - \beta^2)^{-3/2} \, d\beta.$$

After inserting these expressions for m and dm in Eq. (9–8) and simplifying, we get

$$dK = m_0c^2(1 - \beta^2)^{-3/2}\beta \, d\beta.$$

Hence

$$K = m_0c^2 \int (1 - \beta^2)^{-3/2}\beta \, d\beta$$
$$= m_0c^2(1 - \beta^2)^{-1/2} + C, \tag{9–9}$$

where C is an integration constant. But when $v = 0$ the kinetic energy

$$K = 0 \qquad \text{and} \qquad \beta = \frac{v}{c} = 0.$$

Hence

$$0 = m_0c^2 + C,$$

and the constant of integration is

$$C = -m_0c^2.$$

In the first term on the right side of Eq. (9–9) we may replace $m_0(1 - \beta^2)^{-1/2}$ by m, so finally

$$\boxed{K = mc^2 - m_0c^2 = (m - m_0)c^2.} \tag{9–10}$$

That is, the kinetic energy equals the increase in mass over the rest mass, multiplied by the square of the velocity of light. The kinetic energy is in ergs if m is in grams and c is in cm/sec.

When the velocity is small compared with the velocity of light, i.e., when $\beta = v/c$ is small compared with 1, the expression above reduces to the familiar form $\frac{1}{2}mv^2$. Let us write it

$$K = m_0c^2(1 - \beta^2)^{-1/2} - m_0c^2.$$

Now expand $(1 - \beta^2)^{-1/2}$ by the binomial theorem.

$$(1 - \beta^2)^{-1/2} = 1 + \tfrac{1}{2}\beta^2 + \text{terms in } \beta^4, \beta^6, \text{ etc.}$$

Since by hypothesis β is a small quantity, higher powers than the second can be neglected. Then

$$K = m_0c^2(1 + \tfrac{1}{2}\beta^2) - m_0c^2 = m_0c^2 \times \tfrac{1}{2}\beta^2 = \tfrac{1}{2}m_0v^2,$$

and at small velocities the kinetic energy equals one-half the product of the rest-mass and the square of the velocity.

We shall illustrate the general form of the kinetic energy equation by an example taken from the field of nuclear physics. When the nucleus of a lithium atom is struck by a rapidly moving proton (the nucleus of a hydrogen atom), a momentary union of the two nuclei takes place, after which the compound nucleus breaks up into two alpha particles. (Alpha particles are the nuclei of helium atoms.) The alpha particles recoil in almost opposite directions (conservation of momentum) and move initially with very high velocities. Their combined kinetic energy is much greater than the kinetic energy of the original proton. The source of this kinetic energy is the so-called "binding energy" of the nuclear particles. That is, the potential energy of the assemblage of protons and neutrons that makes up the unstable composite nucleus is larger than the potential energy when the same number of particles are combined in the form of two helium nuclei. A crude analogy is that of two masses forced apart by a compressed spring, but tied together by a cord. If the cord is cut, the potential energy of the spring is transformed into kinetic energy of the recoiling masses.

The rest mass of a proton is 1.6715×10^{-24} gm. The rest mass of a lithium nucleus is 11.6399×10^{-24} gm, and that of an alpha particle is 6.6404×10^{-24} gm. Although the proton is moving when it collides with the lithium nucleus, its velocity is not great and we may assume its mass equal to its rest mass. Hence the mass of the original system is

$$(1.6715 + 11.6399) \times 10^{-24} = 13.3114 \times 10^{-24} \text{ gm.}$$

The rest mass of the two alpha particles is

$$2 \times 6.6404 \times 10^{-24} = 13.2808 \times 10^{-24} \text{ gm.}$$

The alpha particles must therefore be traveling with such velocity that their (combined) masses are increased from 13.2808×10^{-24} gm to 13.3114×10^{-24} gm. Then from Eq. (9–10) their (combined) kinetic energy is

$$K = (13.3114 - 13.2808) \times 10^{-24} \times (3 \times 10^{10})^2$$
$$= 2.75 \times 10^{-5} \text{ erg.}$$

Let us compare this with the energy released in a typical chemical reaction. When 2 moles of hydrogen combine with one mole of oxygen to form 2 moles of water in the reaction

$$2H_2 + O_2 = 2H_2O,$$

116,000 calories are released. This energy is shared among the H_2O molecules, of which there are $2 \times 6.02 \times 10^{23} = 12 \times 10^{23}$ molecules. (One mole of any substance contains 6.02×10^{23} molecules.) Since 1 calorie = 4.2 joules = 4.2×10^7 ergs, 1.16×10^5 cal = 4.87×10^{12} ergs, and the energy per molecule is

$$\frac{4.87 \times 10^{12}}{12 \times 10^{23}} = 4 \times 10^{-12} \text{ erg/molecule.}$$

The energy released in the nuclear reaction above is roughly 10 million times as great.

The computation of the energy released in the nuclear reaction is verified by observing the distance the alpha particles travel in air at atmospheric pressure before being brought to rest by collisions with other molecules. This distance is found to be 8.31 cm. (One way of making such measurements is with the help of a cloud chamber, illustrated in Fig. 23–10.) A series of independent experiments is then performed in which the range of alpha particles of known energy is measured. These experiments show that in order to travel 8.31 cm, an alpha particle must have an initial kinetic energy of 1.38×10^{-5} ergs. The energy of the two alphas together is therefore

$$2 \times 1.38 \times 10^{-5} = 2.76 \times 10^{-5} \text{ erg.}$$

This is in excellent agreement with the energy computed from the excess mass.

9–8 The principles of jet propulsion. The flight of a rocket, driven upward by the jet of rapidly moving gas ejected from its tail, is a sight familiar to everyone. Recent applications of the principles of jet propulsion to projectiles and airplanes indicate the increasing importance of this type of motive power. We shall consider briefly the physical principles involved.

A jet propulsion motor, in principle, is merely a combustion chamber in which solid or liquid fuel is burned, and which has an opening or jet to direct the gaseous products of combustion in the desired direction. For concreteness, let us consider the flight of a rocket. The momentum of the rocket is initially zero. When its charge of fuel is ignited, the stream of exhaust gases acquires a momentum in the downward direction, and since momentum is conserved, the rocket acquires an equal and oppositely directed momentum. From the viewpoint of the forces involved, the gas in the combustion chamber pushes downward on the gases in the jet, and upward on the body of the rocket.

We must consider more than the beginning of the motion, however. At the start of its flight, while the rocket is moving slowly, the rocket motor is a very inefficient device. Practically all of the *energy* developed at this stage is used to give kinetic energy to the rapidly moving exhaust gases and very little energy is acquired by the rocket itself. But as the rocket gains velocity the exhaust gases, which are expelled with a certain velocity *relative to the rocket,* move more and more slowly relative to the earth. When the rocket has acquired a velocity relative to the earth, equal to the velocity with which the exhaust gases are expelled from it, these gases as they leave the rocket (or better, as the rocket leaves them) are at rest relative to the earth and their kinetic energy is therefore zero. Hence at this velocity all of the energy developed by the fuel is imparted to the rocket. The energy which a rocket carries in its charge of fuel can therefore be utilized much more effectively if the rocket is initially given a "boost" by some auxiliary means.

It should be noted that a rocket does not depend on the atmosphere for its propulsion, but would actually perform better in the absence of an atmosphere because of lessened air resistance. A helicopter is able to rise vertically only because its propeller sets a stream of air into downward motion. The downward force on the air is equal to the rate of change of downward momentum of the air stream, and the equal and opposite reaction to this force supports the helicopter. The rocket motor, on the other hand, pushes down on its own products of combustion and does not depend on the presence of an external atmosphere.

Although no details of the most recent developments in the jet propulsion of airplanes have been made public, it can readily be seen why this type of motive power is eminently suitable for stratosphere flight at high velocity. In the stratosphere, where the density of the air is small, it is difficult for a conventional propeller to get a "bite" of the air so as to produce rearward momentum, but this is no handicap to a jet propulsion motor since it reacts on its own exhaust. Furthermore, the efficiency of jet propulsion is at maximum when the forward velocity of the motor is sufficiently great so that it equals the (relative) velocity of ejection of the exhaust gases.

PROBLEMS

9–1. (a) What is the momentum of a 10-ton truck whose velocity is 30 mi/hr? At what velocity will a 5-ton truck have (b) the same momentum? (c) the same kinetic energy?

9–2. A block of mass 100 gm, initially at rest on a horizontal frictionless surface, is acted on by a horizontal force $F = 10^4 + 3 \times 10^3 t$, where F is in dynes and t is in seconds. There is no friction. (a) How far does the block move in 5 sec? (b) What is its velocity when $t = 5$ sec? (c) How much work is done by the force in 5 sec?

9–3. A bullet having a mass of .05 kgm, moving with a velocity of 400 m/sec, penetrates a distance of 0.1 m in a wooden block firmly attached to the earth. Assume the decelerating force constant. Compute (a) the deceleration of the bullet, (b) the decelerating force, (c) the time of deceleration, (d) the impulse of the collision. Compare the answer to part (d) with the initial momentum of the bullet.

9–4. A baseball weighs $5\frac{1}{2}$ oz. If the velocity of a pitched ball is 80 ft/sec, and after being batted it is 120 ft/sec in the opposite direction, find the change in momentum of the ball and the impulse of the blow. If the ball remains in contact with the bat for 0.002 sec, find the average force of the blow.

9–5. A 20,000-lb engine coasts along a straight track at a speed of 30 mi/hr. It collides and locks with a similar engine, initially at rest. (a) What is the momentum of the system before collision? (b) What is the velocity of the engines after impact? (c) How much

kinetic energy is lost during the collision?

9–6. A truck weighing 6400 lb is traveling 5 ft/sec when it bumps the rear of a car traveling at 2 ft/sec in the same direction. (a) If, after the collision, the truck has a velocity of 3 ft/sec and the car a velocity of 6 ft/sec (still in the same direction), what is the weight of the car? (b) Calculate the total kinetic energy of the system (truck plus car) before and after the collision.

9–7. On a frictionless table, a 3-kgm block moving 4 m/sec to the right collides with an 8-kgm block moving 1.5 m/sec to the left. (a) If the two blocks stick together what is the final velocity? (b) If the two blocks make a perfectly elastic head-on collision, what are their final velocities? (c) How much mechanical energy is converted into heat in the collision of part (a)?

9–8. (a) Two blocks of mass 300 gm and 200 gm are moving toward one another along a horizontal frictionless surface with velocities of 50 cm/sec and 100 cm/sec, respectively. Find the velocity of the center of mass of the system. (b) If the blocks collide and stick together, find their final velocity. (c) Find the kinetic energy of the system before the collision, with respect to an origin moving with the center of mass. (d) Find the loss of kinetic energy during the collision.

9–9. Find the average recoil force on a machine gun firing 120 shots per minute. The weight of each bullet is 0.025 lb and the muzzle velocity is

2700 ft/sec. Hint: average force equals average rate of change of momentum.

9–10. A 160-lb man standing on skates on ice throws a 6-oz ball horizontally with a speed of 80 ft/sec. (a) With what speed and in what direction will the man begin to move? (b) If the man throws 4 such balls every 3 sec, what is the average force acting on him? (c) What acceleration does the man have in (b), assuming that the ice is frictionless and that the decrease in mass due to throwing the balls is negligible?

9–11. A 75-mm gun fires a projectile weighing 16 lb with a muzzle velocity of 1900 ft/sec. By how many mi/hr is the velocity of a plane mounting such a gun decreased when a projectile is fired directly ahead? The plane weighs 32,000 lb.

9–12. The projectile of a 16-inch seacoast gun weighs 2400 lb, travels a distance of 38 ft in the bore of the gun, and has a muzzle velocity of 2250 ft/sec. The gun weighs 300,000 lb. (a) Compute the initial recoil velocity. (b) What is the acceleration of the projectile while in the gun barrel? Assume it to be constant. (c) How long a time is required for the projectile to travel the length of the gun barrel? (d) If the angle of elevation of the gun is 30°, compute the range of the projectile and the maximum height reached, in miles, neglecting air resistance.

9–13. When a bullet of mass 10 gm strikes a ballistic pendulum of mass 2 kgm, the center of mass of the pendulum is observed to rise a vertical distance of 10 cm. The bullet remains embedded in the pendulum. Calculate the velocity of the bullet.

9–14. A bullet weighing 0.01 lb is shot through a 2-lb wooden block suspended on a string 5 ft long. The block is observed to swing through an angle of 5°. Find the speed of the bullet as it emerges from the block, if its initial speed is 1,000 ft/sec.

9–15. A bullet of mass 2 gm, traveling in a horizontal direction with a velocity of 500 m/sec, is fired into a wooden block of mass 1 kgm, initially at rest on a level surface. The bullet passes through the block and emerges with its velocity reduced to 100 m/sec. The block slides a distance of 20 cm along the surface from its initial position. (a) What was the coefficient of sliding friction between block and surface? (b) What was the decrease in kinetic energy of the bullet? (c) What was the kinetic energy of the block, at the instant after the bullet passed through it?

9–16. A body initially at rest on a level frictionless surface is acted on by a constant horizontal force for 8 sec. During this time 480 ergs of work are done on the body by the force and the body acquires 120 gm-cm/sec of momentum. Find (a) the mass of the body, (b) its speed at the end of the 8-sec interval, (c) the distance moved during the 8 sec, (d) the force.

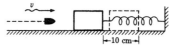

FIGURE 9–5

9–17. A rifle bullet of mass 10 gm strikes and embeds itself in a block of mass 990 gm which rests on a horizontal frictionless surface and is attached to a coil spring as shown in Fig. 9–5. The impact compresses the spring 10 cm. Calibration of the spring shows that a force of 100,000 dynes is required to compress the spring 1 cm. (a) Find the maximum potential energy of the spring. (b) Find the velocity of the block just after the impact. (c) What was the initial velocity of the bullet?

9–18. 100 gm of putty, traveling with a velocity of 700 cm/sec, collides with and sticks to a block of mass 900 gm. The block rests on a table and is attached to a spring, originally under no compression. The coefficient of friction between the block and the table is 0.25. The spring constant k is 98,000 dynes/cm and the spring has negligible mass. How far is the spring compressed?

9–19. A neutron of mass m moving with velocity v makes a head-on collision with a boron atom of mass $M = 10m$. (a) If the collision is completely inelastic, how much energy is lost in the collision? (b) If there is no loss of mechanical energy in the collision, what fraction of its original energy does the neutron transfer to the boron atom?

9–20. A body of mass 600 gm is initially at rest. It is struck by a second body of mass 400 gm initially moving with a velocity of 125 cm/sec toward the right along the X-axis. After the collision the 400-gm body has a velocity of 100 cm/sec at an angle of 37° above the X-axis in the first quadrant. Both bodies move on a horizontal frictionless plane. (a) What is the magnitude and direction of the velocity of the 600-gm body after the collision? (b) What is the loss of kinetic energy during the collision?

9–21. A railroad handcar is moving along straight frictionless tracks. In each of the following cases the car initially has a total weight (car and contents) of 500 lb and is traveling with a velocity of 10 ft/sec. Find the final velocity of the car in each of the three cases. (a) A 50-lb weight is thrown sideways out of the car with a velocity of 8 ft/sec relative to the car. (b) A 50-lb weight is thrown backward out of the car with a velocity of 10 ft/sec relative to the car. (c) A 50-lb weight is thrown into the car with a velocity of 12 ft/sec relative to the ground and opposite in direction to the velocity of the car.

9–22. An open-topped freight car weighing 10 tons is coasting without friction along a level track. It is raining very hard, with the rain falling vertically down. The car is originally empty and moving with a velocity of 2 ft/sec. (a) What is the velocity of the car after it has traveled long enough to collect one ton of rain water? (b) What would be its velocity if it had a small vertical drain pipe in the floor which allowed the one ton of water to leak out of the bottom as fast as it came in?

9–23. A 10-lb shell is projected vertically upward and explodes at its maximum height of 1600 ft into two fragments of 2 lb and 8 lb, respectively. Both fragments strike the ground at the same instant. The 8-lb fragment lands at a horizontal distance of 1000 ft from the point from which it was fired. (a) How long does it take the fragments to strike the ground? (b) What are the initial velocities of the two fragments? (c) What impulse acted on the 8-lb fragment? (d) If the explosion occurred in 0.001 sec, what average force acted on the 8-lb fragment? (e) What is the velocity of the 8-lb fragment when it strikes?

9–24. A ball is dropped from rest onto a fixed horizontal surface and rebounds to a height which is 64% of its original height. (a) What is the coefficient of restitution? (b) With what vertical velocity must the ball strike the surface to rebound to a height of 25 ft?

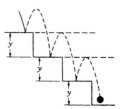

FIGURE 9–6

9–25. A 1-lb rubber ball bounces down a flight of stairs, each time rising to the height y of the step above, as shown in Fig. 9–6. (a) Find the coefficient of restitution. (b) If the height of each step, y, is 1 ft, what is the maximum kinetic energy of the ball?

9–26. A squash ball, dropped from a height of 8 ft, rebounds to a height of 3 ft. With what velocity must it be projected horizontally against a vertical wall, at a height of 6 ft above the floor, in order that it may rebound a horizontal distance of 15 ft?

9–27. A golf ball is dropped on a hard surface from a height of 1 meter and rebounds to a height of 64 cm. (a) What is the height of the second bounce? of the nth bounce? After (about) how many bounces is the height reduced to 1 cm? (b) What is the time of the first bounce, i.e., between the first and second contacts with the surface? What is the time of the nth bounce? (c) What is the coefficient of restitution?

9–28. The rocket motor of a V-2 type rocket is a combustion chamber which ejects hot gases through a nozzle at the rear of the rocket. This creates a fairly constant force against the rocket while the motor is operating. Following are some typical data: Weight of rocket, 30,000 lb; Force due to motor, 60,000 lb for 50 seconds. (a) If the motor ejects 384 lb (12 slugs) of gas per second, at what speed (relative to the rocket) must this gas leave the rocket to produce the 60,000-lb force? (b) If the rocket did not become lighter as it burned fuel, how high would it go if pointed straight up? (c) Will the actual height reached be greater or less than that computed in (b)?

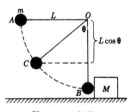

FIGURE 9–7

9–29. A ball of mass m is attached to a cord of length L, pivoted at point O, as shown in Fig. 9–7. The ball is released from rest at point A, swings down and makes a partially elastic collision with a block of mass M, and rebounds to point C. Neglect friction between the block and the horizontal surface. In terms of m, M, L, θ, and g, find the following: (a) the velocity v_1 of the ball just before it collides with the block, (b) the velocity v_2 of the ball just after the collision with the block, (c) the velocity V acquired by the block.

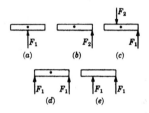

FIGURE 9–8

9–30. Fig. 9–8 shows five identical uniform rods, each weighing 2 lb, resting on a horizontal frictionless surface. Horizontal forces $F_1 = 10$ lb and $F_2 = 4$ lb are applied as shown. (a) What will be the acceleration of the center of mass of each rod? (b) In which cases will there be no rotation of the rod?

9–31. One pound of coal, when burned, develops 13,000 Btu (1 Btu = 778 ft-lb). What is the decrease in mass of an atomic pile, operating at constant temperature, when it develops a quantity of heat equal to that obtained by burning 1 ton of coal?

9–32. An atomic bomb containing 20 kgm of plutonium explodes. The rest mass of the products of the explosion is less than the original rest mass by one ten-thousandth of the original rest mass. (a) How much energy is released in the explosion? (b) If the explosion takes place in one microsecond, what is the average power developed by the bomb? (c) How much water could the released energy lift to a height of one mile?

CHAPTER 10

CIRCULAR MOTION

10–1 Introduction. The concepts of velocity and acceleration were introduced in Chapter 4 in connection with linear motion. However, one has only to glance at some piece of machinery such as a lathe or an automobile engine to realize that rotational motion is of much more common occurrence than is motion in a straight line. The concepts of linear displacement, and average and instantaneous linear velocity and acceleration, have their exact counterparts in rotational motion.

We shall begin this part of the subject by discussing motion of rotation about a fixed axis, for example, the motion of a grinding wheel or the flywheel of a stationary engine. The center line of the shaft on which the wheel is mounted is called the *axis*. All points on the axis remain stationary during the motion, while other points in the body move in circles concentric with the axis and in planes perpendicular to it. Hence such motion is called *circular motion*.

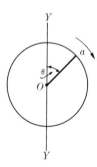

FIGURE 10–1

The position of every point in a body rotating in this way is evidently completely specified if the angular position of any radius of the body is known, relative to some fixed direction. That is, if the rotating body is represented by the circle of Fig. 10–1, and Oa is a radius fixed in the body, the angle θ which Oa makes with the Y-axis is sufficient to determine the position of every point in the body. If the radius Oa is vertical at the start of the motion, and the body rotates in a clockwise direction, the angle θ increases continually as the motion proceeds. It will be seen that the angle θ, in circular motion, corresponds to the coordinate x in linear motion. Angles measured in one direction from the fixed axis are considered positive, those in the opposite direction, negative. It is found convenient to express angles in radians rather than in degrees.

One radian is the angle subtended at the center of a circle by an arc of length equal to the radius of the circle. (Fig. 10–2a.) Since the radius is contained 2π times

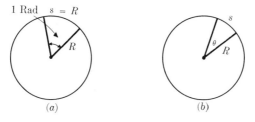

(a) (b)

FIGURE 10–2

$(2\pi = 6.28 \ldots)$ in the circumference, there are 2π or $6.28 \ldots$ radians in one complete revolution or $360°$. Hence

$$1 \text{ radian} = \frac{360}{2\pi} = 57.3 \ldots \text{ degrees.}$$

$$360° = 2\pi \text{ radians} = 6.28 \ldots \text{ radians}$$
$$180° = \pi \qquad `` \qquad = 3.14 \ldots \quad ``$$
$$90° = \pi/2 \quad `` \qquad = 1.57 \ldots \quad ``$$
$$60° = \pi/3 \quad `` \qquad = 1.05 \ldots \quad ``$$

and so on.

In general (Fig. 10–2b), if θ represents any arbitrary angle subtended by an arc of length s on the circumference of a circle of radius R, then θ (in radians) is equal to the length of the arc s divided by the radius R.

$$\theta = \frac{s}{R}, \qquad s = R\theta.$$

An angle in radians, being defined as the ratio of a length to a length, is a pure number.

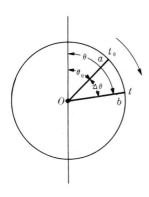

FIGURE 10–3

10–2 Angular velocity. Fig. 10–3 represents a body rotating about an axis through O perpendicular to the plane of the diagram. Oa is the position of some radius in the body at time t_0 and Ob is the position of the same radius at a later time t. The angular coordinates of the radius, measured from the vertical reference line, are θ_0 and θ. The *angular displacement* of the radius (which is the same for all radii) is $\theta - \theta_0$ or $\Delta\theta$. The *average angular* *velocity,* represented by $\bar{\omega}$ (omega), is defined as *the ratio of the angular displacement to the elapsed time.*

$$\text{Average angular velocity} = \frac{\text{angular displacement}}{\text{elapsed time}}.$$

$$\bar{\omega} = \frac{\theta - \theta_0}{t - t_0} = \frac{\Delta\theta}{\Delta t}.$$

Angular velocity is expressed in radians per second.

Instantaneous angular velocity, ω, is the limiting ratio of angular displacement to elapsed time when both are extremely small, or, it is *the instantaneous rate of change of angular displacement.*

$$\omega = \lim_{\Delta t \to 0} \frac{\Delta\theta}{\Delta t} = \frac{d\theta}{dt}.$$

Although the equations of angular motion take a somewhat simpler form when angular velocities are expressed in rad/sec, it is more common engineering practice to express them in revolutions per second (rps) or revolutions per minute (rpm). Since there are 2π radians in one complete revolution the number of rad/sec equals 2π times the number of rps and $2\pi/60$ times the number of rpm.

There are two common methods for measuring angular velocity. In the first, a revolution counter is held against the end of a rotating shaft and the number of revolutions in a measured time interval is noted. The angular displacement and the time interval are thus measured directly and their ratio gives the average angular velocity. The second method employs a tachometer (see page 189) which reads instantaneous angular velocity directly, although most tachometers are calibrated to read rpm rather than rad/sec. The common automobile speedometer is a tachometer whose readings are proportional to the instantaneous angular velocity of the drive shaft to which it is connected. Since the linear velocity of the car is proportional to the angular velocity of the drive shaft, the tachometer can be calibrated to read miles per hour rather than rpm or rad/sec.

When a body rotates with *constant* angular velocity its instantaneous angular velocity is equal to its average angular velocity, whatever the duration of the time interval. This sort of motion is well illustrated by the armature of a synchronous motor or the hands of a Telechron clock. If the angular velocity is constant we may write

$$\omega = \frac{\theta - \theta_0}{t - t_0},$$

where ω is the constant angular velocity and the time interval may be of any length. Then

$$\theta - \theta_0 = \omega(t - t_0),$$

and if t_0 and θ_0 are both zero,

$$\theta = \omega t. \qquad (10\text{--}1)$$

Equation (10–1) is exactly analogous to the equation

$$x = vt$$

for a body moving with constant linear velocity.

10–3 Angular acceleration. While a rotating body is speeding up or slowing down, that is, while its angular velocity is changing, it is said to have an angular acceleration. Let ω_0 represent its instantaneous angular velocity at a time t_0, and ω its angular velocity at a later time t. The *average angular acceleration*, represented by $\bar{\alpha}$ (alpha), is defined as *the ratio of the change in angular velocity to the elapsed time.*

$$\text{Average angular acceleration} = \frac{\text{change in angular velocity}}{\text{elapsed time}}.$$

$$\bar{\alpha} = \frac{\omega - \omega_0}{t - t_0} = \frac{\Delta\omega}{\Delta t}.$$

If the angular velocity is in radians per second and the time in seconds, the angular acceleration will be in radians per second, per second, or rad/sec^2.

Instantaneous angular acceleration, α, is the ratio of the change in angular velocity to the elapsed time during an extremely short time interval, or, it is *the instantaneous rate of change of angular velocity.*

$$\alpha = \lim_{\Delta t \to 0} \frac{\Delta\omega}{\Delta t} = \frac{d\omega}{dt}. \qquad (10\text{--}2)$$

Since $\omega = d\theta/dt$, this can also be written

$$\alpha = \frac{d}{dt}\left(\frac{d\theta}{dt}\right) = \frac{d^2\theta}{dt^2}. \qquad (10\text{--}3)$$

EXAMPLE. The following set of readings of the tachometer of an airplane engine were taken at two-second intervals:

Time (sec)	0	2	4	6	8	10	12	14	16	18
Angular velocity (rpm)	1000	1000	1500	2000	2500	3000	3500	3800	4000	4000

Compute the average angular acceleration, in rad/sec², during each two-second interval. Was the angular acceleration constant during the entire time? During any part of the time?

In the interval between 0 and 2 sec there was no change in angular velocity. Hence the angular acceleration was zero during this interval. In the interval from 2 to 4 sec the angular velocity increased from 1000 to 1500 rpm, or from 105 to 157 rad/sec. The increase in angular velocity was therefore $157 - 105 = 52$ rad/sec, and since the time interval was 2 sec the average angular acceleration was $52/2 = 26$ rad/sec².

The remainder of the example is left as an exercise.

10–4 Constant angular acceleration. When the angular velocity of a body changes by equal amounts in equal intervals of time, the angular acceleration is constant. Under these circumstances the average and instantaneous angular accelerations are equal, whatever the duration of the time interval. One may therefore write

$$\alpha = \frac{\omega - \omega_0}{t - t_0}$$

or

$$\omega = \omega_0 + \alpha(t - t_0), \qquad (10\text{–}4)$$

where α is the constant instantaneous angular acceleration.

If $t_0 = 0$,

$$\boxed{\omega = \omega_0 + \alpha t.} \qquad (10\text{–}5)$$

Equation (10–5) has precisely the same form as Eq. (4–12) on page 58 for linear motion with constant acceleration and may be interpreted in the same way.

The angular displacement of a rotating body, or the angle turned through by the body, corresponds to the linear displacement of a body moving along a straight line. The expression for the angular displacement can be found with the help of the average angular velocity. If the angular acceleration is constant, the angular velocity increases at a uniform rate and its average value during any time interval equals half the sum of its values at the beginning and end of the interval. That is,

$$\bar{\omega} = \frac{\omega_0 + \omega}{2}$$

$$= \frac{\omega_0 + [\omega_0 + \alpha(t - t_0)]}{2}$$

$$= \omega_0 + \tfrac{1}{2}\alpha(t - t_0) \qquad (10\text{–}6)$$

and since by definition

$$\bar{\omega} = \frac{\theta - \theta_0}{t - t_0},$$ (10-7)

we may equate the right sides of Eqs. (10–6) and (10–7), obtaining

$$\theta - \theta_0 = \omega_0(t - t_0) + \tfrac{1}{2}\alpha(t - t_0)^2.$$

If t_0 and θ_0 are both zero,

$$\theta = \omega_0 t + \tfrac{1}{2}\alpha t^2,$$ (10-8)

which has the same form as Eq. (4–17).

On eliminating t between Eqs. (10–5) and (10–8) we obtain

$$\omega^2 = \omega_0^2 + 2\alpha\theta,$$ (10-9)

which corresponds to Eq. (4–19).

The equations of angular motion with constant acceleration can easily be derived by the methods of calculus. From the definition of angular acceleration, we have

$$\alpha = \frac{d\omega}{dt}.$$

Hence

$$d\omega = \alpha \, dt,$$

$$\omega = \int \alpha \, dt + C_1,$$

and if α is constant and $\omega = \omega_0$ when $t = 0$,

$$\omega = \omega_0 + \alpha t.$$

From the definition of angular velocity

$$\omega = \frac{d\theta}{dt}.$$

Hence

$$d\theta = \omega \, dt,$$

$$\theta = \int \omega \, dt + C_2.$$

$$= \int (\omega_0 + \alpha t) \, dt + C_2,$$

and if $\theta = 0$ when $t = 0$,

$$\theta = \omega_0 t + \tfrac{1}{2}\alpha t^2.$$

Equation (10–9) may be obtained by the substitution

$$\alpha = \frac{d\omega}{dt} = \frac{d\omega}{d\theta}\frac{d\theta}{dt} = \omega\frac{d\omega}{d\theta}.$$

Then

$$\alpha\, d\theta = \omega\, d\omega,$$

$$\omega^2 = \omega_0^2 + 2\alpha\theta.$$

10–5 Angular velocity and acceleration as vectors. The vector nature of a quantity like force or linear velocity is obvious, and it seems natural to represent such quantities by arrows. It is also true, although not as obvious, that angular velocity and angular acceleration are also vectors and, like force and linear velocity, can be represented by arrows. The vector representing an angular velocity (or acceleration) is drawn along

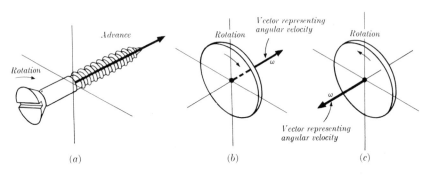

Fig. 10–4. Angular velocity may be represented by a vector along the axis.

the axis of rotation. Its length, to some chosen scale, represents the magnitude of the angular velocity (or acceleration). Imagine now the axis to be a screw with a right-hand thread. The sense of the vector, by convention, is that in which the screw would advance when turned in the direction of the angular velocity (or acceleration). (See Fig. 10–4.) Any quantity associated with an axis can be represented in this way by a vector, and we shall make use of such vectors later in connection with gyroscopic motion.

10–6 Tangential velocity. The angular displacement, angular velocity, and angular acceleration of a rotating body are characteristic of the body as a whole. We wish to consider next the displacement, velocity, and acceleration of some specified point in a rotating body. Every point in a body rotating about a fixed axis moves in a circle with center on the axis. The circle in Fig. 10–5(a) represents the path of such a point. The *displacement* of the point as it moves from p to q is defined as the vector drawn from p to q. The *length of path* is the length of the arc s. It will be seen that these definitions are generalizations of the corresponding definitions on page 62 for a body in linear motion.

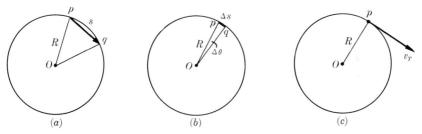

FIG. 10–5. The limiting ratio of Δs to Δt is the instantaneous tangential velocity.

The *average velocity* of the point is defined as the ratio of its displacement to the length of the time interval between p and q. Its *average speed* is the ratio of length of path to the time interval.

$$\text{Average velocity (a vector)} = \frac{\text{displacement (a vector)}}{\text{elapsed time (a scalar)}}.$$

$$\text{Average speed (a scalar)} = \frac{\text{length of path (a scalar)}}{\text{elapsed time (a scalar)}}.$$

The direction of the average velocity vector is the same as that of the displacement. Since the lengths of the arc s and the chord pq are different, the average velocity and the average speed, in Fig. 10–5(a), are not numerically equal.

The *instantaneous velocity* of the point, at p, is defined as its average velocity over an extremely short displacement which includes p. The *instantaneous speed* is the average speed over a short displacement. However, if the displacement is extremely small as in Fig. 10–5(b), the lengths of the vector pq are the arc Δs become practically equal. The instantaneous velocity and the instantaneous speed are therefore numerically equal. In other words, the *magnitude* of the instantaneous velocity is equal to the instantaneous speed.

The direction of an extremely short displacement at point p is the same as the direction of the circle at p; that is, the displacement is in the direction of the tangent at p and is perpendicular to the radius Op. The instantaneous velocity at p is therefore tangent to the circle at p, and it is often called the tangential velocity and written v_T. It is represented by the vector v_T in Fig. 10–5(c).

An important and useful relation connects the angular velocity of a rotating body and the tangential velocity of any point in the body. The angle $\Delta\theta$, in Fig. 10–5(b), is given by

$$\Delta\theta = \frac{\Delta s}{R}.$$

When both sides are divided by Δt we get

$$\frac{\Delta\theta}{\Delta t} = \frac{1}{R}\frac{\Delta s}{\Delta t},$$

and in the limit, when $\Delta t \to 0$,

$$\lim_{\Delta t \to 0}\frac{\Delta\theta}{\Delta t} = \frac{1}{R}\lim_{\Delta t \to 0}\frac{\Delta s}{\Delta t}.$$

But $\lim_{\Delta t \to 0}\Delta\theta/\Delta t$ is the instantaneous angular velocity ω, and $\lim_{\Delta t \to 0}\Delta s/\Delta t$ is the magnitude of the tangential velocity v_T. Hence

$$\omega = \frac{v_T}{R}, \qquad v_T = R\omega. \tag{10–10}$$

The tangential velocity of any point in a rotating body is therefore equal to the product of the angular velocity of the body and the distance of the point from the axis.

Equation (10–10) can also be derived as follows. From the definition of an angle in radians,

$$s = R\theta.$$

Hence

$$\frac{ds}{dt} = R\frac{d\theta}{dt}.$$

But ds/dt is the magnitude of the tangential velocity v_T and $d\theta/dt$ is the angular velocity ω. Therefore

$$v_T = R\omega.$$

If ω, in Eq. (10–10), is in rad/sec, v_T will be in ft/sec when R is expressed in feet, m/sec when R is in meters, and cm/sec when R is in centimeters.

EXAMPLE. An airplane motor and propeller are set up in a test block. The propeller blades are each 6 ft long. (a) When the propeller is rotating at 1200 rpm compute the tangential velocity of the blade tips. (b) What is the tangential velocity of a point on the blade, halfway between axis and tip?

(a) $\omega = 1200 \text{ rpm} = 1200 \times \dfrac{2\pi}{60} = 40\pi \text{ rad/sec},$

$$v_T = R\omega = 6 \times 40\pi = 240\pi = 755 \text{ ft/sec}$$
$$= 514 \text{ mi/hr.}$$

(b) $v_T = 3 \times 40\pi = 120\pi = 378 \text{ ft/sec}$
$$= 257 \text{ mi/hr.}$$

10–7 Acceleration of a point in circular motion. The general definition of acceleration is the rate of change of velocity. Velocity, however, is a vector quantity, involving magnitude and direction. The velocity of a moving point will therefore change if either the magnitude or the direction of its velocity changes. Of course, both may change simultaneously. Hence a moving point may have an acceleration arising either from a change in the magnitude or in the direction of its velocity, or both.

The circle in Fig. 10–6(a) represents the path of a point in a body rotating about a fixed axis through O. Let ω_0 be the angular velocity of the body when the point is at p. The corresponding tangential velocity is $v_0 = R\omega_0$. We shall assume that the rotating body has an angular acceleration. Then when the point under consideration reaches q, the angular velocity will have increased to a larger value ω and the tangential velocity to a value $v = R\omega$. (The subscript T is omitted from the tangential velocity for simplicity.)

The *average acceleration* between points p and q is defined as the change in velocity divided by the time interval between p and q. The change in velocity must now be considered as a *vector* change, or the *vector difference* between v and v_0. This vector difference, which is itself a vector, may be found by either of the methods for subtracting vectors explained in Sec-

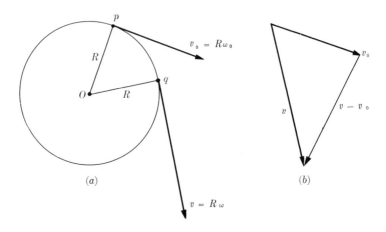

FIG. 10–6. The vector $v - v_0$ is the vector change in velocity.

tion 1–9. In Fig. 10–6(b), the vectors v and v_0 have been transferred to a common origin, keeping them parallel to their directions in Fig. 10–6(a), and the change in velocity or the vector difference $v - v_0$ has been found by the triangle method. The average acceleration is

$$\text{Average acceleration (a vector)} = \frac{\text{change in velocity (a vector)}}{\text{elapsed time (a scalar)}}.$$

$$\bar{a} = \frac{v - v_0 \text{ (vector difference)}}{t - t_0}.$$

where t_0 and t are the times at points p and q. The direction of the average acceleration is the same as that of the vector $v - v_0$.

The *instantaneous acceleration* is found by letting points p and q approach more and more closely as in Fig. 10–7(a). For simplicity, we shall consider

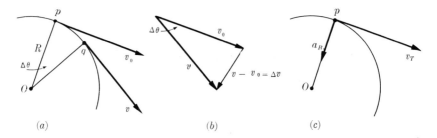

FIG. 10–7. Tangential velocity and radial acceleration.

first a special case in which the angular velocity ω is constant. The *magnitude* of the tangential velocity is then constant also, although its direction is continually changing. The change in velocity, $v - v_0$ or Δv, is found in Fig. 10–7(b), where the vectors v_0 and v have been transferred to a common origin. Notice carefully that although the magnitude of the velocity is constant and the vectors v_0 and v are of the same length, there has nevertheless been a change in velocity because of the change in the direction of motion.

The instantaneous acceleration at point p is the limiting ratio of the vector change in velocity to the elapsed time.

$$a = \lim_{\Delta t \to 0} \frac{\Delta v}{\Delta t}.$$

It is useful to correlate this acceleration with the angular velocity of the rotating body. Since the angle $\Delta \theta$ in Fig. 10–7(b) is small, its approximate value in radians is

$$\Delta \theta = \frac{\Delta v}{v}, \qquad \text{or} \qquad \Delta v = v \, \Delta \theta \text{ (approximately)}.$$

When both sides of this equation are divided by the time interval Δt we obtain

$$\frac{\Delta v}{\Delta t} = v \frac{\Delta \theta}{\Delta t} \text{ (approximately)}.$$

When $\Delta t \to 0$ the approximation becomes exact and

$$\lim_{\Delta t \to 0} \frac{\Delta v}{\Delta t} = v \lim_{\Delta t \to 0} \frac{\Delta \theta}{\Delta t}.$$

But $\lim_{\Delta t \to 0} \Delta v / \Delta t$ is the instantaneous acceleration, and since the angle $\Delta \theta$ in Fig. 10–7(b) is equal to the angle $\Delta \theta$ in Fig. 10–7(a) (their sides are mutually perpendicular) the term $\lim_{\Delta t \to 0} \Delta \theta / \Delta t$ is the instantaneous angular velocity ω. Therefore

$$a = v\omega. \tag{10–11}$$

That is, the magnitude of the acceleration of the point equals the product of its tangential and angular velocities.

The direction of the acceleration, as has been pointed out, is the same as that of the change in velocity Δv. As the angle $\Delta \theta$ becomes smaller, the vectors v_0 and v come more and more nearly into coincidence and the angle between their direction and that of the vector Δv approaches a right angle. In the limit, the vector Δv (or dv) is exactly at right angles to v. Hence the instantaneous acceleration is at right angles to the tangential velocity and is directed inward toward the center, or along the radius. For this reason it is often called a *radial* acceleration and written a_R. The instantaneous tangential velocity and radial acceleration, at the instant the point is at p, are shown in Fig. 10–7(c).

Note that although the vector Δv becomes vanishingly small in the limiting case, the time interval by which it must be divided to obtain the instantaneous acceleration becomes small also. The quotient, a_R, is not necessarily a small quantity.

Finally, from Eq. (10–11) and the relation $v_T = R\omega$, we obtain these useful relations:

$$a_R = v_T\omega = R\omega^2 = \frac{v_T^2}{R}. \qquad (10\text{–}12)$$

From its definition as change in velocity divided by elapsed time, radial acceleration is expressed in ft/sec^2, m/sec^2, or cm/sec^2.

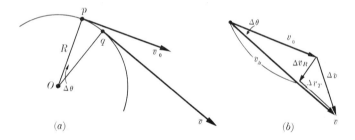

FIG. 10–8. Radial and tangential components of acceleration.

Consider next the more general case, illustrated in Fig. 10–8, in which the rotating body has an angular acceleration α. Then the magnitude of the angular velocity is not constant, and the vector v is longer than the vector v_0 as well as being in a different direction. The change in velocity, found by the usual method, is the vector Δv in Fig. 10–8(b). This vector may be resolved into the components Δv_R and Δv_T. The component Δv_R corresponds exactly to the vector Δv in Fig. 10–7(b). The component Δv_T is equal to the difference in *length* between the vectors v and v_0. That is, this component represents the change in velocity brought about by a change in the *magnitude* of the tangential velocity, while the component Δv_R is the change arising from a change in *direction*.

In the limit, as $\Delta\theta \to 0$, the directions of v and v_0 come more nearly into coincidence. The vector Δv_T, in the limit, coincides with the direction of either and therefore lies along the tangent, whence the subscript T. The vectors Δv_T and Δv_R may be considered as *rectangular* components of Δv, resolved along the tangent and the radius instead of parallel to the X- and Y-axes.

The limiting ratio of the vector Δv_T to the elapsed time is the *instantaneous tangential acceleration*. It is conveniently expressed as follows: Let ω_0 and ω represent the initial and final angular velocities in Fig. 10–8(a), corresponding to the positions p and q. The lengths of the vectors v_0 and v are then $v_0 = R\omega_0$ and $v = R\omega$. Since Δv_T is the difference in length of these vectors,

$$\Delta v_T = R\omega - R\omega_0 = R(\omega - \omega_0) = R\,\Delta\omega.$$

When the first and last terms are divided by Δt we obtain, in the limit,

$$\lim_{\Delta t \to 0} \frac{\Delta v_T}{\Delta t} = R \lim_{\Delta t \to 0} \frac{\Delta\omega}{\Delta t}.$$

The term on the left is, by definition, the tangential acceleration, and $\lim_{\Delta t \to 0} \Delta\omega/\Delta t$ is the instantaneous angular acceleration α. Hence

$$\boxed{a_T = R\alpha.} \tag{10–13}$$

The relation between tangential and angular acceleration can also be obtained by differentiation of the equation

$$v_T = R\omega.$$

This gives

$$\frac{dv_T}{dt} = R\frac{d\omega}{dt}.$$

But dv_T/dt is the rate of change of the magnitude of the tangential velocity, or the tangential acceleration, and $d\omega/dt$ is the angular acceleration. Therefore

$$a_T = R\alpha.$$

Like radial acceleration, tangential acceleration is expressed in ft/sec^2, m/sec^2, or cm/sec^2.

The expressions for the radial and tangential components of the acceleration of a point in circular motion may now be combined to obtain the resultant acceleration a.

$$a = \sqrt{a_R^2 + a_T^2} = \sqrt{\omega^2 v_T^2 + R^2\alpha^2}.$$

The concepts of tangential and radial acceleration are not restricted to motion in a circle but may be applied to motion along any curve. Consider, for example, the parabolic trajectory of a body in free flight. At

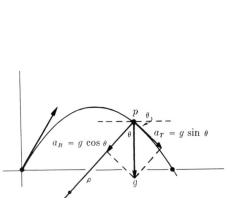

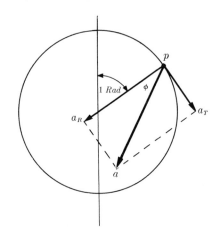

Fig. 10–9. The acceleration of gravity, g, may be resolved into tangential and radial components.

Figure 10–10

all points on the trajectory the magnitude of the acceleration is g and its direction is vertically down. At any point such as p, in Fig. 10–9, where the trajectory makes an angle θ with the horizontal, the acceleration g may be resolved into a tangential component $a_T = g \sin \theta$ and a radial component $a_R = g \cos \theta$.

A sufficiently small portion of any curve may be considered an arc of a circle and the radius of this circle is called the radius of curvature, ρ. (See Thomas, *Calculus and Analytic Geometry*, page 406.) Then if v_T is the velocity of the body at point p,

$$a_R = g \cos \theta = \frac{v_T^2}{\rho}.$$

Summary

(1) When a point moves in a circle its length of path s, its tangential velocity v_T, and its tangential acceleration a_T are related to its angular

displacement θ, its angular velocity ω, and its angular acceleration α by the equations

$$s = R\theta,$$
$$v_T = R\omega,$$
$$a_T = R\alpha.$$

(2) Radial acceleration is the rate of change of velocity arising from a change in direction of this velocity.   Its direction is radially inward toward the center and it is related to the angular and tangential velocities by

$$a_R = \omega v_T = R\omega^2 = \frac{v_T^2}{R}.$$

(3) Tangential acceleration is the rate of change of velocity arising from a change in the magnitude of this velocity.   It is related to the angular acceleration by

$$a_T = R\alpha.$$

EXAMPLE. A disk of radius 10 cm starts from rest and accelerates about a horizontal axis through its center with a constant angular acceleration of 2 rad/sec². A point p on the rim of the disk is located vertically above the center at the start. At the end of 1 sec find (a) the position of the point, (b) its radial acceleration, (c) its tangential acceleration, (d) its resultant acceleration.

(a) $\theta = \omega_0 t + \frac{1}{2}\alpha t^2 = \frac{1}{2} \times 2 \times 1^2 = 1$ rad.

Hence the point is located as in Fig. 10–10.

(b) $a_R = R\omega^2$,

$$\omega^2 = \omega_0^2 + 2\alpha\theta = 2 \times 2 \times 1 = 4 \left(\frac{\text{rad}}{\text{sec}}\right)^2,$$

$$a_R = 10 \times 4 = 40 \text{ cm/sec}^2.$$

(c) $a_T = R\alpha = 10 \times 2 = 20 \text{ cm/sec}^2.$

(d) $a = \sqrt{a_R^2 + a_T^2} = \sqrt{40^2 + 20^2} = 45.7 \text{ cm/sec}^2,$

$$\tan \phi = \frac{a_T}{a_R} = \frac{20}{40} = 0.5, \qquad \phi = 26.5°.$$

The relation between the linear and angular aspects of circular motion is illustrated in the multiflash photograph of Fig. 10–11. A cord is wrapped around the outside of a circular disk whose horizontal axis is supported in ball bearings. A weight hangs from the end of the cord. One radius is

marked on the disk and is horizontal at the start of the motion. When the disk is released, the weight moves down with constant linear acceleration and the disk rotates clockwise with constant angular acceleration. (The dynamics of the problem are discussed in Chapter 11.)

The angle between any two successive positions of the radius, divided by the flash interval, equals the average angular velocity during that interval. Evidently the angles become progressively larger as the motion proceeds, showing that the angular velocity is increasing. Careful measurements show the increases to be equal between each consecutive pair of flashes, or, in other words, the angular acceleration is constant.

The distance moved by the descending weight during any time interval is the same as the circumferential distance moved by any point on the rim of the disk in the same interval. The velocity and acceleration of the weight are therefore numerically equal to the tangential velocity and acceleration of a point on the rim of the disk. It is evident that the weight moves with increasing velocity and careful measurements show the rate of increase to be constant.

Since the angular displacement, velocity, and acceleration can be found from measurements on the disk, and tangential displacement, velocity, and acceleration from

FIG. 10–11. Multiflash photograph showing the relation between tangential and angular acceleration.

measurements on the falling weight, and the radius of the disk can be measured, the relations $s = R\theta$, $v_T = R\omega$, and $a_T = R\alpha$ can all be verified.

10–8 Centripetal and centrifugal forces. Everyone has at some time or other performed the experiment of tying a stone or weight to a cord, and whirling the stone in a circle. While the stone is revolving it can be felt to pull outward on one's hand, and conversely the hand must exert an inward pull on the stone.

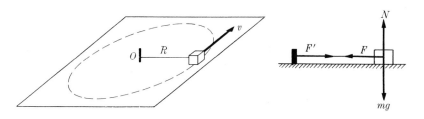

FIGURE 10–12   FIG. 10–13. Force F is the centripetal force. Force F', the reaction to F, is the centrifugal force.

To reduce the problem to its essential terms let us imagine a pin O set into a horizontal frictionless table top as in Fig. 10–12. A small body of mass m is attached to the pin by a cord of length R, and set revolving about it with an angular velocity ω, a tangential velocity v_T, and a radial acceleration $a_R = v_T^2/R = \omega^2 R$. According to Newton's second law, a force must be exerted on the body to produce this radial acceleration, and the direction of this force must be the same as the direction of the acceleration or toward the center of the circle. It is therefore called a central or *centripetal* force. (The term "centripetal" means literally "seeking a center.") Since

$$F = ma \qquad \text{and} \qquad a = v_T^2/R = \omega^2 R,$$

the magnitude of the centripetal force is

$$F = mv_T^2/R = m\omega^2 R. \tag{10–14}$$

This inward force is provided by the cord, which is evidently in tension and which therefore exerts an outward force, equal and opposite to the centripetal force, on the pin at the center. This outward force is called a *centrifugal* force. (The term "centrifugal" means literally "fleeing a center.") The force diagram of the system is given in Fig. 10–13, where F and F' are the equal and opposite forces exerted by the cord on the bodies to which its ends are attached. Force F is the centripetal, force F' the centrifugal force. Centripetal and centrifugal forces always constitute an action-and-reaction pair, the former being the resultant inward force on the revolving body and the latter the reaction to this force.

Unfortunately, there exists much confusion of thought with respect to centrifugal forces. A current notion is that centrifugal force is an outward force exerted *on a revolving body*, causing it to "fly out from the center." There is also the impression that centripetal and centrifugal forces are in some way different from pushes and pulls exerted by sticks and strings, and that they constitute a third class of forces in addition to contact and action-at-a-distance forces. This is not the case. Centripetal forces, like other forces, are pushes or pulls exerted on some material body by some other material body, and their designation as "centripetal" refers only to the effect they produce (a change in direction) and not to something inherently different in their nature.

If the revolving body is not sufficiently small to be considered a point, the radius R in Eq. (10–14) must be taken as the radius of the circle in which the center of mass revolves, and v_T as the tangential velocity of the center of mass.

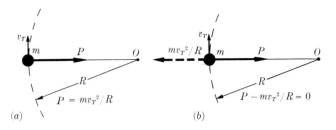

Fig. 10–14. Force P is the centripetal force. The fictitious D'Alembert force, mv_T^2/R, is sometimes called a "centrifugal" force.

The D'Alembert principle (see page 87) may be applied to circular motion as well as to motion in a straight line. Figure 10–14 represents a body of mass m moving with a tangential velocity v_T in a circular path of radius R about a center at O. The Newtonian viewpoint, in (a), is that the connecting cord exerts a resultant inward force P on the body, and this resultant force equals the product of mass and radial acceleration, mv_T^2/R. The D'Alembert viewpoint, in (b), is that the body is in equilibrium under the combined action of the force P and the fictitious outward force mv_T^2/R. When the D'Alembert principle is used, the fictitious outward force mv_T^2/R is called a "centrifugal" force. It is probably this use of the term "centrifugal" for a *fictitious* outward force which is in part responsible for the mistaken impression that a *real* outward force acts on a body in circular motion.

10–9 The banking of curves. Figure 10–15 is a front view of the truck of a railway car of mass m approaching the reader with velocity v, and rounding a curve of radius R whose center is at the right of the diagram. In order to maintain the motion in a curved path, it is necessary that a centripetal force, equal to mv^2/R, shall be exerted on the truck. The direction of this force is toward the center of the circle, or, in this case, toward the right. The centripetal force is provided by the outer rail pushing toward the right against the flange of the outer wheel, and is represented by P in Fig. 10–15(a). The other forces exerted on the truck are its weight, mg, and the upward push of the rails, N. For simplicity these forces are shown as if they all acted at the center of mass. (See problems 31 and 32 at the end of this chapter for a more complete solution.) The resultant force exerted on the truck by the rails is shown by the dotted vector.

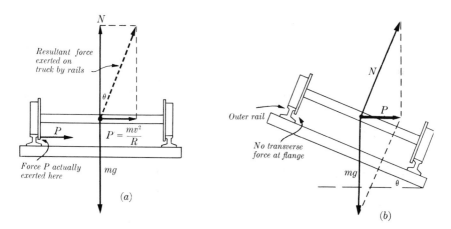

Fig. 10–15. The banking of curves.

If, now, the rails, instead of being level, are "banked" as in Fig. 10–15(b), so as to be perpendicular to the direction of the force which they must exert on the truck, this force becomes a normal force, and the pressure of the rails against the wheel flanges need no longer be relied on to keep the truck moving in a circle. The *vertical component* of the normal force now supports the weight of the truck, and its *horizontal component* provides the centripetal force. The resultant of the entire set of forces is the same in both Fig. 10–15(a) and (b), namely, the centripetal force P.

The banking angle θ which the roadbed makes with the horizontal is equal to the angle θ in Fig. 10–15(a). Hence

$$\tan \theta = \frac{mv^2/R}{mg} = \frac{v^2}{Rg}. \tag{10–15}$$

It can be seen from this equation that the correct angle of banking is proportional to the square of the velocity and inversely proportional to the radius of the curve. For a given radius no one angle is correct for all velocities. Hence in the design of highways and railroads, curves are banked for the average velocity of the traffic over them, and will be somewhat too steep for velocities lower than the average and vice versa.

The same considerations determine the correct banking angle of a plane when making a turn. The angle should be such that the resultant of the lift and the centripetal force is perpendicular to the wing surfaces. (Fig. 10–16.)

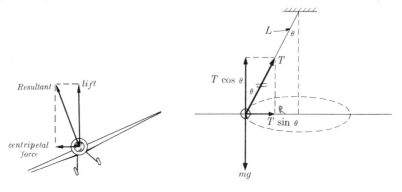

FIGURE 10–16

FIG. 10–17. The conical pendulum.

10–10 The conical pendulum. Figure 10–17 represents a small body of mass m revolving in a horizontal circle at constant angular velocity ω at the end of a light cord of length L. We shall ignore the interval during which the motion was being started and consider only the state of affairs after the mass has been set in motion. If θ is the angle which the cord makes with the vertical, the radius R of the circle in which it is moving is

$$R = L \sin \theta.$$

As the body swings around its path, the cord sweeps over the surface of a cone. Hence the device is called a *conical pendulum*.

The forces exerted on the body when in the position shown are its weight, mg; the tension T in the cord, and nothing else. There is a great temptation to add an outward "centrifugal" force to the diagram, but, as we have seen, it does not belong in the set of forces acting on the body. We know furthermore that the acceleration of the body is directed toward the center of the horizontal circle in which it is moving. Hence we choose axes in this direction and at right angles to it, and resolve the tension T into components as shown. The resultant Y-force is $T \cos \theta - mg$ and the resultant X-force is $T \sin \theta$. Then from the second law

$$\Sigma Y = T \cos \theta - mg = ma_y,$$

$$\Sigma X = T \sin \theta = ma_x.$$

But $a_y = 0$, since the elevation of the body does not change, and $a_x = v^2/R = \omega^2 R$. Therefore

$$T \cos \theta = mg,$$

$$T \sin \theta = m\omega^2 R. \tag{10–16}$$

Since $R = L \sin \theta$,

$$T \sin \theta = m\omega^2 L \sin \theta,$$

$$T = m\omega^2 L.$$

When this value of T is inserted in Eq. (10–16) we obtain

$$m\omega^2 L \cos \theta = mg,$$

$$\cos \theta = \frac{g}{\omega^2 L}. \tag{10–17}$$

That is, this relation must hold between the angular velocity, the length of the supporting cord, and the angle θ. Hence for a given angular velocity and a cord of given length, there is a definite angle θ which the cord must make with the vertical. This equation explains why the ball revolves in a circle of larger radius as its angular velocity is increased. If ω increases, $\cos \theta$ must decrease and the angle θ must increase, since the cosine of an angle between 0 and 90° decreases as the angle increases.

A useful technical application of this effect is found in a common type of tachometer, illustrated in Fig. 10–18. Shaft S is connected by a flexible drive shaft to the device whose angular velocity is to be measured. The shaft carrying the flyweights W is coupled to shaft S by a gear and pinion. The flyweights are connected by a linkage mechanism to collars F and C. Collar F is fixed to the shaft but collar C is free to move up or down. The collars are forced apart by the coil spring.

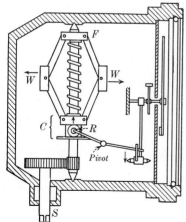

FIG. 10–18. A common type of tachometer. (Courtesy of Pioneer Instrument Co.)

When the shaft S rotates, the flyweights move out, compressing the coil spring until a position is reached where the force exerted by the spring through the linkage provides the requisite centripetal force. The motion of collar C is transmitted by the roller R and a system of levers and gears to the pointer on the dial of the instrument.

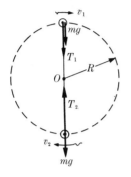

FIG. 10–19. Motion in a vertical circle.

10–11 Motion in a vertical circle. Figure 10–19 represents a small body attached to a cord of length R and whirling in a vertical circle about a fixed point O to which the other end of the cord is attached. The motion, while circular, is not one of constant angular velocity, since the body accelerates on the way down and decelerates on the way up.

Let v_1 represent the velocity of the body as it passes the highest point. The forces acting on it at this point are its weight mg and the tension T_1 in the cord, both acting downward. The resultant force is $T_1 + mg$, and therefore at this point

$$T_1 + mg = \frac{mv_1^2}{R},$$

or

$$T_1 = \frac{mv_1^2}{R} - mg. \qquad (10\text{–}18)$$

Since by definition the centripetal force is the resultant inward force on a body in circular motion, the centripetal force in this instance is provided partly by the body's weight and partly by the tension in the cord. Similarly, at the lowest point of the circle

$$T_2 - mg = \frac{mv_2^2}{R},$$

or

$$T_2 = \frac{mv_2^2}{R} + mg \qquad (10\text{–}19)$$

where T_2 and v_2 represent the tension and the velocity at this point, and the upward direction has been taken as positive.

With motion of this sort, it is a familiar fact that there is a certain

FIG. 10–20. Multiflash photographs of a ball looping-the-loop in a vertical circle.

FIGURE 10–20(a)

FIGURE 10–20(b)

FIGURE 10–20(c)

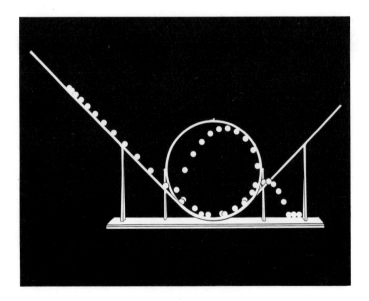

FIGURE 10–20(d)

critical velocity below which the cord becomes slack at the highest point. To find this velocity, set $T_1 = 0$ in Eq. (10–18).

$$0 = \frac{mv_1^2}{R} - mg,$$

$$v_1 = \sqrt{gR}. \tag{10–20}$$

The corresponding velocity at the lowest point may be found from energy considerations. The decrease in potential energy between top and bottom is $2mgR$, and this must equal the increase in kinetic energy. Hence

$$2mgR = \tfrac{1}{2}mv_2^2 - \tfrac{1}{2}mv_1^2$$

$$= \tfrac{1}{2}mv_2^2 - \tfrac{1}{2}mgR,$$

$$v_2 = \sqrt{5gR}.$$

That is, the body must have at least this velocity at the bottom of the circle if it is to get over the top without having the string slacken.

The multiflash photographs of Fig. 10–20 illustrate another case of motion in a vertical circle: a small ball "looping-the-loop" on the inside of a vertical circular track. The inward normal force exerted on the ball by the track takes the place of the tension T in Fig. 10–19.

In Fig. 10–20(a) the ball is released from such an elevation that its velocity at the top of the track is greater than the critical velocity, $\sqrt{gR}$. In Fig. 10–20(b) the ball starts from a lower elevation and reaches the top of the circle with a velocity such that its own weight is larger than the requisite centripetal force. In other words, the track would have to pull outward to maintain the circular motion. Since this is impossible, the ball leaves the track and moves for a short distance in a parabola. This parabola soon intersects the circle, however, and the remainder of the trip is completed successfully.

In Fig. 10–20(c) the start is made from a still lower elevation, the ball leaves the track sooner, and the parabolic path is clearly evident. In Fig. 10–20(d), while the ball eventually returns to the track, the collision is so nearly at right angles that it bounces a few times and finally rolls off.

10–12 Effect of the earth's rotation on weight. The weight of a body is defined as the force of gravitational attraction between the body and the earth, and we have stated that this force may be measured by suspending the body from a spring balance. The latter statement is not strictly correct unless the weighing operation is carried out at one of the poles.

Figure 10–21(a) is a view of the earth looking down on the north pole and showing a (greatly enlarged) body at the equator, hanging from a spring balance. The forces exerted on the hanging body are shown in Fig. 10–21(b), where w is the true force of gravitational attraction and P is the upward pull exerted on the body by the spring balance. From Newton's third law a downward force equal and opposite to P is exerted on the balance and hence the balance reading equals the force P.

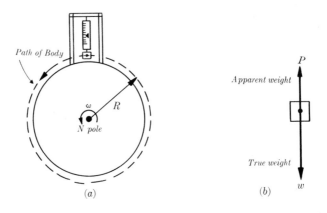

FIG. 10–21. True and apparent weight.

If the suspended body were in equilibrium, the forces w and P would be equal by Newton's first law and the balance reading would equal the true weight. Actually, since the body is carried along by the earth's rotation, it is not in equilibrium but has a radial acceleration toward the earth's center. The force w must, therefore, be slightly larger than the force P in order to provide the necessary centripetal force. If R is the earth's radius and ω its angular velocity, then

$$w - P = m\omega^2 R = \frac{w}{g}\omega^2 R,$$

or

$$P = w - \frac{w}{g}\omega^2 R,$$

$$P = w\left(1 - \frac{\omega^2 R}{g}\right). \qquad (10\text{–}21)$$

If we call the balance reading P the "apparent weight" and w the "true weight" of the body, it will be seen that the apparent weight is somewhat smaller than the true weight. Approximate numerical values of R and ω are

$$R = 6400 \text{ km} = 6.4 \times 10^6 \text{ meters,}$$

$$\omega = 1 \text{ rev/day} = 7.3 \times 10^{-5} \text{ rad/sec,}$$

and hence

$$\frac{\omega^2 R}{g} = \frac{(7.3 \times 10^{-5})^2 \times 6.4 \times 10^6}{9.8} = 0.0034,$$

and $P = 0.9966w$.

The calculations above were made for a body at the equator. At points north or south of the equator, bodies revolve in circles of smaller radius and the difference between true and apparent weight is less, becoming zero at the poles.

10–13 The centrifuge. A centrifuge is a device for whirling an object with a high angular velocity. The consequent large radial acceleration is equivalent to increasing the value of g, and such processes as sedimentation, which would otherwise take place only slowly, can be greatly accelerated in this way. Very high speed centrifuges, called ultracentrifuges, have been operated at angular velocities as high as 180,000 rpm, and small experimental units have been driven as fast as 1,300,000 rpm.

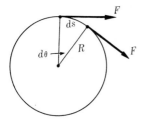

FIGURE 10–22

10–14 Work and power in circular motion. Suppose a force F acts as shown in Fig. 10–22 at the rim of a pivoted wheel of radius R, while the wheel rotates through an angle $d\theta$ radians. By definition, the work done by the force is

$$dW = F \, ds.$$

But $ds = R \, d\theta$, so that

$$dW = FR \, d\theta.$$

But FR is the torque, τ, due to the force F, so we have finally

$$dW = \tau \, d\theta \qquad (10\text{–}22)$$

as the equation corresponding to $dW = F \, dx$ in linear motion.

The work done in a finite angular displacement is

$$W = \int_{\theta_1}^{\theta_2} \tau \, d\theta,$$

and if the torque is constant

$$W = \tau(\theta_2 - \theta_1).$$

That is, the work done by a constant torque equals the product of the torque and the angular displacement.

If τ is expressed in lb·ft, the work is in ft·lb. If τ is in meter-newtons the work is in joules, and if τ is in cm-dynes the work is in ergs.

When both sides of Eq. (10–22) are divided by dt we obtain

$$\frac{dW}{dt} = \tau \frac{d\theta}{dt}.$$

But dW/dt is the rate of doing work or the power, and $d\theta/dt$ is the angular velocity. Hence

$$P = \tau\omega. \tag{10–23}$$

That is, the instantaneous power developed by a torque equals the product of the torque and the instantaneous angular velocity.

EXAMPLE. The drive shaft of an automobile rotates at 3600 rpm and transmits 80 hp from the engine to the rear wheels. Compute the torque developed by the engine.

$$\omega = 3600 \times \frac{2\pi}{60} = 120\pi \text{ rad/sec,}$$

$$80 \text{ hp} = 44,000 \text{ ft·lb/sec,}$$

$$\tau = \frac{P}{\omega} = \frac{44,000}{120\pi} = 117 \text{ lb·ft.}$$

Problems

10–1. (a) What angle in radians is subtended by an arc 6 ft in length, on the circumference of a circle whose radius is 4 ft? (b) What angle in radians is subtended by an arc of length 78.54 cm on the circumference of a circle of diameter 100 cm? What is this angle in degrees? (c) The angle between two radii of a circle is 0.60 radians. What length of arc is intercepted on the circumference of a circle of radius 200 cm? of radius 200 ft?

10–2. Compute the angular velocity, in rad/sec, of the crankshaft of an automobile engine rotating at 4800 rpm.

10–3. (a) A cylinder 6 inches in diameter rotates in a lathe at 750 rpm. What is the tangential velocity of the surface of the cylinder? (b) The proper tangential velocity for machining cast iron is about 2 ft/sec. At how many rpm should a piece of stock 2 inches in diameter be rotated in a lathe?

10–4. An electric motor running at 1800 rpm has on its shaft three pulleys, of diameters 2, 4, and 6 in respectively. Find the linear velocity of the surface of each pulley, in ft/sec. The pulleys may be connected by a belt to a similar set on a countershaft; the 2 in to the 6 in, the 4 in to the 4 in, and the 6 in to the 2 in. Find the three possible angular velocities of the countershaft, in rpm.

10–5. A wheel 2.4 ft in diameter starts from rest and accelerates. uniformly to an angular velocity of 100 rad/sec in 20 sec. Find the angular acceleration and the angle turned through.

10–6. The angular velocity of a flywheel decreases uniformly from 1000 rpm to 400 rpm in 5 sec. Find the angular acceleration and the number of revolutions made by the wheel in the 5 sec interval. How many more seconds are required for the wheel to come to rest?

10–7. A flywheel requires 3 sec to rotate through 234 radians. Its angular velocity at the end of this time is 108 rad/sec. Find its constant angular acceleration.

10–8. A flywheel whose angular acceleration is constant and equal to 2 rad/sec², rotates through an angle of 100 radians in 5 sec. How long had it been in motion at the beginning of the 5 sec interval if it started from rest?

10–9. (a) Distinguish clearly between tangential and radial acceleration. (b) A flywheel rotates with constant angular velocity. Does a point on its rim have a tangential acceleration? a radial acceleration? (c) A flywheel is rotating with constant angular acceleration. Does a point on its rim have a tangential acceleration? a radial acceleration? Are these accelerations constant in magnitude?

10–10. At time $t = 0$ a body is moving east at 10 cm/sec. At time $t = 2$ sec it is moving 25° north of east at 14 cm/sec. Find graphically its change in velocity during this time and its average acceleration.

10–11. A wheel 30 inches in diameter is rotating about a fixed axis with an initial angular velocity of 2 revolutions per sec. The acceleration is 3 rev/sec². (a) Compute the angular velocity after 6 sec. (b) Through what angle has the wheel turned in this time interval? (c) What is the tangential velocity of a point on the rim of the wheel at $t = 6$ sec? (d) What is the resultant acceleration of a point on the rim of the wheel at $t = 6$ sec?

10–12. A point on the rim of a wheel of radius 16 cm has its angular position given as a function of time by $\theta = 12 - 9t - 3t^2 + t^3$, where θ is in radians, t in seconds, and the counterclockwise direction is positive. (a) Find the equations for the angular velocity and the angular acceleration of the point as a function of time. (b) At what time is the resultant acceleration in the direction of the radius? What is the value of this acceleration? (c) At what time is the resultant acceleration tangent to the circle? What is the value of this acceleration?

10–13. A wheel rotates about a fixed axis in such a way that its angular velocity at any time t is given by the relation $\omega = 2 + 6t^2$, where ω is in radians/sec and t in seconds. (a) What is the angular acceleration of the wheel at times $t = 0$ and $t = 2$ sec? (b) What is the angle through which the wheel rotates in the interval between $t = 0$ and $t = 2$ sec?

10–14. The flywheel of an electric motor has an angular acceleration which increases linearly with time, so that $\alpha = 10t$ rad/sec^2. The wheel is at rest at time $t = 0$. (a) Obtain an expression for the angular velocity at any time. What is the angular velocity at $t = 2$ sec? (b) Find the time required for the wheel to make its first complete revolution.

10–15. A racing car is driven with a velocity of 90 ft/sec, constant in magnitude, around a circular track whose circumference is 3600 ft. Find graphically the magnitude and direction of the average acceleration of the car in an interval of (a) 8 sec, (b) 4 sec, (c) 2 sec. (d) Compare the answer to (c) with the magnitude and direction of the instantaneous radial acceleration.

10–16. A body moves in the X-Y plane according to the law $x = R \cos \omega t$, $y = R \sin \omega t$, where x and y are the coordinates of the body, t is the time, and R and ω are constants. (a) Eliminate t between these equations to find the equation of the curve in which the body moves. What is this curve? (b) Differentiate the given equations to find the X- and Y-components of the velocity of the body. Combine these expressions to obtain the magnitude and direction of the resultant velocity. (c) Differentiate again to obtain the magnitude and direction of the resultant acceleration.

This problem illustrates an alternate method of deriving the expressions for tangential velocity and radial acceleration.

10–17. Find the required angular velocity of an ultracentrifuge, in rpm, in order that the radial acceleration of a point 1 cm from the axis shall equal 300,000 g (i.e., 300,000 times the acceleration of gravity).

10–18. The pilot of a dive bomber who has been diving at a velocity of 400 mi/hr pulls out of the dive by changing his course to a circle in a vertical plane. What is the minimum radius of the circle in order that the acceleration at the lowest point shall not exceed "$7g$." How much does a 180-lb pilot apparently weigh at the lowest point of the pull-out?

10–19. A wheel starts from rest and accelerates uniformly to an angular velocity of 900 rpm in 20 sec. (a) Find the position, at the end of 1 sec, of a point originally at the top of the wheel. (b) Compute and show in a diagram the magnitude and direction of the tangential and radial components of its acceleration at this instant. The distance of the point from the axis is 6 in.

10–20. (a) Prove that when a body starts from rest and rotates about a fixed axis with constant angular acceleration, the radial acceleration of a point in the body is directly proportional to its angular displacement. (b) Through what angle will the body have turned when the resultant acceleration makes an angle of 60° with the radial acceleration?

10–21. A small sphere of mass m is fastened to a weightless string of length 2 ft to form a pendulum. The pendulum is swinging so as to make a maximum angle of 60° with the vertical. Compute and show in a diagram the magnitude and direction of the resultant acceleration of the sphere when the string makes an angle of (a) 60° and (b) 37° with the vertical.

10–22. Find the centripetal force exerted on a 4-oz bolt at the rim of the flywheel of an engine, 18 inches in diameter, rotating at (a) 2000 rpm; (b) 4000 rpm.

10–23. A block of mass M rests on a

FIGURE 10–23

turntable which is rotating at constant angular velocity ω. A smooth cord runs from the block through a hole in the center of the table down to a hanging block of mass m. The coefficient of friction between the first block and the turntable is μ. (See Fig. 10–23.) Find the largest and smallest values of the radius r for which the first block will remain at rest relative to the turntable.

10–24. A stone of mass 1 kgm is attached to a string 1 m long, of breaking strength 500 newtons, and is whirled in a horizontal circle. With what speed will the stone fly off if the angular velocity is just great enough so that the string breaks?

10–25. A bicycle and rider, weighing together 160 lb, loop-the-loop in a circular track of radius 8 ft. The velocity at the lowest point is $32\sqrt{2}$ ft/sec. (a) Find the radial acceleration at the highest point. Assume that the bicycle "coasts" without friction. (b) Show in a diagram all of the forces acting on the bicycle and rider at the highest point, and compute the force with which the track pushes against the bicycle. (c) With what force does the bicycle press against the track? (d) What is the minimum velocity the bicycle can have at the highest point without leaving the track?

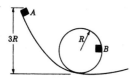

FIGURE 10–24

10–26. A small body of mass m slides without friction around the loop-the-

loop apparatus shown in Fig. 10–24. It starts from rest at point A at a height $3R$ above the bottom of the loop. When it reaches point B at the end of a horizontal diameter of the loop, compute (a) its radial acceleration, (b) its tangential acceleration, (c) its resultant acceleration. Show these accelerations in a diagram, approximately to scale. (See Problem 17 in Chapter 12.)

10–27. Prove that when a body attached to a cord is revolving in a vertical circle, the tension in the cord when the body is at the lowest point exceeds the tension when it is at the highest point by six times the weight of the body.

FIGURE 10–25

10–28. A boy riding a bicycle at 12 ft/sec rounds a curve and finds he must incline his bicycle at an angle of 53° with the horizontal in order to keep from tipping over. The combined weight of the boy and bicycle is 180 lb and the combined center of gravity is as shown in Fig. 10–25. (a) Draw a diagram showing all the forces acting on the bicycle and rider. (b) Calculate the radius of the curve. (c) Calculate the friction force between the road and bicycle tire.

10–29. A car moves in a curve of radius of curvature R. The width between wheels is b, and the height of the center of mass from the ground is h. With what velocity must the car move in order that the vertical force on the inside wheels shall be reduced to zero?

10–30. A curve of 600 ft radius on a level road is banked at the correct angle for a velocity of 30 mi/hr. If an automobile rounds this curve at 60 mi/hr, what is the minimum coefficient of friction between tires and road so that the automobile will not skid? Assume all forces to act at the center of mass.

10–31. An airplane is flying at 120 mi/hr, in a horizontal circle of radius 5000 ft. (a) What is the banking angle, θ, of the plane? (b) What is the magnitude and direction of the resultant force on the airplane? (c) What is the magnitude of the aerodynamic force on the wing?

FIGURE 10–26

10–32. A packing case 8 ft high and 4 ft square weighs 500 lb and its center of gravity is at its center. The case stands on one end on the floor of a delivery truck. (a) Show in a diagram all of the forces acting on the case when the truck is rounding a curve with constant speed. (b) What is the maximum speed with which the truck can round a curve whose radius is 64 ft without causing the case to tip over? Assume sufficient friction so that the case does not slide. (c) What is the minimum value the coefficient of friction can have?

10–33. A motorcycle velodrome is 40 ft in diameter, and the coefficient of friction between tires and boards is 0.4. (a) Draw a diagram of the forces acting on the motorcycle when it is moving in a horizontal circle around the vertical wall of the velodrome. (b) Compute the minimum speed of the motorcycle. (c) If the motorcycle and rider weigh 320 lb, find the vertical and horizontal forces acting on the track at this speed.

10–34. A body is projected with an initial velocity of 120 ft/sec at an angle of 53° above the horizontal. Find the radius of curvature of its trajectory (a) at the highest point, (b) 4 sec after the motion begins. Illustrate by a diagram approximately to scale.

FIGURE 10–27

10–35. A two-ton coaster rolls on a track as shown in Fig. 10–27. It starts at S with zero velocity, rolls down to A (100 ft below S) where the radius of curvature is 50 ft, and then up to B (50 ft below S) where the radius of curvature is again 50 ft. (a) What force is exerted on the track as the roller coasts by point A? (b) By point B? (c) Is the track adequately designed? (It is built to take a force of 20 tons.)

FIGURE 10–28

10–36. A small sphere is placed in a grooved vertical circular track of radius 30 cm. When the track is spinning about a vertical axis with a constant angular velocity of 7 rad/sec, the sphere takes a position given by the angle θ as shown in Fig. 10–28. (a) Draw a diagram showing the direction and magnitude of all the forces acting on the sphere. (b) Find the angle θ.

10–37. The 8-lb block in Fig. 10–29 is attached to a vertical rod by means of two strings. When the system rotates about the axis of the rod with an angular velocity of 4 rad/sec the strings are extended as shown in the diagram. What is the tension in (a) the upper string, (b) the lower string?

Marlow

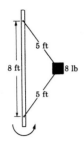

FIGURE 10–29

10–38. The mechanism in Fig. 10–30 rotates about the vertical axis with an angular velocity of 20 rad/sec. The mass m of each flyweight is 500 gm, and

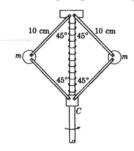

FIGURE 10–30

the masses of the other elements of the system are negligible. Collar C slides without friction on the vertical shaft. With how great a force is the spring compressed?

10–39. The apparent weight of a man at the equator is 240 lb. By how many ounces does this differ from his true weight?

10–40. A plane flies from east to west above the equator, parallel to the curvature of the earth, at a velocity of 300 mi/hr relative to the earth. Find the difference between its apparent weight and its true weight of 32,000 lb.

10–41. (a) Compute the torque developed by an airplane engine whose output is 2000 hp at an angular velocity of 2400 rpm. (b) If a drum 18 inches in diameter were attached to the motor shaft, and the power output of the motor were used to raise a weight hanging from a rope wrapped around the

shaft, how large a weight could be lifted? (c) With what velocity would it rise?

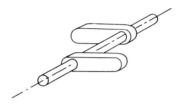

FIGURE 10–31

10–42. The system shown in Fig. 10–31 is said to be "statically balanced" but "dynamically unbalanced." Each counterweight weighs 8 lb, the center of mass of each is 4 inches from the axis, and the counterweights are 8 inches apart. The shaft is 18 inches long and is mounted in bearings at its ends. (a) Find the forces exerted on the shaft by the bearings when the system is rotating at 900 rpm, and show in a diagram the directions of these forces at an instant when the system is in the position shown in the figure. (b) The system is to be dynamically balanced by the addition of two other counterweights weighing 16 lb each and whose centers of mass are 6 inches from the axis. Describe how these weights should be attached to the shaft.

10–43. An automobile engine delivers 20 horsepower at 1200 rpm to the transmission of a car. The ratio of engine speed to that of the driveshaft is 1:1 in high gear and 3:1 in low gear. The ratio of drive shaft speed to rear axle speed is 4:1. Each rear wheel has a diameter of 28 inches (6.00 × 16 tire). If the overall efficiency of the power transmitting system is 80% at all gear ratios, find (a) the power delivered to the rear wheels when in high gear and when in low gear, (b) the torque delivered to the rear wheels when in high gear and when in low gear, and (c) the tangential force exerted by the rear wheels on the road when in high gear and when in low gear.

CHAPTER 11

MOMENT OF INERTIA

11-1 Moment of inertia. In the preceding chapter we discussed motion of rotation about a fixed axis without inquiring into the "causes" of the motion. If we go back to fundamental principles, the motion of each particle of matter in a rotating body is determined by Newton's second law. That is, the resultant force exerted on a particle is at every instant equal to the product of the mass of the particle and its acceleration, and is in the same direction as the acceleration. It turns out, however, that simplification is possible if instead of working with the accelerations of the individual particles of a rotating body we consider the *angular acceleration* of the body, which is the same for all particles, and instead of dealing with the forces on the particles we consider the *resultant torque* on the body as a whole. That is, we look for a relation between resultant torque and angular acceleration which will correspond to Newton's second law connecting resultant force with linear acceleration.

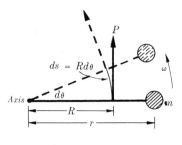

FIGURE 11-1

Let us take as our first example the motion of a light, rigid rod, having a point mass m attached to one end and pivoted at its other end about an axis at right angles to its length, as in Fig. 11-1. An external force P is exerted on the rod at a distance R from the axis, and we shall assume that P remains at right angles to the rod as it rotates. The system rests on a level frictionless surface so that gravitational forces are balanced by the upward push of the surface.

The external force P is not exerted directly on the mass m. The motion of this mass is brought about by the *internal* forces exerted on it by the rod to which it is attached. The complexities that arise when the problem is analyzed in terms of internal forces can be avoided by the use of work-energy relations. That is, the work done on the system by the external force P equals the increase in kinetic energy of the mass m. (The rod itself is of negligible mass.)

For generality, let the system be in rotation about the axis with an angular velocity ω. In a short time interval dt the rod rotates through an angle $d\theta$ and the angular velocity increases by $d\omega$. The work dW done by the force P in the angular displacement is

$$dW = P\,ds = PR\,d\theta \qquad (11\text{–}1)$$

or, since PR is the torque about the axis,

$$dW = \tau\,d\theta.$$

The velocity of the mass m, when the angular velocity of the system is ω, is

$$v = r\omega,$$

and its kinetic energy is

$$\text{KE} = \tfrac{1}{2}mv^2 = \tfrac{1}{2}(mr^2)\omega^2. \qquad (11\text{–}2)$$

When the angular velocity increases by $d\omega$ the increase in kinetic energy is

$$d\,(\text{KE}) = (mr^2)\omega\,d\omega.$$

Since the work done equals the increase in kinetic energy,

$$dW = d\,(\text{KE}),$$

$$\tau\,d\theta = (mr^2)\omega\,d\omega,$$

$$\tau = (mr^2)\omega\,\frac{d\omega}{d\theta}.$$

But we have shown that the angular acceleration, α, can be expressed as $\omega\,(d\omega/d\theta)$. Hence

$$\tau = (mr^2)\alpha. \qquad (11\text{–}3)$$

This equation is the relation we set out to find between the resultant torque about the axis and the angular acceleration. When we compare it with Newton's second law,

$$F = ma,$$

we see that the two are of exactly the same form. Resultant torque is the analogue of resultant force, angular acceleration is the analogue of linear acceleration, and the quantity (mr^2), the product of the mass and the square of its distance from the axis, plays the same role as does mass or inertia in linear motion. This product mr^2 is called the *moment of inertia* of the mass about the axis (the name was first suggested by Euler in 1765) and is represented by the letter I. Equation (11–3) may therefore be written

$$\boxed{\tau = I\alpha.} \qquad (11\text{–}4)$$

The kinetic energy of the body can also be expressed in terms of its moment of inertia. From Eq. (11–2),

$$KE = \tfrac{1}{2}I\omega^2, \qquad (11\text{–}5)$$

a relation which is exactly analogous to

$$KE = \tfrac{1}{2}mv^2$$

in linear motion.

The extension of the concept of moment of inertia to any number of rigidly connected point masses is obvious. We must multiply each mass by the square of its distance from the axis and add the products. That is, for a number of rigidly connected point masses,

$$I = \Sigma mr^2. \qquad (11\text{–}6)$$

The moment of inertia of a mass point depends only on its mass and its radial distance from the axis, not on its angular position. Thus each of the arrangements in Fig. 11–2 has the same moment of inertia, $m_1r_1^2 + m_2r_2^2$.

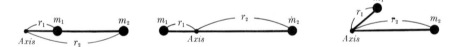

FIG. 11–2. The moment of inertia in each case equals $m_1r_1^2 + m_2r_2^2$.

It is evident from its definition that moment of inertia is expressed in slug·ft^2 in the English system, kgm·m^2 in the mks system, and gm·cm^2 in the cgs system. A point mass of 1 slug, 1 ft from an axis, has a moment of inertia of 1 slug·ft^2 about that axis, and similar statements may be made regarding the kgm·m^2 and the gm·cm^2.

EXAMPLES. (1) A simple pendulum consists of a small lead sphere of mass 100 gm at the end of a cord 1 meter long. What is its moment of inertia about an axis through the upper end of the cord, perpendicular to its length?

Consider the sphere to be a point mass. In mks units

$$I = 0.10 \times (1)^2 = 0.10 \text{ kgm·m}^2.$$

In cgs units

$$I = 100 \times (100)^2 = 10^6 \text{ gm·cm}^2.$$

(2) A rod one meter long has three 10-gm blocks clamped to it as in Fig. 11–3. Find the moment of inertia of the system (a) about an axis through one end, (b) about an axis through the center. Neglect the moment of inertia of the rod itself.

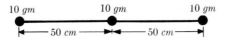

$$10 \ gm \qquad 10 \ gm \qquad 10 \ gm$$

$$\leftarrow\!\!-\!\!-\!\!-\ 50 \ cm \ -\!\!-\!\!-\!\!\rightarrow\!\!\leftarrow\!\!-\!\!-\!\!-\ 50 \ cm -\!\!-\!\!-\!\!\rightarrow$$

FIGURE 11–3

If the axis passes through one end,

$$I = \Sigma mr^2 = 10(0)^2 + 10(50)^2 + 10(100)^2$$
$$= 125{,}000 \text{ gm·cm}^2.$$

If the axis passes through the center,

$$I = \Sigma mr^2 = 10(50)^2 + 10(0)^2 + 10(50)^2$$
$$= 50{,}000 \text{ gm·cm}^2.$$

This example illustrates an extremely important fact, namely, that the moment of inertia of a body, unlike its mass, is not a unique property of the body but depends on the location of the axis about which the moment of inertia is computed. Thus in this example, the moment of inertia of the system about an axis through one end is $2\frac{1}{2}$ times as great as its moment of inertia about an axis through the center.

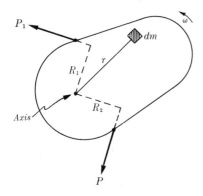

FIGURE 11–4

11–2 Moment of inertia. General case. Consider next a body of any arbitrary shape (Fig. 11–4) rotating about a fixed axis and acted on by any number of external forces P_1, P_2, etc. The work done by the external forces in a small angular displacement is

$$dW = P_1 R_1 \, d\theta + P_2 R_2 \, d\theta + \cdots$$
$$= (\Sigma PR) \, d\theta$$
$$= (\Sigma \tau) \, d\theta.$$

That is, the work is the product of the resultant external torque about the axis and the angular displacement.

The kinetic energy of a small mass dm, at a distance r from the axis, is

$$\tfrac{1}{2} \, dm \, v^2 = \tfrac{1}{2} \, dm r^2 \omega^2,$$

and the total kinetic energy of the body is

$$KE = \int \tfrac{1}{2} \, dm \, r^2\omega^2 = \tfrac{1}{2}(\int r^2 \, dm)\omega^2,$$

where the limits of integration must be chosen so as to include the entire body.

The increase in kinetic energy as a result of the work done is

$$d(KE) = (\int r^2 \, dm)\omega \, d\omega.$$

Hence

$$(\Sigma \tau) \, d\theta = (\int r^2 \, dm)\omega \, d\omega,$$

and since

$$\alpha = \frac{d\omega}{dt} \frac{d\theta}{d\theta} = \frac{d\theta}{dt} \frac{d\omega}{d\theta} = \omega \frac{d\omega}{d\theta},$$

$$\Sigma \tau = (\int r^2 \, dm)\alpha. \qquad (11\text{-}7)$$

Hence, in the general case, the term mr^2 or the sum Σmr^2 is replaced by $\int r^2 \, dm$ and the general definition of moment of inertia is*

$$\boxed{I = \int r^2 \, dm.} \qquad (11\text{-}8)$$

That is, *the moment of inertia of a body about any axis is found by multiplying each element of mass in the body by the square of its distance from the axis, and summing the products over the entire body.*

Equation (11–7) becomes

$$\Sigma \tau = I\alpha. \qquad (11\text{-}9)$$

Obviously, if the angular acceleration is zero, $\Sigma \tau = 0$, and the "third condition of equilibrium," mentioned in Chapter 3, is simply a special case of Eq. (11–9) when the angular acceleration is zero.

***

EXAMPLES. (1) Compute the moment of inertia of a slender rod of constant cross section about an axis perpendicular to the rod and passing through one of its ends.

Let L represent the length of the rod, M its mass, A its cross section and ρ its density. Then (see Fig. 11–5),

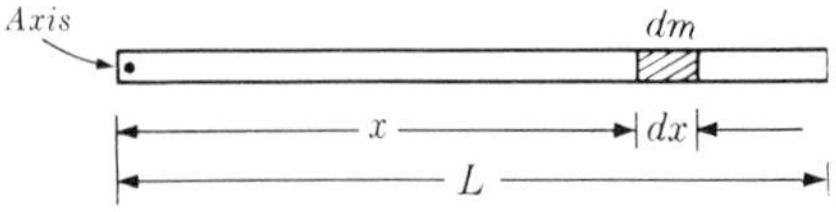

FIGURE 11–5

$$dm = \rho \, dV = \rho A \, dx,$$

$$I = \int r^2 \, dm = \int_0^L x^2 \rho A \, dx = \rho A \int_0^L x^2 \, dx$$

$$= \tfrac{1}{3}\rho A L^3.$$

***

*See Thomas, *Calculus and Analytic Geometry,* p. 199.

Or, since the mass M of the rod is equal to $\rho A L$,

$$I = \tfrac{1}{3}ML^2.$$

(2) Compute the moment of inertia of an annular disk about an axis perpendicular to the plane of the disk through its center (Fig. 11–6).

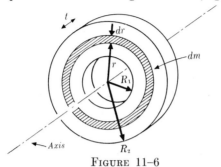

FIGURE 11–6

All parts of the shaded element in Fig. 11–6 are at the same distance r from the axis. Hence the moment of inertia of this element is

$$dI = r^2\, dm.$$

If t is the thickness of the disk and ρ its density,

$$dm = \rho\, dV = \rho \times 2\pi r\, dr \times t,$$

$$dI = 2\pi \rho t r^3\, dr,$$

$$I = 2\pi \rho t \int_{R_1}^{R_2} r^3\, dr = \tfrac{1}{2}\pi \rho t (R_2^4 - R_1^4). \tag{11–10}$$

The moment of inertia is more conveniently expressed in terms of the total mass, M. The volume of the disk is $\pi t(R_2^2 - R_1^2)$ and hence

$$M = \pi \rho t(R_2^2 - R_1^2).$$

Eq. (11-10) can be factored and written

$$I = \tfrac{1}{2}\pi \rho t(R_2^2 - R_1^2)(R_2^2 + R_1^2),$$

and therefore

$$I = \tfrac{1}{2}M(R_2^2 + R_1^2).$$

It is left as an exercise for the reader to deduce from this expression that the moment of inertia of a solid cylinder, about an axis through its center, is

$$I = \tfrac{1}{2}MR^2,$$

and the moment of inertia of a thin-walled cylinder or a hoop is

$$I = MR^2.$$

Moments of inertia of a few bodies of simple geometrical shapes are given in Fig. 11–7.

(3) A rope is wrapped around the surface of a flywheel 2 ft in radius and a 10-lb weight hangs from the rope (Fig. 11–8). The wheel is free to rotate about a

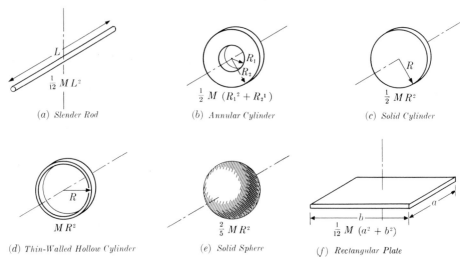

(a) Slender Rod   *(b) Annular Cylinder*   *(c) Solid Cylinder*

(d) Thin-Walled Hollow Cylinder   *(e) Solid Sphere*   *(f) Rectangular Plate*

FIG. 11–7. Moments of inertia.

horizontal axis through its center. Compute its angular acceleration and the tension in the rope if the moment of inertia of the wheel is 2 slug·ft^2.

Isolate the flywheel and the hanging weight. The force diagrams are given in Fig. 11–8(b) and (c). The forces at the center of the flywheel are omitted, since their moment about the axis is zero. From Fig. 11–8(b)

$$\tau = I\alpha,$$
$$2T = 2\alpha,$$

and from Fig. 11–8(c)

$$F = ma,$$
$$10 - T = \frac{10}{32} a.$$

Also, since the linear acceleration of the weight equals the tangential acceleration of the surface of the flywheel,

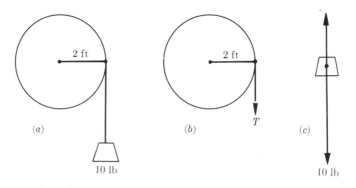

(a)   *(b)*   *(c)*

FIGURE 11–8

Simultaneous solution of these equations gives

$$a = 12.3 \text{ ft/sec}^2,$$
$$\alpha = 6.2 \text{ rad/sec}^2,$$
$$T = 6.2 \text{ lb}.$$

It is left as an exercise to show that the loss in potential energy of the weight equals the *combined* kinetic energies of the weight and of the flywheel.

11-3 Radius of gyration. Whatever the shape of a body, it is always possible to find a radial distance from any given axis at which the mass of the body could be concentrated without altering the moment of inertia of the body about that axis. This distance is called the *radius of gyration* of the body about the given axis, and is represented by k.

If the mass M of the body actually were concentrated at this distance, the moment of inertia would be that of a point mass M at a distance k from an axis, or Mk^2. Since this equals the actual moment of inertia, I, then

$$Mk^2 = I,$$

$$k = \sqrt{\frac{I}{M}}. \tag{11–11}$$

Equation (11–11) may be considered the definition of radius of gyration.

EXAMPLE. What is the radius of gyration of a slender rod of mass M and length L about an axis perpendicular to its length and passing (a) through one end? (b) through the center?

(a) The moment of inertia about an axis through one end is $I = \frac{1}{3}ML^2$. Hence

$$k = \sqrt{\frac{\frac{1}{3}ML^2}{M}} = \frac{L}{\sqrt{3}} = 0.577L.$$

(b) The moment of inertia about an axis through the center is $I_0 = \frac{1}{12}ML^2$. Hence

$$k_0 = \sqrt{\frac{\frac{1}{12}ML^2}{M}} = \frac{L}{2\sqrt{3}} = 0.289L.$$

The radius of gyration, like the moment of inertia, depends on the location of the axis.

Note carefully that, in general, the mass of a body can *not* be considered as concentrated at its center of mass, for the purpose of computing its moment of inertia. For example, when a rod is pivoted about its center, the distance from the axis to the center of mass is zero, although the radius of gyration is $L/2\sqrt{3}$.

11–4 The parallel-axis theorem. The parallel-axis theorem is a useful relation which enables one to find the moment of inertia of a body about any axis whatever if its moment of inertia about some parallel axis is known. The theorem states that *the moment of inertia of a body about any axis is equal to its moment of inertia about a parallel axis through the center of mass, plus the product of the mass of the body and the square of the distance between the axes.* The theorem was first derived by Lagrange in 1783.

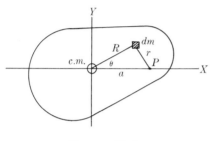

FIGURE 11–9

In Fig. 11–9, point P represents any arbitrary point in the body. The X-axis has been taken along the line joining P and the center of mass. The moment of inertia about an axis through the center of mass, perpendicular to the plane of the the diagram, is

$$I_0 = \int R^2 \, dm.$$

The moment of inertia about a parallel axis through P is

$$I = \int r^2 \, dm.$$

But

$$r^2 = R^2 + a^2 - 2aR \cos \theta$$

or, since $R \cos \theta$ is simply the X-coordinate of the mass dm,

$$r^2 = R^2 + a^2 - 2ax.$$

When this expression for r^2 is substituted in the second integral we get

$$I = \int R^2 \, dm + a^2 \int dm - 2a \int x \, dm.$$

The first term on the right is I_0. The second is Ma^2, where $M = \int dm$ is the total mass of the body. The third term is zero, as may be seen by making use of the expression for the X-coordinate of the center of mass,

$$\bar{x} = \frac{\int x \, dm}{\int dm}.$$

In this case the center of mass is at the origin so $\bar{x} = 0$ and $\int x \, dm = 0$. Then finally

$$\boxed{I = I_0 + Ma^2.} \qquad (11\text{–}12)$$

EXAMPLE. Use the parallel-axis theorem to find the moment of inertia of a slender rod about an axis through one end, given that the moment of inertia about an axis through the center of mass is $\frac{1}{12}ML^2$.

$$I = I_0 + Ma^2 = \frac{1}{12}ML^2 + M\left(\frac{L}{2}\right)^2$$
$$= \tfrac{1}{3}ML^2,$$

which agrees with the result on page 207.

11–5 Forces at the axis. In addition to the equation

$$\Sigma\tau = I\alpha, \tag{11-13}$$

which expresses the relation between the resultant external *torque*, the moment of inertia, and the angular acceleration of a body pivoted about a fixed axis, we also have the equations

$$F = m\bar{a} \quad\text{or}\quad \Sigma X = m\bar{a}_x, \quad \Sigma Y = m\bar{a}_y, \tag{11-14}$$

which relate the resultant external *force*, the mass, and the linear acceleration of the center of mass of the body. The latter equations must be used to obtain a complete analysis of the forces on a pivoted body, since any forces whose lines of action pass through the axis have no moment about that axis and do not appear in Eq. (11–13).

In the special case where the fixed axis passes through the center of mass, the acceleration of the center of mass is zero and Eq. (11–14) reduces to

$$\Sigma X = 0, \quad \Sigma Y = 0.$$

In all other cases the center of mass moves in a circular path about the axis and, in general, has an acceleration whose radial component is $\omega^2\bar{r}$ and whose tangential component is $\alpha\bar{r}$, where ω and α have their usual meanings and $\bar{r}$ is the radial distance from the center of mass to the axis.

Figure 11–10 represents a body of arbitrary shape pivoted about an axis perpendicular to the plane of the diagram. The path of the center of mass is indicated by the dotted circular arc. The X-axis is taken along the line joining the center of mass and the axis. If the angular velocity and angular acceleration are both clockwise, the radial and tangential components of the acceleration of the center of mass have the directions shown. Then

$$\bar{a}_x = \omega^2\bar{r}, \quad \bar{a}_y = \alpha\bar{r},$$

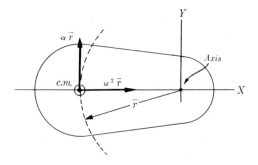

FIGURE 11–10

and we may write as the complete set of equations of motion

$$\Sigma\tau = I\alpha,$$

$$\Sigma X = m\bar{a}_x = m\omega^2\bar{r},$$

$$\Sigma Y = m\bar{a}_y = m\alpha\bar{r}.$$

EXAMPLE. A slender rod of mass m and length L is pivoted about a horizontal axis through one end and released from rest at an angle of 30° above the horizontal. Find the horizontal and vertical components of the force exerted on the rod by the pivot at the instant when the rod is passing through a horizontal position (Fig. 11–11).

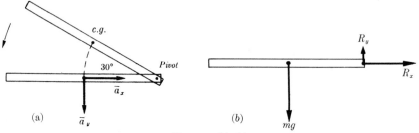

FIGURE 11–11

The loss in potential energy, $mg(L/4)$, equals the gain in kinetic energy, $\frac{1}{2}I\omega^2$.

$$mg\frac{L}{4} = \frac{1}{2}I\omega^2 = \frac{1}{2}\left(\frac{1}{3}mL^2\right)\omega^2,$$

$$\omega^2 = \frac{3}{2}\frac{g}{L},$$

$$\bar{a}_x = \omega^2\bar{r} = \omega^2\frac{L}{2} = \frac{3}{4}g.$$

The torque about the axis when the rod is horizontal is $mg(L/2)$, and from the relation $\Sigma\tau = I\alpha$ we obtain

$$mg\,\frac{L}{2} = \frac{1}{3}\,mL^2\alpha,$$

$$\alpha = \frac{3}{2}\frac{g}{L},$$

$$\bar{a}_y = \alpha\bar{r} = \frac{3}{4}\,g.$$

(Consider the downward direction positive.)
From the force diagram, Fig. 11–11(b),

$$\Sigma X = R_x = m\bar{a}_x = \tfrac{3}{4}mg$$

$$\Sigma Y = mg - R_y = m\bar{a}_y = \tfrac{3}{4}mg$$

$$R_y = \tfrac{1}{4}mg.$$

Hence R_x equals three-fourths of the weight of the rod and is directed toward the right, while R_y is upward and equals one-fourth of the weight.

PROBLEMS

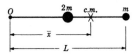

FIGURE 11–12

11–1. A rigid rod of length L and negligible mass has a point mass $2m$ attached to its mid-point and a mass m attached at one end. The rod is pivoted at point O (see Fig. 11–12). (a) Find the distance $\bar{x}$ from the pivot O to the center of mass, in terms of L. (b) Find the moment of inertia I of the system about an axis through point O, in terms of m and L. (c) Find the moment of inertia I_0 about an axis through the center of mass, in terms of m and L. (d) Find the radius of gyration about an axis through point O, in terms of L.

11–2. Find the moment of inertia of a rod 4 cm in diameter and 2 m long, of mass 8 kgm, (a) about an axis perpendicular to the rod and passing through its center; (b) about an axis perpendicular to the rod and passing through one end; (c) about a longitudinal axis through the center of the rod.

11–3. Starting from the definition of moment of inertia as $\int r^2\,dm$, compute the moment of inertia of a slender homogeneous rod about an axis passing through a point one-quarter of its length from the center of the rod and perpendicular to its length. Express the answer in terms of the mass and length of the rod.

11–4. A slim rod has a density which varies proportionally to the distance from one end. The density at one end is twice that at the other. Thus the density at a point x is $\rho = \rho_0[1 + (x/L)]$. By integration calculate an expression for the moment of inertia of this rod about an axis passing through one end (the one of smaller density) and perpendicular to the length of the rod. Express your answer in terms of

M, the mass of the rod, and L, its length.

11–5. The inner radius of a hollow cylinder is 3 inches, the outer radius is 4 inches, and the length is 6 inches. What is the radius of gyration of the cylinder about its axis?

11–6. What is the moment of inertia of a sphere about an axis tangent to the sphere?

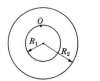

FIGURE 11–14   FIGURE 11–15

R and M, where M is the mass of the remaining portion of the disk.

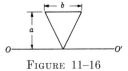

FIGURE 11–16

11–10. Find the moment of inertia about the axis OO' of a thin triangular board of uniform density, with height a and base b, and total mass m (Fig. 11–16). Express the answer in terms of the mass and dimensions of the board.

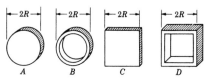

FIGURE 11–13

11–7. The four bodies shown in Fig. 11–13 have equal masses M. Body A is a solid cylinder of radius R. Body B is a hollow thin cylinder of radius R. Body C is a solid square with length of side $= 2R$. Body D is the same size as C, but hollow (i.e., made up of four thin sticks). The bodies have axes of rotation perpendicular to the page and through the center of mass of each body. (a) Which body has the smallest moment of inertia? Which body has the largest moment of inertia? (b) Calculate the moment of inertia of body D, about the given axis assuming the rim is very thin compared to R. Give the result in terms of total mass M and length R.

11–8. (a) Compute by integration the moment of inertia I_0 of a hollow cylinder of mass M, inner radius R_1, outer radius R_2, about a longitudinal axis through the center of the cylinder. (b) What is the moment of inertia of the cylinder about an axis through point O, Fig. 11–14, parallel to the first axis?

11–9. A circular disk of *diameter* R is cut from a solid disk of *radius* R as in Fig. 11–15. Find the moment of inertia of the remaining portion of the large disk, about an axis through point O perpendicular to its plane, in terms of

FIGURE 11–17

11–11. Derive by integration a formula for the moment of inertia of a semicircular disk of mass M and radius R about an axis perpendicular to the disk and passing through the point O (Fig. 11–17).

11–12. A grinding wheel 6 inches in diameter, weighing 4 lb, is rotating at 3600 rpm. What is its kinetic energy? How far would it have to fall to acquire the same kinetic energy?

11–13. The flywheel of a motor weighs 640 lb and has a radius of gyration of 4 ft. The motor develops a constant torque of 1280 lb·ft, and the flywheel starts from rest. (a) What is the angular acceleration of the flywheel? (b) What will be its angular velocity after making 4 revolutions? (c) How much work is done by the motor during the first 4 revolutions?

11–14. The flywheel of a gasoline engine is required to give up 380 ft·lb of kinetic energy while its angular velocity decreases from 600 rpm to 540 rpm. What moment of inertia is required?

11–15. A grindstone in the form of a solid cylinder weighs 80 lb and is 2 ft in diameter. What force applied at right angles to the end of a crank 9 in long will bring it up to an angular velocity of 120 rpm in 5 sec?

11–16. The flywheel of a punch press has a moment of inertia of 15 slug·ft^2 and it runs at 300 rpm. The flywheel supplies all the energy needed in a quick punching operation. (a) Find the speed in rpm to which the flywheel will be reduced by a sudden punching operation requiring 4500 ft·lb of work. (b) What must be the constant power supply to the flywheel in horsepower to bring it back to its initial speed in 5 sec?

11–17. A constant torque of 20 newton-meters is exerted on a pivoted wheel for 10 sec, during which time the angular velocity of the wheel increases from zero to 100 rpm. The external torque is then removed and the wheel is brought to rest by friction in its bearings in 100 sec. Compute (a) the moment of inertia of the wheel, (b) the friction torque, (c) the total number of revolutions made by the wheel.

11–18. A flywheel whose moment of inertia is 1.5 slug·ft^2 is rotating at 1800 rpm. A variable retarding torque given by $\tau = 0.18\,\pi t^2$ (where τ is in lb·ft and t is in sec) is exerted on the wheel. (a) How long a time is required to bring the wheel to rest? (b) How many revolutions does it make during this time?

11–19. A dumbbell consists of two solid spheres, each of radius 10 cm and mass 1000 gm, connected by a light rigid rod so that their centers are 30 cm apart. A vertical axis is passed through the center of one of the spheres, perpendicular to the rod (Fig. 11–18). (a) What is the moment of inertia of the dumbbell around this vertical axis? (b) A torque of 98,000 cm·dyne is ap-

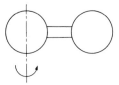

FIGURE 11–18

plied to the dumbbell, which is initially at rest. What is the angular velocity after one complete revolution?

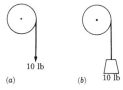

10 lb

(a)   (b)   10 lb

FIGURE 11–19

11–20. A cord is wrapped around the rim of a flywheel 2 ft in radius and a steady pull of 10 lb is exerted on the cord as in Fig. 11–19(a). The wheel is mounted in frictionless bearings on a horizontal shaft through its center. The moment of inertia of the wheel is 2 slug·ft^2. (a) Compute the angular acceleration of the wheel. (b) Show that the work done in unwinding 20 ft of cord equals the gain in kinetic energy of the wheel. (c) If a 10-lb weight hangs from the cord as in Fig. 11–19(b), compute the angular acceleration of the wheel. Why is this not the same as in part (a)?

11–21. A flywheel of mass 400 kgm and radius of gyration 0.5 meter is accelerated from rest by a motor which develops a constant power of 2 kilowatts. (a) How long a time is required to reach an angular velocity of 600 rpm? (b) How many revolutions does the flywheel make during this time?

11–22. The Atwood machine shown in Fig. 11–20 has a solid cylindrical pulley of mass 4 slugs. The masses m_1 and m_2 are 5 slugs and 3 slugs, respectively, and $R = 0.5$ ft. Find (a) the acceleration of mass m_1, (b) the ten-

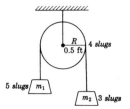

FIGURE 11–20

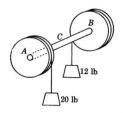

FIGURE 11–22

sion in the cord connecting the two weights and (c) the tension in the cord supporting the pulley.

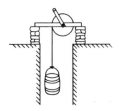

FIGURE 11–21

11–23. A bucket of water weighing 64 lb is suspended by a rope wrapped around a windlass in the form of a solid cylinder 1 ft in diameter, also weighing 64 lb. The bucket is released from rest at the top of a well and falls 64 ft to the water. (a) What is the tension in the rope while the bucket is falling? (b) With what velocity does the bucket strike the water? (c) What was the time of fall? Neglect the weight of the rope.

11–24. A 16-lb block rests on a horizontal frictionless surface. A cord attached to the block passes over a pulley, whose diameter is 6 inches, to a hanging block which also weighs 16 lb. The system is released from rest, and the blocks are observed to move 16 ft in 2 sec. (a) What was the moment of inertia of the pulley? (b) What was the tension in each part of the cord?

11–25. In Fig. 11–22 disks A and B, each 2 ft in radius and weighing 64 lb, are rigidly attached at the ends of shaft C, which is supported in a horizontal position in frictionless bearings (not shown). Cords are wrapped around the

outside of each disk, with a 12-lb weight hanging from one and a 20-lb weight hanging from the other. The moment of inertia of the shaft can be neglected. The system is released from rest. (a) Find the linear velocity of each weight after the 20-lb weight has descended 6 ft. (b) Find the torque exerted by shaft C at the instant after the motion starts. (c) Will this torque increase, decrease, or remain constant as the motion proceeds?

11–26. A slender vertical rod of length L and mass M, pivoted about an axis through its lower end perpendicular to its length, is released from rest in a vertical position. (a) Find its angular velocity when it reaches a horizontal position. (b) Find its angular acceleration in the horizontal position. (c) Find the horizontal and vertical components of the acceleration of its center of mass in the horizontal position. (d) Find the horizontal and vertical components of the force exerted on the rod by the pivot, in the horizontal position. Show these forces in a diagram.

11–27. A light rigid rod 100 cm long has a small block of mass 50 gm attached at one end. The other end is pivoted and the rod rotates in a vertical circle. At a certain instant the rod makes an angle of 53° with the vertical, and the tangential speed of the block is 400 cm/sec. (a) What are the horizontal and vertical components of the velocity of the block? (b) What is the moment of inertia of the system? (c) What is the radial acceleration of the block? (d) What is the tangential acceleration of the block? (e) What is the tension or compression in the rod?

FIGURE 11–23

11–28. A wheel of radius 1 ft consists of a 10-lb rim and one 10-lb spoke (see Fig. 11–23). (a) What is the moment of inertia of the wheel about a perpendicular axis through the center? (b) Where is the center of mass of the wheel? (c) If the wheel is pivoted at the center and released from a position in which the spoke is horizontal, what is the angular velocity of the wheel when the spoke reaches the bottom position? (d) What is the force exerted on the axis of the wheel when the spoke is horizontal? (e) What is the force on the axis when the spoke is at the bottom? (f) What is the force on the axis when the spoke is at a 45° position?

11–29. A disk of mass M and radius R is pivoted about a horizontal axis through its center, and a small body also of mass M is attached to the rim of the disk. If the disk is released from rest with the small body at the end of a horizontal radius, find: (a) the angular velocity when the small body is at the bottom, (b) the force exerted on the disk by the pivot at the same instant.

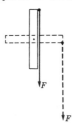

FIGURE 11–24

11–30. A bar of mass M and length L is pivoted at its center. A constant vertical force $F = Mg/2$ is exerted by a cord attached to the top end of the bar. The system is released from rest when the bar is in the vertical position (Fig. 11–24). (a) Find the linear velocity of the end of the bar at the instant that the bar is in a horizontal position.

(b) What are the vertical and horizontal components of force exerted on the bar by the pivot at the same instant?

FIGURE 11–25

11–31. A solid disk of mass M and radius R is pivoted about a horizontal axis through the point A on its rim (Fig. 11–25). It is released from rest with its center at the same height as point A. (a) Immediately after it is released, what is the tangential acceleration of the point B, opposite point A? (b) What is the angular velocity of the disk when it reaches the lowest point in its swing?

11–32. If the moment of inertia of the spool in Fig. 7–9 about an axis through its center is I_0, and the radius of the central shaft is r, find the acceleration with which the free ends of the cords must be moved upward in order that the center of mass of the spool shall remain at rest.

11–33. Show that the axis about which the moment of inertia of a body is a minimum must pass through the center of mass.

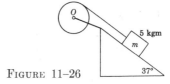

FIGURE 11–26

11–34. A block of mass $m = 5$ kgm slides down a surface inclined 37° to the horizontal as shown in Fig. 11–26. The coefficient of sliding friction is 0.25. A string attached to the block is wrapped around a flywheel on a fixed axis at O. The flywheel has a mass $M = 20$ kgm, an outer radius $R = 0.2$ m, and a radius of gyration with respect to the axis $k_0 = 0.1$ m. (a) What is the acceleration of the block down the plane? (b) What is the tension in the string?

CHAPTER 12

ROTATION AND TRANSLATION

12–1 The general equations of motion. We have shown in Chapter 7 that when a body is acted on by a resultant external force F, the acceleration of the center of mass of the body is given by the equation $F = m\bar{a}$, or in component form,

$$\Sigma X = m\bar{a}_x, \qquad \Sigma Y = m\bar{a}_y. \qquad (12\text{–}1)$$

We have also shown in Chapter 11 that when a body is pivoted about a fixed axis, its angular acceleration is given by

$$\Sigma \tau = I\alpha, \qquad (12\text{–}2)$$

where $\Sigma \tau$ is the resultant torque about the fixed axis and I is the moment of inertia about the same axis.

In this chapter we are to combine these relations to obtain the general equations of motion of a rigid body, whether it is pivoted about an axis or not. The body may be acted on by any number of external forces, having arbitrary directions, and applied at arbitrary points of the body. We shall restrict the discussion to co-planar forces, but the principles involved are the same in any case. By the use of methods already explained, the set of external forces can be combined to obtain their resultant, and when this is done we find that there are three, and only three, possible cases to consider, namely, the original set of forces reduces to either (1) a single force whose line of action passes through the center of mass, (2) a couple, i.e., two equal and oppositely directed forces not in the same straight line, or (3) a single force whose line of action does not pass through the center of mass.

Case (1), that of a single force whose line of action passes through the center of mass, we have already discussed in Chapter 7 and have shown that the motion is one of pure translational acceleration. We shall consider next case (2), where the set of external forces reduces to a couple, and then show that case (3), a single force not passing through the center of mass, is simply a combination of cases (1) and (2).

Figure 12–1(a) represents a body of irregular shape on which is exerted a couple consisting of the equal and opposite forces of magnitude P, located

218

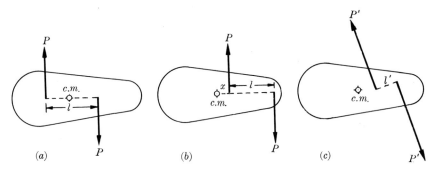

Fig. 12-1. Effect of a couple.

symmetrically with respect to the center of mass, and separated by the perpendicular distance l. It is obvious that ΣX and ΣY are both zero and therefore from Eq. (12-1) the acceleration of the center of mass of the body is zero. Hence, if the body is initially at rest, *its center of mass remains at rest.*

The resultant torque τ_0 about an axis through the center of mass is $P \times l/2 + P \times l/2 = P \times l$, and therefore

$$\Sigma \tau_0 = P \times l = I_0 \alpha.$$

Since the center of mass remains at rest, the motion will be one of angular acceleration about an axis through the center of mass, *even though there is no physical axis through the body at this point.*

In Fig. 12-1(b) the couple is applied "off center." ΣX and ΣY are still zero, and the center of mass remains at rest. The torque about the center of mass is $P \times (l + x) - P \times x = P \times l$, and hence the angular acceleration is the same as in Fig. 12-1(a). The location of the couple relative to the center of mass is immaterial—the angular acceleration depends only on the product $P \times l$. Hence if the couple $P' \times l'$ in Fig. 12-1(c) is exerted on the body, the angular acceleration will still be the same as in (a) and (b) provided only that $P' \times l' = P \times l$.

To summarize, then, we may say that the sole effect of a couple is to produce an angular acceleration about an axis through the center of mass. No physical axis for the body to rotate about needs to be provided, since the couple produces no acceleration of the center of mass.

We shall next show that a single force whose line of action does not pass through the center of mass is equivalent to a single force through the center of mass, combined with a couple. The force F, in Fig. 12-2, represents the resultant of the set of external forces. Let us apply at the center of mass the forces F' and F'', each equal in magnitude to F and parallel

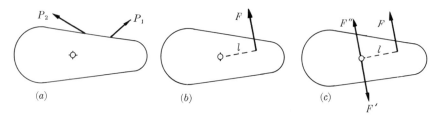

FIG. 12–2. Any force can be replaced by an equal force through the center of mass, together with a couple.

and antiparallel to it. The resultant of F' and F'' is zero, so these forces do not affect the motion of the body.

The forces F and F' constitute a couple of moment $F \times l$. The original set of forces therefore reduces to the force F'' (equal to F) acting at the center of mass, and the couple of moment $F \times l$. Hence the motion of the body can be broken down into two parts. One is a linear acceleration of the center of mass, produced by the force F''. The other is an angular acceleration about an axis through the center of mass produced by the couple FF'. The force F'' is equal to the resultant force F, and the moment of the couple, $F \times l$, is equal to the moment of the resultant force about an axis through the center of mass. The equations of motion are therefore

$$F = m\bar{a}, \quad \text{or} \quad \Sigma X = m\bar{a}_x, \quad \Sigma Y = m\bar{a}_y \quad (12\text{–}3)$$

and

$$\Sigma \tau_0 = I_0 \alpha. \quad (12\text{–}4)$$

F is the resultant of the set of forces, ΣX and ΣY are its X and Y components, and $\Sigma \tau_0$ is the moment of the original set of forces about an axis through the center of mass.

These two effects of the external forces, translation and rotation, can be treated independently. That is, the rotational effect can be ignored and every point in the body assumed to acquire a linear acceleration equal to that of the center of mass. Second, the translational effect can be ignored and every point assumed to acquire an angular acceleration about an axis through the center of mass.

The multiflash photographs on page 107 illustrate these two effects of an "off-center" force. In (a) and (b) the rod receives both a linear and an angular acceleration. In (c), where the line of action of the force passes through the center of mass, all points receive a linear acceleration only. The motion in (a) and (b) can be considered to be that of (c), on which an angular motion is superposed.

The kinetic energy of the body can also be considered as made up of two parts, kinetic energy of translation assuming each point to have the same linear velocity as the center of mass, and kinetic energy of rotation about an axis through the center of mass.

$$KE = \tfrac{1}{2}mv^2 + \tfrac{1}{2}I_0\omega^2.$$

EXAMPLE. A horizontal force P, Fig. 12–3, is exerted on a rod of mass m and length L at a distance from one end equal to one-quarter of the length of the rod. The rod rests on a horizontal frictionless surface. Find the initial linear acceleration of the center of mass, the angular acceleration about an axis through the center of mass, and the linear accelerations of the ends of the rod.

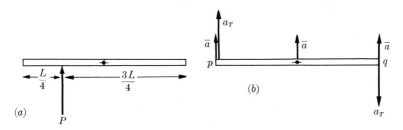

FIGURE 12–3

The linear acceleration of the center of mass is the same as though the force P were applied at that point. Hence

$$\bar{a} = \frac{P}{m}.$$

The resultant torque about the center of mass is

$$\tau_0 = P \times \frac{L}{4},$$

and since the moment of inertia about the center is

$$I_0 = \tfrac{1}{12}\, mL^2,$$

the angular acceleration is

$$\alpha = \frac{\tau_0}{I_0} = \frac{3}{L}\frac{P}{m}.$$

The accelerations of the ends and the center of the rod are shown in Fig. 12–3(b). Every point on the rod has a component of linear acceleration $\bar{a}$, equal to that of the center of mass. In addition, every point has a tangential acceleration given by $a_T = r\alpha$, where α is the angular acceleration and r is the distance of that point from the center. The tangential acceleration of the center is, of course, zero, while that of the ends is

$$a_T = \frac{L}{2}\,\alpha = \frac{L}{2} \times \frac{3}{L}\frac{P}{m} = \frac{3}{2}\frac{P}{m} = \frac{3}{2}\,\bar{a}.$$

$M\,a_T = M R \dfrac{d\omega}{dt} \qquad \underline{T O R q u e}$

The resultant acceleration of end p is therefore

$$\bar{a} + a_T = \frac{5}{2}\frac{P}{m},$$

while that of end q is

$$\bar{a} - a_T = -\frac{1}{2}\frac{P}{m}.$$

12-2 Rolling. The motion of a body along a surface on which it rolls without slipping affords a number of illustrations of the general equations of motion. The condition that a body shall roll without slipping imposes a definite relation on the relative values of its linear and angular displacement, velocity, and acceleration. Consider a cylinder of radius R rolling

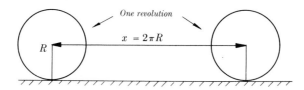

FIG. 12–4. Linear and angular displacement of a rolling body.

on a plane without slipping, as in Fig. 12–4. The cylinder must make one complete revolution in the time that its center advances a distance equal to the circumference of the cylinder. That is,

$$\theta = 2\pi, \qquad x = 2\pi R,$$

or,

$$x = R\theta.$$

If the cylinder rolls with constant angular velocity ω, while its center moves with constant linear velocity v, and t is the time to make one revolution,

$$\theta = \omega t = 2\pi, \qquad x = vt = 2\pi R,$$

and

$$v = R\omega.$$

If the cylinder starts from rest and rolls with constant angular acceleration α, while its center moves with constant linear acceleration a, then

$$\theta = \tfrac{1}{2}\alpha t^2 = 2\pi, \qquad x = \tfrac{1}{2}at^2 = 2\pi R,$$

and

$$a = R\alpha.$$

Although derived for the special cases of constant velocity and constant acceleration, the equations above are true in general. That is, when a body rolls without slipping, the relations between the linear displacement, velocity, and acceleration of the center of the body and the angular displacement, velocity, and acceleration about the center, are of the same form as those connecting tangential and angular quantities when the rotation takes place about a fixed axis.

EXAMPLES. (1) Figure 12–5(a) represents a solid cylinder of mass m and radius R being pulled along a horizontal surface, on which it rolls without slipping, by a horizontal force P applied at its center. Find the linear acceleration of the center of mass.

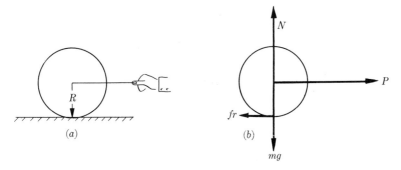

FIGURE 12–5

It is evident that sufficient friction must exist between the cylinder and surface in order that the cylinder may accelerate without slipping. If the surfaces were frictionless, the cylinder would simply slide. The forces on the cylinder are shown in Fig. 12–5(b). From the general equations of motion,

$$\Sigma X = P - fr = m\bar{a}_x,$$

$$\Sigma Y = N - mg = m\bar{a}_y = 0,$$

$$\Sigma \tau_0 = fr \times R = I_0\alpha.$$

From the condition for rolling without slipping

$$\bar{a}_x = R\alpha.$$

The moment of inertia of a solid cylinder about an axis through its center of mass is $\frac{1}{2}mR^2$, and on combining the preceding equations we find

$$\bar{a}_x = \frac{2}{3}\frac{P}{m}.$$

That is, the acceleration is only $\frac{2}{3}$ as great as if the surface were frictionless.

(2) If the mass of the cylinder is $\frac{1}{2}$ slug, its radius 6 in, and the applied force P is 4 lb, find the velocity of point p, Fig. 12–6, 3 sec after the motion begins.

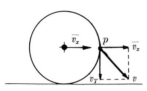

FIG. 12–6. The velocity v of point p is the resultant of the translational velocity $\bar{v}_x$ and the tangential velocity v_T about the center of mass.

When these numerical values are introduced in the results of the preceding example we get

$$\bar{a}_x = 5\tfrac{1}{3} \text{ ft/sec}^2,$$

$$\alpha = 10\tfrac{2}{3} \text{ rad/sec}^2,$$

$$fr = \tfrac{4}{3} \text{ lb.}$$

After 3 sec, the velocity of the center of mass is

$$\bar{v}_x = \bar{a}_x t = 16 \text{ ft/sec}$$

and the angular velocity is

$$\omega = \alpha t = 32 \text{ rad/sec.}$$

Point p, in common with every other point in the cylinder, has a velocity component equal to $\bar{v}_x$. In addition, every point has a tangential velocity due to the rotation about the moving center of mass. For point p the magnitude of the tangential velocity is

$$v_T = R\omega = 16 \text{ ft/sec,}$$

and its direction is vertically downward.

The resultant velocity of p is therefore $16\sqrt{2}$ ft/sec at an angle of 45° below the horizontal.

(3) Find the acceleration of the center of mass of a solid cylinder which rolls without slipping down a plane inclined at an angle θ with the horizontal (Fig. 12–7).

Comparison of Figs. 12–7 and 12–5 will show that the situation here is essentially the same as in Example (1). The force $mg \sin \theta$ corresponds to P, and $mg \cos \theta$ corresponds to mg. From the general equations of motion,

$$\Sigma X = mg \sin \theta - fr = m\bar{a}_x,$$

$$\Sigma Y = N - mg \cos \theta = m\bar{a}_y,$$

$$fr \times R = I_0 \alpha,$$

and

$$\bar{a}_x = R\alpha.$$

Since

$$I_0 = \tfrac{1}{2} m R^2,$$

$$a_x = \tfrac{2}{3} g \sin \theta.$$

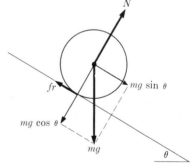

FIGURE 12–7

That is, the acceleration of the center of mass is only two-thirds as great as that of a body *sliding* down a frictionless plane of the same slope.

(4) The same result can be obtained from energy considerations. No work is done on the cylinder as it rolls down the plane and no heat is developed. (Although there is a friction force, there is no sliding along the surface.) Hence the

kinetic energy of the cylinder at the foot of the incline, if it starts from rest, equals its loss of potential energy as it descends. The kinetic energy must now be thought of as consisting of two parts—kinetic energy of translation, $\frac{1}{2}m\bar{v}^2$, and kinetic energy of rotation about the center of mass, $\frac{1}{2}I_0\omega^2$.

In descending a distance x along the plane, (Fig. 12–8) the center of gravity drops vertically a distance $h = x \sin \theta$. The loss of potential energy is therefore $mgx \sin \theta$. The kinetic energy at the bottom is $\frac{1}{2}m\bar{v}^2 + \frac{1}{2}I_0\omega^2$. Therefore

$$mgx \sin \theta = \tfrac{1}{2}m\bar{v}^2 + \tfrac{1}{2}I_0\omega^2$$

and since $I_0 = \frac{1}{2}mR^2$ and $\omega^2 = \bar{v}^2/R^2$,

$$mgx \sin \theta = \tfrac{1}{2}m\bar{v}^2 + \tfrac{1}{2} \times \tfrac{1}{2}mR^2 \times \frac{\bar{v}^2}{R^2}$$

$$= \tfrac{3}{4}m\bar{v}^2,$$

$$\bar{v}^2 = \tfrac{4}{3}gx \sin \theta. \qquad (12\text{–}5)$$

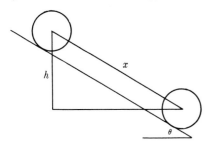

FIGURE 12–8

Let $\bar{a}_x$ be the linear acceleration of the center of mass. It follows from the equations of motion with constant acceleration that

$$\bar{v}^2 = 2\bar{a}_x x. \qquad (12\text{–}6)$$

From Eqs. (12–5) and (12–6) we find

$$\bar{a}_x = \tfrac{2}{3}g \sin \theta,$$

which agrees with the result previously derived.

12–3 Instantaneous axis. The viewpoint of the preceding sections is that the general motion of a body can be considered as compounded of two motions—one a translation of the body as a whole and the other a rotation about an axis through the center of mass. It is also possible to consider any type of motion as a succession of *pure rotations only*, but about an axis which in general does not pass through the center of mass, and may not even pass through the body. This axis is called the *instantaneous axis*.

To see how the position of the instantaneous axis is determined, let us first consider Fig. 12–9(a), which shows a cylinder rotating about a *fixed* axis through its center O. The velocities of points P, Q, and R can be found by drawing the radii r_1, r_2, and r_3 from O to P, Q, and R, and then constructing the velocity vectors v_P, v_Q, and v_R at right angles to these radii and of lengths $v_P = \omega r_1$, $v_Q = \omega r_2$, and $v_R = \omega r_3$.

Conversely, if the directions of the velocities of any two points such as P and Q were given, the position of the axis could be determined by constructing lines through P and Q at right angles to these velocities and find-

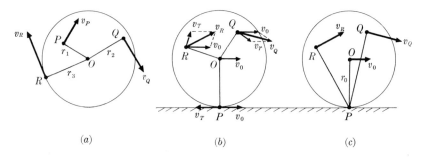

Fig. 12–9. (a) Velocities of points in a body rotating about a fixed axis. (b) Velocities of points in a rolling body. (c) The instantaneous axis passes through point P.

ing their point of intersection. The magnitude of the angular velocity could then be found if the magnitude of the velocity of any one point were known, together with its distance from the axis. Thus if v_P and r_1 are known,

$$\omega = \frac{v_P}{r_1}.$$

Consider now a cylinder rolling on a plane without slipping, as in Fig. 12–9(b) and (c). The velocity of the center of mass is v_0. The velocities of points P, Q, and R have been found in Fig. 12–9(b) by the method illustrated in Fig. 12–6, that is, by finding the resultant of the velocity v_0 and the tangential velocity v_T about an axis through the center of mass. Note that the velocity of point P is zero.

To avoid confusion, the same velocities are shown again in Fig. 12–9(c). Now if we use the method described above for locating the axis of a rotating body, that is, if we construct lines through the points O, Q, and R at right angles to the velocities of these points, we find that all such lines intersect at point P. In other words, the motion of the rolling cylinder, at any instant, is the same as if it were rotating about an axis through P perpendicular to the plane of the diagram. This axis is the *instantaneous axis*. It differs from a fixed axis, or from an axis through the moving center of mass, in that those points of the body which lie on the instantaneous axis are continually changing. Thus in Fig. 12–10, where the line a–a' is the line of tangency between cylinder and surface at the instant shown, all points on the line a–a' lie on the instantaneous axis at this instant. A moment later line b–b' will make contact with the surface and, at the instant it does so, all points on this line lie on the instantaneous axis. At a later moment points on c–c' lie on the instantaneous axis, and so on.

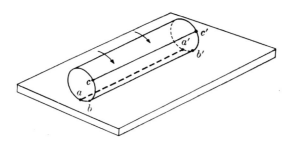

Fɪɢ. 12–10. As a body rolls, the instantaneous axis passes through different points on its surface.

We next show that the angular velocity (and angular acceleration) about the instantaneous axis are the same as about a moving axis through the center of mass. It was pointed out above, in connection with rotation about a fixed axis, that the angular velocity could be found if the linear velocity of any one point were known, together with its distance from the axis. But in Fig. 12–9(c) we know that the velocity v_0 of the center of mass, for a body rolling without slipping, is

$$v_0 = \omega r_0,$$

where r_0 is the outside radius of the cylinder. Also, the distance of the center of mass from the instantaneous axis through P is r_0. Let ω' represent the angular velocity about P. Then

$$\omega' = \frac{v_0}{r_0} = \frac{\omega r_0}{r_0} = \omega,$$

and the angular velocity about the instantaneous axis equals that about a moving axis through the center of mass. Since the angular velocities are equal, the angular accelerations (the time derivatives of the angular velocities) are equal also.

Since the motion of the body at any instant is one of pure rotation about the instantaneous axis, it follows that the resultant torque *about the instantaneous axis* equals the product of the angular acceleration and the moment of inertia *about the instantaneous axis*. Hence, instead of the three equations (12–3) and (12–4), the motion can be expressed by a single equation,

$$\tau_{IA} = I_{IA}\alpha,$$

where the subscript IA means that the torque and moment of inertia are computed about the instantaneous axis rather than about an axis through the center of mass. This simplifies the solution of many problems.

The kinetic energy can also be considered as purely rotational, about the instantaneous axis. Thus instead of writing

$$KE = \tfrac{1}{2}mv_0^2 + \tfrac{1}{2}I_0\omega^2,$$

we have the single equation

$$KE = \tfrac{1}{2}I_{IA}\omega^2.$$

EXAMPLE. Compute the acceleration of a solid cylinder rolling down an inclined plane. Consider the motion to be pure rotation about an instantaneous axis. Refer to Fig. 12–7.

The torque about the instantaneous axis is

$$\tau = mgR \sin\theta.$$

The moment of inertia about this axis is

$$I = I_0 + mR^2 = \tfrac{3}{2}mR^2.$$

Hence, from $\tau = I\alpha$,

$$\alpha = \frac{2}{3}\frac{g\sin\theta}{R}.$$

The acceleration of the center of mass is now considered as a tangential acceleration about the instantaneous axis. Hence it is

$$\bar{a} = R\alpha,$$

and therefore

$$\bar{a} = \tfrac{2}{3}g\sin\theta,$$

which agrees with the result on page 224.

It will be left as an exercise to show that the same expression is obtained from energy considerations, treating the entire kinetic energy as rotational about the instantaneous axis.

12–4 Angular momentum and angular impulse. Figure 12–11(a) represents a small body of mass m moving in the plane of the diagram with a velocity v and a momentum mv. We define its *moment of momentum*, about an axis through O perpendicular to the plane of the diagram, as the product of its linear momentum and the perpendicular distance from the axis to its line of motion. That is,

moment of momentum $= mvr.$

It will be seen that moment of momentum is defined in the same way as the moment of a force. Moment of momentum is also called *angular momentum*.

Figure 12–11(b) represents a body of finite size rotating in the plane of the diagram about an axis through O. The velocity v of a small element of the body is related to the angular velocity of the body by $v = \omega r$. The

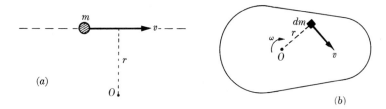

FIG. 12–11. Moment of momentum.

angular momentum of the element is therefore $vr\,dm = \omega r^2\,dm$, and the total angular momentum of the body is $\int \omega r^2\,dm = \omega \int r^2\,dm$. But $\int r^2\,dm$ is the moment of inertia of the body about its axis of rotation. Hence the angular momentum can be written as $I\omega$. In this form it is completely analogous to linear momentum mv.

The general form of Newton's second law (see page 160) is: *resultant force equals rate of change of momentum*. The analogue of this law for angular motion is: *resultant torque equals rate of change of angular momentum*. More precisely, the resultant torque about any axis equals the rate of change of angular momentum about that axis.

$$\tau = \frac{d}{dt}(I\omega). \tag{12–9}$$

If the moment of inertia is constant, this equation reduces to

$$\tau = I\frac{d\omega}{dt} = I\alpha,$$

just as the equation $F = (d/dt)(mv)$ reduces to $F = ma$ when the mass m is constant.

Equation (12–9) may be written

$$\tau\,dt = d(I\omega), \tag{12–10}$$

$$\int_{t_0}^{t} \tau\,dt = I\omega - (I\omega)_0. \tag{12–11}$$

The integral of a torque over the time interval it acts is called the *angular impulse* of the torque. (Compare with the definition of the impulse of a force on page 152.) In terms of angular impulse we may state Eq. (12–11) as: *the angular impulse about any axis equals the change in angular momentum about that axis.*

If angular momentum is represented by a single letter G, Eqs. (12–10) and (12–11) become

$$\tau \, dt = dG, \tag{12–12}$$

$$\int_{t_0}^{t} \tau \, dt = G - G_0. \tag{12–13}$$

It follows at once from Eq. (12–13) that *if the resultant external torque on a system is zero, the angular momentum of the system remains constant,* and hence any interaction between the parts of a system cannot alter its total angular momentum. This is the principle of the *conservation of angular momentum,* and it ranks with the principles of conservation of linear momentum and conservation of energy as one of the most fundamental of physical laws.

EXAMPLES. (1) A slender rod of mass m and length L rests on a horizontal frictionless surface. A pivot passes through one end of the rod. Compute the impulse of the force exerted on the rod by the pivot when the rod is struck a blow of impulse J, perpendicular to the rod, at the other end (Fig. 12–12).

Let J' represent the unknown impulse at the pivot. Since impulse equals change in linear momentum,

$$J + J' = m\bar{v},$$

where $\bar{v}$ is the linear velocity acquired by the center of mass. Also, since angular impulse equals change in angular momentum,

$$JL = I\omega,$$

where I is the moment of inertia of the rod about an axis through the pivot and ω is the angular velocity acquired. But

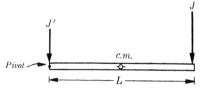

FIGURE 12–12

$$I = \frac{1}{3}mL^2 \text{ and } \bar{v} = \omega \times \frac{L}{2}.$$

Hence

$$J = \frac{I\omega}{L} = \frac{2}{3} \, m\bar{v},$$

or

$$m\bar{v} = \tfrac{3}{2}J.$$

When this value of $m\bar{v}$ is inserted in the first equation we get

$$J' = m\bar{v} - J = \tfrac{3}{2}J - J = \tfrac{1}{2}J.$$

Hence the impulse exerted on the rod at the pivot is one-half the impulse of the blow.

Note that the impulse J' exists only during the short interval while the rod is being struck. The forces at the pivot *after* the rod begins to rotate are found by the methods explained in Section 11–5.

(2) A man stands at the center of a turntable, holding his arms extended horizontally with a 10-lb weight in each hand. He is set rotating about a vertical axis with an angular velocity of one revolution in 2 sec. Find his new angular velocity if he drops his hands to his sides. The moment of inertia of the man may be assumed constant and equal to 4 slug·ft². The original distance of the weights from the axis is 3 ft, and their final distance is 6 in.

If friction in the turntable is neglected, no external torques act about a vertical axis and the angular momentum about this axis is constant. That is,

$$I\omega = (I\omega)_0 = I_0\omega_0,$$

where I and ω are the final moment of inertia and angular velocity, and I_0 and ω_0 are the initial values of these quantities.

$$I = I_{\text{man}} + I_{\text{weights}},$$

$$I = 4 + 2(\tfrac{10}{32})(\tfrac{1}{2})^2 = 4.3 \text{ slug·ft}^2,$$

$$I_c = 4 + 2(\tfrac{10}{32})(3)^2 = 9.6 \text{ slug·ft}^2,$$

$$\omega_0 = \pi \text{ rad/sec},$$

$$\omega = \omega_0 \frac{I_0}{I} = 2.2\pi \text{ rad/sec}.$$

That is, the angular velocity is more than doubled. This experiment is easily performed with a piano stool as a turntable. The results are most surprising.

12–5 Vector representation of angular quantities. It was mentioned briefly on page 174 that any quantity associated with an axis, such as angular velocity, angular acceleration, etc., could be represented by a vector along the axis. The sense of the vector is usually considered to be that in which a nut would advance along the axis if threaded on it with a right-hand thread and rotated in the direction of the angular quantity to be represented. Evidently torque, angular impulse, and angular momentum can all be represented in this way.

EXAMPLE. A couple consisting of the two forces P and P', each equal to 4 lb, is applied for 2 sec to a disk of radius 2 ft and moment of inertia 20 slug·ft², pivoted about an axis through its center as in Fig. 12–13. The initial angular velocity of the disk is 5 rad/sec. Show in a vector diagram the torque, the initial angular momentum, and the final angular momentum.

From the preceding discussion, the vectors representing the torque τ due to the couple, and the initial angular momentum G_0, are directed as in Fig. 12–13. The magnitude of the initial angular momentum is

$$G_0 = I\omega_0 = 20 \times 5 = 100 \text{ slug·ft}^2/\text{sec}.$$

Let us write Eq. (12–12) in the form

$$\tau \, \Delta t = \Delta G.$$

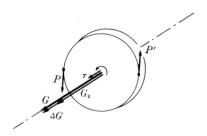

FIG. 12–13. Vector ΔG is the change in angular momentum produced by the couple $P - P'$.

Then

$$\Delta G = (4 \times 4) \times 2 = 32 \text{ slug·ft}^2/\text{sec}.$$

This increase has been represented by the vector ΔG in Fig. 12–13.

The final angular momentum, G, is the vector sum of G_0 and ΔG. Since both are in the same direction, the vector sum is simply the arithmetic sum. That is,

$$G = G_0 + \Delta G = 132 \text{ slug·ft}^2/\text{sec}.$$

12–6 Precession. The disk shown in Fig. 12–14 is rotating about a shaft which coincides with the X-axis. A couple consisting of the forces P and P' is applied to the *shaft*, and we wish to find the resulting motion of the disk. The problem will be solved by the same method used in the preceding example. The initial angular momentum is represented by the vector G_0. The effect of the forces P and P' is to produce a torque τ about the Z-axis, perpendicular to the X-Y plane in which the forces act. In the time interval Δt, the change in angular momentum produced by this torque is $\tau \Delta t$ and this change is represented by the vector ΔG in the figure. The final angular momentum is the vector sum of G_0 and ΔG, and is represented by the vector G.

It will be seen that the situation is exactly analogous to that of Fig. 12–13, the only difference being that in the first case the direction of the applied torque vector is the same as the initial angular momentum vector, so that the vectors G_0 and ΔG are in the same line, while in the second case the torque vector is perpendicular to the initial angular momentum vector, and hence G_0 and ΔG are at right angles to each other. The new angular momentum, G, will be seen to lie in the horizontal (X-Z) plane, but displaced from the original angular momentum by an angle $\Delta \theta$.

Finally, since an angular momentum vector in a given direction implies rotation in a plane perpendicular to that direction, it follows that the plane

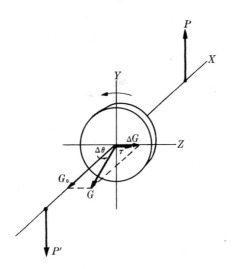

Fig. 12–14. Vector ΔG is the change in angular momentum produced by the couple $P - P'$. Compare with Fig. 12–13.

of the spinning disk must also have rotated through an angle $\Delta\theta$. That is, the vector G must lie along the new axis of rotation. This leads to the very unexpected result that the forces P and P', instead of forcing the further end of the shaft up and the nearer end down, as would be the case if the disk were not spinning, cause the further end of the shaft to move to the left and the nearer end to move to the right, perpendicular to the directions in which P and P' are acting.

If we assume that the forces P and P' follow the motion of the shaft as it turns in the X-Z plane, the shaft and spinning disk will rotate about the vertical Y-axis with a uniform angular velocity, say Ω (capital ω). This type of motion is called *precession*, and Ω is the precessional velocity.

Precessional effects have an important influence on the maneuvering of an airplane. Let the disk of Fig. 12–14 represent the motor and propeller of an airplane, the plane traveling along the X-axis toward the reader. If the pilot wishes to turn to the left, so as to change the direction of the axis of rotation from that of G_0 to that of G, the bearings of the motor and propeller system must exert a couple on the shaft equivalent to that produced by P and P'. From Newton's third law the shaft then exerts a couple on the bearings in such a direction as to push the nose of the plane up and its tail down, an effect which must be counteracted by adjustment of the elevators. It is easy to see that a turn to the right will produce the opposite effect, and that a change from level flight to climbing or descending will deflect the plane to the left or right.

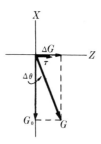

FIGURE 12–15

Figure 12–15 is a top view of the vector diagram of Fig. 12–14. This diagram is not strictly correct. At the first instant when P and P' start to act, the directions of the torque and the angular momentum change produced by it are along the Z-axis, as shown. But this change in angular momentum immediately causes a swing of the axis toward the direction of G, and if the forces move with the axis, the direction of the torque vector swings also. By the time the axis has moved to the direction of G, the torque vector is at right angles to this new direction. Hence Fig. 12–15 must be considered to apply to an extremely short time interval only, while the axis swings from its initial position through a very small angle $\Delta\theta$. If the angle is small, the length of G is practically the same as that of G_0, and the only effect of the change is to alter the direction of G_0 slightly. In other words, the couple $P - P'$ changes the direction of the angular momentum but not its magnitude. (This is exactly analogous to the effect produced by centripetal force in circular motion. The centripetal force changes the direction of the tangential velocity but not its magnitude.)

If the angle $\Delta\theta$ in Fig. 12–15 is small, its value in radians is very nearly

$$\Delta\theta = \frac{\Delta G}{G} = \frac{\tau\,\Delta t}{G}.$$

(Since G_0 and G are numerically equal, let G represent the magnitude of either.)

The precessional velocity, Ω, is

$$\Omega = \frac{\Delta\theta}{\Delta t} = \frac{\tau}{G},$$

or

$$\tau = \Omega G.$$

Finally, replacing G by $I\omega$, we obtain

$$\boxed{\tau = I\omega\Omega.} \qquad (12\text{–}14)$$

τ is the torque necessary to produce an angular velocity of precession Ω in a rotating system of angular momentum $I\omega$. Both Ω and ω are to be expressed in rad/sec, and τ and I in appropriate units.

Equation (12–19) may help to make clear, in part, the stabilizing properties of the gyroscope. If we write it as $\Omega = \tau/I\omega$, it is seen that for a given torque τ the precessional velocity Ω will be small if the product $I\omega$ is large, and this product can be made very large indeed, even in a gryoscope of small mass and dimensions, by running it at a high angular velocity ω. Hence, for a couple at right angles to its axis of spin, the gyroscope behaves like a body of very large inertia.

EXAMPLE. The moment of inertia of an airplane motor and propeller is 15 slug·ft^2 and the motor turns at 2000 rpm. Compute the torque causing the plane to nose up when it is flying in a horizontal circle of radius 1000 ft at a speed of 300 mi/hr.

The tangential velocity is 300 mi/hr = 440 ft/sec, and since $R = 1000$ ft,

$$\Omega = \frac{v_T}{R} = \frac{440}{1000} = 0.44 \text{ rad/sec.}$$

The angular momentum is

$$G = I\omega = 15 \times \frac{2000 \times 2\pi}{60}$$

$$= 3140 \text{ slug·ft}^2/\text{sec.}$$

Hence the torque is

$$\tau = I\omega\Omega = 1400 \text{ lb·ft.}$$

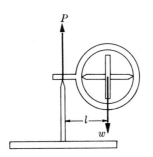

FIG. 12–16. The forces P and w, equal and oppositely directed, constitute a couple.

12–7 The gyroscope. Another illustration of precessional motion is afforded by the gyroscope. As well as being an interesting and puzzling toy, it finds important technical applications in the gyrocompass, the directional gyro, the artificial horizon, the turn indicator, and in the stabilization of ships.

The usual mounting of a toy gyroscope is shown in Fig. 12–16. The forces acting on the gyroscope are its weight w and the upward push of the pivot P. If these forces are equal they constitute a couple, and since the effect of a couple is the same wherever it may be applied, the arrangement of forces in Fig. 12–16 is entirely equivalent to that in Fig. 12–14. The gyroscope will not "fall" under these circumstances because the resultant vertical force acting on it is zero!

The torque produced by the couple is wl. Hence the precessional velocity is

$$\Omega = \frac{wl}{I\omega}.\tag{12-15}$$

If the gyroscope is held with its axis of rotation at rest, say by supporting the free end with the finger, the upward push P and the upward force exerted by the finger will each equal $w/2$. If the finger is suddenly removed, the forces on the gyroscope do not constitute a couple, since the downward force is w and the upward force only $w/2$. The center of mass of the gyroscope therefore starts to fall, and at the same time precession begins, but with a smaller precessional velocity than that given by Eq. (12–15), since the resultant torque is smaller. The effect of the motion of the gyroscope is such as to increase the magnitude of P to a value larger than w. This causes the axis to rise again, after which the motion repeats itself. The resulting motion is one of precession combined with an up and down oscillation of the axis of rotation. This motion is called *nutation* and its complete analysis is too lengthy to be carried out here.

If it is desired to start the gyroscope off with a motion of pure precession after releasing it, it is necessary to give the free end a push in the direction in which it will naturally precess. The effect of the horizontal push is to cause the pivoted end to bear down on the pivot, thus increasing the force P. At the instant when P has increased to equal w, the precessional velocity of the gyroscope will have reached its proper value. The outer end may then be released and the motion will continue.

The gyro-stabilizer, used to decrease the rolling motion of a ship, consists of a large motor-driven gyroscope spinning about a vertical axis and pivoted in trunnions so that the axis of spin may be tipped in a fore-and-aft-direction. Figure 12–17(a) shows a fore-and-aft, and Fig. 12–17(b) a transverse view of the stabilizer.

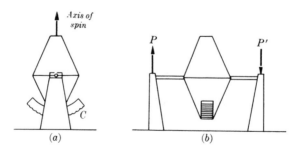

Fig. 12–17. The stabilizing gyroscope.

At the instant when a roll starts, a smaller gyroscope serving as a control closes the circuit to a motor which engages the sector C and forcibly tips the top of the gyroscope axis forward or aft. As a result of this precession, forces P and P' are exerted on the bearings, producing a couple in the proper direction to counteract the roll. The reaction of the sector C against the driving motor results in an equal couple which forces the bow of the ship up or down. The effect of the stabilizer is thus to transform a "roll" to a "pitch," but since the length of a ship enables it to offer a greater resistance to pitching than to rolling, the resulting motion is smaller than it otherwise would be.

PROBLEMS

12–1. The driving wheel of a locomotive is 6 ft in diameter and the locomotive is traveling at 60 ft/sec. Find the magnitude and direction of the velocity of the following points of the wheel: (a) its center, (b) the point of contact between wheel and rail, (c) the point at the top of the wheel, (d) a point at the end of a horizontal diameter, (e) a point on the flange 4 in below the instantaneous axis. (f) Locate a point on the rim of the wheel which has a velocity of the same magnitude as that of the center of the wheel. What is the direction of this velocity?

12–2. If the wheel in problem 12–1 weighs 1 ton and its radius of gyration about an axis through its center is 2.5 ft, compute (a) its kinetic energy of translation, (b) its kinetic energy of rotation about its center, (c) its moment of inertia about the instantaneous axis, (d) its kinetic energy of rotation about the instantaneous axis.

12–3. A symmetrical wheel of mass M, radius R, and radius of gyration k_0 about its center, rolls without slipping on a horizontal surface. The velocity of its center is v. (a) What is the kinetic energy of the wheel, considering its motion as a combined translation of the center of mass and rotation about an axis through the center of mass? (b) If k represents the radius of gyration

about the instantaneous axis, what is the expression for the kinetic energy of the wheel, considering its motion as pure rotation about the instantaneous axis? (c) Use the answers to (a) and (b) to prove that $k^2 = k_0^2 + R^2$.

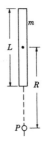

FIGURE 12–18

12–4. A uniform stick of length L and mass m undergoes pure rotation about a point P which is at a distance R from the center of mass of the stick, as shown in Fig. 12–18. (a) Considering the motion as pure rotation about P, determine the kinetic energy of the stick in terms of m, L, R, and the angular velocity of the stick about P. (b) Repeat the calculation of kinetic energy by considering the motion to be made up of translational motion of the center of mass of the stick plus rotation of the stick about its center of mass. (c)

Show that the two results are equivalent.

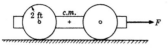

FIGURE 12–19

12–5. A cart has four disk-shaped wheels, each of radius 2 ft and weight 20 lb. The total weight of cart plus wheels is 200 lb, and its center of mass is at its center, as shown in Fig. 12–19. The cart is initially at rest when a constant horizontal force, F, whose line of action passes through the center of mass, starts pulling it forward. It moves 36 ft in 3 sec. (a) What force F is necessary if the surface is frictionless? (b) What force F is necessary if the coefficient of friction is large enough so that the wheels roll without sliding?

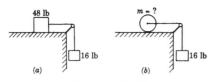

FIGURE 12–20

12–6. In part (a) of Fig. 12–20, a 16-lb block drags a 48-lb block along a horizontal frictionless surface. In part (b) the same 16-lb block pulls a solid cylinder along a level surface on which the cylinder rolls without slipping. The acceleration of the 16-lb block is the same in both diagrams. (a) Find the acceleration of the 16-lb block. (b) Find the tension in the cord. (c) Find the mass of the cylinder. Any effect of the small pulley can be neglected.

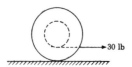

FIGURE 12–21

12–7. A wheel consists of two disks, each weighing 20 lb and of radius 6 inches joined by an axle of negligible mass and of radius 3 inches. The wheel is pulled along a horizontal plane by a force of 30 lb exerted by a string wrapped around the axle as shown in Fig. 12–21. The coefficients of static and sliding friction each equal 0.25. (a) Find the acceleration of the center of mass. (b) Find the angular acceleration. (c) Find the distance the center of mass moves during one revolution.

12–8. Show that the general expression for the acceleration of a symmetrical body of circular cross section, rolling down an inclined plane, is

$$a = \left(\frac{1}{1 + \dfrac{k_0^2}{R^2}}\right) g \sin \theta,$$

where R is the outside radius of the body and k_0 is the radius of gyration about an axis through the center of mass.

12–9. A solid cylinder rolls without slipping down a plane inclined 37° to the horizontal. (a) What is the minimum coefficient of friction between cylinder and surface which will allow the cylinder to roll without slipping? (b) If the cylinder starts from rest, how far must it roll down the plane to acquire a velocity of 280 cm/sec?

12–10. A solid cylinder is placed on end on an inclined plane. It is found that the plane can be tipped at an angle θ before the cylinder starts to slide. When turned on its side and allowed to roll down the plane, it is found that ϕ is the steepest angle at which the cylinder will roll without slipping. Compute the ratio of $\tan \phi$ to $\tan \theta$.

12–11. A block slides down a frictionless inclined plane, starting from rest. A solid cylinder rolls the same distance down a plane of similar geometry without slipping. (a) Find the ratio of the velocity of the block to the velocity of the center of mass of the cylinder when each body has reached the bottom of the plane. (b) What is the ratio of the final kinetic energies of the two bodies?

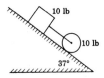

FIGURE 12–22

12–12. A 10-lb solid cylindrical wheel of radius 1 ft rolls down a 37° incline, as shown in Fig. 12–22. The wheel drags along a 10-lb block by means of a rope tied to the center of the wheel. The coefficient of sliding friction between the block and the incline is 0.5. (a) What is the acceleration of the system? (b) What is the tension in the rope?

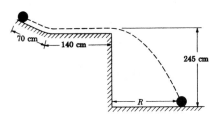

FIGURE 12–23

12–13. A sphere starts from rest at the top of an inclined plane of slope angle 30° (Fig. 12–23) and rolls down the incline and along a horizontal surface without slipping. (a) What forces act on the sphere while it is on the horizontal surface? (b) How many seconds after the start does it strike the floor? (c) What is the distance R?

12–14. A solid cylinder, a thin-walled hollow cylinder, and a sphere are released at such distances from the foot of an inclined plane that all three reach the foot of the plane at the same instant. Find their relative initial distances from the foot of the plane.

12–15. A freight car loaded with iron pipe is to be unloaded by rolling the pipes down two skids leading from the car door to the ground. The skids are 80 ft long and the floor of the car is 10 ft above the lower end of the skids. The distance between any two sections of pipe is not to be less than 20 ft. If the car contains 200 sections of pipe, what is the shortest time in which it can be unloaded?

12–16. A 14-lb sphere, initially rolling on a horizontal surface with a translational velocity of 16 ft/sec, comes to a 37° inclined plane up which it rolls without slipping. (a) How far up the plane does it roll? (b) How long a time does it remain on the plane?

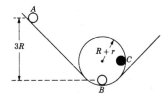

FIGURE 12–24

12–17. A small sphere of mass m and radius r is released from rest at point A in the loop-the-loop apparatus in Fig. 12–24. The radius of the loop is $R + r$, so the center of the sphere moves around the loop in a circle of radius R. The sphere rolls without slipping. Find the radial and tangential components of the acceleration of the center of the sphere as it passes point C, at the same elevation as the center of the circle. (See Problem 26 in Chapter 10.)

12–18. What is the minimum height at which the sphere in problem 12–17 can be released so that it will remain in contact with the circular track at the highest point?

12–19. A sphere of 1 cm radius and 100 gm mass is released from rest at a height h_1 above the end of a curved track, as shown in Fig. 12–25. It rolls without slipping, leaves the track at an angle of 37° with the horizontal, and

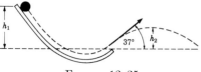

FIGURE 12–25

then follows a parabolic trajectory rising to a maximum height h_2 of 20.4 cm. (a) What is the linear velocity at the end of the track? (b) Compute the translational and rotational kinetic energies at the end of the track. (c) What is the height h_1?

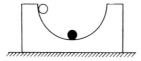

FIGURE 12–26

12–20. A uniform cylinder 1.5 ft in radius and weighing 8 lb is released from rest at the top of a semicircular track of radius 4 ft, cut in a block weighing 32 lb. The block rests on a horizontal frictionless surface, and the cylinder rolls without slipping (see Fig. 12–26). (a) How far has the block moved when the cylinder reaches the bottom of the track? (b) How fast is the block moving when the cylinder reaches the bottom of the track?

12–21. A solid sphere of radius r is placed on top of a large fixed sphere of radius R and rolls on it without slipping. If the rolling sphere starts from rest at the highest point, compute (a) the velocity of the center of the sphere when it is a distance h below its starting point, (b) the angular velocity of the sphere at this position. (c) For what value of h will the small sphere leave the surface of the large one?

12–22. A spool consists of two disks, each of mass 1 kgm and each 10 cm in radius, joined by a solid cylinder of mass 1 kgm and 4 cm in radius. The spool rolls without slipping on a hori-

zontal table. A string is wrapped around the central cylinder and one end hangs down through a slot in the table. A constant downward pull P of 5 newtons is applied to the string, which is kept vertical as the spool moves. (a) Find the linear acceleration of the center of mass of the spool. (b) Find the horizontal and vertical components of the force exerted on the spool by the table.

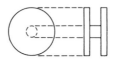

FIGURE 12–27

12–23. A "Yo-yo" is constructed of two flat disks, each of mass 100 gm and radius 10 cm, separated by a connecting axle of mass 50 gm and radius 2 cm, as shown in Fig. 12–27. A string is wound around the axle and the system is released from rest, the upper end of the string being held fixed. (a) Compute the linear acceleration of the center of mass and the angular acceleration of the system. (b) What is the kinetic energy of the system after it has dropped 40 cm? (c) What is the speed of the center of mass at the instant when the disk has made one complete revolution?

12–24. If the coefficient of sliding friction between the surface and the cylinder in Fig. 12–5 is μ, what is the maximum acceleration the center of mass may have without causing the cylinder to slide? If the applied force P, in the same example, is 4 μmg, how far does the center of mass move while the cylinder makes one complete revolution about its axis?

12–25. An 8-lb block slides on a horizontal table as in Fig. 12–28. The coefficient of sliding friction between the block and the table is 0.25. A string attached to the block passes over a light frictionless pulley and is wrapped

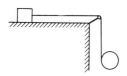

FIGURE 12–28

around the outside of a solid cylinder whose weight is 8 lb and whose radius is 2 ft. The cylinder is released, unrolling the string as it falls. Compute the acceleration of the block and the tension in the string.

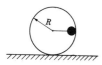

FIGURE 12–29

12–26. A solid disk of mass m and radius R has a small counterweight of mass m (equal to that of the disk) attached close to its rim as in Fig. 12–29. (a) Compute the initial linear acceleration of the center of the disk if it is released from rest with the attached mass at the same elevation as the center of the disk. The disk rolls without slipping. (b) Compute the horizontal and vertical forces exerted on the disk by the surface at the instant after the disk is released.

12–27. A man sits on a piano stool holding a pair of dumbbells at a distance of 3 ft from the axis of rotation of the stool. He is given an angular velocity of 2 rad/sec, after which he pulls the dumbbells in until they are but 1 ft distant from the axis. The moment of inertia of the man about the axis of rotation is 3 slug·ft^2 and may be considered constant. The dumbbells weigh 16 lb each and may be considered point masses. Neglect friction. (a) What is the initial angular momentum of the system? (b) What is the angular velocity of the system after the dumbbells are pulled in toward the axis? (c)

Compute the kinetic energy of the system, before and after the dumbbells are pulled in. Account for the difference, if any.

FIGURE 12–30

12–28. A block of mass 50 gm is attached to a cord passing through a hole in a horizontal frictionless surface as in Fig. 12–30. The block is originally revolving at a distance of 20 cm from the hole with an angular velocity of 3 rad/sec. The cord is then pulled from below, shortening the radius of the circle in which the block revolves to 10 cm. The block may be considered a point mass. (a) What is the new angular velocity? (b) Find the change in kinetic energy of the block.

12–29. A uniform rod of mass 30 gm and 20 cm long rotates in a horizontal plane about a fixed vertical through its center. Two small bodies, each of mass 20 gm, are mounted so that they can slide along the rod. They are initially held by catches at positions 5 cm on either side of the center of the rod, and the system is rotating at 15 rpm. Without otherwise changing the system, the catches are released and the masses slide outward along the rod and fly off at the ends. (a) What is the angular velocity of the system at the instant when the small masses reach the ends of the rod? (b) What is the angular velocity of the rod after the small masses leave it?

12–30. A turntable rotates about a fixed vertical axis, making one revolution in 10 sec. The moment of inertia of the turntable about this axis is 720 slug·ft^2. A man weighing 160 lb, initially standing at the center of the turntable, runs out along a radius. What is the angular velocity of the turntable when the man is 6 ft from the center?

12–31. A man weighing 160 lb stands at the rim of a turntable of radius 10 ft and moment of inertia 2500 slug·ft^2, mounted on a vertical frictionless shaft at its center. The whole system is initially at rest. The man now walks along the outer edge of the turntable with a velocity of 2 ft/sec, relative to the earth. With what angular velocity and in what direction does the turntable rotate? Through what angle will it have rotated when the man reaches his initial position on the turntable? Through what angle will it have rotated when he reaches his initial position relative to the earth?

FIGURE 12–31

12–32. Disks A and B are mounted on a shaft SS and may be connected or disconnected by clutch C, as in Fig. 12–31. The moment of inertia of disk A is one-half that of disk B. With the clutch disconnected, A is brought up to an angular velocity ω_0. The accelerating torque is then removed from A and it is coupled to disk B by the clutch. Bearing friction may be neglected. It is found that 3000 ft·lb of heat are developed in the clutch when the connection is made. What was the original kinetic energy of disk A?

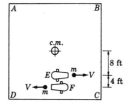

FIGURE 12–32

12–33. The rectangle $ABCD$ of Fig. 12–32 is a top view of a large wooden platform resting on the frictionless surface of a frozen lake. Two guns, E and F, fastened to the platform and pointed in opposite directions, simultaneously fire two cannon balls of mass $m = 5$ slugs with velocities V of 800 ft/sec. The moment of inertia of platform plus guns, about an axis through the center of mass and perpendicular to the diagram, is 8000 slug·ft^2. (a) What is the linear velocity of the center of mass of the platform after the guns are fired? (b) What is the angular velocity of the platform after the guns are fired?

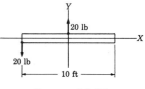

FIGURE 12–33

12–34. A 64-lb plank, 10 ft long, rests on a horizontal frictionless surface. Equal and opposite forces of 20 lb are applied as shown in Fig. 12–33, for 0.5 sec. Describe the motion of the plank after the forces have ceased to act.

12–35. A slender rod whose mass is 100 gm and whose length is 120 cm is struck a blow of impulse 1000 dyne·sec. The blow is applied at one end of the rod in a direction at right angles to its length. (a) If the rod is pivoted at its center, find the magnitude and direction of the impulse exerted on the rod by the pivot, and the kinetic energy acquired by the rod. (b) What would have been the kinetic energy of the rod if it had not been pivoted, the impulse and its point of application being the same? (c) How do you account for the fact that although the impulse is the same in both cases the work done by the blow is not the same?

12–36. A uniform rod of mass M and length L stands on the edge of a frictionless table. A blow of impulse J is

FIGURE 12–34

applied horizontally at the lower end of the rod (Fig. 12–34). (a) What is the initial horizontal velocity of the lower end of the rod relative to the table? (b) How far does the center of mass of the rod move horizontally in the time required to make one revolution?

12–37. A uniform rod of length 100 cm and mass 400 gm hangs from a pivot through the upper end. A bullet of mass 2 gm traveling horizontally with an initial velocity of 30,000 cm/sec, passes through the lower end of the rod and emerges with a velocity of 10,000 cm/sec. (a) What is the impulse of the force exerted on the bullet by the rod as it passes through the rod? (b) What is the angular velocity of the rod just after the bullet has passed through?

12–38. The stabilizing gyroscope of a ship weighs 50 tons, its radius of gyration is 5 ft, and it rotates about a vertical axis with an angular velocity of 900 rpm. (a) How long a time is required to bring it up to speed, starting from rest, with a constant power input of 100 hp? (b) Find the righting moment exerted on the ship, in lb·ft, when the

axis is forced to precess in a vertical fore-and-aft plane at the rate of 1 degree/sec.

FIGURE 12–35

12–39. The mass of the rotor of a toy gyroscope is 150 gm and its moment of inertia about its axis is 1500 gm·cm². The mass of the frame is 30 gm. The gyroscope is supported on a single pivot as in Fig. 12–35 with its center of gravity distant 4 cm horizontally from the pivot, and is precessing in a horizontal plane at the rate of 1 revolution in 6 sec. (a) Find the upward force exerted by the pivot. (b) Find the angular velocity with which the rotor is spinning about its axis, expressed in rpm. (c) Copy the diagram, and show by vectors the angular velocity of the rotor and the angular velocity of precession.

12–40. If the projection that rests on the pivot in Fig. 12–35 were extended to the left, at what horizontal distance from the pivot should a 200-gm body be hung to cause an angular velocity of precession of 1 revolution in 10 sec, in a direction opposite to that in the figure?

CHAPTER 13

ELASTICITY

13–1 Introduction. In the preceding chapters we have developed the principles by which the engineer can compute the tensile or compressive forces in the various members of a structure. It is not enough, however, to know how much force each part of a structure will exert. One must also know how large a cable or strut is needed to withstand this force, and how much the structure will distort under load. The subject of elasticity is the study of the way in which actual materials such as wood, steel, concrete, etc., are changed in shape by forces applied to them. In engineering work this part of mechanics is called "Strength of Materials."

All real substances are found to yield somewhat under the influence of a force. Some materials return to their original form when the force is removed, while others remain more or less distorted. A *perfectly elastic* material is one which returns exactly to its original form when the distorting force is removed; a *perfectly inelastic* material is one which does not return at all. Many substances are nearly perfectly elastic up to a certain maximum distortion but do not recover completely if distorted beyond this point, which is known as the *elastic limit*.

The elastic properties of materials are described in terms of two concepts known as *stress* and *strain*. These terms are used loosely in everyday life, often as synonyms, but like such terms as force and work, they are given a very restricted meaning in physics.

13–2 Stress. Figure 13–1(a) represents a bar subjected at its ends to equal and opposite pulls of magnitude F. The bar is in equilibrium under the action of these forces, and hence every part of it is also in equilibrium. Imagine the bar to be cut at the dotted section, and consider the portion of the bar at the left of the cut. (Fig. 13–1b.) Since this portion was in equilibrium before the cut was made, the portion of the bar at the right of the section must have been exerting a force on it,

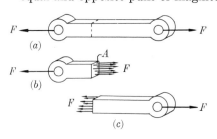

(a)

(b)

(c)

Fig. 13–1. Tensile stress.

equal to F and directed toward the right. If the cut is not too near the
end of the bar, this force will be distributed uniformly over the cross section, as indicated by the small arrows.

Let A represent the cross-sectional area of the bar. The ratio of the
distributed force to the cross-sectional area is called the *stress* in the bar,
and the bar is said to be in a state of stress (in this particular case, in
tensile stress). Since the cut may be made at any point along the bar, the
entire bar is in a state of stress.

Stress is a force per unit area, and the units of stress in our three systems
are lb/ft^2, $newtons/m^2$, and $dynes/cm^2$. It is almost universal engineering
practice, however, to express a stress in lb/in^2.

Some engineering texts use the term "stress" for the *total* force F
acting across a section, and the term "unit stress" for the force per unit
area.

It is evident that the portion of the bar shown in Fig. 13–1(b) must
itself be exerting a force toward the left, on the right-hand portion of the
bar. This force, shown in Fig. 13–1(c), is the reaction to the distributed
force in 13–1(b) and hence is also equal in magnitude to F. The concept
of stress is considered to include both of these distributed forces. If, for
example, the force F is 1000 pounds and the area A is two square inches,
the stress at the section is $1000/2 = 500$ lb/in^2. One cannot say, however, in which direction the stress acts. The portion of the bar at the left
of the section is being pulled toward the right, and that at the right of the
section is being pulled toward the left.

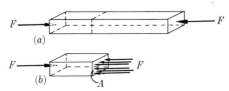

FIG. 13–2. Compressive stress.

If a member, such as a column or strut, is subject to compression, a
similar state of affairs exists at every cross section, except that a push
instead of a pull acts across the section. The portion of the member at
either side of the section exerts a push on that portion at the other side,
as in Fig. 13–2. The stress in the member is defined in the same way, as
the ratio of the force to the area, and is called a *compressive* stress.

A third type of stress is illustrated in Fig. 13–3. The lower face of the
block is being pulled to the left and the upper face to the right. The
force which the portion above any horizontal plane exerts on the portion
below that plane is shown in Fig. 13–3(b). This force is also distributed

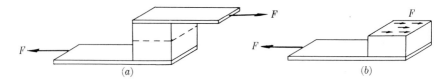

FIG. 13–3. Shearing stress.

over the cross section of the block, but it is parallel to the plane of the cross section instead of being perpendicular to it as in tensile or compressive stress. A stress of this type is called a *shearing stress* or simply a *shear*.

Since the lower half of the block is in equilibrium,* the distributed force at its top face must equal the force F, and, as before, the shearing stress is defined as the ratio of this force to the cross-sectional area.

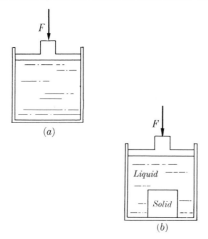

FIG. 13–4. Both liquid and solid are under a hydrostatic pressure.

Still another type of stress is illustrated in Fig. 13–4. Figure 13–4(a) represents a cylinder provided with a tightly fitting piston and filled with a liquid. A force F presses down on the piston, compressing the liquid in the cylinder. Let A be the cross section of the piston. The *pressure* exerted on the liquid by the piston is defined as the ratio of the force F to the area A. It is well known (see Chapter 16) that the side walls and bottom of the cylinder press inward on the liquid with a pressure equal to that exerted by the piston, so that the liquid is subjected to a uniform inward pressure over its entire surface.

If a solid block is placed in the liquid and pressure applied as in Fig. 13–4(b), the same uniform inward pressure is exerted over the entire surface of the block.

A stress of this sort is called a *hydrostatic pressure*, and, like other types of stress, is expressed in force units per unit of area. Like other stresses

---

*The two forces F in Fig. 13–3(a) constitute a couple, and by themselves would produce clockwise rotation. Since a couple can only be balanced by another couple of opposite sign, there must be other forces on the block for complete equilibrium, but we shall ignore them here.

also, it is not confined to the surface of the liquid or solid. Across any imagined area within either one, a force is exerted by that part of the body at one side of the area on the part at the other side.

13–3 Strain. The term *strain* refers to the relative change in dimensions or shape of a body which is subjected to stress. Associated with each type of stress described in the preceding section is a corresponding type of strain.

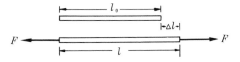

Fig. 13–5. Longitudinal strain.

Figure 13–5 shows a bar whose natural length is l_0 and which elongates to a length l when equal and opposite pulls are exerted at its ends. The elongation, of course, does not occur at the ends only, but every element of the bar stretches in the same proportion as does the bar as a whole. The *tensile strain* in the bar is defined as the ratio of the increase in length to the original length.

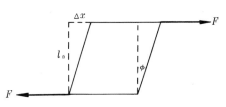

Fig. 13–6. Shearing strain.

$$\text{Tensile strain} = \frac{l - l_0}{l_0} = \frac{\Delta l}{l_0}.$$

The *compressive strain* of a bar in compression is defined in the same way, as the ratio of the decrease in length to the original length.

Figure 13–6 illustrates a shearing strain. The block whose original rectangular shape is shown by the dotted lines is subjected to a shearing stress and becomes distorted as shown. In this type of strain it is the change in shape, rather than the change in size, which is of interest, and the strain is defined as the angle ϕ. In all cases of practical interest the angle is small and hence is nearly equal to the ratio of the displacement Δx to the dimension l_0.

$$\text{Shearing strain} = \phi \text{ (radians)} = \frac{\Delta x}{l_0}.$$

In both longitudinal and shearing strain, the displacements Δl or Δx and the dimension l_0 are to be expressed in the same unit. (Any unit may be used.) Since a strain is the ratio of one length to another, both expressed in the same unit, it is a pure number.

The strain produced by a hydrostatic pressure is defined as the ratio of the change in volume to the original volume. If we call the original volume V_0 and the change in volume ΔV, the strain is the ratio $\Delta V/V_0$. Like other strains it is a pure number.

13–4 Elastic modulus. The ratio of a stress to the corresponding strain is called an elastic modulus and, provided the elastic limit is not exceeded, this ratio is found experimentally to be constant, characteristic of a given material. In other words, the stress is directly proportional to the strain, or is a linear function of the strain (within the elastic limit). This linear relationship between stress and strain is called *Hooke's law*.

Let us first consider longitudinal (i.e., tensile or compressive) stresses and strains. Experiment shows that with a given material, a given longitudinal stress produces a strain of the same magnitude whether the stress is a compression or a tension. Hence the ratio of tensile stress to tensile strain, for a given material, equals the ratio of compressive stress to compressive strain. This ratio is called the *stretch modulus* or *Young's modulus* of the material and will be denoted by Y.

$$Y = \frac{\text{tensile stress}}{\text{tensile strain}} = \frac{\text{compressive stress}}{\text{compressive strain}},$$

or

$$Y = \frac{F/A}{\Delta l/l_0}. \tag{13–1}$$

Since a strain is a pure number, the units of Young's modulus are the same as those of stress, namely, force per unit area. Tabulated values are usually in lb/in^2 or $dynes/cm^2$. Some typical values are listed in Table III.

TABLE III

Elastic Constants

(Representative values)

Material	Young's Modulus		Shear Modulus		Bulk Modulus	
	dynes/cm^2	lb/in^2	dynes/cm^2	lb/in^2	dynes/cm^2	lb/in^2
Aluminum	7×10^{11}	10×10^6	2.4×10^{11}	3.4×10^6	7×10^{11}	10×10^6
Brass	9	13	3.5	5.1	6.1	8.5
Copper	10–12	14–18	4	6	12	17
Iron, cast	8–10	12–14			9.6	14
Iron, wrought	18–20	26–29			15	21
Lead	1.5	2.3	0.5	0.8	0.8	1.1
Steel	19–21	27–30	8	12	16	23

TABLE IV

COMPRESSIBILITIES OF LIQUIDS

Liquid	Compressibility (atm^{-1})
Carbon disulphide	66×10^{-6}
Ethyl alcohol	112
Glycerine	22
Mercury	3.8
Water	50

The ratio of a shearing stress to the corresponding shearing strain is called the *shear modulus* of a material and will be represented by M. It is also called the modulus of rigidity or the torsion modulus.

$$M = \frac{\text{shearing stress}}{\text{shearing strain}}$$
$$= \frac{F/A}{\phi} = \frac{F/A}{\Delta x/l_0}. \tag{13-2}$$

(Refer to Fig. 13–6 for the meanings of ϕ, Δx, and l_0.) The shear modulus of a material is also expressed in force per unit area. For most materials it is one-half to one-third as great as Young's modulus.

The modulus relating an increase in hydrostatic pressure to the corresponding decrease in volume is called the *bulk modulus* and we shall represent it by B.

$$B = -\frac{p}{\Delta V/V_0}. \tag{13-3}$$

The minus sign is included in the definition of B since an increase of pressure always causes a decrease in volume. That is, if p is positive ΔV is negative. By including a minus sign in its definition the bulk modulus itself is a positive quantity.

The reciprocal of the bulk modulus is called the *compressibility*, k. Tables of physical constants often list the compressibility rather than the bulk modulus. From its definition,

$$k = \frac{1}{B} = -\frac{1}{p}\frac{\Delta V}{V_0}, \tag{13-4}$$
$$\Delta V = -kV_0 p. \tag{13-5}$$

The ratio $\Delta V/V_0$ is the fractional change in volume. Hence the compressibility of a substance may be defined as its fractional change in volume per unit increase in pressure.

The units of a bulk modulus, from Eq. (13–3), are the same as those of pressure, and the units of compressibility, from Eq. (13–4), are those of a reciprocal pressure. In tabulating compressibilities, the pressure is often expressed in atmospheres. (1 atmosphere = 14.7 lb/in^2.) The corresponding units of compressibility are therefore "reciprocal atmospheres" or atm^{-1}. For example, the statement that the compressibility of water is 50×10^{-6} atm^{-1}, or 50×10^{-6} per atmosphere, means that the volume decreases by 50 one-millionths of the original volume for each atmosphere increase in pressure.

Strictly speaking, an elastic modulus is defined as the ratio of an infinitesimal change in stress to the corresponding change in strain. Thus if the force stretching a bar is increased from F to $F + dF$, causing an increase in length from l to $l + dl$, Young's modulus is defined as

$$Y = \frac{dF/A}{dl/l} = \frac{l}{A}\frac{dF}{dl}. \tag{13–6}$$

Similarly, the shear modulus is defined as

$$M = \frac{dF/A}{dx/l} = \frac{l}{A}\frac{dF}{dx}. \tag{13–7}$$

The bulk modulus becomes

$$B = -\frac{dp}{dV/V} = -V\frac{dp}{dV}, \tag{13–8}$$

and the compressibility

$$k = -\frac{1}{V}\frac{dV}{dp}. \tag{13–9}$$

We shall need to use these general forms later in connection with the compressibility of gases and the propagation of sound waves.

When a metal rod is subjected to an increasing tensile stress, the strain is found to change as in Fig. 13–7. The first part of the curve, from O to A, is a straight line. That is, in this region there is a linear relationship between stress and strain and the material obeys Hooke's law. If the stress is not carried beyond that corresponding to point A, the specimen returns to its original length when the stress is removed. In other words, the portion of the curve from O to A is the region of perfect elasticity.

If the stress is increased to a value corresponding to point B and then removed, the specimen does not return to its original length but retains a *permanent strain* or a *set*. Point A is called the *elastic limit* or the *yield point* of the material. (Actual materials may show some small irregularities at this point which are omitted for simplicity.) Finally, when the stress is increased sufficiently, the specimen breaks at point C.

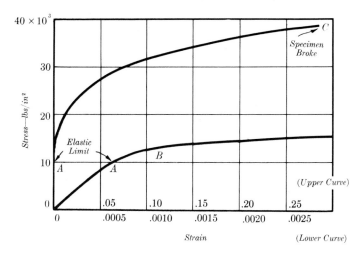

FIG. 13–7. Stress-strain diagram.

EXAMPLES. (1) In an experiment to measure Young's modulus, a load of 1000 lb hanging from a steel wire 8 ft long, of cross section 0.025 in^2, was found to stretch the wire 0.12 in. above its no-load length. What were the stress, the strain, and the value of Young's modulus for the steel of which the wire was composed?

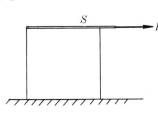

FIGURE 13–8

$$\text{Stress} = \frac{F}{A} = \frac{1000}{0.025} = 40{,}000 \text{ lb/in}^2.$$

$$\text{Strain} = \frac{\Delta l}{l_0} = \frac{0.010}{8} = 0.00125.$$

$$Y = \frac{\text{stress}}{\text{strain}} = \frac{40{,}000}{0.00125} = 32 \times 10^6 \text{ lb/in}^2.$$

(2) A brass plate 2 ft square and $\frac{1}{4}$ in. thick is rigidly fastened to the floor along one edge as in Fig. 13–8. A flat strip S is brazed to its top edge. How large a force is needed to pull the top edge a distance of 0.01 in. to the right? The shear modulus for this brass is 5×10^6 lb/in^2.

The force F is applied to the strip over an area of $24 \times \frac{1}{4} = 6$ in^2. Hence

$$\text{Shearing stress} = F/6 \text{ lb/in}^2.$$

$$\text{Also,} \quad \Delta x = 0.01 \text{ in}, \quad l_0 = 24 \text{ in}.$$

$$\text{Shearing strain} = \Delta x/l_0 = \frac{0.01}{24} = 0.000417.$$

$$\text{Shear modulus } M = \frac{\text{stress}}{\text{strain}},$$

$$5 \times 10^6 = \frac{F/6}{0.000417},$$

$$F = 12{,}500 \text{ lb}.$$

(3) The volume of oil contained in a certain hydraulic press is 5 ft³. Find the decrease in volume of the oil when subjected to a pressure of 2000 lb/in². The compressibility of the oil is 20×10^{-6} per atm.

The volume decreases by 20 parts per million for a pressure increase of one atm. Since 2000 lb/in² = 136 atm, the volume decrease is $136 \times 20 = 2720$ parts per million. Since the original volume is 5 ft³, the actual decrease is

$$\frac{2720}{1{,}000{,}000} \times 5 = 0.0136 \text{ ft}^3 = 23.5 \text{ in}^3,$$

or, from Eq. (13–5),

$$\Delta V = -kV_0 p = -20 \times 10^{-6} \times 5 \times 136$$

$$= -0.0136 \text{ ft}^3.$$

13–5 Poisson's ratio. When a rod or bar is subjected to a tensile stress, it not only elongates in the direction of the stress but its transverse dimensions decrease. If subjected to a compressive stress the transverse dimen-

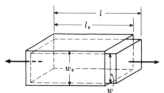

sions increase. There is thus set up a *transverse strain* in addition to the longitudinal strain. This effect is shown (greatly exaggerated) in Fig. 13–9. If w_0 represents an original transverse dimension and Δw the change in this dimension, the trans-

FIG. 13–9. A longitudinal stress produces both longitudinal and transverse strains.

verse strain is $\Delta w/w_0$. (The ratio $\Delta w/w_0$ is the same for *all* transverse dimensions, except for nonisotropic substances such as certain crystals whose properties are different in different directions.)

The ratio of the transverse to the longitudinal strain is called Poisson's ratio and is represented by σ.

$$\sigma = -\frac{\Delta w/w_0}{\Delta l/l_0}.$$

The minus sign is introduced since an increase in length always results in a decrease in transverse dimensions and vice versa. Hence if Δl is positive Δw is negative and σ is a positive quantity. Its magnitude, for most metals, is about 0.3.

Figure 13–10 represents a block in the form of a rectangular parallelepiped whose unstressed dimensions are a_0, b_0, and c_0. Suppose that compressive stresses F/A are exerted as shown on the upper and lower faces, producing a decrease Δa in the dimension of original length a_0. From the definition of Young's modulus,

$$Y = \frac{F/A}{-\Delta a/a_0},$$

and hence $\quad \dfrac{\Delta a}{a_0} = -\dfrac{F}{AY}.$ (13–10)

The negative sign is introduced since Δa is a negative quantity. Then from the definition of Poisson's ratio, the strains in the other two dimensions of the block are

$$\frac{\Delta b}{b_0} = -\frac{\sigma\,\Delta a}{a_0} = \frac{\sigma F}{AY}, \quad (13\text{–}11)$$

$$\frac{\Delta c}{c_0} = -\frac{\sigma\,\Delta a}{a_0} = \frac{F\sigma}{AY}. \quad (13\text{–}12)$$

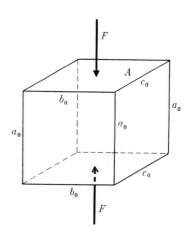

FIGURE 13–10

The quantities Δb and Δc are positive, that is, the transverse dimensions increase.

The volume V of the stressed block is

$$
\begin{aligned}
V &= (a_0 + \Delta a)(b_0 + \Delta b)(c_0 + \Delta c) \\
&= (a_0 b_0 c_0) + (a_0 b_0\,\Delta c + b_0 c_0\,\Delta a + c_0 a_0\,\Delta b) \\
&\quad + (a_0\,\Delta b\,\Delta c + b_0\,\Delta c\,\Delta a + c_0\,\Delta a\,\Delta b) \\
&\quad + (\Delta a\,\Delta b\,\Delta c).
\end{aligned}
$$

Since Δa, Δb, and Δc are small, we can neglect the last two parentheses. The original unstressed volume V_0 is equal to $a_0 b_0 c_0$, so the change in volume, $\Delta V = V - V_0$, is

$$\Delta V = (a_0 b_0\,\Delta c + b_0 c_0\,\Delta a + c_0 a_0\,\Delta b).$$

Divide the left side by V_0, and each term on the right by $a_0 b_0 c_0$. Then

$$\frac{\Delta V}{V_0} = \frac{\Delta a}{a_0} + \frac{\Delta b}{b_0} + \frac{\Delta c}{c_0},$$

and from Eqs. (13–10), (13–11), and (13–12),

$$\frac{\Delta V}{V_0} = \frac{\Delta a}{a_0} - \frac{\sigma \, \Delta a}{a_0} - \frac{\sigma \, \Delta a}{a_0},$$

$$\frac{\Delta V}{V_0} = (1 - 2\sigma)\frac{\Delta a}{a_0}. \qquad (13\text{–}13)$$

Hence the fractional change in volume equals $(1 - 2\sigma)$ times the fractional change in length. It is found by experiment that Poisson's ratio is always less than $\frac{1}{2}$. Hence the term $(1 - 2\sigma)$ is a positive quantity always. In our example $\Delta a/a_0$ is negative, so $\Delta V/V_0$ is negative. In other words, when a body is subjected to a compressive stress, its volume always decreases, the decrease in length more than compensating for the increase in cross section. Had the body in Fig. 13–10 been subjected to a tensile stress, $\Delta a/a_0$ would be positive and the volume would increase.

13–6 Relations between elastic constants. It was shown in the preceding section that when a block is subjected to a compressive stress, the fractional change in its volume is

$$\frac{\Delta V}{V_0} = (1 - 2\sigma)\frac{\Delta a}{a_0}.$$

Making use of Eq. (13–10), this can be written

$$\frac{\Delta V}{V_0} = -(1 - 2\sigma)\left(\frac{F}{AY}\right). \qquad (13\text{–}14)$$

Now suppose that we set up the same compressive stress F/A on the other pairs of faces of the block in Fig. 13–10, so that the block is subjected to an inward stress F/A on all six of its faces. In other words, it is under a uniform hydrostatic pressure $p = F/A$. The stress on each pair of opposite faces produces the same fractional volume change as that on the first pair, so the total volume change is three times as great as that due to a compressive stress alone. That is,

$$\frac{\Delta V}{V_0} = -3(1 - 2\sigma)\frac{F}{AY},$$

or, since $F/A = p$,

$$\frac{\Delta V}{V_0} = -3(1 - 2\sigma)\frac{p}{Y}. \qquad (13\text{–}15)$$

But from the definition of bulk modulus B in Eq. (13–3),

$$B = - \frac{p}{\Delta V / V_0}.$$ (13–16)

When this is combined with Eq. (13–15), we get

$$B = \frac{Y}{3(1 - 2\sigma)}.$$ (13–17)

By applying the same type of reasoning to a block subjected to a compressive stress over one pair of opposite faces and simultaneously to a tensile stress over another pair of faces, it can be shown that

$$M = \frac{Y}{2(1 + \sigma)},$$ (13–18)

where M is the shear modulus. Hence, of the four elastic constants Y, B, M, and σ, only two are independent. If any two are known, the others may be computed.

The statement above is true for homogeneous isotropic substances only. The elastic properties of crystalline (i.e., nonisotropic) substances are, in general, different in different directions and the treatment of their elastic properties is considerably more complicated.

13–7 Torsion. When equal and opposite couples are exerted at the ends of a rod or shaft as in Fig. 13–11 (a), the rod twists as shown, a line such as AB being displaced to the position $A'B'$. The rod is then said to be in *torsion*. This situation exists in every shaft which is transmitting power, as, for instance, the drive shaft of an automobile. The couples at the ends of such a shaft are provided by the engine at one end and the resistance of the load at the other end.

Consider the thin outer shell of a shaft in torsion. Imagine it to be separated from the rest of the shaft, split along the line AB, and spread out flat as in Fig. 13–11(b). Originally its shape was the thin rectangle $AABB$. After the twist, it takes the shape $A'A'B'B'$ and is hence in shear. In the same way, each successive inner shell into which the shaft can, in imagination, be divided, is also in shear, so that torsion and shear are equivalent types of strain.

The couple necessary to twist one end of a shaft through a certain angle with respect to the other end can be found by dividing the shaft into thin shells as described above, computing the torque due to each

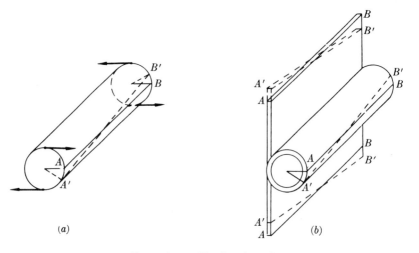

FIG. 13–11. Torsional strain.

shell, and summing these to find the resultant torque. The result is

$$\tau = \frac{\pi M R^4 \theta}{2L},\qquad(13\text{–}19)$$

where in the English system τ is the torque in pound-feet due to the couple at either end of the shaft, M is the shear modulus of the material of which the shaft is composed expressed in pounds per square *foot*, R is the radius of the shaft in feet, θ is the angle in radians through which one end of the shaft is twisted relative to the other, and L is the length of the shaft in feet.

To derive Eq. (13–19) let Fig. 13–12 be a "cut-away" view of a solid shaft of length L and radius R, unstressed in (a) and under torsion in (b). The upper face is twisted relative to the lower by an angle θ. The twisting forces, distributed over the top end, are shown by the small arrows. Similar but oppositely directed forces are distributed over the lower end. A thin shell of radius r and thickness dr is shown shaded. The upper edge of this shell is displaced relative to the lower by a distance $r\theta$, which corresponds to AA' in Fig. 13–11 or to Δx in Fig. 13–6. The strain of the shell is therefore $r\theta/L$. Let dF represent the sum of the tangential forces on the top edge of the shell. The area of this edge is $2\pi r\,dr$, and the stress is therefore $dF/2\pi r\,dr$. Hence, from the definition of shear modulus,

$$M = \frac{dF/2\pi r\,dr}{r\theta/L},$$

or

$$dF = \frac{2\pi M\theta}{L}r^2\,dr.$$

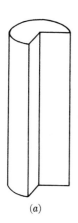
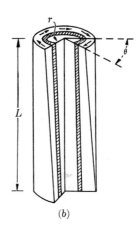

(a) (b)

FIGURE 13–12

The torque due to this force is

$$d\tau = r\,dF = \frac{2\pi M\theta}{L} r^3\,dr,$$

and the torque to twist the entire rod is found by integrating the above from $r = 0$ to $r = R$.

$$\tau = \int d\tau = \frac{2\pi M\theta}{L} \int_0^R r^3\,dr,$$

which leads to Eq. (13–19).

13–8 Bending of a beam. A beam usually means a horizontal portion of a structure, designed to support transverse loads. The beam itself may be supported at its ends, as in Fig. 13–13(a) or at one end only as in Fig. 13–13(b). The latter type is called a *cantilever*.

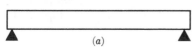

(a) (b)

FIG. 13–13. (a) A beam supported at its ends. (b) A cantilever.

When subjected to a load w as in Fig. 13–14 or 13–15, the beam becomes distorted as shown. The bending is, of course, greatly exaggerated. In either case the center line of the beam is unchanged in length.* If the

---

*This discussion is only approximately correct. The complete analysis of the bending of a beam is beyond the scope of this text.

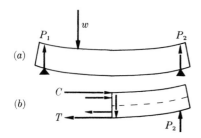

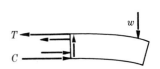

FIG. 13–14. Stresses in a beam.   FIG. 13–15. Stresses in a cantilever.

beam is supported at its ends, that part above the center line is shortened and is in compression, while the part below the center line is stretched and is in tension. The reverse is true of the cantilever, where the upper portion is in tension and the lower in compression. The longitudinal stress in each case is zero at the center line, and increases in either direction away from the center.

If an imaginary transverse cut is made at the center of the beam supported at its ends, the stress distribution at the cut is as shown in Fig. 13–14(b). The compressive and tensile stresses C and T give rise to a couple in a clockwise direction. Hence, since the beam is in equilibrium, there must be an equal counterclockwise couple exerted on it. This couple is provided by the upward force P_2, and a downward force at the cut, exerted by the portion of the beam at the left of the cut. That is, a shearing stress, as well as longitudinal stresses, must exist at the section, of such magnitude that the shearing *force* is equal to P_2.

Figure 13–14 may help to explain why the I-beam is such a rigid member for its weight, since a large amount of material is concentrated at the upper and lower surfaces where the tensile and compressive stresses are greatest.

If an imaginary cut is made in the cantilever beam at the point where it enters the wall, as in Fig. 13–15, it will be seen that tensile, compressive, and shearing stresses are also exerted at this section. In fact, Fig. 13–15 is the same as Fig. 13–14 turned upside down.

13–9 The force constant. The various elastic moduli are quantities which describe the elastic properties of a particular *material* and do not directly indicate how much a given rod, cable, or spring constructed of the material will distort under load. If Eq. (13–1) is solved for F one obtains

$$F = \frac{YA}{l_0} \Delta l,$$

or if YA/l_0 is replaced by a single constant k and the elongation Δl is represented by x,

$$F = kx. \qquad (13\text{--}11)$$

In other words, the elongation above its no-load length of a body in tension is directly proportional to the stretching force. Hooke's law was originally stated in this form, rather than in terms of stress and strain.

When a helical spring is stretched, the actual distortion of the wire composing it is a combination of stretching, bending, and torsion. It is nevertheless true that the elongation of the spring as a whole is directly proportional to the stretching force, provided the elastic limit is not exceeded. That is, an equation of the form $F = kx$ still applies, although the proportionality constant k cannot be simply expressed in terms of elastic moduli.

The constant k, or the ratio of the force to the elongation, is called the *force constant* or the *coefficient of stiffness* of the spring, and is expressed in pounds per foot, newtons per meter, or dynes per centimeter. It is equal numerically to the force required to produce unit elongation.

Problems

13–1. A steel wire 10 ft long and 0.1 square inch in cross section, is found to stretch 0.01 ft under a tension of 2500 lb. What is Young's modulus for this steel?

13–2. The elastic limit of a steel elevator cable is 40,000 lb/in². Find the maximum upward acceleration which can be given a 2-ton elevator when supported by a cable whose cross section is one-half a square inch, if the stress is not to exceed ¼ of the elastic limit?

13–3. A copper wire 12 ft long and 0.036 inch in diameter was given the test below. A load of 4.5 lb was originally hung from the wire to keep it taut. The position of the lower end of the wire was read on a scale.

Added load (lb)	Scale reading (in)
0	3.02
2	3.04
4	3.06
6	3.08
8	3.10
10	3.12
12	3.14
14	3.65

Make a graph of these values, plotting the increase in length horizontally and the added load vertically. Calculate the value of Young's modulus. What was the stress at the elastic limit?

13–4. A steel wire has the following properties:

Length = 10 ft
Cross section = .01 in²
Young's modulus = 30,000,000 lb/in²
Shear modulus = 10,000,000 lb/in²
Elastic limit = 60,000 lb/in²
Breaking stress = 120,000 lb/in²

The wire is fastened at its upper end and hangs vertically. (a) How great a load can be supported without exceeding the elastic limit? (b) How much will the wire stretch under this load? (c) What is the maximum load that can be supported?

13–5. A copper rod of length 3 ft and cross-sectional area 0.5 in² is fastened end-to-end to a steel rod of length L and cross-sectional area 0.2 in². The compound rod is subjected to equal and opposite pulls of magnitude 6000 lb at its ends. (a) Find the length L of the steel rod if the elongations of the two rods are equal. (b) What is the stress in each rod? (c) What is the strain in each rod?

13–6. A copper wire 320 inches long and a steel wire 160 inches long, each of cross section 0.1 in², are fastened end-to-end and stretched with a tension of 100 lb. (a) What is the change in length of each wire? (b) What is the elastic potential energy of the system?

13–7. A 32-lb weight, fastened to the end of a steel wire of unstretched length 2 ft, is whirled in a vertical circle with an angular velocity at the bottom of the circle of 2 rps. The cross section of the wire is 0.01 in². Calculate the elongation of the wire when the weight is at the lowest point of its path.

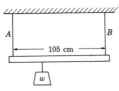

FIGURE 13–16

13–8. A rod 105 cm long, whose weight is negligible, is supported at its ends by wires A and B of equal length. The cross section of A is 1 mm², that of B is 2 mm². Young's modulus for wire A is 30×10^6 lb/in² and for B it is 20×10^6 lb/in². At what point along the bar should a weight w be suspended in order to produce (a) equal stresses in A and B, (b) equal strains in A and B?

13–9. A bar of length L, cross-sectional area A, Young's modulus Y, is subjected to a tension F. Represent the stress in the bar by S and the strain by P. Derive the expression for the elastic potential energy, per unit volume, of the bar in terms of S and P.

13–10. A 16-lb bar of square cross section 2 inches on a side, and 1 ft long, is pulled along a smooth horizontal sur-

face by a force applied uniformly over one end. The block has a constant acceleration of 8 ft/sec². What is the stress at a transverse section of the block, perpendicular to its length, (a) at a point 1 inch from the back end of the bar, (b) at the center of the bar.

13–11. Two strips of metal are riveted together at their ends by four rivets, each of diameter 0.25 inch. What is the maximum tension that can be exerted by the riveted strip if the shearing stress on the rivets is not to exceed 10,000 lb/in²? Assume each rivet to carry one-quarter of the load.

13–12. The compressibility of sodium is to be measured by observing the displacement of the piston in Fig. 13–4(b) when a force is applied. The sodium is immersed in an oil which fills the cylinder below the piston. Assume that the piston and walls of the cylinder are perfectly rigid, that there is no friction, and no oil leak. Compute the compressibility of the sodium in terms of the applied force F, the piston displacement x, the piston area A, the initial volume of the oil V_0, the initial volume of the sodium v_0, and the compressibility of the oil k_0.

13–13. Find the weight-density of ocean water at a depth where the pressure is 4700 lb/ft². The weight-density at the surface is 64 lb/ft³.

13–14. Compute the compressibility of steel, in reciprocal atmospheres, and compare with that of water. Which material is more readily compressed?

13–15. A bar of length l, width w, and thickness t, is subjected to a tensile strain of 0.001. Poisson's ratio is 0.30. Find the fractional increase or decrease in cross section of the bar, and the fractional increase or decrease in area of a face initially of area wl.

13–16. A steel post 6 inches in diameter and 10 ft long is placed vertically and is required to support a load of 20,000 lb. (a) What is the stress in the post? (b) What is the strain in the post? (c) What is the change in length of the post? (d) What is the change in the area of a cross section?

13–17. A cubical block of steel is 10 cm on a side. Compute the fractional change in length of each edge of the block, (a) if the block is subjected to a compressional stress of 10^5 lb/in^2 by forces applied at one pair of opposite faces, (b) if the block is subjected to a hydrostatic pressure of 10^5 lb/in^2.

FIGURE 13–17

13–18. A steel cube 2 inches on a side is subjected to four shearing forces of 2.4×10^3 lb each, as shown in Fig. 13–17. Calculate the shear strain.

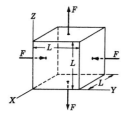

FIGURE 13–18

13–19. A cube of material of length L on each side is acted on by four equal forces F, two of which are tensile and two compressive. (The forces are actually distributed uniformly over the four faces. They are shown as single forces in Fig. 13–18 for simplicity.) (a) Find the fractional change in length of the edges of the cube parallel to each of the three coordinate axes X, Y, Z. Express your answer in terms of F, L, Young's modulus Y, and Poisson's ratio σ. (b) Find the fractional change in volume of the cube.

13–20. Compressive forces are applied to two opposite surfaces of a rec-

tangular block of volume $V = abc$. The resulting fractional decrease in the length of the block is .001. The fractional volume decrease is .0005. Determine Poisson's ratio for the material of the block.

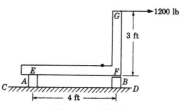

FIGURE 13–19

13–21. In Fig. 13–19, A and B are two short steel rods, of 0.5 in^2 cross-sectional area. The lower ends of A and B are welded to the fixed plate CD. The upper end of A is welded to the L-shaped piece EFG, which can slide without friction on the upper end of B. A horizontal pull of 1200 lb is exerted at G. Neglect the weight of EFG. (a) Compute the shearing stresses in rods A and B. (b) Compute the longitudinal stress in A. Is the stress a tension or a compression? (c) Compute the longitudinal stress in B. Is it a tension or a compression?

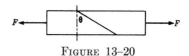

FIGURE 13–20

13–22. A bar of cross section A is subjected to equal and opposite tensile forces F at its ends. Consider a plane through the bar making an angle θ with a plane at right angles to the bar (Fig. 13–20). (a) What is the tensile (normal) stress at this plane, in terms of F, A, and θ? (b) What is the shearing (tangential) stress at the plane, in terms of F, A, and θ? (c) For what value of θ is the tensile stress a maximum? (d) For what value of θ is the shearing stress a maximum?

13–23. Calculate the torque in newton-meters necessary to hold a hollow cylinder of inside radius 2 cm and wall thickness 1 mm, twisted through an angle of 1 degree. The length of the cylinder is 1 meter and the shear modulus is 6×10^{11} dynes/cm^2.

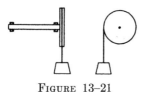

FIGURE 13–21

13–24. A steel rod 1 meter long and 0.20 cm in radius is rigidly clamped at one end. (Fig. 13–21.) A disk 20 cm in radius is attached to the other end, which is free to rotate. When a load of 500 gm is hung from a cord wrapped around the disk, the load is observed to descend 10 cm. (a) What is the shear modulus of the material of which the rod is composed? (b) What is the decrease in potential energy of the load? (c) What is the elastic potential energy of the twisted rod?

13–25. A thin-walled circular tube of mean radius 10 cm and thickness .05 cm is melted up and recast into a solid rod of the same length. In each case the rod is to be twisted by a torque τ through an angle θ such that $\tau = k\theta$. Find the ratio of the values of k for the two cases.

13–26. A steel wire 5 ft long and 0.1 inch diameter is stretched by a force of 100 lb and then twisted about its axis by a torque of 15 lb·in. Compute the elastic potential energy of the stressed wire.

13–27. The drive shaft of an automobile rotates at 3000 rpm and transmits 10 hp from the engine to the rear wheels. (a) What is the torque acting on the shaft? (b) Through what angle is one end of the shaft twisted with respect to the other if the shaft is 1 inch in radius, 5 ft long, and has a torsion modulus of $12^3 \times 10^6$ lb/ft^2?

13–28. A hollow shaft of length 3 ft, inner and outer radii 1.5 and 2 in, shear modulus 11×10^6 lb/in^2, is used to transmit power from an 8 hp motor to a drum of radius 2 ft. A cable wrapped around the drum supports an elevator. Through what angle is the shaft twisted when the elevator is drawn upward with a constant velocity of 2 ft/sec?

CHAPTER 14

HARMONIC MOTION

14–1 Introduction. The motion of a body when acted upon by a constant force was considered in detail in Chapters 5 and 6. The motion is one of constant acceleration, and it was found useful to derive expressions for the position and velocity of the body at any time, and for its velocity in any position. In the present chapter we are to study the motion of a body when the resultant force on it is not constant, but varies during the motion. Naturally, there are an infinite number of ways in which a force may vary and hence no general expressions can be given for the motion of a body when acted on by a variable force, except that the acceleration at each instant must equal the force at that instant divided by the mass of the body. There is, however, one particular mode of variation which is met with in practice so frequently that it is worth while to develop formulas for this special case. The force referred to is an elastic restoring force, brought into play whenever a body is distorted from its normal shape. When released, the body will be found to vibrate about its equilibrium position.

Examples of this sort of motion are the up-and-down motion which ensues when a weight hanging from a spring is pulled down and released, the vibrations of the strings or air columns of musical instruments, the vibration of a bridge or building under impact loads, and the oscillation of the balance wheel of a watch or of a clock pendulum. Furthermore, many reciprocating motions such as those of the crosshead in a steam engine or the piston of an automobile engine, while not exactly of this type, do approximate it quite closely.

It turns out that the equations of motion involve sines or cosines, and the term *harmonic* is applied to expressions containing these functions. This type of vibratory motion is therefore called *harmonic motion*.

14–2 Elastic restoring forces. It has been shown in Chapter 13 that when a body is caused to change its shape, the distorting force is proportional to the amount of the change, provided the elastic limit is not exceeded. The change may be in the nature of an increase in length, as of a rubber band or a coil spring, or a decrease in length, or a bending as of a flat spring, or a twisting of a rod about its axis, or of many other forms. The term "force" is to be interpreted liberally as the force, or

263

torque, or pressure, or whatever may be producing the distortion. If we restrict the discussion to the case of a push or a pull, where the distortion is simply the displacement of the point of application of the force, the force and displacement are related by Hooke's law,

$$F = kx,$$

where k is a proportionality constant called the force constant and x is the displacement from the equilibrium position.

In this equation, F stands for the force which must be exerted *on* an elastic body to produce the displacement x. We shall find it more convenient to work with the reaction to this force, that is, the force with which the distorted body pulls back. This force is called the *restoring force* and is given by

$$F = -kx.$$

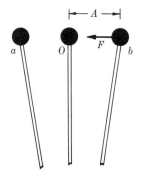

FIG. 14-1. Motion under an elastic restoring force.

14-3 Definitions. To fix our ideas, suppose that a flat strip of steel such as a hacksaw blade is clamped vertically in a vise and a small mass is attached to its upper end as in Fig. 14-1. We shall assume that the strip is sufficiently long and the displacement sufficiently small so that the motion is essentially along a straight line. The mass of the strip itself is negligible.

Let the top of the spring be pulled to the right a distance A as in Fig. 14-1 and released. The attached mass is then acted on by a restoring force exerted by the steel strip and directed toward the equilibrium position O. It therefore accelerates in the direction of this force, and moves in toward the center with increasing speed. The *rate* of increase (i.e., the acceleration) is not constant, however, since the accelerating force becomes smaller as the body approaches the center.

When the body reaches the center the restoring force has decreased to zero, but because of the velocity which has been acquired, the body "overshoots" the equilibrium position and continues to move toward the left. As soon as the equilibrium position is passed the restoring force again comes into play, directed now toward the right. The body therefore decelerates, and at a rate which increases with increasing distance from O.

It will therefore be brought to rest at some point to the left of O, and repeat its motion in the opposite direction.

Both experiment and theory show that the motion will be confined to a range $\pm A$ on either side of the equilibrium position, each to-and-fro movement taking place in the same length of time. If there were no loss of energy by friction the motion would continue indefinitely once it had been started. This type of motion, under the influence of an elastic restoring force and in the absence of all friction, is called *simple* harmonic motion, often abbreviated *SHM*.

Any sort of motion which repeats itself in equal intervals of time is called *periodic*, and if the motion is back and forth over the same path it is also called *oscillatory*.

A *complete vibration* or *oscillation* means one round trip, say from a to b and back to a, or from O to b to O to a and back to O.

The *periodic time*, or simply the *period* of the motion, represented by T, is the time required for one complete vibration.

The *frequency*, f, is the number of complete vibrations per unit time. Evidently the frequency is the reciprocal of the period, or

$$T = \frac{1}{f}.$$

The *displacement*, x, at any instant, is the distance away from the equilibrium position or center of the path at that instant.

The *amplitude*, A, is the maximum displacement. The total range of the motion is therefore $2A$.

14–4 Equations of simple harmonic motion. We now wish to find expressions for the displacement, velocity, and acceleration of a body moving with *SHM*, just as we found those for a body moving with constant acceleration. It must be emphasized that the equations of motion with *constant* acceleration cannot be applied, since the acceleration is continually changing.

Figure 14–2 represents the vibrating body of Fig. 14–1 at some instant when its displacement is x. The resultant force on it is simply the elastic restoring force, $-kx$, and from Newton's second law,

$$F = -kx = ma,$$

or

$$a = -\frac{k}{m}x, \tag{14-1}$$

where m is the mass of the body.

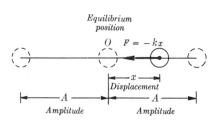

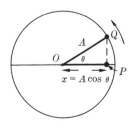

FIGURE 14–2

FIG. 14–3. Coordinate of a body in simple harmonic motion.

Since k and m are both constants, the ratio k/m is constant. Hence the acceleration is directly proportional to the displacement and, because of the minus sign, is in the opposite direction to the displacement. That is, when the body is at the right of its equilibrium position, its acceleration is toward the left and vice versa. Another way of stating this is that the acceleration is always directed toward the center of the path. With the help of calculus, the equation above can be solved at once to find the desired expressions for displacement and velocity. However, we shall first make use of a simple geometrical method of deducing these equations.

Consider a type of motion determined as follows. Let Q, Fig. 14–3, be a point revolving in a circle of radius A with a constant angular velocity of ω rad/sec. Let P be a point on the horizontal diameter of the circle, directly below Q. Point P is called the projection of Q onto the diameter. Point Q is referred to as the *reference point,* and the circle in which it moves as the *reference circle.*

As the reference point revolves, the point P moves back and forth along a horizontal line, keeping always directly below (or above) Q. We shall show that the motion of P is the same as that of a body moving under the influence of an elastic restoring force in the absence of friction.

The displacement of P at any time t is the distance OP or x, and if θ represents the angle which OQ makes with the horizontal diameter,

$$x = A \cos \theta.$$

The angle θ is called the *phase angle,* or simply the *phase* of the motion.

If point Q is at the extreme right-hand end of the diameter at time $t = 0$, the angle θ may be written

$$\theta = \omega t.$$

Hence

$$x = A \cos \omega t.$$

Now ω, the angular velocity of Q in radians per second, is related to f, the number of complete revolutions of Q per second, by

$$\omega = 2\pi f,$$

since there are 2π radians in one complete revolution. Furthermore, the point P makes one complete vibration for each revolution of Q. Hence f may also be interpreted as the number of vibrations per second or the frequency of vibration of point P. Replacing ω by $2\pi f$, we have

$$x = A \cos 2\pi ft. \tag{14-2}$$

Equation (14–2) gives the displacement of point P at any time t after the start of the motion, and thus corresponds to

$$x = v_0 t + \tfrac{1}{2}at^2$$

for a body moving with constant acceleration.

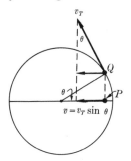

FIG. 14–4. Velocity in simple harmonic motion.

The instantaneous velocity of P may be found with the aid of Fig. 14–4. The reference point Q moves with a tangential velocity

$$v_T = \omega A = 2\pi f A.$$

Since point P is always directly below or above the reference point, the velocity of P at each instant must equal the X-component of the velocity of Q. That is, from Fig. 14–4,

$$v = v_T \sin \theta = -2\pi f A \sin \theta,$$

or

$$v = -2\pi f A \sin 2\pi ft. \tag{14-3}$$

The minus sign is introduced since the direction of the velocity is toward the left. When Q is below the horizontal diameter, the velocity of P will be toward the right, but since $\sin \theta$ is negative at such points, the minus sign is still needed. Equation (14–3) gives the velocity of point P at any time, and corresponds to

$$v = v_0 + at$$

for motion with constant acceleration.

Since $\sin\theta = \sqrt{1 - \cos^2\theta}$ and $\cos\theta = x/A$, Eq. (14–3) may be written

$$v = \pm 2\pi f A \sqrt{1 - \frac{x^2}{A^2}},$$

$$v = \pm 2\pi f \sqrt{A^2 - x^2}. \tag{14–4}$$

The symbol $\pm$ is required, since at any given displacement x the point may be moving either toward the right or left. Equation (14–4) gives the velocity of P at any *displacement*. It thus corresponds to

$$v^2 = v_0^2 + 2ax$$

for constant acceleration.

Finally, the acceleration of point P may be found, making use again of the fact that since P is always directly below or above Q its acceleration must equal the X-component of the acceleration of Q. Point Q, since it moves in a circular path with a constant angular velocity ω, has at each instant an acceleration toward the center given by

$$a_R = \omega^2 A = 4\pi^2 f^2 A.$$

From Fig. 14–5, the X-component of this acceleration is

$$a = a_R \cos\theta,$$

or

$$a = -4\pi^2 f^2 A \cos 2\pi f t, \tag{14–5}$$

and since $A \cos 2\pi f t = x$,

$$a = -4\pi^2 f^2 x. \tag{14–6}$$

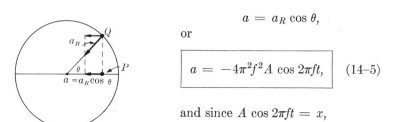

Fig. 14–5. Acceleration in simple harmonic motion.

The minus sign is introduced since the acceleration is toward the left. When Q is at the left of the center the acceleration of P is toward the right, but since $\cos\theta$ is negative at such points, the minus sign is still required. Equations (14–5) and (14–6) give the acceleration of P at any time and at any displacement. There are no corresponding equations for motion with constant acceleration except the extremely simple one

$$a = \text{constant.}$$

Equation (14–6) states that the acceleration of point P, moving in the manner described, is proportional to the displacement x and is in the direction opposite to x. But this is just the condition which must be fulfilled by a body moving under the influence of an elastic restoring force (Eq. 14–1). Hence it may be concluded that Eqs. (14–2) to (14–6) describe the motion of a body when acted on by such a force, and are the equations which we set out to find.

In order that the motion of point P may coincide in all respects with that of a given vibrating body, the radius A of the reference circle must equal the amplitude A of the actual vibration, and the frequency of revolution of Q must be the same as the frequency of the actual vibration. The proper value of the latter may be found by combining Eq. (14–1),

$$a = -\frac{k}{m}x,$$

which gives the acceleration of the vibrating body at any displacement, and Eq. (14–6),

$$a = -4\pi^2 f^2 x,$$

which gives the acceleration of point P at any displacement. Since these accelerations must be the same,

$$\frac{k}{m} = 4\pi^2 f^2,$$

and therefore

$$f = \frac{1}{2\pi}\sqrt{\frac{k}{m}}. \qquad (14\text{–}7)$$

Equation (14–7) may be used to find the vibration frequency of a body of given mass when vibrating under the influence of an elastic restoring force of given force constant.

Since the period is the reciprocal of the frequency, Eq. (14–7) may also be written

$$T = 2\pi\sqrt{\frac{m}{k}}. \qquad (14\text{–}8)$$

In using Eqs. (14–7) or (14–8), m must be expressed in slugs, kilograms, or grams, and k in lb/ft, newtons/meter, or dynes/cm. The frequency f

will then be in vibrations per second, and the period T in seconds per vibration.

A somewhat unexpected conclusion to be drawn from these equations is that the period does not depend on the amplitude, but on the mass and force constant alone.

The equations of simple harmonic motion may be deduced by the methods of calculus as follows. Equation (14–1), in calculus notation, is

$$\frac{d^2x}{dt^2} = -\frac{k}{m}\,x. \tag{14–9}$$

That is, x is a function of t such that its second derivative with respect to t is equal to x multiplied by a negative constant. This suggests that x is a sine or cosine function of t. We therefore try $x = A\cos\omega t$. Then

$$\frac{dx}{dt} = -\omega A \sin \omega t, \tag{14–10}$$

and

$$\frac{d^2x}{dt^2} = -\omega^2 A \cos \omega t = -\omega^2 x. \tag{14–11}$$

Hence the differential equation is satisfied if $\omega^2 = k/m$ or $\omega = \sqrt{k/m} = 2\pi f$, which is the same as Eq. (14–7).

The constant A may have any value so far as the differential equation is concerned. It evidently must be set equal to the amplitude of the motion to satisfy the initial condition that x is equal to the amplitude of the motion when t is zero.

Equation (14–10) is evidently the same as Eq. (14–3), and Eq. (14–11) is the same as Eqs. (14–5) and (14–6).

It is helpful in visualizing harmonic motion to represent the position, velocity, and acceleration of the vibrating body graphically. Graphs of these quantities against time are given in Fig. 14–6, which may be considered as a graph of Eqs. (14–2), (14–3), and (14–5). Note that the velocity is a maximum when the displacement is zero, that is, at the center, while the velocity is zero when the displacement is a maximum. The acceleration, on the other hand, is zero at the center and a maximum at the ends of the path.

Figure 14–7 is a multiflash photograph of the motion of a mass suspended from a coil spring and set into vertical vibration. The camera was rotated about a vertical axis while the photographs were taken so that each image is displaced laterally from the preceding image. In effect this introduces

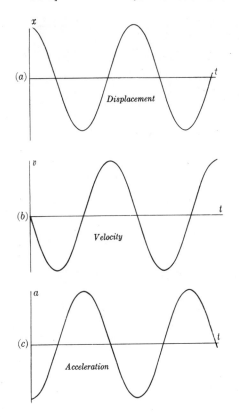

(a)

Displacement

(b)

Velocity

(c)

Acceleration

FIG. 14–6. Graphs of position, velocity, and acceleration.

a horizontal time scale into the motion and the body traces out its own sinusoidal displacement-time graph corresponding to the upper diagram in Fig. 14–6. By comparing the *vertical* separation of successive images, it is seen that the velocity is greatest at the center of the path and zero at the ends, while the acceleration is greatest at the ends and zero at the center.

If the body is not at the extreme end of its path when $t = 0$ the equations of motion must be generalized. Figure 14–8, representing the reference circle of a harmonic vibration, shows the reference point at an angle θ_0 when $t = 0$. After time t, the reference point has advanced through an angle ωt or $2\pi ft$, and its angular position is

$$\theta = \theta_0 + \omega t = 2\pi ft + \theta_0.$$

The coordinate of its projection on the X-axis is therefore

$$x = A \cos \theta = A \cos (2\pi ft + \theta_0).$$

The term θ_0 corresponds to the initial position x_0 in Eq. (4–16) on page 59, and is often called the *epoch angle*.

The velocity is given by

$$v = -2\pi fA \sin (2\pi ft + \theta_0),$$

and the acceleration is

$$a = -4\pi^2 f^2 A \cos (2\pi ft + \theta_0).$$

For the special case where $\theta_0 = \pi/2$, that is, when the initial position is at the center of the path,

$$x = A \cos (2\pi ft + \pi/2) = A \sin 2\pi ft,$$
$$v = 2\pi fA \cos 2\pi ft,$$
$$a = -4\pi^2 f^2 A \sin 2\pi ft,$$

and the equations are the same as those for motion starting at the end of the path except that sines and cosines are interchanged.

EXAMPLES. (1) A flat steel strip is mounted as in Fig. 14–1. By attaching a spring balance to the end of the strip and pulling it sidewise, it is found that a force of 1 lb will produce a deflection of 6 in. A 4-lb body is attached to the end of the strip, pulled aside a distance of 8 in., and released.

FIG. 14–7. Simple harmonic motion of a mass suspended from a spring.

(a) Compute the force constant of the spring.

A force of 1 lb produces a displacement of 6 in. or $\frac{1}{2}$ ft. Hence

$$k = \frac{F}{x} = \frac{1}{\frac{1}{2}} = 2 \text{ lb/ft.}$$

(b) Compute the period of vibration.

$$T = 2\pi \sqrt{\frac{m}{k}} = 2\pi \sqrt{\frac{4/32}{2}} = \frac{\pi}{2} \text{ sec.}$$

(c) Compute the maximum velocity attained by the vibrating body.

The maximum velocity occurs at the center when the displacement is zero. From Eq. (14–4),

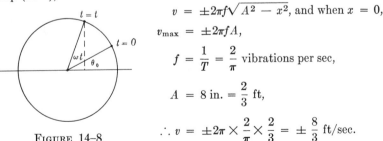

FIGURE 14–8

$v = \pm 2\pi f \sqrt{A^2 - x^2}$, and when $x = 0$,

$$v_{\max} = \pm 2\pi f A,$$

$$f = \frac{1}{T} = \frac{2}{\pi} \text{ vibrations per sec,}$$

$$A = 8 \text{ in.} = \frac{2}{3} \text{ ft,}$$

$$\therefore v = \pm 2\pi \times \frac{2}{\pi} \times \frac{2}{3} = \pm \frac{8}{3} \text{ ft/sec.}$$

(d) Compute the maximum acceleration.

The maximum acceleration occurs at the ends of the path, where $x = \pm A$. From Eq. (14–6),

$$a_{max} = \mp 4\pi^2 f^2 A$$

$$= \mp 4\pi^2 \times \left(\frac{2}{\pi}\right)^2 \times \frac{2}{3}$$

$$= \mp \frac{32}{3} \text{ ft/sec}^2.$$

(e) Compute the velocity and acceleration when the body has moved halfway in toward the center from its initial position.

At this point,

$$x = \frac{A}{2} = \frac{1}{3} \text{ ft,}$$

$$v = -2\pi \times \frac{2}{\pi} \sqrt{\left(\frac{2}{3}\right)^2 - \left(\frac{1}{3}\right)^2} = -\frac{4}{\sqrt{3}} \text{ ft/sec,}$$

$$a = -4\pi^2 \times \left(\frac{2}{\pi}\right)^2 \times \frac{1}{3} = -\frac{16}{3} \text{ ft/sec}^2.$$

(f) How long a time is required for the body to move halfway in to the center from its initial position?

Note that the motion is neither one of constant velocity nor of constant acceleration. The simplest method of handling a problem involving the time required to move from one point to another in harmonic motion is to make use of the reference circle. While the body moves halfway in, the reference point revolves through an angle of 60° (Fig. 14–9). Since the reference point moves with constant angular velocity and makes one complete revolution in $\pi/2$ sec (in this particular example), the time to rotate through 60° is $\frac{1}{6} \times \pi/2 = \pi/12$ sec.

The time may also be computed directly from the equation

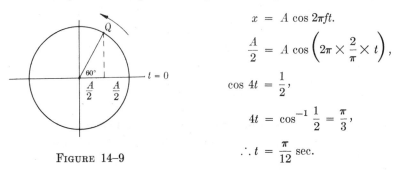

$$x = A \cos 2\pi f t.$$

$$\frac{A}{2} = A \cos \left(2\pi \times \frac{2}{\pi} \times t\right),$$

$$\cos 4t = \frac{1}{2},$$

$$4t = \cos^{-1}\frac{1}{2} = \frac{\pi}{3},$$

$$\therefore t = \frac{\pi}{12} \text{ sec.}$$

FIGURE 14–9

14–5 Energy relations in harmonic motion. If friction effects are neglected, the total mechanical energy of a vibrating spring-mass system remains constant and, once started, the motion will continue indefinitely.

Since the velocity and position of the vibrating body are continually changing, the kinetic and potential energies change also, but their sum must have the same value at all times.

It has been shown in Chapter 8 that the potential energy of a stretched spring is

$$PE = \tfrac{1}{2}kx^2.$$

The kinetic energy is

$$KE = \tfrac{1}{2}mv^2,$$

and since

$$v = 2\pi f\sqrt{A^2 - x^2}$$

and

$$f = \frac{1}{2\pi}\sqrt{\frac{k}{m}},$$

the kinetic energy may be written

$$KE = \tfrac{1}{2}k(A^2 - x^2).$$

The total energy is therefore

$$PE + KE = \tfrac{1}{2}kx^2 + \tfrac{1}{2}k(A^2 - x^2)$$
$$= \tfrac{1}{2}kA^2,$$

which is a constant, and equal to the potential energy at either end of the path. The equation may be interpreted to mean that the system is given an amount of potential energy $\tfrac{1}{2}kA^2$ at the start, by displacing it a distance A, and that thereafter the total energy is equal to this initial value.

It will be left as an exercise to show that the total energy is also equal to the kinetic energy of the vibrating body as it passes through its equilibrium position.

14–6 The simple pendulum. A simple pendulum consists of a mass of small dimensions suspended by an inextensible weightless string. When pulled to one side of its equilibrium position and released, the pendulum bob vibrates about this position with motion which is both periodic and oscillatory. We wish to discover whether the motion is simple harmonic.

The necessary condition for simple harmonic motion is that the restoring force shall be directly proportional to the displacement and oppositely directed. The path of the bob is, of course, not a straight line, but the arc of a circle of radius L, where L is the length of the supporting cord.

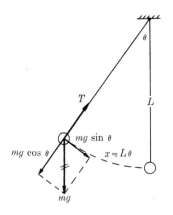

The displacement refers to distances measured along this arc. (See Fig. 14–10.) Hence if $F = -kx$ the motion will be simple harmonic, or since $x = L\theta$, the requirement may be written $F = -kL\theta$.

Figure 14–10 shows the forces on the bob at an instant when its displacement is x. Choose axes tangent to the circle and along the radius, and resolve the weight into components. The restoring force F is

FIG. 14–10. The simple pendulum.

$$F = -mg \sin \theta. \quad (14\text{–}12)$$

The restoring force is therefore *not* proportional to θ but to $\sin \theta$, so the motion is *not* simple harmonic. However, *if the angle θ is small*, $\sin \theta$ is very nearly equal to θ and Eq. (14–12) becomes

$$F = -mg\theta = -mg \frac{x}{L},$$

or

$$F = -\frac{mg}{L} x.$$

The restoring force is then proportional to the displacement *for small displacements*, and the constant mg/L represents the force constant k. The period of a simple pendulum when its amplitude is small is therefore

$$T = 2\pi \sqrt{\frac{m}{k}} = 2\pi \sqrt{\frac{m}{mg/L}},$$

or

$$T = 2\pi \sqrt{\frac{L}{g}}. \quad (14\text{–}13)$$

What constitutes a "small" amplitude? It can be shown that the general equation for the time of swing, when the maximum angular displacement is α, is

$$T = 2\pi \sqrt{\frac{L}{g}} \left(1 + \frac{1}{4} \sin^2 \frac{\alpha}{2} + \frac{9}{64} \sin^4 \frac{\alpha}{2} + \cdots \right).$$

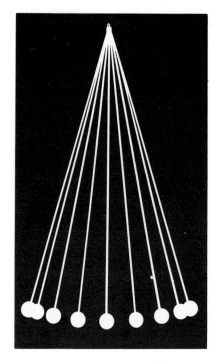

FIG. 14–11. A single swing of a simple pendulum.

The time may be computed to any desired degree of precision by taking enough terms in the infinite series. When $\alpha = 15°$ (on either side of the central position), the true period differs from that given by the approximate Eq. (14–13) by less than one-half of one percent.

The utility of the pendulum as a timekeeper is based on the fact that the period is practically independent of the amplitude. Thus, as a clock runs down and the amplitude of the swings becomes slightly smaller, the clock will still keep very nearly correct time.

The simple pendulum is also a precise and convenient method of measuring the acceleration of gravity, g, without actually resorting to free fall, since L and T may readily be measured. More complicated pendulums find considerable application in the field of geophysics. Local deposits of ore or oil, if their density differs from that of their surroundings, affect the local value of "g," and precise measurements of this quantity over an area which is being prospected often furnish valuable information regarding the nature of underlying deposits.

Figure 14–11 is a multiflash photograph of a single swing of a simple pendulum. The motion is evidently of the simple harmonic type with maximum speed at the center and maximum acceleration at the ends of the swing.

14–7 Lissajous' figures. The path traced out by a point which simultaneously executes simple harmonic motion in two directions at right angles to each other is called a *Lissajous' figure*. Given the amplitude and frequency of each component vibration and the phase relation between them, the resulting figure can be found by the method shown in Fig. 14–12. Line AB represents one component vibration together with its reference circle. Line CD represents the other vibration. The figure is constructed for the special case in which the component vibrations have equal frequen-

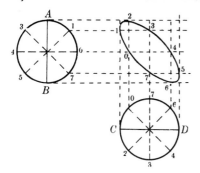

Fig. 14–12. Method of constructing Lissajous' figures.

cies and amplitudes, and where the phase difference is 45° or $\pi/4$ radians. At the start, the reference point is at position 0 in each circle. Points in the Lissajous' figure are obtained by projecting horizontally and vertically from corresponding points on the reference circles. The resultant motion in this case is an ellipse with its axes at 45° with either component vibration. Patterns obtained with other frequency ratios and phase relations are shown in Fig. 14–13.

Lissajous' figures are a very useful method for comparing two frequencies, for example, when calibrating an electrical oscillator. A known frequency is applied to one pair of plates of a cathode-ray oscillograph and the unknown frequency to the other pair of plates. If the two frequencies are exactly equal, the pattern will have one of the forms of the sequence of the top line in Fig. 14–13. If the frequencies are nearly but not exactly equal, the pattern migrates slowly through the sequence. If one frequency is twice that of the other, a pattern from the sequence of the second line is formed, and so on.

14–8 Damped harmonic motion. The effect of friction has been neglected in the preceding discussion, and the equations of motion predict, as would be expected, a vibration which continues indefinitely with the same amplitude. Actually, a weight on a spring (or a pendulum) does not oscillate indefinitely, but as a result of friction its amplitude decreases regularly and it finally comes to rest. The motion is said to be *damped out* by friction and is called *damped harmonic motion*.

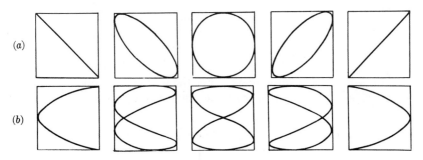

Fig. 14–13. Lissajous' figures. (a) Frequencies equal. (b) Frequency ratio 2:1.

In most cases of practical interest, the friction force is due to air resistance and internal friction in the spring and so is of the viscous type. In other words, the friction force is not constant but is proportional to the velocity and oppositely directed. We must therefore add to the elastic restoring force a friction force proportional to the negative of the velocity, obtaining

$$\text{Resultant force} = -kx - \mu v = ma,$$

or in calculus notation

$$\frac{d^2x}{dt^2} + \frac{\mu}{m}\frac{dx}{dt} + \frac{k}{m}x = 0.$$

Provided the damping is not too great, the solution of this equation is

$$x = \left[\frac{f_0}{f'} A \epsilon^{-\mu t/2m}\right] \cos(2\pi f't + \phi), \tag{14-14}$$

where A is the initial displacement and

$$f_0 = \frac{1}{2\pi}\sqrt{\frac{k}{m}}, \qquad f' = \frac{1}{2\pi}\sqrt{\frac{k}{m} - \frac{\mu^2}{4m^2}}, \qquad \cos\phi = \frac{f'}{f_0}.$$

If the friction coefficient μ is equal to zero, then

$$f' = f_0, \qquad \epsilon^{-\mu t/2m} = 1, \qquad \phi = 0,$$

and Eq. (14-14) reduces to that for undamped harmonic motion. When friction is present the frequency is reduced, or the periodic time is increased, as shown by the appearance of the term $-\mu^2/4m^2$. The term in brackets in Eq. (14-14) can be considered as the amplitude of the vibration, and since the exponential factor steadily decreases with time, the motion eventually dies out. The natural logarithm of the ratio of the amplitudes of two successive swings is called the *logarithmic decrement* of the motion.

If the friction is so great that $\mu^2/4m^2 > k/m$, the radial becomes imaginary and the motion is not oscillatory. The ball or pendulum, if pulled aside, returns slowly to its initial position without overshooting. The motion is said to be "overdamped." For the particular friction force which separates one type of motion from the other, namely, when

$$\frac{\mu^2}{4m^2} = \frac{k}{m},$$

the motion just fails to be oscillatory and is called "critically damped."

Figure 14-14 is a series of multiflash photographs of the motion of a

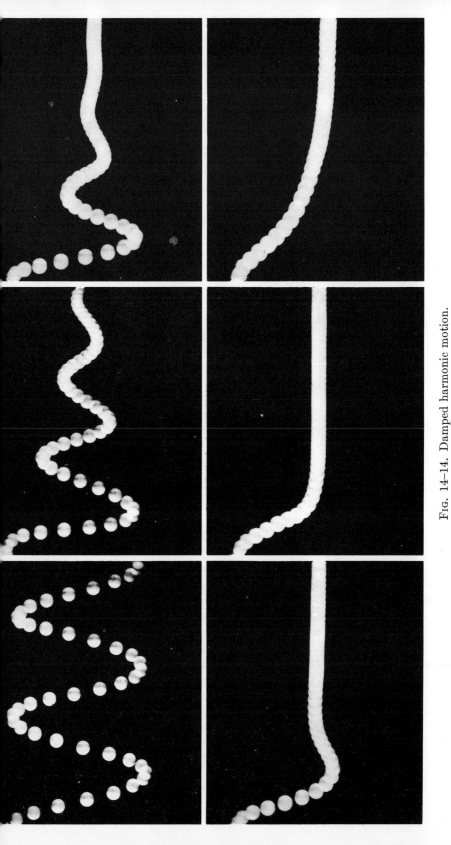

Fig. 14-14. Damped harmonic motion.

spring-weight system with an increase in damping between successive pictures. The next to the last photograph shows critically damped motion, and in the last view the motion is overdamped.

14–9 Forced harmonic motion. Resonance. The type of motion analyzed in the preceding sections results when a spring-weight system is displaced from its equilibrium position and left to itself thereafter. Another case of interest arises when such a system is subjected to a series of periodic impulses. A succession of relatively small impulses, if properly timed, will set up a vibration of relatively large amplitude. Everyone who uses a swing discovers this very quickly by experiment.

For mathematical simplicity we shall assume the applied force to vary sinusoidally according to the law

$$F \cos 2\pi f t,$$

where F is the maximum force and f is its frequency. The force on the vibrating body is the resultant of the periodic driving force, the elastic restoring force, and the friction force. Hence from Newton's second law,

$$\text{Resultant force} = F \cos 2\pi f t - kx - \mu v = ma,$$

or in calculus notation

$$\frac{d^2 x}{dt^2} + \frac{\mu}{m} \frac{dx}{dt} + \frac{k}{m} x = \frac{F}{m} \cos 2\pi f t.$$

The complete solution of this equation is rather complex and will not be given. After the external force has been in operation for a sufficiently long time, the system settles down to a state of undamped harmonic motion, not of its own natural frequency but with the same frequency as the driving force. The *amplitude* of the vibration is smaller the greater the friction, and depends also on the frequency of the driving force. Figure 14–15 shows four curves, each corresponding to a certain friction coefficient. The frequency of the driving force is plotted horizontally and the amplitude of the resulting vibration is plotted vertically. The natural frequency in the absence of damping is shown by the vertical line.

The particular frequency which results in the maximum amplitude of vibration is called the *resonant frequency*. If the system is frictionless, the resonant frequency equals the natural frequency. If friction is present, the natural frequency, as we have seen, is slightly less than it is for the same system without friction and the resonant frequency is somewhat smaller still. Except when there is considerable friction, however, the resonant frequency and natural frequency are nearly the same.

If there is no friction (upper curve) the amplitude becomes infinite

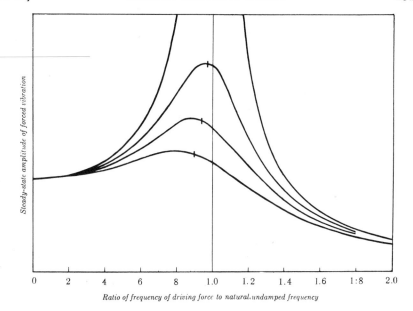

FIG. 14–15. Amplitude of forced harmonic motion as a function of frequency of driving force.

when the frequency of the driving force equals the natural frequency. The lower curves show the response for three different friction coefficients. The resonant frequency is that corresponding to the maximum of each curve. The natural frequency of each is indicated by the short vertical dash.

The friction coefficients of the three lower curves are such that if the system is displaced and released so that it oscillates as in Fig. 14–14, the amplitudes of successive swings decrease to 50%, 25%, and 2% respectively.

14–10 Angular harmonic motion. Angular harmonic motion results when a body which is pivoted about an axis experiences a restoring torque proportional to the angular displacement from its equilibrium position. This type of vibration is very similar to linear harmonic motion, and the corresponding equations may be written down immediately from the analogies between linear and angular quantities. The oscillatory motion of the balance wheel of a watch is a common example of angular harmonic motion.

A restoring torque proportional to angular displacement is expressed by

$$\tau = -k\theta. \tag{14–15}$$

The moment of inertia of the pivoted body corresponds to the mass of a body in linear motion. Hence the period formula for angular harmonic motion is

$$T = 2\pi \sqrt{\frac{I}{k}}, \tag{14-16}$$

where k is the constant in Eq. (14-15).

The equations for angular displacement, angular velocity, and angular acceleration can be obtained by comparison with the corresponding equations in Section 14-4.

14-11 The physical pendulum. A so-called "physical" pendulum is any real pendulum, as contrasted with a simple pendulum in which all of the mass is assumed to be concentrated at a point. Let Fig. 14-16 represent a body of irregular shape pivoted about a horizontal frictionless axis and displaced from the vertical by an angle θ. The distance from the pivot to the center of gravity is l, the moment of inertia of the pendulum about an axis through the pivot is I, and the mass of the pendulum is m. The restoring torque in the position shown in the figure is

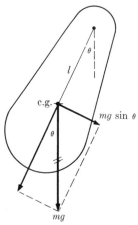

$$\tau = -mgl \sin \theta.$$

If θ is small we may replace $\sin \theta$ by θ, and

$$\tau = -mgl\theta.$$

Hence the pendulum is acted on, in effect, by an elastic restoring torque with $k = mgl$. The period of vibration is therefore

FIG. 14-16. The physical pendulum.

$$T = 2\pi \sqrt{\frac{I}{k}} = 2\pi \sqrt{\frac{I}{mgl}} \text{ (for small amplitudes).} \tag{14-17}$$

EXAMPLES. (1) Let the body in Fig. 14-16 be a meter stick pivoted at one end. Then if L stands for the length of 1 meter,

$$I = \frac{1}{3} mL^2, \qquad l = \frac{L}{2}, \qquad g = 9.8 \text{ m/sec}^2,$$

$$T = 2\pi \sqrt{\frac{\frac{1}{3}mL^2}{mg(L/2)}} = 2\pi \sqrt{\frac{2}{3}\frac{L}{g}}$$

$$= 2\pi \sqrt{\frac{2}{3}\frac{1}{9.8}} = 1.65 \text{ sec.}$$

(2) Equation (14–16) may be solved for the moment of inertia I, giving

$$I = \frac{T^2 mgl}{4\pi^2}.$$ (14–18)

The quantities on the right of the equation are all directly measurable. Hence the moment of inertia of a body of any complex shape may be found by suspending the body as a physical pendulum and measuring its period of vibration. The location of the center of gravity can be found by balancing. Since T, m, g, and l are known, I can be computed. For example, Fig. 14–17 illustrates a connecting rod pivoted about a horizontal knife edge. The connecting rod weighs 4 lb and its center of gravity has been found by balancing to be 8 in. below the knife's edge. When set into oscillation, it is found to make 100 complete vibrations in 120 sec, so that $T = 120/100 = 1.2$ sec. Therefore

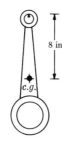

8 in.

$$I = \frac{(1.2)^2 \times 4 \times \frac{2}{3}}{4\pi^2}$$

$$= 0.097 \text{ slug·ft}^2.$$

From the parallel-axis theorem,

$$I_0 = I - ma^2 = 0.097 - \frac{1}{8}(\frac{2}{3})^2$$

$$= 0.042 \text{ slug·ft}^2.$$

FIGURE 14–17

14–12 Center of oscillation. It is always possible to find an *equivalent* simple pendulum whose period is equal to that of a given physical pendulum. If L_0 is the length of the equivalent simple pendulum,

$$T = 2\pi \sqrt{\frac{L_0}{g}} = 2\pi \sqrt{\frac{I}{mgl}},$$

or

$$L_0 = \frac{I}{ml}.$$ (14–19)

Thus, so far as its period of vibration is concerned, the mass of a physical pendulum may be considered to be concentrated at a point whose distance from the pivot is $L_0 = I/ml$. This point is called the *center of oscillation* of the pendulum.

Figure 14–18 shows a body pivoted about an axis through P and whose center of oscillation is at point C. The center of oscillation and the point of support have the following interesting property, namely, if the pendulum is pivoted about a new axis through point C its period is unchanged and point P becomes the new center of oscillation. The point of support and the center of oscillation are said to be *conjugate* to one another.

The length of the equivalent simple pendulum when the pivot is at P is

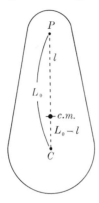

$$L_0 = \frac{I_P}{ml},$$

where I_P is the moment of inertia about an axis through P.

The length L_0' of the equivalent simple pendulum when the body is pivoted at C is

$$L_0' = \frac{I_C}{m(L_0 - l)}, \quad (14\text{-}20)$$

FIG. 14–18. Point C is the center of oscillation.

where I_C is the moment of inertia about an axis through C.

From the parallel-axis theorem,

$$I_P = I_0 + ml^2, \qquad I_C = I_0 + m(L_0 - l)^2.$$

When I_0 is eliminated between these equations, we get

$$I_C = I_P + mL_0^2 - 2mlL_0,$$

and since $I_P = mlL_0$, it follows that $I_C = mL_0(L_0 - l)$. Then from Eq. (14–20),

$$L_0' = \frac{mL_0(L_0 - l)}{m(L_0 - l)} = L_0.$$

The equivalent length when the pivot passes through C is therefore the same as when the pivot passes through P.

Let us return to Eq. (14–19) and replace I by mk^2, where k is the radius of gyration about an axis through the pivot. Then

$$L_0 = \frac{k^2}{l}, \qquad L_0 l = k^2. \quad (14\text{-}21)$$

Equation (14–21) contains three different distances, L_0, k, and l, at which the mass of the pendulum "may be considered to be concentrated." When computing linear accelerations, the point is the center of mass at a distance l

from the pivot. When computing angular accelerations, the point is at the terminus of the radius of gyration, k. When computing the period of vibration, the point is the center of oscillation, distant L_0 from the pivot. For example, for a meter stick pivoted at one end,

$$l = 50 \text{ cm},$$

$$k = \sqrt{\frac{I}{m}} = \sqrt{\frac{\frac{1}{3}mL^2}{m}} = \frac{100}{\sqrt{3}} = 58 \text{ cm},$$

$$L_0 = \frac{I}{ml} = \frac{\frac{1}{3}mL^2}{ml} = 67 \text{ cm}.$$

It is of interest to correlate these three distances by returning to fundamental definitions and writing l, k^2, and L_0 as follows:

$$l = \frac{\int x \, dm}{\int dm},$$

$$k^2 = \frac{\int x^2 \, dm}{\int dm},$$

$$L_0 = \frac{\int x^2 \, dm}{\int x \, dm}.$$

14–13 Center of percussion. Everyone knows that there is one point on a baseball bat or golf club where the ball may be struck and no "sting" will be felt. This point is called the *center of percussion* and its position may be found as follows. Let Fig. 14–19 represent a baseball bat held or pivoted at the point O and struck at its center of percussion with a blow of impulse J. Let l be the distance from O to the center of mass, and L the distance from O to the center of percussion.

In order that no sting may be felt, it is necessary that there shall be no tendency for point O to move to the right or left while the blow is struck. In other words, the bat must start to pivot about point O.

The linear momentum acquired by the bat is found from the relation

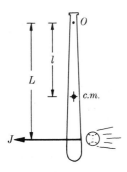

Fig. 14–19. Center of percussion.

$$J = m\bar{v},$$

and the angular momentum about an axis through O from the relation

$$JL = I\omega.$$

Since the instantaneous axis passes through O,

$$\bar{v} = \omega l.$$

When these three equations are solved for L, we get

$$L = \frac{I}{ml}. \tag{14–22}$$

Comparison with Eq. (14–19) shows that the center of percussion coincides with the center of oscillation. Note, however, that the position of either center depends on the location of the pivot.

It is easy to show that the pivot and the center of percussion are interchangeable. That is, if the bat in Fig. 14–19 were to be held at the point at which it is being struck, the proper place to hit a ball would be at O.

Figure 14–20 is a series of multiflash photographs illustrating the motion of a body suspended by a cord when the body is struck a horizontal blow. The *center of mass* is marked by a black band. In (a), the body is struck at its center of percussion relative to a pivot at the upper end of the cord, and it starts to swing smoothly about this pivot. In (b), the body is struck at its center of mass. Note that it does not start to rotate about the pivot, but that its initial motion is one of pure translation. That is, the center of percussion does not coincide with the center of mass. In (c), the body is struck above, and in (d) below its center of percussion.

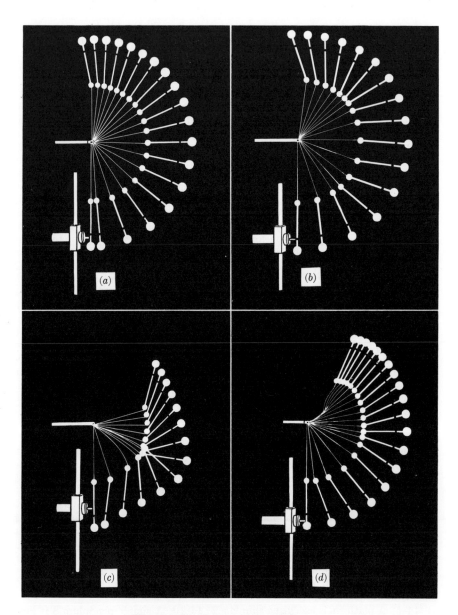

Fig. 14–20. Center of percussion.

PROBLEMS

14-1. Draw carefully on the same time axis two diagrams showing the velocity and acceleration of a point moving with simple harmonic motion, for a time of at least one complete cycle. Show clearly on the diagram the instants when the point is at the ends and at the center of its path.

14-2. Three force equations are given below. F is the force applied to a body, k is a positive constant, c is a constant, and x is the displacement of the body. (a) $F = -kx^2$, (b) $F = -k(x + c)$, (c) $F = kx$. Which of these forces will give rise to simple harmonic motion? State your reasons.

14-3. (a) Write an equation for the position of a particle of mass $m = 15$ gm moving along the X-axis at any time t if it is executing simple harmonic motion. Let its equilibrium position be at $x_0 = 10$ cm and the amplitude of motion be 5 cm. It takes 2 seconds for the particle to go through a complete cycle of the motion. (b) Write an equation for the force that must be acting on the particle to cause this motion. (c) For what value of x does the particle have maximum velocity? (d) For what values of x does the particle have maximum acceleration?

14-4. (a) A block rests on a horizontal surface which is executing simple harmonic motion in a horizontal plane at the rate of 2 oscillations/sec. The coefficient of static friction between block and plane is 0.5. How large can the amplitude be without slipping between block and surface? (b) A platform vibrates vertically with simple harmonic motion of amplitude 2 inches. What is the maximum frequency for which a weight will remain continuously on the platform?

14-5. A platform is executing simple harmonic motion in a vertical direction, with an amplitude of 5 cm and a frequency of $10/\pi$ vibr/sec. A block is placed on the platform at the lowest point of its path. (a) At what point will the block leave the platform? (b)

How far will the block rise above the highest point reached by the platform? (c) Optional. At what point will the block return to the platform? Hint: solve graphically.

14-6. A body of mass 100 gm hangs from a long spiral spring. When pulled down 10 cm below its equilibrium position and released, it vibrates with a period of 2 seconds. (a) What is its velocity as it passes through the equilibrium position? (b) What is its acceleration when it is 5 cm above the equilibrium position? (c) When it is moving upward, how long a time is required for it to move from a point 5 cm below its equilibrium position to a point 5 cm above it? (d) How much will the spring shorten if the body is removed?

14-7. The top end of a light spring of unstretched length 20 cm is held fixed and masses of 40 gm and 80 gm are hung from the spring, the 80-gm mass below the other. The length of the stretched spring is 26 cm. If the bottom mass falls off, compute (a) the frequency of the ensuing simple harmonic motion of the 40-gm mass, (b) the maximum kinetic energy of the 40-gm mass.

14-8. A block suspended from a spring vibrates with simple harmonic motion. At an instant when the displacement of the block is equal to one-half the amplitude, what fraction of the total energy of the system is kinetic and what fraction is potential?

14-9. Two bodies of equal mass are hung from separate springs having force constants k_1 and k_2. k_2 is greater than k_1. The two bodies oscillate with amplitudes such that the maximum velocities are equal. For which system is the amplitude of motion greater? Give reasons.

14-10. A simple pendulum 8 ft long swings with an amplitude of 1 ft. (a) Compute the velocity of the pendulum at its lowest point. (b) Compute its acceleration at the ends of its path.

14-11. Two springs, each of un-stretched length 20 cm but having dif-

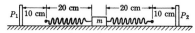

FIGURE 14-20

ferent force constants k_1 and k_2, are attached to opposite ends of a block of mass m on a level frictionless surface. The outer ends of the springs are now attached to the two pins P_1 and P_2, 10 cm from the original positions of the springs. Let $k_1 = 1000$ dynes/cm, $k_2 = 3000$ dynes/cm, $m = 100$ gm. (See Fig. 14-20.) (a) Find the length of each spring when the block is in its new equilibrium position, after the springs have been attached to the pins. (b) Find the period of vibration of the block if it is slightly displaced from its new equilibrium position and released.

14-12. A simple pendulum with a supporting steel wire of cross-sectional area 0.01 cm^2 is observed to have a period of 2 sec when a 10 kgm lead bob is used. The lead bob is replaced by an aluminum bob of the same dimensions having a mass of 2 kgm, and the period is remeasured. (a) What was the length of the pendulum with the lead bob? (b) By what fraction is the period changed when the aluminum bob is used? Is it an increase or decrease?

14-13. A pendulum clock which keeps correct time at a point where $g = 980.0$ cm/sec^2 is found to lose 10 sec per day at a higher altitude. Find the value of g at the new location.

14-14. A simple pendulum vibrates through a total angle of 60°, 30° on each side of the vertical. Find the difference between its period and the period for small oscillations, expressed as a percentage of the latter.

14-15. The balance wheel of a watch vibrates with an angular amplitude of π radians and with a period of 0.5 sec. (a) Find its maximum angular velocity. (b) Find its angular velocity when its displacement is one-half its amplitude. (c) Find its angular acceleration when its displacement is 45°.

14-16. A pendulum is constructed of a spherical bob 2 cm in radius attached to a cord 98 cm long. How far is the center of oscillation below the center of the sphere?

14-17. A load of 320 lb suspended from a wire whose unstretched length L_0 is 10 ft, is found to stretch the wire by 0.12 inch. The cross-sectional area of the wire, which can be assumed constant, is 0.016 square inch. (a) If the load is pulled down a small additional distance and released, find the frequency at which it will vibrate. (b) Compute Young's modulus for the wire.

14-18. A disk of mass $m = 1$ kgm and radius $r = 10$ cm is supported by a steel wire of length $L = 100$ cm and radius $R = 1$ mm to form a torsion pendulum, as shown in Fig. 14-21. The period is found to be $T = 1.25$ sec. What is the shear modulus M for steel?

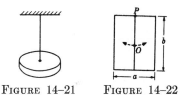

FIGURE 14-21   FIGURE 14-22

14-19. A thin uniform rod of length $L = 100$ cm is to be pivoted at some point along the rod and allowed to swing as a physical pendulum. (a) At what distance from the center of the rod should it be pivoted to give the smallest period? (b) What is then the period?

14-20. The uniform rectangular plate in Fig. 14-22 is suspended at point P. It swings in the plane of the paper about an axis through P. At what other point between P and O along the line PO could it be suspended to give the same period?

14-21. A monkey wrench is pivoted at one end and allowed to swing as a physical pendulum. The period is 0.9 sec and the pivot point is 6 inches from the center of mass. (a) What is the radius of gyration of the wrench about an axis through the pivot? (b) If the wrench was initially displaced 0.1 radian from its equilibrium position, what is the angular velocity of the wrench as it passes through the equilibrium position?

FIGURE 14–23

14–22. A section of thin-walled tube of radius R and mass m has a rod of small cross section and mass m soldered on its inner surface. The tube is placed on a level table top, as shown in Fig. 14–23 and given a small angular displacement from its equilibrium position. Find its period of vibration if it rolls without slipping.

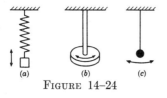

FIGURE 14–24

14–23. In Fig. 14–24, (a) is a mass suspended from a spring, (b) is a disk supported by a rod at its center (a torsion pendulum), and (c) is a simple pendulum. When set oscillating as shown by the double arrows, all three are found to have the same frequency. Suppose they are set in vibration at a second point where the acceleration of gravity is 4/9ths as great as at the first. How many vibrations will the torsion pendulum and the simple pendulum make while the mass on the spring makes 60 vibrations?

14–24. A steel rod of unstretched length 10 ft and square cross section of area 1 in^2 is suspended from one end. A disk of 1 ft radius and weight 300 lb is attached to the end of the rod as shown in Fig. 14–25. Neglect friction. Neglect the weight of the rod, except in

part (e). (a) Compute the change in the length of the rod due to the addition of the disk. (b) Compute the corresponding change in the cross-sectional area of the rod. (c) If the weight is pulled down an additional distance equal to the answer to part (a) and then released, find the period with which it will vibrate. (d) What is the total elastic potential energy of the rod for the lowest position of the weight in part (c)? (e) If, instead of pulling the weight down, it is swung aside in the plane of the diagram through a small angle, what is the period of vibration of the system of disk and rod?

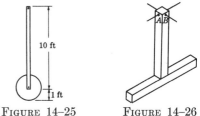

FIGURE 14–25   FIGURE 14–26

14–25. A slender rod of length 1 meter hangs vertically, supported by a string 40 cm long. At what point should the rod be struck a transverse blow so that the system starts swinging smoothly about the upper end of the string?

14–26. A point executes simple harmonic motion in two mutually perpendicular directions. The amplitudes of the vibrations are equal, their frequencies are in the ratio 3:2, and the initial displacement in both vibrations is zero. Construct the Lissajous' figure.

14–27. Two slender rods, each of length L and mass m, are fastened together as shown in Fig. 14–26, in the form of a T. Two holes, at right angles to each other, are bored through one of the rods near one end as shown. Find the period of oscillation if the system is set into vibration as a physical pendulum (a) about a horizontal axis through the hole in face A, (b) about a horizontal axis through the hole in face B. Assume small amplitudes.

CHAPTER 15

GRAVITATION

15–1 Newton's law of universal gravitation. Throughout our study of mechanics we have been continually encountering forces due to gravitational attraction between the earth and bodies on its surface, forces which are called the *weights* of the bodies. We now wish to study this phenomenon of gravitation in somewhat more detail.

The law of universal gravitation was discovered by Sir Isaac Newton, and was first announced by him in the year 1686. It may be stated:

Every particle of matter in the universe attracts every other particle with a force which is directly proportional to the product of the masses of the particles and inversely proportional to the square of the distance between them.

$$F \propto \frac{mm'}{r^2}.$$

The proportion above may be converted to an equation on multiplication by a constant γ (gamma) which is called the gravitational constant.

$$F = \gamma \frac{mm'}{r^2}. \qquad (15\text{–}1)$$

There seems to be no certain evidence that Newton was led to deduce this law from speculations concerning the fall of an apple to the earth. His first published calculations to justify its correctness had to do with the motion of the moon around the earth.

The numerical value of the constant γ depends on the units in which force, mass, and distance are expressed. Its magnitude can be found experimentally by measuring the force of gravitational attraction between two bodies of known masses m and m', at a known separation. For bodies of moderate size the force is extremely small, but it can be measured without too much difficulty by the Cavendish balance, adapted for this purpose by Sir Henry Cavendish in the year 1798 from a similar balance invented by Coulomb for studying forces of electrical attraction and repulsion.

The Cavendish balance consists of two small spheres of mass m (Fig. 15–1), usually of gold or platinum, mounted at opposite ends of a light horizontal rod which is supported at its center by a fine vertical fibre

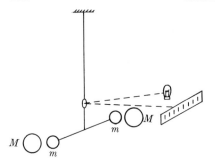

FIG. 15–1. Principle of the Cavendish balance.

such as a quartz thread. A small mirror fastened to the fibre reflects a beam of light onto a scale. To use the balance, two large spheres of mass M, usually of lead, are brought up to the positions shown. The forces of gravitational attraction between the large and small spheres result in a couple which twists the fibre and mirror through a small angle, thereby moving the light beam along the scale.

By using an extremely fine fibre the deflection of the light beam may be made sufficiently large so that the gravitational forces may be measured quite accurately. The numerical value of the gravitational constant, as measured in this way, is found to be

$$\gamma = 6.670 \times 10^{-8} \text{ dyne·cm}^2/\text{gm}^2,$$

or

$$\gamma = 6.670 \times 10^{-11} \text{ newton·m}^2/\text{kgm}^2.$$

EXAMPLE. Compute the force of gravitational attraction between the large and small spheres of a Cavendish balance, if $m = 1$ gm, $M = 500$ gm, $r = 5$ cm. (We shall show later that two spheres attract each other as if the mass of each were concentrated at its center.)

$$F = \frac{6.67 \times 10^{-8} \times 1 \times 500}{(5)^2} = 1.33 \times 10^{-6} \text{ dyne,}$$

or about one-millionth of a dyne!

15–2 The mass of the earth. Since the constant γ in Eq. (15–1) can be found from measurement in the laboratory, the mass of the earth may be computed. From measurements on freely falling bodies, we know that the earth attracts a one-gram mass at its surface with a force of (about) 980 dynes. The distance between the centers of the masses is the radius of the earth, 6380 km or 6.38×10^8 cm. Therefore,

$$980 = \frac{6.67 \times 10^{-8} \times 1 \times M}{(6.38 \times 10^8)^2},$$

where M is the mass of the earth. Hence

$$M = 5.98 \times 10^{27} \text{ grams.}$$

It is rather absurd to speak of the "weight of the earth," in the sense in which we have defined weight, but if we use the relations between the gram and pound which hold at the earth's surface, this corresponds to 13.2×10^{24} lb or 6.6×10^{21} tons.

The volume of the earth is

$$V = \frac{4}{3}\pi R^3 = 3.84 \times 10^{22} \text{ cu ft},$$

and its average weight-density is

$$\frac{13.2 \times 10^{24}}{3.84 \times 10^{22}} = 344 \text{ lb/cu ft}.$$

Its average specific gravity is therefore

$$\frac{344}{62.5} = 5.5.$$

This is considerably larger than the average specific gravity of the material near the earth's surface, so that the interior of the earth must be of much higher density.

15–3 Variations in "g." The acceleration of gravity, g, is the acceleration imparted to a body by its own weight. Its weight, however, can be written

$$w = \gamma \frac{mM}{R^2},$$

where m is the mass of the body, M is the mass of the earth, and R is the distance to the earth's center. Then since $w = mg$,

$$mg = \gamma \frac{mM}{R^2},$$

$$g = \frac{\gamma M}{R^2}. \qquad (15\text{--}2)$$

Since γ and M are constants, the acceleration of gravity should decrease with increasing distance from the center of the earth. In other words, it should be smaller at high altitudes. This is illustrated by the data in Table V.

TABLE V

Station	Elevation, meters	g, cm/sec^2	g, ft/sec^2
Cambridge, Mass.	14	980.398	32.1652
Worcester, Mass.	170	980.324	32.1628
Denver, Col.	1638	979.609	32.1393

The earth is not a perfect sphere but an oblate spheroid, slightly flattened at the poles. Hence the distance from sea level to the earth's center decreases slightly as one proceeds north or south from the equator, and g at sea level should increase with increasing north or south latitude. Some data illustrating this are given in Table VI.*

Table VI

Station	Latitude	g, cm/sec^2	g, ft/sec^2
Canal Zone	9° 00′	978.243	32.0944
Jamaica	17° 58′	978.591	32.1059
Bermuda	32° 21′	979.806	32.1548
Cambridge	42° 23′	980.398	32.1652
Greenland	70° 27′	982.534	32.2353

Local deposits of ore, oil, or other substances whose density is greater or less than the average density of the earth will cause local variations in g for points at the same latitude and elevation. Conversely, from a knowledge of such variations, conclusions can be drawn as to the presence of deposits of ore or oil beneath the earth's surface. Hence the precise measurement of g is one of the methods of geophysical prospecting. Such measurements are made with a physical pendulum of special construction.

Because of the decrease of g with altitude, it is only approximately correct to state that a body falls toward the earth with constant acceleration. Actually, the acceleration continually increases as the body approaches the earth, air resistance being neglected. For most purposes this variation is negligible, as is shown below.

EXAMPLE. At what height above the earth's surface has g diminished by one-tenth of one percent?

A problem such as this, where a small fractional change in one quantity is given and we wish to find the corresponding small change in another, is best handled by calculus methods.

From Eq. (15–2),

$$g = \frac{\gamma M}{R^2},$$

and by differentiation we obtain

$$dg = -\frac{2\gamma M}{R^3} \, dR,$$

where R is interpreted, not necessarily as the earth's radius, but as a distance from the earth's center. Dividing the second equation by the first gives

$$\frac{dg}{g} = -2 \frac{dR}{R}.$$

---

*The variation of g with latitude is also due in part to the earth's rotation. See page 193.

While the relation above is strictly true only if dg and dR are infinitesimal, it will be very nearly true if dg and dR are small compared with g and R, which is the case in our example. We may therefore consider dg to represent a small but finite change in g, when the distance from the earth's center is increased by dR. The ratio dg/g is the *fractional* change in g, which was given as 0.1% or 0.001, or, since g *decreases*, the fractional change is −0.001. Hence

$$-0.001 = -2\,\frac{dR}{R},$$

$$dR = \frac{0.001 \times R}{2} = 0.0005 \times 6380 = 3.2 \text{ km or about 2.5 miles.}$$

The motion of a body falling through a very great height, when the variation of g cannot be neglected, is discussed on page 305.

The English gravitational force unit, the pound, was defined on page 2 as the force of the earth's attraction on a standard pound at sea level and 45° lat. This definition has been made more precise by agreeing that the pound of force shall equal the weight of the standard pound at any point where the acceleration of gravity is 980.665 cm/sec^2 or 32.170 ft/sec^2. (This figure is sometimes called the "standard" acceleration of gravity.) This is equivalent to defining the pound of force as that force which imparts to a standard pound an acceleration of 32.170 ft/sec^2. The definition is thus independent of any gravitational phenomenon and if it is used, the term "gravitational" ceases to have any significance when applied to the English system of units.

15–4 The gravitational field. The concept of a *field of force* is extremely useful when one is dealing with any type of action-at-a-distance force, that is, a force due to gravitational, electrical, or magnetic causes. While the concept is used chiefly in electricity and magnetism, the same ideas and terms are applicable to gravitation and a brief discussion of gravitational fields will be given.

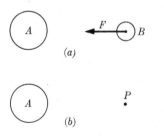

(a)

(b)

Fig. 15–2. Body A sets up a gravitational field at point P.

We know by experiment that a gravitational force of attraction (F in Fig. 15–2a) is exerted on body B by body A, even though both bodies are in a vacuum and not connected in any way. How is this possible? No one knows. That is the way the world is made. However, we may think of the phenomenon in some such fashion as this. Let the body B be removed as in Fig. 15–2(b), where P designates the point at which body B was formerly located. We imagine that the presence of body A modifies the state of affairs at the point P in such a way that *if* a body is placed at P this state of strain, or whatever it may be, is ready and waiting to influence B and urge it toward A. Since a force will be exerted on B toward A, wherever B may be placed, the whole of space around A must be considered to be affected in this way. We say that a *gravitational field* exists at all points of space around A.

A gravitational field is said to exist at a point if a force due to gravitational causes is exerted on a body at the point.

When making use of the field concept, we adopt the viewpoint that the gravitational force on a body is exerted *by the gravitational field* in which the body is located, rather than by the body (or bodies) that set up the field. Thus in Fig. 15–2 one would say that body A produces or sets up a gravitational field at point P and when body B is placed at P a force is exerted on it by this gravitational field. Of course, body B sets up its own gravitational field, and when B is placed at P body A experiences a force toward B due to B's field.

The *intensity* of the gravitational field at a point is defined as the ratio of the force exerted by the field to the mass of the body on which the force is exerted. Field intensity may therefore be described as the force per unit mass, and its magnitude and direction at any point may be determined experimentally by simply placing a body (called a test body) at the point, measuring the force on it, and dividing the force by the mass of the test body.

In Fig. 15–2, body B of mass m' can be considered the test body. If the field exerts a force F on this body, the intensity of the gravitational field at P, which we shall represent by G, is

$$\text{Gravitational field intensity at } P = \frac{\text{Force on mass } m' \text{ at } P}{m'},$$

$$G = \frac{\text{Force on } m'}{m'}. \tag{15–3}$$

Gravitational field intensity is expressed in pounds per slug, newtons per kilogram, or dynes per gram. For example, at the surface of the earth the earth's gravitational field exerts a force of 9.8 newtons on a 1-kgm mass, or a force of 32.2 pounds on a 1-slug mass. Hence the gravitational field intensity at the earth's surface is 9.8 newtons/kgm or 32.2 lb/slug.

If the gravitational field at a point is set up by a single point mass, such as body A in Fig. 15–2, the force on body B may be computed from the law of gravitation,

$$F = \frac{\gamma m m'}{r^2},$$

where m is the mass of body A and m' the mass of body B. Then, since $G = F/m'$,

$$G = \frac{\gamma m}{r^2}. \tag{15–4}$$

Since field intensity is force per unit mass, it has a definite direction (the direction of the force on a test body) and is therefore a vector quantity. If more than one body contributes to the gravitational field at a point, the resultant field must therefore be found by the usual methods for combining vectors.

$$G = \gamma \, \Sigma \left(\frac{m}{r^2}\right) \text{ (vector sum).} \tag{15–5}$$

Either Eq. (15–3) or (15–5) may be considered the definition of G. It should be noted carefully that m' in Eq. (15–3) represents the mass of a test body at the point P, while the m's in Eq. (15–5) are the masses of bodies located at points other than P and which exert forces on the test mass m' at P.

EXAMPLES. (1) Compute the magnitude and direction of the gravitational field at the points P and Q in Fig. 15–3 due to the masses m_1 and m_2.

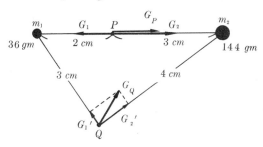

FIGURE 15–3

The field intensity at point P due to m_1 alone is

$$G_1 = \frac{\gamma m_1}{r_1^2} = \gamma \, \frac{36}{4} = 9\gamma \text{ dynes/gm.}$$

Its direction is toward m_1.
    The field due to m_2 is

$$G_2 = \frac{\gamma m_2}{r_2^2} = \gamma \, \frac{144}{9} = 16\gamma \text{ dynes/gm.}$$

Its direction is toward m_2.

The resultant field G_P at P is therefore $16\gamma - 9\gamma = 7\gamma$ dynes/gm directed toward m_2.

The field intensity at Q due to m_1 alone is

$$G_1' = \gamma \frac{36}{9} = 4\gamma \text{ dynes/gm}$$

directed toward m_1.

The field due to m_2 is

$$G_2' = \gamma \frac{144}{16} = 9\gamma \text{ dynes/gm}$$

directed toward m_2.

The resultant field G_Q is the vector sum of G_1' and G_2', and by the usual methods we find it to be

$$G_Q = 9.85\gamma \text{ dynes/gm}$$

at an angle of 24° with the line joining Q and m_2.

(2) What gravitational force would be experienced by a 20-gm body at points P and Q?

The gravitational field intensity at a point is the force per unit mass. Hence a 20-gm body at point P would experience a force of $20 \times 7\gamma = 140\gamma$ dynes toward m_2, and at point Q it would experience a force of $20 \times 9.85\gamma = 197\gamma$ dynes in the direction of G_Q.

In the foregoing discussion it has been tacitly assumed that the masses of the bodies were concentrated at a point. We shall now consider briefly the field due to a distributed mass.

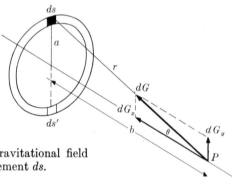

FIG. 15–4. Vector dG is the gravitational field intensity at point P due to the element ds.

Figure 15–4 represents a ring or hoop of small cross section and of radius a. At a point P on its axis, a short element of the ring of length ds and mass dm sets up a field of intensity dG in the direction shown, where

$$dG = \frac{\gamma \, dm}{r^2}.$$

This field may be resolved into components dG_x and dG_y. For every element of the ring such as ds, there is a corresponding one ds' at the opposite end of a diameter, which sets up at point P a field whose Y-component is equal and opposite to dG_y, but whose X-component is equal to and in the same direction as dG_x. The Y-components of all the elements of the ring will therefore mutually cancel, and the resultant field at P is the arithmetic sum or integral of all the X-components.

The magnitude of dG_x is

$$dG_x = dG \cos \theta = \gamma \frac{dm}{r^2} \frac{b}{r} = \frac{\gamma b}{r^3} dm.$$

If the ring is homogeneous, the mass of an element is proportional to its length, so that if m is the mass of the entire ring,

$$\frac{dm}{m} = \frac{ds}{2\pi a}, \quad \text{or} \quad dm = \frac{m}{2\pi a} ds.$$

Hence

$$dG_x = \frac{\gamma mb}{2\pi a r^3} ds,$$

and

$$G_x = \int dG_x = \frac{\gamma mb}{2\pi a r^3} \int ds$$

$$= \frac{\gamma mb}{2\pi a r^3} 2\pi a,$$

or, finally,

$$G = \frac{\gamma mb}{r^3} = \frac{\gamma mb}{(a^2 + b^2)^{3/2}}. \tag{15–6}$$

(The subscript may be dropped, since $G_y = 0$.)

When $b = 0$, $G = 0$. That is, the gravitational field of a ring is zero at the center of the ring, as would be expected.

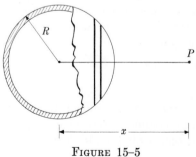

FIGURE 15–5

Consider next a thin homogeneous spherical shell of radius R (Fig. 15–5). The shell may be subdivided into narrow zones as shown, each of which corresponds to the ring of Fig. 15–4. From Eq. (15–6) we know the field intensity at P due to any one ring, and by integrating over the shell, an expression can be obtained for the resultant field at P.

The details will not be given, but it turns out that if P is *outside* the shell the field intensity is

$$G = \frac{\gamma M}{x^2},$$

where M is the mass of the shell and x the distance from P to its center, while if P is *inside* the shell $G = 0$. In other words, a thin homogeneous spherical shell of matter attracts bodies outside it as though all of its mass were concentrated at its center, while at internal points it exerts no attractive force at all.

The fact that the field is zero at internal points can also be shown by the following simple argument. Let P in Fig. 15–6 be any point within a thin spherical shell

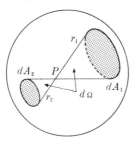

FIGURE 15–6

of matter. Consider the two small cones having a common apex at P and intercepting the areas dA_1 and dA_2 of the shell. The solid angle $d\Omega$ is the same in both cones and hence the areas dA_1 and dA_2 are proportional to the squares of the distances r_1 and r_2. The masses of the elements are proportional to their areas. The gravitational fields at P, due to these elements, therefore cancel each other, since the masses of the elements are proportional to the squares of the distances r_1 and r_2 and their gravitational fields are inversely proportional to the squares of r_1 and r_2. The entire shell can be subdivided in this way, and complete cancellation of internal fields results.

The extension to a solid sphere follows at once. Figure 15–7 is a sectional view. Since each shell into which the sphere may be divided acts at outside points as though all of its mass were at its center, the same is true for the sphere as a whole. This justifies the construction of Fig. 15–2, etc., where the distances between the spherical bodies were taken as the distances between their centers.

At internal points such as P in Fig. 15–7, those shells *outside* P contribute nothing to the field, which is simply that at the surface of a sphere of radius r. If m is the mass of this smaller sphere,

$$G_r = \frac{\gamma m}{r^2}.$$

If the sphere is homogeneous the mass of this sphere stands in the same ratio to that of the whole sphere as does its volume to that of the whole sphere. If the mass of the whole sphere is represented by M, then

Marlow

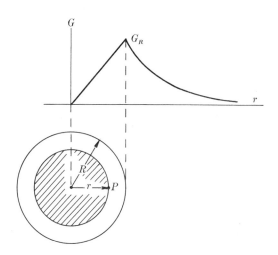

FIG. 15–7. Gravitational field intensity of a sphere.

$$\frac{m}{M} = \frac{\frac{4}{3}\pi r^3}{\frac{4}{3}\pi R^3}, \qquad \text{or} \qquad m = M\,\frac{r^3}{R^3}.$$

Hence

$$G_r = \frac{\gamma M}{R^3}\, r.$$

At the surface of the large sphere, where $r = R$,

$$G_R = \frac{\gamma M}{R^2}.$$

Hence

$$G_r = G_R\, \frac{r}{R}.$$

The field at internal points in a homogeneous sphere is therefore zero at the center and increases linearly with distance as one proceeds outward. Outside the sphere it is inversely proportional to the square of the distance from the center. A graph of G is given in the upper part of Fig. 15–7.

15–5 Gravitational potential energy. In Chapter 8 we showed that when a body was raised from one elevation to another, the increase in its potential energy was equal to the work done in lifting it. We considered only cases in which the change in elevation was so small relative to the diameter of the earth that the weight of the body remained essentially constant. In the general case, however, when two bodies are pulled apart against the force of gravitational attraction between them, the inverse square variation of force must be taken into account in computing the work done and the potential energy.

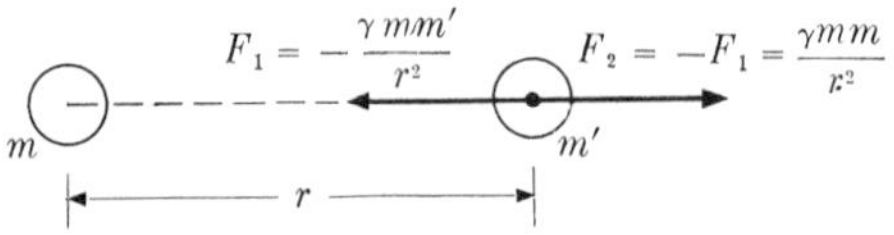

$$F_1 = -\frac{\gamma mm'}{r^2} \qquad F_2 = -F_1 = \frac{\gamma mm'}{r^2}$$

FIGURE 15-8

Figure 15-8 shows two masses m and m', separated by a distance r. Consider mass m fixed at the origin. The magnitude of the attractive force exerted on m' by m is $\gamma mm'/r^2$, and since it is directed toward the left, we shall write it as $F_1 = -\gamma mm'/r^2$. The external force F_2 that must be exerted on m' to pull it away from m is $F_2 = -F_1 = \gamma mm'/r^2$. The work done by this force when m' is moved from a separation r_1 to a separation r_2 is

$$W = \int_{r_1}^{r_2} F_2\,dr = \gamma mm' \int_{r_1}^{r_2} \frac{dr}{r^2} = \gamma mm' \left(\frac{1}{r_1} - \frac{1}{r_2}\right).$$

Since the work done equals the increase in potential energy,

$$W = \mathrm{PE}_{r_2} - \mathrm{PE}_{r_1} = \gamma mm' \left(\frac{1}{r_1} - \frac{1}{r_2}\right), \qquad (15\text{-}7)$$

where PE_{r_1} and PE_{r_2} are the potential energies at the separations r_1 and r_2.

In engineering work it is convenient to take the reference level of gravitational potential energy at sea level or at the lowest point in any particular problem. In the general theory of gravitation, however, simplification often results when the reference level is at infinity. That is, the gravitational potential energy of a body is assumed zero when it is far removed from all other bodies. Accordingly, let r_1 in Eq. (15-7) be ∞ and $\mathrm{PE}_{r_1} = 0$. Then

$$\mathrm{PE}_{r_2} - 0 = \gamma mm' \left(\frac{1}{\infty} - \frac{1}{r_2}\right),$$

and since r_2 may be any distance, the subscript can be dropped and

$$\mathrm{PE}_r = -\frac{\gamma mm'}{r}. \qquad (15\text{-}8)$$

This is then the expression for the potential energy of a mass m' at a distance r from a mass m.* The fact that the potential energy has a negative sign simply means that the potential energy at a finite separation r is less than it is at infinity.

———————

EXAMPLE. What is the gravitational potential energy of a 100-gm mass at sea level when the reference level of potential energy is at infinity?

If M is the mass of the earth, R its radius, and m the mass of a body at its surface, Eq. (15–8) becomes

$$PE = -\frac{\gamma M m}{R}.$$

We have shown that $g = \gamma M/R^2$, so that $\gamma M/R = Rg$ and

$$PE = -mgR$$
$$= -100 \times 980 \times 6.38 \times 10^8$$
$$= -6.25 \times 10^{13} \text{ ergs.}$$

15–6 Gravitational potential. The gravitational field intensity at a point is defined as the gravitational force per unit mass on a test body at the point. It is found useful to define a similar quantity, the *potential energy per unit mass*, called the *gravitational potential* and represented by V. If we denote by m' the mass of a test body at a point P of a gravitational field, then

$$\text{Gravitational potential at } P = \frac{\text{Potential energy of mass } m' \text{ at } P}{m'},$$

$$V = \frac{\text{PE of } m'}{m'}. \tag{15–9}$$

Gravitational potential is evidently expressed in foot-pounds per slug, joules per kilogram, or ergs per gram. If the potential at a point is known, the potential energy of a mass at the point can be found at once from Eq. (15–9).

Since energy is a scalar quantity, potential is a scalar also, unlike the intensity of a gravitational field, which is a vector. Hence if a number of bodies contribute to the potential at a point, the combined potential equals the *arithmetic* sum of the potentials due to the bodies individually.

———————

*As explained earlier, the potential energy is a property of the *system* and should not be assigned to either mass separately. Nevertheless it is convenient to speak of "the potential energy of the mass m'."

Since the potential energy of a point mass m' at a distance r from a point mass m is

$$PE = -\frac{\gamma mm'}{r},$$

the potential at a distance r from a point mass m is

$$V = \frac{PE \text{ of } m'}{m'} = -\frac{\gamma m}{r}, \qquad (15\text{-}10)$$

and the potential due to a number of point masses is

$$V = -\gamma \Sigma \left(\frac{m}{r}\right). \qquad (15\text{-}11)$$

Either Eq. (15–9) or (15–11) may be considered the definition of the gravitational potential at a point. Compare with the corresponding definitions of gravitational field intensity.

The *difference of potential* between any two points in a gravitational field is defined as the potential at one point minus the potential at the other, and is numerically equal to the work necessary to take a unit mass from one point to the other along any path whatever. Hence the work required to take a mass m from one point to the other is m times the difference of potential between the points.

EXAMPLES. (1) Compute the gravitational potential at the points P and Q in Fig. 15–3.

The potential at point P due to m_1 is

$$V_1 = -\frac{\gamma m_1}{r_1} = -\gamma \frac{36}{2} = -18\gamma \text{ ergs/gm.}$$

The potential due to m_2 is

$$V_2 = -\gamma \frac{m_2}{r_2} = -\gamma \frac{144}{3} = -48\gamma \text{ ergs/gm.}$$

The combined potential V_P is

$$V_P = -18\gamma - 48\gamma = -66\gamma \text{ ergs/gm.}$$

The potential at point Q due to m_1 is

$$V_1' = -\gamma \frac{36}{3} = -12\gamma \text{ ergs/gm.}$$

The potential due to m_2 is

$$V_2' = -\gamma \frac{144}{4} = -36\gamma \text{ ergs/gm.}$$

The combined potential V_Q is

$$V_Q = -12\gamma - 36\gamma = -48\gamma \text{ ergs/gm.}$$

(2) How much work is required to move a 20-gm mass from point P to point Q? The difference in gravitational potential between the points is

$$V_Q - V_P = -48\gamma - (-66\gamma) = +18\gamma \text{ ergs/gm.}$$

This is the increase in potential energy of a unit mass when moved from P to Q, or the work required to move a unit mass from P to Q. The work required to move a mass of 20 gm is therefore

$$\text{Work} = 20 \times 18\gamma = 360\gamma \text{ ergs,}$$

and the work is the same for all paths between P and Q.

The motion of a body in a gravitational field when the intensity is not the same at all points is best treated with the help of the concept of gravitational potential or potential energy. If friction forces are neglected, the basic relation is simply the principle of the conservation of energy

$$\text{PE} + \text{KE} = \text{constant,}$$

or

$$\text{PE}_1 + \text{KE}_1 = \text{PE}_2 + \text{KE}_2,$$

where the subscripts 1 and 2 refer to two points on the path of the body.

Let us restrict the discussion to that of the motion of a mass m in the gravitational field of a single fixed point mass or homogeneous sphere of mass M. Then

$$\text{KE} = \tfrac{1}{2}mv^2, \qquad \text{PE} = mV = -\frac{\gamma mM}{r},$$

where r is the distance of mass m from the center of the mass M. Hence

$$-\frac{\gamma mM}{r_1} + \tfrac{1}{2}mv_1^2 = -\frac{\gamma mM}{r_2} + \tfrac{1}{2}mv_2^2,$$

$$v_2^2 = v_1^2 + 2\gamma M \left(\frac{1}{r_2} - \frac{1}{r_1}\right).$$

EXAMPLES. (1) With what velocity must a body be projected vertically upward from the earth's surface to rise to a height equal to the earth's radius? Neglect air resistance.

$r_1 = R, r_2 = 2R$, and if the body just reaches this point, $v_2 = 0$. Hence

$$v_1^2 = 2\gamma M \left(\frac{1}{R} - \frac{1}{2R}\right) = \frac{\gamma M}{R}.$$

This result may be conveniently evaluated with the aid of Eq. (15–2),

$$g = \frac{\gamma M}{R^2},$$

where g is the acceleration of gravity at the earth's surface. Combining the equations above, we find

$$v_1 = \sqrt{Rg} = \sqrt{6.38 \times 10^6 \times 9.8}$$
$$= 7.9 \times 10^3 \text{ m/sec} = 7.9 \text{ km/sec},$$

or about 6 mi/sec.

(2) What velocity of projection is necessary for the body to escape from the earth's gravitational field?
$v_2 = 0, \qquad r_2 = \infty.$

$$v_1^2 = 2\gamma M \left(\frac{1}{R} - \frac{1}{\infty} \right) = \frac{2\gamma M}{R},$$
$$v_1 = \sqrt{2Rg} = 11 \times 10^3 \text{ m/sec} = 11 \text{ km/sec}.$$

15–7 Planetary motion. To a first approximation, the planets move around the sun in orbits which are circles with the sun at the center. The mass of the sun is so much greater than the mass of any planet that the sun may be considered to remain at rest. The gravitational force between sun and planet provides the requisite centripetal force to maintain the circular motion. In Fig. 15–9, M is the mass of the sun and m the mass of a planet revolving in a circle of radius r with an angular velocity ω. The gravitational force F is given by

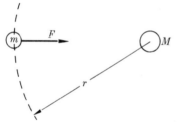

$$F = \frac{\gamma mM}{r^2},$$

and hence

$$\frac{\gamma mM}{r^2} = m\omega^2 r,$$

or

Fig. 15–9. The force of gravitational attraction provides the centripetal force.

$$\omega^2 = \frac{\gamma M}{r^3}. \qquad (15\text{–}12)$$

The angular velocity is thus independent of the mass of the planet and depends on the mass of the sun and the radius of the planetary orbit only. In other words, a planet of any mass at the same distance from the sun as is the earth, would revolve about the sun in the same time as does the earth.

Since the gravitational constant γ is known, the mass of the sun may be computed from Eq. (15–12) and data on any one of the planets. One justification of the law of gravitation is provided by the fact that using the values of ω and r for any planet, the same solar mass is found. See problem 16 on page 309.

Figure 15–9 and Eq. (15–12) would apply equally well to a planet and one of its satellites, such as the earth and the moon, or Jupiter and one of its moons.   Hence one can compute the mass of any planet which has a satellite whose angular velocity and distance from the planet can be measured.

PROBLEMS

15–1. The mass of the moon is approximately 6.7×10^{22} kgm and its distance from the earth is about 250,000 mi. (a) What is the gravitational force of attraction between the earth and the moon, in tons? (b) If the moon were to be retained in its orbit by a steel cable, instead of by gravity, what should be the diameter of the cable if the stress in it is not to exceed 60,000 lb/in²?

15–2. In an experiment using the Cavendish balance to measure the gravitational constant γ, it is found that a sphere of mass 800 gm attracts another sphere of mass 4 gm with a force of 13×10^{-6} dyne, when the distance between the centers of the spheres is 4 cm.   The acceleration of gravity at the earth's surface is 980 cm/sec², and the radius of the earth is 6400 km.   Compute the mass of the earth from these data.

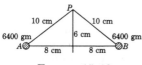

FIGURE 15–10

15–3. Two spheres, each of mass 6400 gm, are fixed at points A and B (Fig. 15–10). (a) Find the magnitude and direction of the initial acceleration of a sphere of mass 10 gm, if released from rest at point P and acted on only by forces of gravitational attraction of the spheres at A and B. (b) Find the velocity of the 10-gm sphere after it has traveled 6 cm.

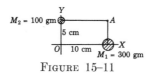

FIGURE 15–11

15–4. (a) What is the magnitude of the force exerted by the masses in Fig. 15–11 on a 10-gm mass placed at the origin?   What is the gravitational potential due to M_1 and M_2 at (b) the origin, (c) point A? (d) What is the work required to bring a 10-gm mass from point A to the origin?

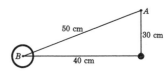

FIGURE 15–12

15–5. A solid sphere and a thin spherical shell are located with their centers 40 cm apart, as shown in Fig. 15–12. The solid sphere is 2 cm in radius and has a mass of 3600 gm. The spherical shell has a radius of 5 cm and a mass of 25 kgm. Calculate the gravitational field and gravitational potential at points A and B. Express the results in terms of the gravitational constant γ.

15–6. In (a) and (b) of Fig. 15–13 all bodies have identical masses m and are separated from each other by the distance a, as shown. Calculate (a) the gravitational potential and the gravitational field intensity at a point x cm to

FIGURE 15-13

the right of body 1 in system (a). (b) Make the same calculations for system (b) at a point equidistant from bodies 1 and 2 on the line joining their centers.

15-7. In round numbers, the distance from the earth to the moon is 250,000 miles, the distance from the earth to the sun is 93 million miles, the mass of the earth is 6×10^{27} gm and the mass of the sun is 2×10^{33} gm. Approximately, what is the ratio of the gravitational pull of the sun on the moon to that of the earth on the moon?

FIGURE 15-14

15-8. A point mass m is released from rest at a distance $3R$ from the center of a thin-walled hollow sphere of radius R and mass M. The hollow sphere is fixed in position, and the only force on the point mass is the gravitational attraction of the hollow sphere. There is a very small hole in the hollow sphere through which the point mass falls, as shown in Fig. 15-14. (a) What is the initial acceleration of the point mass? (b) What is its velocity just as it reaches the hole in the hollow sphere? (c) What is its velocity when it passes through the center of the hollow sphere? All answers may be expressed in terms of the gravitational constant γ.

15-9. The mass of the moon is approximately 6.7×10^{22} kgm and its radius is approximately 16×10^5 meters. (a) Compute the velocity with

which an object must be projected vertically from the surface of the moon so that it will rise above the surface to a height equal to the moon's radius. (b) How far will a body fall toward the moon in 1 sec if released from rest at a point near the moon's surface? (c) If a man can raise his center of gravity 4 ft vertically in a high jump at the earth's surface, how high could he jump on the moon's surface if he exerted the same impulse? (d) What would be the period of vibration, at the moon's surface, of a pendulum whose period is 1 sec at the surface of the earth?

15-10. With what velocity must a body be projected upward from the earth's surface to reach a height of 400 mi? What percent error would be made if g were assumed constant and equal to its value at the earth's surface?

15-11. What velocity would a body acquire in falling freely from a height h to the surface of the earth? Neglect friction. Express your answer in terms of the acceleration at the surface of the earth g, and the radius of the earth R. In this problem h is assumed to be so great that the variation of gravity with altitude must be taken into account.

15-12. (a) Compute the fractional change in the period of a pendulum when it is moved from sea level to an altitude of 10 mi. (Approximate finite changes by differentials.) (b) If a pendulum clock keeps correct time at sea level, how many seconds will it gain or lose each day at an altitude of 10 mi?

15-13. At what distance from the earth's center, outside the earth, is the intensity of the earth's gravitational field equal to its value at a point within the earth, halfway between the earth's surface and its center?

15-14. If a tunnel could be bored through the earth, passing through the earth's center, (a) what type of motion would be performed by a body dropped into one opening of the tunnel? Neglect friction. (b) How long a time would be required for the trip from one end of the tunnel to the other?

(c) What would be the velocity at the earth's center?

FIGURE 15-15

15-15. In Fig. 15-15 $ABCD$ is a square 10 cm on a side. A 100-gm mass is placed at A, and a 50-gm mass is placed at C. (a) What is the force of attraction between the two masses? (b) What is the magnitude of the gravitational field intensity at D? Show its direction approximately on a diagram. (c) What potential energy would a 10-gm mass have if placed at the point B? (d) If the 10-gm mass were thrown from point B with a velocity of 6 cm/sec, and in such a way that it reached D, what would its velocity be at point D? Leave answers in terms of γ.

15-16. Compute the mass of the sun from the data that the earth's orbit (assumed circular) has a radius of 93 million miles, and the time of revolution is about 365 days.

15-17. One of Jupiter's moons makes a complete revolution about Jupiter in 1.53×10^6 sec. Its orbit can be considered a circle of radius 4.21×10^8 meters. Compute the mass of Jupiter, and compare with that of the earth.

15-18. What is the weight on Jupiter of a man who weighs 200 lb on the earth?

15-19. Calculate the gravitational intensity produced by a uniform thin rod of length L and mass M at a point

P situated on the long axis of the rod at a distance A from one end.

15-20. (a) Prove that the gravitational field intensity at a point on the axis of a thin hoop of mass m and radius R, at a distance x from the center of the hoop, is

$$G = \gamma m \frac{x}{(x^2 + R^2)^{3/2}}.$$

(b) At what distance from the center of the hoop is the field a maximum?

15-21. The gravitational field intensity at a point along the axis of a circular hoop is given in problem 15-20. (a) Prove that in the absence of any other gravitational field but that due to the hoop, if a small mass m' is placed on the axis of the hoop at a distance x small compared with R and released, the resulting motion of m' will be simple harmonic. (b) Derive an expression for the frequency of the oscillations.

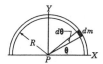

FIGURE 15-16

15-22. A slender rod of mass m is bent into a semicircle of radius R (Fig. 15-16). (a) What is the gravitational potential at point P? (b) What is the magnitude of the gravitational intensity dG at P, due to the element dm? Show the vector dG in a diagram. (c) What are the X- and Y-components of the vector dG? (d) Find the resultant gravitational field intensity at P.

CHAPTER 16

HYDROSTATICS AND SURFACE TENSION

16–1 Introduction. The term "hydrostatics" is applied to the study of fluids at rest, and "hydrodynamics" to fluids in motion. The special branch of hydrodynamics relating to the flow of gases and of air in particular is called "aerodynamics."

A fluid is a substance which can flow. Hence the term includes both liquids and gases. A liquid, while it adapts its shape to that of the containing vessel, has a definite volume. A gas, on the other hand, completely fills the volume of any container, however large, in which it may be placed. Liquids and gases differ markedly in their compressibilities, a gas being easily compressed while a liquid is practically incompressible. The small volume changes of a liquid under pressure can usually be neglected in this part of the subject.

Fluids also differ from one another in viscosity, a term which refers qualitatively to the readiness with which they flow. The viscosity of a gas is extremely small. Liquids like water, alcohol, and kerosene have much smaller viscosities than those like glycerine, molasses, or heavy oil. A substance like pitch is on the borderline between a liquid and a solid. A lump of pitch will fracture under a blow like a solid, but in the course of time a lump of it placed on a horizontal surface will spread out into a thin sheet. It may be described as a liquid of very high viscosity. For the present, we shall assume that liquids are nonviscous and incompressible.

16–2 Pressure in a fluid. A fluid (liquid or gas) confined in a vessel exerts forces against the walls of the vessel, and by Newton's third law the walls exert oppositely directed forces on the confined fluid. The magnitude of *the pressure at any point* is defined as the ratio of the force dF exerted on a small area dA including the point, to the area dA.

$$p = \frac{dF}{dA}, \qquad dF = p\,dA. \qquad (16\text{–}1)$$

Pressure is expressed in lb/ft^2, $newtons/m^2$, or $dynes/cm^2$ in our three systems of units.

If a fluid is at rest, the force exerted by it against any infinitesimal wall area is at right angles to that area, and the force exerted by the wall on the fluid is also at right angles to the wall. That this is so is evident when we realize that a fluid cannot permanently support a shearing stress. Any sidewise or tangential force exerted on the fluid by the walls would constitute a shearing stress and cause a flow parallel to the wall. If the liquid is at rest, there is no such flow, hence there is no tangential force and the force is normal to the surface at every point.

The same is true for any imagined area within the fluid. A small cube of the fluid, in any orientation, is subjected to inward forces at right angles to all of its faces. Conversely, at each face of the cube the fluid within it exerts a force at right angles to the face on the surrounding fluid. A state of stress, similar to the longitudinal stress in a bar under compression, exists therefore throughout the fluid, with the important difference that the forces at any imagined plane in the fluid, whatever the orientation of the plane, are normal to that plane. This is what is meant by the statement sometimes made that "the pressure in a fluid acts in all directions." Pressure is not a vector quantity and no direction can be assigned to it, but the *force* exerted by the fluid at one side of a plane, on the fluid at the other side of the plane, is at right angles to the plane whatever the orientation of the plane.

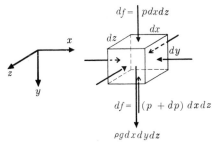

FIG. 16–1. Forces on an element of fluid.

It is a familiar fact that atmospheric pressure decreases with increasing altitude, and that the pressure in a lake or the ocean increases with increasing depth below the surface. We next derive the relation between pressure and elevation in a fluid. Any small cube of the fluid (Fig. 16–1) is in equilibrium under the action of the inward forces exerted on its faces by the surrounding fluid, together with its weight. The horizontal forces against opposite faces are evidently equal and opposite, but the upward force on the bottom face of the cube must be enough larger than the downward force on its top face to balance the weight of the fluid in the cube. Let p represent the pressure at the upper face, $p + dp$ the pressure at the lower face, dx, dy, and dz the dimensions of the cube, and ρ the density of the fluid. Then from the conditions of equilibrium,

$$(p + dp)\, dx\, dz = p\, dx\, dz + \rho g\, dx\, dy\, dz.$$

After expanding and canceling, this becomes

$$dp = \rho g\, dy, \qquad (16\text{--}2)$$

which is the fundamental equation governing change in pressure with elevation. (The coordinate y is considered positive when measured downward.) Note that the product ρg is the *weight* per unit volume of the fluid or its "weight density."

Let us now apply this equation to compute the pressure at a point

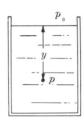

below the surface of a liquid in an open tank. (Fig. 16–2.) If the liquid is incompressible, its density, ρ, is a constant independent of p and y and the equation can readily be integrated, giving

$$p = \rho g y + C,$$

FIGURE 16–2

where C is a constant of integration. If we let p_0 represent the pressure at the top surface of the liquid, where $y = 0$, then $p_0 = C$ and

$$p = p_0 + \rho g y. \qquad (16\text{--}3)$$

That is, the pressure p at any depth y below the surface of a liquid equals the pressure p_0 at the surface plus the product of "weight density" times depth. Notice that the shape of the containing vessel does not affect the pressure, and that the pressure is the same at all points at the same depth. It also follows from Eq. (16–3) that if the pressure p_0 is increased in any way, say by inserting a piston on the top surface and pressing down on it, the pressure p at any depth must increase by exactly the same amount. This fact was stated by the French scientist Blaise Pascal (1623–1662) in 1653 and is called "Pascal's law." It is often stated: "Pressure applied to an enclosed fluid is transmitted undiminished to every portion of the fluid and the walls of the containing vessel." We can see now that it is not an independent principle but a necessary consequence of the laws of mechanics.

Let us next apply Eq. (16–2) to the earth's atmosphere. It will be more convenient to measure y *upward* from the earth's surface, so Eq. (16–2) becomes

$$dp = -\rho g\, dy.$$

A gas is *not* an incompressible fluid, but its density varies with pressure and temperature according to the gas law (see Section 22–3).

$$\rho = \frac{pM}{RT},$$

where p is the pressure, M the molecular weight, T the absolute temperature, and R a universal constant. Then for a gas,

$$dp = -\frac{pMg}{RT}\, dy,$$

$$\frac{dp}{p} = -\frac{Mg}{RT}\, dy.$$

If we assume the temperature of the atmosphere to remain constant with height (not a very good assumption), then

$$\ln p = -\frac{Mgy}{RT} + C.$$

If p_0 represents atmospheric pressure at the earth's surface, where $y = 0$, then $\ln p_0 = C$ and $\ln p/p_0 = -Mgy/RT$, or

$$p = p_0\epsilon^{-Mgy/RT}. \tag{16–4}$$

This is sometimes called the "barometric equation."

16–3 Pressure gauges. The simplest type of pressure gauge is the open tube manometer, illustrated in Fig. 16–3. It consists of a U-shaped tube containing a liquid, one end of the tube being at the pressure p which it is desired to measure, while the other end is open to the atmosphere. The lowest point of the U may be thought of as the bottom of either column of the U. The pressure due to the left column is

$$p + \rho gx,$$

while that due to the right column is

$$p_0 + \rho g(x + h)$$

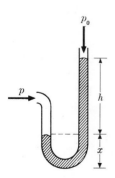

FIG. 16–3. The open-tube manometer.

(ρ is the density of the liquid in the manometer). Since these pressures both refer to the same point, they are equal. Hence

$$p + \rho g x = p_0 + \rho g(x + h),$$

and

$$p - p_0 = \rho g h. \tag{16-5}$$

The difference in height between the liquid columns is therefore proportional to the difference between the pressure p and the atmospheric pressure p_0. This difference, $p - p_0$, is called the *gauge pressure*, while the pressure p is the *absolute pressure*.

The *mercurial barometer* is simply a U-tube with one arm sealed off and evacuated, so that the pressure at the top of that arm is zero (Fig. 16-4). It is easy to show that

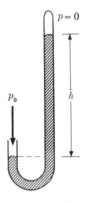

$$p_0 = \rho g h, \tag{16-6}$$

where p_0 is the atmospheric pressure and h the difference in level between the tops of the mercury columns in the barometer arms. Since the pressure is proportional to the height h, it is customary to express atmospheric pressure (and other pressures also) as so many "inches of mercury" or "centimeters of mercury." Note, however, that an "inch of mercury" is *not* a unit of pressure. (Pressure is the ratio of force to area.)

FIG. 16-4. The barometer.

EXAMPLE. Compute the atmospheric pressure on a day when the height of the barometer is 76.0 cm.

The height of the mercury column, from Eq. (16-6), depends on ρ and g as well as on the atmospheric pressure. Hence both the density of mercury and the local acceleration of gravity must be known. The density varies with the temperature, and g with the latitude and elevation above sea level. All accurate barometers are provided with a thermometer and with a table or chart from which corrections for temperature and elevation can be found. If we assume $g = 980$ cm/sec^2 and $\rho = 13.6$ gm/cm^3,

$$\begin{aligned} p_0 = \rho g h &= 13.6 \times 980 \times 76 \\ &= 1{,}013{,}000 \text{ dynes/cm}^2 \end{aligned}$$

(about a million dynes per square centimeter).

In English units,

$$76 \text{ cm} = 30 \text{ in} = 2.5 \text{ ft},$$
$$\rho g = 850 \text{ lb/ft}^3,$$
$$p_0 = 2120 \text{ lb/ft}^2 = 14.7 \text{ lb/in}^2.$$

A pressure of 1.013×10^6 dynes/cm^2, or 14.7 lb/in^2, is called *one atmosphere*. A pressure of exactly one million dynes per square centimeter is called one *bar*, and a pressure one one-thousandth as great is one *millibar*. Evidently, 1 bar = 1000 millibars. Atmospheric pressures are of the order of 1000 millibars, and are now stated in terms of this unit by the United States Weather Bureau.

Unfortunately, workers in the field of acoustics have adopted the term "bar" to mean a pressure of 1 dyne/cm^2. This need cause little confusion, however, since it is usually obvious which definition is being used.

The Bourdon type pressure gauge is more convenient for most purposes than a liquid manometer. It consists of a flattened brass tube closed at one end and bent into a circular form. The closed end of the tube is connected by a gear and pinion to a pointer which moves over a scale. The open end of the tube is connected to the apparatus, the pressure within which is to be measured. When pressure is exerted within the flattened tube it straightens slightly just as a bent rubber hose straightens when water is admitted. The resulting motion of the closed end of the tube is transmitted to the pointer.

16–4 Archimedes' principle. It is a fact of common experience that a body, when wholly or partly immersed in a fluid, is buoyed up by the fluid. Like Pascal's principle, the explanation of this effect follows directly from the laws of mechanics and does not depend on any special properties of a fluid. Figure 16–5 represents a body in the shape of a right cylinder of height h and cross section A, submerged in a fluid of density ρ. The horizontal forces exerted on the cylinder by the fluid are evidently in equilibrium and are not shown. At the upper end of the cylinder the liquid exerts a downward force F_1 given by

$$F_1 = p_1 A = (p_0 + \rho g x) A,$$

where x is the depth of the upper end below the surface. Similarly,

$$F_2 = p_2 A = [p_0 + \rho g(x + h)]A.$$

The net upward force, or the buoyant force, is

$$F_2 - F_1 = \rho g h A.$$

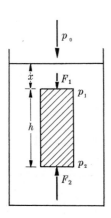

FIG. 16–5. Buoyant force equals weight of fluid displaced.

But hA is the volume of the body, and ρg is the weight per unit volume of the fluid. Hence, $\rho g h A$ is the weight of a volume of the fluid equal to the volume of the body, or, as it is called, the "weight of the displaced fluid." Hence,

A body immersed in a fluid is buoyed up with a force equal to the weight of the displaced fluid.

This is Archimedes' principle. Although derived from the special case of a right cylinder, it can be shown to hold regardless of the shape of the body. According to legend, Archimedes (287–212 B.C.) discovered this relation when given the problem of determining whether a crown of king Herod was or was not pure gold.

If the body is not wholly submerged in the fluid, the buoyant force is equal to the weight of a volume of fluid equal to the volume of the submerged portion of the body. If a body can displace its own weight of fluid before it is completely submerged, it will float. Otherwise it will sink in the fluid.

When making "weighings" with a sensitive analytical balance, correction must be made for the buoyant force of the air if the density of the body being "weighed" is very different from that of the standard "weights," which are usually of brass. For example, suppose a block of wood of density 0.4 gm/cm^3 is balanced on an equal-arm balance by brass "weights" of 20 gm, density 8.0 gm/cm^3. The apparent weight of each body is the difference between its true weight and the buoyant force of the air. If ρ_w, ρ_b, and ρ_a are the densities of the wood, brass, and air, and V_w and V_b are the volumes of the wood and brass, the apparent weights, which are equal, are

$$\rho_w V_w g - \rho_a V_w g = \rho_b V_b g - \rho_a V_b g.$$

The true mass of the wood is $\rho_w V_w$, and the true mass of the standard is $\rho_b V_b$. Hence,

$$\text{True mass} = \rho_w V_w = \rho_b V_b + \rho_a (V_w - V_b)$$

$$= \text{mass of standard} + \rho_a (V_w - V_b).$$

In the specific example cited

$$V_w = \frac{20}{0.4} = 50 \text{ cm}^3 \text{ (very nearly)},$$

$$V_b = \frac{20}{8} = 2.5 \text{ cm}^3, \qquad \rho_a = 0.0013 \text{ gm/cm}^3.$$

Hence

$$\rho_a(V_w - V_b) = 0.0013 \times 47.5 = 0.062 \text{ gm.}$$

$$\text{True mass} = 20.062 \text{ gm.}$$

If measurements are being made to one one-thousandth of a gram, it is obvious that the correction of 62 thousandths is of the greatest importance.

16–5 Stability of a ship. Archimedes' principle gives the magnitude of the buoyant force but not its line of action. It can be shown that the latter passes through the center of gravity of the displaced fluid. This has an important bearing on the stability of a floating object such as a ship. Figure 16–6 represents a section of the hull of a ship when on an even keel and when heeled over. The weight w and the buoyant force B give rise to a couple in such a direction as to right the ship.

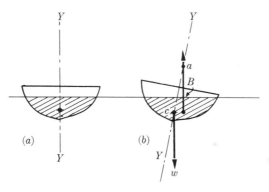

Fig. 16–6. Forces on a ship. Point a is the metacenter.

Point a, at which the line of action of the buoyant force intersects the line YY, is called the *metacenter*, and the distance ca is the *metacentric height*. The greater the metacentric height, the greater will be the stability. The righting couple is the same as though the ship were hung from a pivot at the metacenter.

If the line of action of B intersects YY at a point below the center of gravity, the ship is unstable and will capsize.

16–6 The hydrostatic paradox. If a number of vessels of different shapes are interconnected as in Fig. 16–7, it will be found that a liquid poured into them will stand at the same level in each. Before the principles of hydrostatics were completely understood, this seemed a very puzzling phenomenon and was called the "hydrostatic paradox." It would appear

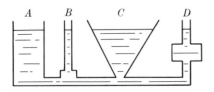

Fig. 16–7. Liquid stands at the same level in each vessel.

Figure 16–8

at first sight, for example, that vessel C should develop a greater pressure at its base than should B, and hence that liquid would be forced from C into B.

Equation (16–3), however, states that the pressure depends only on the depth below the liquid surface and not at all on the shape of the containing vessel. Since the depth of the liquid is the same in each vessel, the pressure at the base of each is the same and hence the system is in equilibrium.

A more detailed explanation may be helpful in understanding the situation. Consider vessel C in Fig. 16–8. The forces exerted against the liquid by the walls are shown by arrows, the force being everywhere perpendicular to the walls of the vessel. The inclined forces at the sloping walls may be resolved into horizontal and vertical components. The weight of the liquid in the sections lettered A is supported by the vertical components of these forces. Hence the pressure at the base of the vessel is due only to the weight of liquid in the cylindrical column B. Any vessel, regardless of its shape, may be treated the same way.

16–7 Forces against a dam. Water stands at a depth h behind the vertical upstream face of a dam (Fig. 16–9). It exerts a certain resultant horizontal force on the dam, tending to slide it along its foundation, and a certain moment tending to overturn the dam about the point O. We wish to find the horizontal force and its moment.

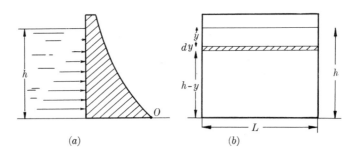

(a)   (b)

Fig. 16–9. Forces on a dam.

Figure 16–9(b) is a view of the upstream face of the dam. The pressure at a depth y is

$$p = \rho g y.$$

(Atmospheric pressure can be omitted since it also acts upstream against the other face of the dam.) The force against the shaded strip is

$$dF = \rho \, dA = \rho g y \times L \, dy.$$

The total force is

$$F = \int dF = \int_0^h \rho g L y \, dy = \rho g L \frac{h^2}{2}.$$

The moment of the force dF about an axis through O is

$$d\tau = dF(h - y) = \rho g L y(h - y) \, dy.$$

The total torque about O is

$$\tau = \int d\tau = \int_0^h \rho g L y(h - y) \, dy$$

$$= \rho g L \frac{h^3}{6}.$$

If H is the height above O at which the total force F would have to act to produce this torque,

$$FH = \rho g L \frac{h^2}{2} H = \rho g L \frac{h^3}{6},$$

$$H = \tfrac{1}{3}h.$$

Hence the line of action of the resultant force is at $\frac{1}{3}$ of the depth above O, or at $\frac{2}{3}$ of the depth below the surface.

16–8 The physics of surfaces. A liquid flowing slowly from the tip of a medicine dropper or a burette does not emerge as a continuous stream but as a succession of drops. A needle placed carefully on a water surface makes a small depression in the surface in which it "floats." When a clean glass tube of small bore is dipped into water, the water rises in the tube, but if the tube is dipped in mercury, the mercury is depressed. All these phenomena, and many others of a similar nature, are associated with the existence of a boundary surface between a liquid and some other

substance. The investigation of surface phenomena is a subject of increasing importance in many branches of pure and applied science.

In order to understand the origin of these surface effects we must know
something about the sizes and separations of the molecules of a liquid, and
the forces between them, although historically the argument has proceeded
in the reverse direction and much of our knowledge of molecules has been
derived from a study of the phenomena we are now considering. In the
first place, we know from many lines of experimental evidence, some of
which will be described in Chapter 24, that the dimensions of molecules are
of the order of 2 or 3×10^{-8} cm. We also know that one gram-mole of
any substance contains about 6×10^{23} molecules, and that at standard
conditions one gram-mole of a gas occupies 22.4 liters or 22,400 cm^3.
The volume per molecule, in a gas at standard conditions, is therefore
$22,400 \div 6 \times 10^{23}$, or about 37×10^{-21} cm^3. We may think of the gas
as divided into cubes of this volume with, on the average, a molecule at
the center of each cube. The average distance between the molecules
would then be equal to the length of one side of the cube, or $\sqrt[3]{37 \times 10^{-21}} =$
3.4×10^{-7} cm, which is about ten times the size of a molecule.

Consider next the distance between the molecules of a liquid. One
gram-mole of water, in the liquid state, occupies a volume of 18 cm^3. The
volume per molecule is $18 \div 6 \times 10^{23} = 30 \times 10^{-24}$ cm^3, and the average intermolecular distance is $\sqrt[3]{30 \times 10^{-24}}$ or very nearly 3×10^{-8} cm,
which is about the size of the molecules themselves. That is, the molecules
of a liquid are practically in contact with one another.

Now, what can we conclude about the forces that act between molecules? Our knowledge of these forces is not sufficient, at the present time,
for us to give a completely satisfactory interpretation of surface phenomena
in terms of them, and although plausible explanations of the experimental
facts can be made, the reader is advised to accept them with reservations.
There is, of course, a force of gravitational attraction between every pair
of molecules, but it turns out that this is negligible in comparison with the
forces we are now considering. The forces that hold the molecules of a
liquid (or solid) together are, in part at least, of electrical origin and do
not follow a simple inverse square law. When the separation of the molecules is large, as in a gas, the force is extremely small and is an attraction.
The attractive force increases as a gas is compressed and its molecules
brought closer together. But since tremendous pressures are needed to
compress a liquid, i.e., to force its molecules closer together than their
normal spacing in the liquid state, we conclude that at separations only
slightly less than the dimensions of a molecule the force is one of repulsion
and is relatively large. The force must then vary with separation in
somewhat the fashion shown in Fig. 16–10. At large separations the

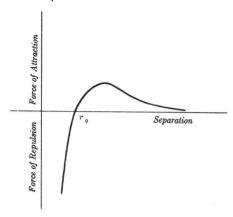

FIG. 16–10. Intermolecular force as a function of separation.

force is one of attraction but is extremely small. The force of attraction at first increases as the separation decreases, then passes through zero and changes to a large force of repulsion when the separation is less than r_0.

A single pair of molecules could remain in equilibrium at a center-to-center spacing equal to r_0 in Fig. 16–10. If they were separated slightly, the force between them would be attractive and they would be drawn together. If they were forced closer together than the distance r_0, the force would be one of repulsion and they would spring apart. If they were either pulled apart or pushed together, and then released, they would oscillate about their equilibrium separation r_0. From the energy viewpoint, the separation r_0, at which the force is zero, is that of minimum potential energy which, as we have seen in Chapter 8, corresponds to an equilibrium position.

While the behavior of two molecules can be visualized without much difficulty, it is not as easy to describe or understand that of a tremendous number of molecules such as those comprising a liquid. The general nature of the problem, however, at least so far as the interior of the liquid is concerned, cannot be very different from that of two molecules. We know, of course, that because of their thermal energy ("heat energy") the molecules of a liquid are continually in motion, and we picture them as performing some sort of vibration about an equilibrium position. The state of affairs near the surface of a liquid (that is, within a few molecular diameters of the surface) is somewhat different. Suppose a molecule happens to find itself in the liquid surface, and moving outward. There are no molecules outside to repel it, so it may move out to a considerably larger separation than that prevailing within the body of the liquid before it is brought to rest and accelerated inward by the attractive force of the molecules it has left behind. In fact, if it happens to have sufficient kinetic energy it may escape from the liquid altogether. This is the process we all recognize as *evaporation*. We are concerned here, however, with those molecules that do not succeed in escaping but form the outer layer of the liquid. These are continually performing a series of excursions out to a distance slightly larger than that of normal separation, and then returning. In other words, most of their time is spent in a region in which there is exerted on them an inward force of attraction. The fact that the environment of those molecules in or very near the surface differs from that of the molecules in the interior gives rise to the surface effects we are now considering.

16–9 Coefficient of surface tension. We turn now from the molecular picture of a liquid and its surface to the experimental facts of the problem. Most of the phenomena associated with the physics of surfaces can be understood with the help of a single measurable property of a surface that is, when the shape of a liquid is altered in such a way that its surface area increases, a definite amount of work per unit area is needed to create fresh surface. Since the work can be recovered when the area decreases, the surface appears capable of storing potential energy.

Energy considerations afford a simple explanation of the fact that a free liquid droplet is spherical. We have seen in Chapter 8 that a system is in stable equilibrium when its potential energy is a minimum. The equilibrium state of a liquid, then, is that in which its surface area is a minimum, consistent with other constraints. Hence a free liquid droplet assumes a spherical shape, since a sphere has a minimum surface for a given volume.

The work that must be done to increase the surface of a liquid is found to be proportional to the increase. The proportionality constant, or the work per unit area, is called the *coefficient of surface tension* of the liquid and is represented by γ. The work dW necessary to increase the area of a liquid surface by dA is therefore

$$dW = \gamma \, dA. \tag{16–7}$$

Surface tension may be expressed in any unit of energy per unit of area. The most common unit is 1 erg/cm^2 or 1 dyne-cm/cm^2, which is evidently equivalent to 1 dyne/cm. The reason for the name "surface tension" will be explained later. Some representative values are listed in Table V.

TABLE V

SURFACE TENSION
(Liquids against air)

Substance	γ (dynes/cm)
Benzene	28.9
Carbon tetrachloride	26.8
Ethyl alcohol	22.3
Mercury	465
Water	72.8

A second property of a liquid surface, which is helpful in understanding many surface effects, is that if a surface is curved, the pressure on the convex side of the curve is less than that on the concave side. The magnitude of this pressure differential across a curved surface can be derived at once from the fact that work must be done to create fresh surface.

Consider the experiment shown in Fig. 16–11. A cylindrical tube of small inside diameter (of the order of 1 mm) is dipped just below the surface of a liquid in a much larger container and a small bubble of air is blown at its lower end. As the air pressure p in the tube is increased, the size of the bubble increases and the area of its surface increases also. This does not mean that the surface already existing is "stretched" in the sense that a rubber balloon stretches as more air is forced into the balloon. The separation of the surface molecules remains constant as the area increases, and molecules which were formerly in the interior of the liquid migrate into the surface layer to keep the intermolecular spacing in this layer always the same.

An air bubble formed at the end of a tube as in Fig. 16–11 goes through a number of complex changes in shape before it becomes unstable and breaks away from the tube. At the instant when it becomes unstable, however, its shape is very nearly a hemisphere of radius R equal to the radius of the tube. The surface area of a hemisphere is

$$A = 2\pi R^2,$$

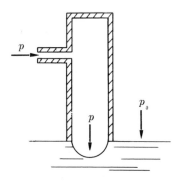

and if the radius of the bubble is increased to $R + dR$, the surface increases by

$$dA = 4\pi R\, dR.$$

The work done in creating this surface is, from Eq. (16–7),

$$dW = 4\pi \gamma R\, dR. \quad (16\text{–}8)$$

Fig. 16–11. A bubble of air blown at the end of a tube.

The work done in increasing the surface of the bubble can also be expressed in terms of the pressures at its convex and concave surfaces, and by equating the two expressions for the work we can deduce the pressure differential across the surface. Let p represent the air pressure within the tube in Fig. 16–11. This is the pressure on the concave side of the surface. The pressure against the convex side is simply the atmospheric pressure p_0, since the bubble is so small that we may neglect the increase of hydrostatic pressure with depth below the surface of the surrounding liquid. Consider a small portion of the hemispherical surface of area dS. The net force against it is

$$(p - p_0) \, dS,$$

and in the process of increasing the size of the bubble this portion moves out radially through a distance dR. The work done is the product of force and distance, or

$$(p - p_0) \, dS \, dR.$$

When this expression is integrated over all the surface elements dS we find, as the total work done by the pressure forces,

$$dW = (p - p_0) 2\pi R^2 \, dR.$$

Hence, from Eq. (16–8),

$$(p - p_0) 2\pi R^2 \, dR = 4\pi\gamma \, R \, dR \,,$$

$$(p - p_0) = \frac{2\gamma}{R} \,, \qquad\qquad (16\text{–}9)$$

$$p_0 = p - \frac{2\gamma}{R} \,.$$

The pressure p_0 on the convex side of a spherical surface is therefore smaller than the pressure p on the concave side by an amount $2\gamma/R$.

The pressure differential across a cylindrical surface can be found in a similar way. Figure 16–11 can be considered to represent two glass plates dipping into a liquid; the semicircle then corresponds to the cross section of a semicylindrical air bubble. (End effects are neglected.) The area of a semicylindrical surface of radius R and length L is

$$A = \pi R L.$$

The increase of area, when R increases by dR, is

$$dA = \pi L \, dR,$$

and the work done in creating fresh surface is

$$dW = \pi\gamma L \, dR.$$

The work done by the pressure forces is

$$dW = (p - p_0)\pi R L \, dR.$$

Hence

$$(p - p_0)\pi RL\,dR = \pi\gamma L\,dR,$$

and

$$(p - p_0) = \frac{\gamma}{R}, \tag{16–10}$$

$$p_0 = p - \frac{\gamma}{R}.$$

The pressure differential is therefore only one-half of that across a spherical surface of the same radius.

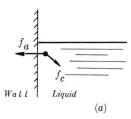

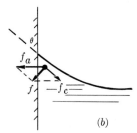

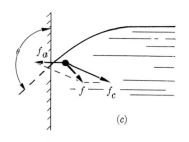

FIG. 16–12. The liquid surface is at right angles to the resultant force.

The expression for the pressure differential across an arbitrary curved surface can be derived by the same method we have used for a sphere and a cylinder. Any small portion of any curved surface can be fitted to a surface shaped like a blow-out patch, that is, having two different radii of curvature in mutually perpendicular directions. If these radii, known as the principal radii of curvature, are called R_1 and R_2, the general expression for the pressure differential is

$$(p - p_0) = \gamma\left(\frac{1}{R_1} + \frac{1}{R_2}\right). \tag{16–11}$$

The relations we have derived for a sphere and a cylinder are special cases of the above. The radii of curvature of a sphere are equal in any two mutually perpendicular directions and each is equal to the radius of the sphere, R. If $R_1 = R_2 = R$, then from Eq. (16–11),

$$(p - p_0) = 2\gamma/R.$$

One of the radii of curvature of a cylinder is infinite and the other equals the radius of the cylinder. Hence for a cylinder

$$(p - p_0) = \gamma/R.$$

16–10 Angle of contact. Consider next the molecules in a liquid surface near the walls of the containing vessel or near some body immersed in the liquid. These molecules are acted on both by forces of *cohesion*, exerted on them by other molecules of the liquid, and by forces of *adhesion* exerted by the molecules of the wall. Figure 16–12(a) represents a liquid in contact with a wall. The dot represents a molecule in the surface layer, the vector f_a the force of adhesion between it and the wall, and the vector f_c the force of cohesion between it and the liquid. The resultant of these forces can be found by the usual method of combining vectors.

If the forces of adhesion and cohesion have the relative magnitudes indicated in Fig. 16–12(b) the resultant force f is in the direction shown, and since a liquid can be in equilibrium only when the force on its surface is at right angles to the surface at any point, the tangent to the surface at the point of contact is perpendicular to the resultant force f. The angle θ is called the *angle of contact*.

If the forces of adhesion and cohesion have the relative magnitudes of Fig. 16–12(c), the liquid surface curves down instead of up and the angle of contact is greater than 90°. Speaking generally, if adhesion is greater than cohesion the contact angle is small, while if the reverse is true the angle is large. The angle of contact between clean water and clean glass is nearly zero, between water and silver it is nearly 90°, and between mercury and glass it is about 140°. Small amounts of impurities, however, may result in large variations from the values quoted.

When the angle of contact between a liquid and a surface is small, the liquid is said to "wet" the surface. If the angle is large, it does not "wet" the surface. Thus water "wets" clean glass but mercury does not. We can see, however, that any angle of contact is possible and there is no sharp distinction between "wetting" and "nonwetting."

16–11 Capillary rise in tubes. One of the most familiar of surface effects is the elevation of a liquid in an open tube of small cross section. In fact, the term "capillarity," which is often used to describe all of these surface effects, originates from the description of such tubes as "capillary" or "hair-like." The process may be considered to take place in two steps.

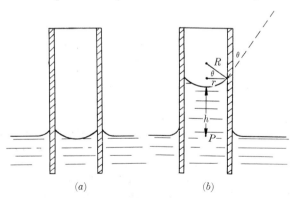

Fig. 16–13. Rise of a liquid in a capillary tube.

Assume first that the liquid level in the tube does not rise, as in Fig. 16–13(a). The forces of cohesion and adhesion cause the liquid surface within the tube to become curved as shown. This curved surface is called a *meniscus*. Since the tube is open at the top, the air pressure on the upper side of the meniscus is the atmospheric pressure, p_0. The pressure differential that we have shown to exist across a curved surface means that the pressure just below the meniscus is less than atmospheric, and hence less than the pressure at points just beneath the flat surface of the surrounding liquid. The system is therefore not in equilibrium, and liquid is forced up the tube by the excess pressure of the liquid around it. This explanation of capillary rise is more satisfactory than to consider that the liquid is "pulled" up the tube, although we shall explain in Section 16–12 how the latter viewpoint may be used.

The elevation of liquid in the tube will evidently go on until the pressure at point P in Fig. 16–13(b) becomes equal to that at the same elevation in the surrounding liquid, which is the atmospheric pressure p_0. The height of rise, h, is found as follows. To a good approximation, the meniscus can be considered a portion of a spherical surface. From the diagram, the radius R of this surface is

$$R = \frac{r}{\cos \theta},$$

where r is the radius of the tube and θ is the angle of contact. The pressure p' just below the meniscus is, from Eq. (16–9), (remember that the pressure on the concave side of the meniscus is now p_0)

$$p' = p_0 - \frac{2\gamma}{R}$$

$$= p_0 - \frac{2\gamma \cos \theta}{r},$$

and the pressure at point P is

$$p' + \rho g h,$$

or

$$p_0 - \frac{2\gamma \cos \theta}{r} + \rho g h.$$

Since in the equilibrium state this equals atmospheric pressure,

$$p_0 = p_0 - \frac{2\gamma \cos \theta}{r} + \rho g h,$$

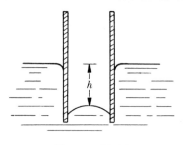

FIGURE 16–14

and finally,

$$h = \frac{2\gamma \cos \theta}{\rho g r}. \qquad (16\text{–}12)$$

This equation then expresses the relation between the capillary rise, the surface tension, and the radius of the tube. Note that the rise is inversely proportional to the tube radius.

The depression of a liquid such as mercury in a capillary tube is illustrated in Fig. 16–14. It is left as an exercise to derive the expression for the depression h in terms of r, γ, and θ. Note that the liquid now lies on the *concave* side of the curved surface.

16–12 Alternate treatment of surface tension. An instructive demonstration of surface effects is afforded by the apparatus shown in Fig. 16–15. The wire frame A is provided with a smoothly sliding crosspiece B. The frame is dipped in a soap solution and a film formed as indicated by the shading. Since the free surface tends toward a minimum, it tends to shrink and pull the crosspiece up. Suppose a force F is exerted on the crosspiece and the latter is moved down a distance dx. The area of fresh surface created in this displacement is $2l\,dx$ (the film has two sides), so the work required is

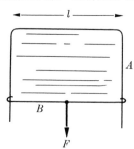

$$dW = 2\gamma l\,dx.$$

The work done by the force F in moving a distance dx is

$$dW = F\,dx.$$

Hence

$$F\,dx = 2\gamma l\,dx,$$

and

$$F = 2\gamma l.$$

FIG. 16–15. The force F, which balances the surface tension force, is independent of the area of the soap film.

The equation above may now be given the following interpretation. The force F, equal to $2\gamma l$, is the same *as if* each surface of the film pulled up on the crosspiece with a force γl, or with a force γ per unit length. The surfaces of the liquid are in a way analogous to elastic membranes in tension, and the quantity γ can be considered as the force per unit length with which they pull on their boundary. The origin of the term "surface tension" is now evident. Note, however, that unlike the tension in an elastic membrane, the force due to surface tension is *independent* of the area of the surface. From this point of view, considering surface tension as force per unit length, the cgs unit of surface tension is 1 dyne/cm, which we have shown is equivalent to 1 erg/cm^2.

Consider again the question of the rise of liquid in a capillary tube. The surface of the liquid makes contact with the tube along a line of length $2\pi r$. The "surface tension force" at every point makes an angle θ with the vertical, so the resultant upward force is (see Fig. 16–16)

$$F = 2\pi r \gamma \cos \theta.$$

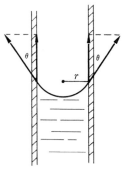

We set this equal to the weight of the liquid column "supported" by the force, or $\pi r^2 \rho g h$. Hence

$$2\pi r \gamma \cos \theta = \pi r^2 \rho g h,$$

$$h = \frac{2\gamma \cos \theta}{\rho g r},$$

which is the same as Eq. (16–12).

FIGURE 16–16

The surface tension of a liquid may be determined by measuring the force required to detach an object from the liquid surface. The duNouy apparatus for the measurement of surface tension, which is widely used because of its simplicity and convenience, measures the vertical force needed to detach a ring of fine platinum wire from a liquid surface. The simple theory of its operation is that the downward force on the ring, just before detachment, equals the surface tension force along a line of length equal to twice the perimeter of the ring. If the radius of the ring is R, and F is the force exerted on it,

$$F = 4\pi R \gamma,$$

$$\gamma = \frac{F}{4\pi R}.$$

Actually, the surface tension is not always given by this simple equation, which may be considerably in error depending on the size and shape of the ring.

16–13 Excess pressure in bubbles. Everyone has at some time blown a soap bubble and, if asked, would undoubtedly reply that in order to increase the size of the bubble he would have to increase the air pressure within it. Actually the reverse is true, as is easily seen.

Figure 16–17 is a sectional view of a portion of the wall of a soap bubble. The film is so thin that both inside and outside surfaces can be considered to have the same radius R. Let p_0 be the external, atmospheric pressure. The liquid layer lies on the *concave* side of the *outer* surface of the bubble. Hence the pressure p_i in the liquid, from Eq. (16–9), is

$$p_i = p_0 + \frac{2\gamma}{R}.$$

But the liquid layer lies on the *convex* side of the *inner* surface of the bubble. Hence the air pressure p inside the bubble is

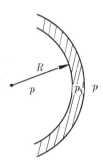

$$p = p_i + \frac{2\gamma}{R},$$

or

$$p = p_0 + \frac{4\gamma}{R},$$

$$p - p_0 = \frac{4\gamma}{R}.$$

Fig. 16–17. Excess pressure in a soap bubble.

The gauge pressure within the bubble, $p - p_0$, is therefore inversely proportional to the radius of the bubble and decreases as the size of the bubble increases. Of course, more air must be forced in to cause the bubble to grow, but, contrary to the process of blowing up a toy balloon, the excess pressure required to force air into the bubble decreases with bubble size.

16–14 Formation of drops. A series of high-speed photographs of successive stages in the formation of a drop of milk at the end of a vertical tube is reproduced in Fig. 16–18. The photographs were taken by Dr. Edgerton of M.I.T. It will be seen that the process, if examined in detail, is exceedingly complex. An interesting feature is the small drop that follows the larger one. Both drops execute a few oscillations after their formation (4, 5, and 6) and eventually assume a spherical shape (7) which would be retained but for the effects of air resistance, as shown in (8) and (9). The drop in (9) has fallen 14 ft.

An approximate relation between the weight of the drop, w, and the radius of the tube from which it falls can be derived by assuming that the limiting size of the drop that can be supported by the tube is reached when

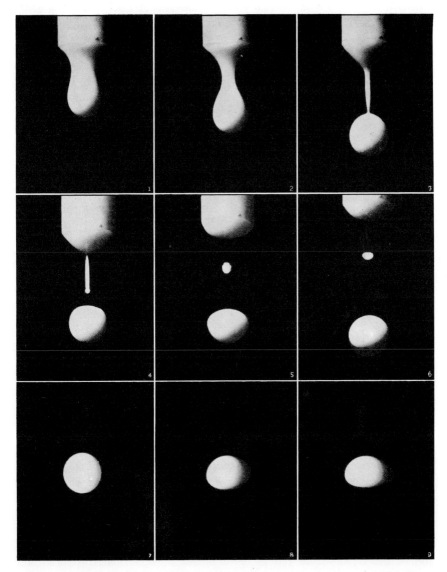

Fig. 16–18. Successive stages in the formation of a drop. (Reproduced from *Flash*, courtesy of Ralph S. Hale & Co.)

the weight of the drop equals the surface tension force around a circle whose diameter equals that of the tube. This leads to the equation

$$w = 2\pi r \gamma.$$

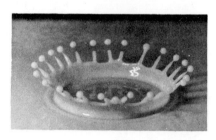

FIG. 16–19. A drop of milk splashes on a hard surface. (Reproduced from *Flash*, courtesy of Ralph S. Hale & Co.)

An examination of Fig. 16–18 shows that the assumption that $w = 2\pi r \gamma$ is seriously in error. For one thing, the edge of the drop at the tip of the tube at the instant the drop breaks away is rarely vertical, and for another only a part of the drop falls. A very thorough study has been made of the relation between drop weight, radius of tube, and surface tension. When the proper corrections are made to the simple formula this method is one of the most satisfactory means of measuring surface tensions.

A beautiful photograph of the splash made by a drop of milk falling on a hard surface is reproduced in Fig. 16–19. It also was taken by Dr. Edgerton.

16–15 Surface tension and surface energy. The coefficient of surface tension, γ, is defined as the work per unit area required to create fresh liquid surface. It would seem, therefore, that the *surface energy* associated with a surface of area A should equal the product γA. This is not correct. It is found that when the surface area of a liquid is increased, as by blowing a larger bubble or stretching a surface film as in Fig. 16–15, the temperature of the film decreases unless heat is supplied to it from some outside source. The phenomenon is similar to the drop in temperature of a gas when allowed to expand from a high to a low pressure. Hence if the temperature is to be kept constant while the surface is increased, heat must be supplied to the film, and the increase in its surface energy is equal to the sum of the work done and the heat supplied. A text on thermodynamics should be consulted for further details. If we let U represent the energy of a surface film of area A, its surface energy per unit area, U/A, is related to its surface tension by the equation

$$\frac{U}{A} = \gamma - T \frac{d\gamma}{dT},$$

where T is the Kelvin temperature (see Section 24–9).

The surface tension almost invariably decreases with increasing temperature. Hence the surface energy per unit area is larger than the surface tension. Graphs of surface energy per unit area and surface tension, as functions of temperature, are given in Fig. 16–20 for water. The temperature of 374°C, where both become zero, is the *critical temperature* (see Section 23–1).

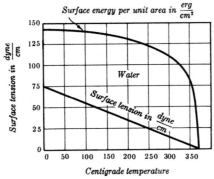

Fig. 16–20. Surface energy and surface tension of water, as functions of temperature.

PROBLEMS

16–1. The piston of a hydraulic automobile lift is 12 inches in diameter. What pressure, in lb/in^2, is required to lift a car weighing 2400 lb?

16–2. The expansion tank of a household hot-water heating system is open to the atmosphere and is 30 ft above a pressure gauge attached to the furnace. What is the gauge pressure, in lb/in^2?

16–3. The submarine Squalus sank at a depth of 240 ft. Compute the absolute pressure at this depth, in lb/in^2 and lb/ft^2. The specific gravity of sea water is 1.025.

16–4. A piece of gold-aluminum alloy weighs 10 lb. When suspended from a spring balance and submerged in water, the balance reads 8 lb. What is the weight of gold in the alloy if the specific gravity of gold is 19.3 and the specific gravity of aluminum is 2.5?

16–5. The densities of air, helium, and hydrogen (at standard conditions) are respectively 0.00129 gm/cm^3, 0.000178 gm/cm^3, and 0.0000899

gm/cm^3. What is the volume in cubic feet displaced by a hydrogen-filled dirigible which has a total "lift" of 10 tons? What would be the "lift" if helium were used instead of hydrogen?

16–6. A hollow sphere of inner radius 9 cm and outer radius 10 cm floats half-submerged in a liquid of specific gravity 0.8. (a) Calculate the density of the material of which the sphere is made. (b) What would be the density of a liquid in which the hollow sphere would just float completely submerged?

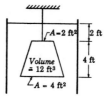

FIGURE 16–21

16–7. An object in the shape of a truncated cone weighs 1000 lb in vacuum and is suspended by a rope in an open tank of liquid of density 2 slugs/ft^3 as in Fig. 16–21. (a) Find the total downward force exerted by the liquid on the top of the object, of area 2 ft^2. (b) Find the total upward force exerted by the liquid on the bottom of the object, of area 4 ft^2. (c) Find the tension in the cord supporting the object.

16–8. A block of balsa wood placed in one scale pan of an equal arm balance is found to be exactly balanced by a 100-gm brass "weight" in the other scale pan. Find the true mass of the balsa wood, if its specific gravity is 0.15.

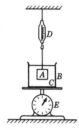

FIGURE 16–22

16–9. Block A in Fig. 16–22 hangs by a cord from spring balance D and is submerged in a liquid C contained in beaker B. The weight of the beaker is 2 lb, the weight of the liquid is 3 lb. Balance D reads 5 lb and balance E reads 15 lb. The volume of block A is 0.1 ft^3. (a) What is the "weight-density" of the liquid? (b) What will each balance read if block A is pulled up out of the liquid?

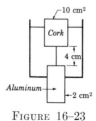

FIGURE 16–23

16–10. A cylindrical cork of mass 15 gm and cross section 10 cm^2 floats in a

pan of water as shown in Fig. 16–23. An aluminum cylinder of mass 25 gm and cross section 2 cm^2 is attached 4 cm below the cork and slides through a watertight, frictionless hole in the bottom of the pan. What is the distance of the bottom of the cork below the surface of the water?

16–11. A hydrometer consists of a spherical bulb and a cylindrical stem of cross section 0.4 cm^2. The total volume of bulb and stem is 13.2 cm^3. When immersed in water the hydrometer floats with 8 cm of the stem above the water surface. In alcohol, 1 cm of the stem is above the surface. Find the density of the alcohol.

16–12. A 3200-lb cylindrical can buoy floats vertically in salt water (specific gravity = 1.03). The diameter of the buoy is 3 ft. Calculate (a) the additional distance the buoy will sink when a 150-lb man stands on top, (b) the period of the resulting vertical simple harmonic motion when the man dives off.

FIGURE 16–24

16–13. A uniform rod AB, 12 ft long, weighing 24 lb, is supported at end B by a flexible cord and weighted at end A with a 12-lb lead weight. The rod floats as shown in Fig. 16–24 with one-half its length submerged. The buoyant force on the lead weight can be neglected. (a) Show in a diagram all of the forces acting on the rod. (b) Find the tension in the cord. (c) Find the total volume of the rod.

16–14. A swimming pool measures $75 \times 25 \times 8$ ft deep. Compute the force exerted by the water against each end and against the bottom.

16–15. The upper edge of a vertical gate in a dam lies along the water surface. The gate is 6 ft wide and is hinged along the bottom edge, which is 10 ft below the water surface. What is the torque about the hinge?

16–16. The upper edge of a gate in a dam runs along the water surface. The gate is 6 ft high and 10 ft wide and is hinged along a horizontal line through its center. Calculate the torque about the hinge.

16–17. The cross section of a certain dam is a rectangle 10 ft wide and 20 ft high. The depth of water behind the dam is 20 ft and the dam is 500 ft long. (a) What is the torque tending to overturn the dam about the bottom edge of the downstream face? (b) If the material of the dam weighs 100 lb/ft³, show whether or not the restoring torque due to the weight of the dam is greater than the torque due to water pressure.

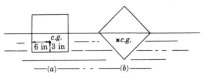

FIGURE 16–25

16–18. A cubical block of wood 1 ft on a side is weighted so that its center of gravity is at the point shown in Fig. 16–25(a), and it floats in water with one-half its volume submerged. Compute the righting moment and the metacentric height when the block is "heeled" at an angle of 45° as in Fig. 16–25(b).

16–19. A capillary tube is dipped in water with its lower end 10 cm below the water surface. Water rises in the tube to a height of 4 cm above that of the surrounding liquid, and the angle of contact is zero. What gauge pressure is required to blow a hemispherical bubble at the lower end of the tube?

16–20. A sealed vessel has inserted in it a capillary tube (open to the atmosphere) of radius 0.1 mm. It contains water under an absolute pressure p of 1.01 atmospheres. How high

FIGURE 16–26

above the surface does the water rise in the capillary tube? (See Fig. 16–26.)

16–21. On a day when the atmospheric pressure is 950 millibars, (a) what would be the height of the mercury column in a barometric tube of inside diameter 2 mm? (b) What would be the height in the absence of any surface tension effects? (c) What is the minimum diameter a barometric tube may have in order that the correction for capillary depression shall be less than 0.01 cm of mercury?

16–22. (a) Derive the expression for the height of capillary rise in the space between two parallel plates dipping in a liquid. (b) Two glass plates, parallel to each other and separated by 0.5 mm, are dipped in water. To what height will the water rise between them? Assume zero angle of contact.

16–23. A tube of circular cross section and outer radius 0.14 cm is closed at one end. This end is weighted and the tube floats vertically in water, heavy end down. The total mass of the tube and weights is 0.2 gm. If the angle of contact is zero, how far below the water surface is the bottom of the tube?

16–24. A glass tube of inside diameter 1 mm and wall thickness 0.5 mm is aligned coaxially within a larger tube of inside diameter 2.5 mm. When the tubes are dipped below the flat surface of a liquid, the liquid rises to a height of 1.2 cm in the inner tube. To what height does it rise in the annular space between the tubes? The angle of contact between the liquid and the glass is 10° and the specific gravity of the liquid is 0.80.

FIGURE 16-27

16-25. A wire frame like that in Fig. 16-27 is forced beneath the surface of the water in a large beaker and prevented from rising by surface tension. Given that the mass of the cork is 3.5 gm and its volume 7.0 cm³, and the surface tension of water is 72 dynes/cm, find the minimum radius r of the ring for which the system will remain submerged. Neglect the weight of the wire frame and the buoyant force on it.

16-26. Mercury is poured out onto a plane horizontal glass surface which it does not wet, and forms a pool 3.7 mm deep, regardless of the size of the pool. The density of mercury is 13.6 gm/cm³. What is the surface tension of mercury?

16-27. Find the gauge pressure, in dynes/cm², in a soap bubble 5 cm in diameter. The coefficient of surface tension is 25 dynes/cm.

16-28. A bubble is made of 0.12 gm of soap solution. It is inflated with hydrogen of density 0.00009 gm/cm³. What is the minimum volume of the bubble which will just float in air?

16-29. Use the concept of surface tension as a force per unit length to derive the expression for the excess pressure in a soap bubble. Hint: imagine the bubble cut along its equatorial plane and consider either half as a body in equilibrium.

16-30. A spherical bubble is blown so that its radius increases at the constant rate of 1 cm/sec. Compute the power required to increase the surface of the bubble when its radius is 2 cm. The surface tension is 25 dynes/cm.

16-31. A soap bubble may be drawn out into a cylinder by touching to it a ring of the same diameter as the tube from which the bubble is blown, and then "stretching" the bubble between the tube and the ring. Derive the expression for the gauge pressure within a cylindrical bubble.

CHAPTER 17

HYDRODYNAMICS AND VISCOSITY

17-1 Streamline flow. Hydrodynamics is the study of fluids in motion. It is one of the most complex branches of mechanics, as will be realized by considering such common examples of fluid flow as a river in flood or a swirling cloud of cigarette smoke. While it must be true that $F = ma$ at each instant for each drop of water or each smoke particle, imagine attempting to write their equations of motion! However, the problem is not as hopeless as it seems at first sight.

When the proper conditions are fulfilled, the flow of a fluid is of a relatively simple type called *streamline* or *steady* flow. Figure 17-1 represents a portion of a pipe in which a fluid is flowing from left to right. If the flow is of the streamline type, every particle passing a point such as a follows exactly the same path as the preceding particles which passed the same point. These paths are called *lines of flow* or *streamlines* and three of them are shown in the figure. If the cross section of the pipe varies from point to point, the velocity of any one particle will vary along its line of flow, but at any fixed point in the pipe the velocity of the particle which happens to be at that point is always the same. The particle which is now at a in the figure will be a moment later at point b, traveling in a different direction with a different speed, and a moment later yet it will be at c, having again changed its velocity. However, if we fix our attention on the point of space marked b, then each successive particle as it passes through b will be traveling in exactly the same direction and with the same speed as is the particle which is at that point now.

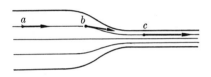

FIG. 17-1. Streamline or steady flow.

Any real fluid, because of its viscosity, will have a higher velocity at the center of the pipe than at the outside. For the present we shall assume the fluid to be nonviscous and the velocity to be the same at all points of a transverse cross section.

The flow of a fluid is of the streamline type provided the velocity is not too great and the obstructions, constrictions, or bends in the pipe are not such as to cause the lines of flow to change their direction too abruptly. If these conditions are not fulfilled, the flow is of a much more complicated type called *turbulent*.

17–2 Bernoulli's equation.   The fundamental equation of hydrodynamics is Bernoulli's equation, which is a relation between the pressure, velocity, and elevation at points along a streamline.   Essentially, it is the form taken by the work-energy equation when applied to fluid flow.

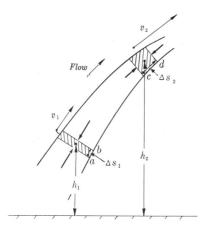

Figure 17–2 represents a portion of a pipeline in which a nonviscous incompressible liquid is flowing with streamline flow.   (In order to maintain the flow in the direction shown there must be a pump in the line or an elevated tank somewhere at the left of the diagram.)   We are to follow a small element of the liquid, indicated by shading, as it passes from one point to another along the pipe.

Let h_1 be the elevation of the first point above some reference level, v_1 the velocity at that point, A_1 the cross section of the pipe, and p_1 the pressure.   All of these quantities

FIG. 17–2. The net work done on the shaded element equals the increase in its kinetic and potential energy.

may vary from point to point, and h_2, v_2, A_2, and p_2 are their values at the second point.

Since the liquid is under pressure at all points, inward forces shown by the heavy arrows are exerted against both faces of the element.   As the element moves from point 1 to point 2, work is done *on* it by the force acting on its left face, and work is done *by* it against the force acting on its right face.   The net work done on the element, or the difference between these quantities, equals the gain in its kinetic energy plus the gain in its potential energy.

If A represents the cross section of the pipe at any point and p the corresponding pressure, the force against a face of the element at any point is pA.   The work done *on* the element in the motion in the diagram is

$$\int_a^c F \, ds = \int_a^c pA \, ds,$$

where ds is any short distance measured along the pipe, and the limits are from a to c, since these are the initial and final positions of the left face. This integral may be written

$$\int_a^c pA \, ds = \int_a^b pA \, ds + \int_b^c pA \, ds. \qquad (17\text{–}1)$$

Similarly, the work done *by* the element in its motion is

$$\int_b^d pA \, ds = \int_b^c pA \, ds + \int_c^d pA \, ds. \tag{17–2}$$

The *net* work done on the element is

$$\int_a^b pA \, ds + \int_b^c pA \, ds - \int_b^c pA \, ds - \int_c^d pA \, ds$$

$$= \int_a^b pA \, ds - \int_c^d pA \, ds.$$

The distances from a to b and from c to d are sufficiently small so that, without appreciable error, the pressures and areas may be considered constant along their extent. Then

$$\int_a^b pA \, ds = p_1 A_1 \, \Delta s_1,$$

and

$$\int_c^d pA \, ds = p_2 A_2 \, \Delta s_2.$$

But $A_1 \, \Delta s_1 = A_2 \, \Delta s_2 = V$, where V is the volume of the element. Hence

$$\text{Net work} = (p_1 - p_2)V. \tag{17–3}$$

Let ρ be the density of the liquid and m the mass of the element. Then $V = m/\rho$, and Eq. (17–3) becomes

$$\text{Net work} = (p_1 - p_2)\frac{m}{\rho}.$$

We now equate the net work done on the element to the sum of the increases in its kinetic and potential energy.

$$(p_1 - p_2)\frac{m}{\rho} = (\tfrac{1}{2}mv_2^2 - \tfrac{1}{2}mv_1^2) + (mgh_2 - mgh_1).$$

After cancelling m and rearranging terms, we obtain

$$p_1 + \tfrac{1}{2}\rho v_1^2 + \rho g h_1 = p_2 + \tfrac{1}{2}\rho v_2^2 + \rho g h_2, \tag{17–4}$$

and since the subscripts 1 and 2 refer to *any* two points along the pipeline, we may write

$$p + \tfrac{1}{2}\rho v^2 + \rho g h = \text{constant}. \tag{17–5}$$

Either Eq. (17–4) or Eq. (17–5) may be considered Bernoulli's equation.

Note carefully: p is the *absolute* (not gauge) pressure and must be expressed in pounds per square foot, newtons per square meter, or dynes per square centimeter. The density ρ must be expressed in slugs per cubic foot, kilograms per cubic meter, or grams per cubic centimeter.

17–3 Discharge rate of a pipe. Figure 17–3 represents the open end of a pipeline of cross section A, out of which is flowing a liquid with velocity v. The quantity of liquid discharged in time t is that contained in a cylinder of area A, extending back from the end of the pipe a distance vt. In other words, each point of the liquid which is now at the dotted section will, in

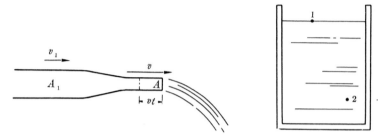

FIG. 17–3. Discharge rate of a pipe.   FIGURE 17–4

time t, have just reached the end of the pipe, and all of the liquid between that section and the end of the pipe will have been discharged in the meantime. Hence a volume of liquid equal to Avt is discharged in a time interval t, and the discharge *rate*, Q, is

$$Q = \frac{Avt}{t} = Av. \qquad (17\text{–}6)$$

Q is expressed in $\mathrm{ft}^3/\mathrm{sec}$, $\mathrm{m}^3/\mathrm{sec}$, or $\mathrm{cm}^3/\mathrm{sec}$ if the appropriate units are used for A and v.

If an incompressible liquid completely fills the pipe at all points, in any given time the same volume of liquid as is discharged from the end of the pipe must pass every cross section. If that were not the case, the volume of liquid between that cross section and the end would be increasing or decreasing. Hence if A_1 and v_1, (Fig. 17–3), are the area and velocity at any other point along the pipe,

$$Q = Av = A_1 v_1 = \text{constant.} \qquad (17\text{–}7)$$

This is known as the *equation of continuity*. A consequence of this relation is that the velocity is greatest at points where the cross section is least and vice versa.

17–4 Applications of Bernoulli's equation. (1) The equations of hydrostatics are special cases of Bernoulli's equation, when the velocity is everywhere zero. For example, the variation of pressure with depth in an incompressible liquid may be found by applying Bernoulli's equation to points 1 and 2 in Fig. 17–4. We have

$$p_1 = p_0 \text{ (atmospheric)}, \quad v_1 = v_2 = 0.$$

Let elevations be measured from the level of point 2. Then

$$h_2 = 0, \quad h_1 = h,$$

and

$$p_0 + \rho gh = p_2,$$

or

$$p_2 = p_0 + \rho gh,$$

which is the same as Eq. (16–3).

(2) *Torricelli's theorem.* Figure 17–5 represents a liquid flowing from an orifice in a tank at a depth h below the surface of the liquid in the tank. Take point 1 at the surface and point 2 at the orifice. The pressure at each point is the atmospheric pressure, p_0, since both are open to the atmosphere. Take the reference level at the elevation of point 2. If the orifice is small, the level of liquid in the tank will fall only slowly. Hence v_1 is small and we shall assume it to be zero. Then

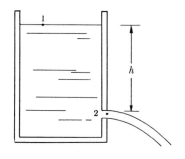

$$p_0 + \rho gh = p_0 + \tfrac{1}{2}\rho v_2^2,$$

or

$$v_2^2 = 2gh. \qquad (17\text{–}8)$$

This is Torricelli's theorem. Note that the velocity of discharge is the same as that which would be acquired by a body falling freely from rest through a height h.

If A is the area of the opening, the discharge rate Q is

FIG. 17–5. Velocity of efflux equals $\sqrt{2gh}$.

$$Q = Av = A\sqrt{2gh}. \qquad (17\text{–}9)$$

Because of the converging of the streamlines as they approach the orifice, the cross section of the stream continues to diminish for a short distance outside the tank. It is the area of smallest cross section, known as the *vena contracta*, which should be used in Eq. (17–9). For a sharp-edged circular opening, the area of the *vena contracta* is about 65% as great as the area of the orifice.

(3) *The Venturi meter.* The Venturi meter, illustrated in Fig. 17–6, consists of a constriction or throat inserted in a pipeline, and having properly designed tapers at inlet and outlet to avoid turbulence and assure

streamline flow. Bernoulli's equation, applied to the wide and constricted portions of the pipe, becomes

$$p_1 + \tfrac{1}{2}\rho v_1^2 = p_2 + \tfrac{1}{2}\rho v_2^2$$

(the "h" terms drop out if the pipe is level).

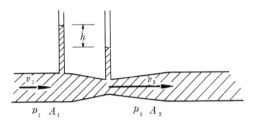

FIG. 17–6. The Venturi meter.

Since v_2 is greater than v_1, it follows that p_2 is less than p_1. That is, the pressure in the throat is smaller than in the main pipeline. The pressure difference may be measured by attaching vertical side tubes as shown in the diagram. If h is the difference in height of the liquid in the tubes, then

$$p_1 - p_2 = \rho g h.$$

The discharge rate, Q, may be obtained by combining Bernoulli's equation with the equation of continuity The result is

$$Q = A_1 A_2 \sqrt{\frac{2(p_1 - p_2)}{\rho(A_1^2 - A_2^2)}} \,.$$

Hence, if the pressure difference $p_1 - p_2$ is measured, and the areas A_1 and A_2 are known, the flow through the meter can be computed.

The reduced pressure at a constriction finds a number of technical applications. Gasoline vapor is drawn into the intake manifold of an internal combustion engine by the low pressure produced in a Venturi throat to which the carburetor is connected. The aspirator pump is a Venturi throat through which water is forced. Air is drawn into the low-pressure water rushing through the constricted portion. The injection pump used on a steam locomotive to draw water from the tender makes use of the same principle.

(4) *The pitot tube.* A pitot tube is shown in Fig. 17–7 as it would be used to measure the velocity of a gas

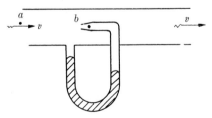

FIG. 17–7. The pitot tube.

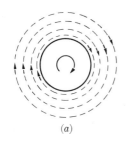

(a)

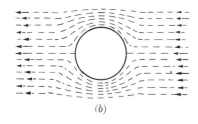

(b)

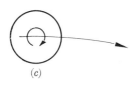

(c)

Fig. 17–8. Curved flight of a spinning ball.

flowing in a tube or pipe. An open-tube manometer is connected as shown to the tube in which the gas is flowing. The pressure at the left arm of the manometer, whose opening is parallel to the direction of flow, is equal to the pressure in the gas stream. The pressure in the right arm, whose opening is at right angles to the stream, may be computed by applying Bernoulli's equation to the points a and b. Let v be the velocity of the stream, ρ the density of the gas, and p_0 the pressure at point a. The velocity at point b, of course, is zero. Then

$$p_b = p_a + \tfrac{1}{2}\rho v^2.$$

Since p_b is greater than p_a, the liquid in the manometer becomes displaced as shown. If ρ_0 is the density of the liquid in the manometer and h the difference in height of the liquid in its arms, then

$$p_b = p_a + \rho_0 g h.$$

When this is combined with the preceding equation, we get

$$\rho_0 g h = \tfrac{1}{2}\rho v^2,$$

from which v may be expressed in terms of measurable quantities.

(5) *The curved flight of a spinning ball.* Figure 17–8(a) represents a top view of a ball spinning about a vertical axis. Because of friction between the ball and the surrounding air, a thin layer of air is dragged around by the spinning ball.

Figure 17–8(b) represents a stationary ball in a blast of air moving from right to left. The motion of the air stream around and past the ball is the same as though the ball were moving through still air from left to right. If the ball is moving from left to right and spinning at the same time, the actual velocity of the air around the ball is the resultant of the velocities in (a) and (b). At the top of the diagram the two velocities are in opposite directions, while the reverse is true at the bottom of the diagram. The top is a region of low velocity and high pressure, while the bottom is a region of high velocity and low pressure. There is therefore an excess pressure forcing the ball down in the diagram, so that if moving from left to right and spinning at the same time, the ball deviates from a straight line as shown in the top view in Fig. 17–8(c).

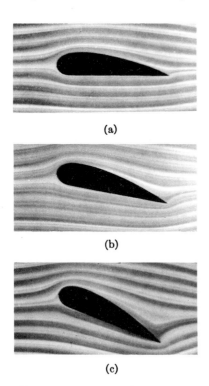

(a)

(b)

(c)

Fig. 17–9. Lines of flow around an airfoil.

(6) *Lift on an airplane wing.* Figure 17–9 is a photograph of streamline flow around a section in the shape of an airplane wing or an airfoil, at three different angles of attack. The apparatus consists of two parallel glass plates spaced about 1 mm apart. The wing section, whose thickness equals the separation of the plates, is inserted between them and alternate streams of clear water and ink flow by gravity between the plates and past the section. The photographs have been turned through 90° to give the effect of horizontal airflow past an airplane wing. Because the fluid is water flowing relatively slowly, the nature of the flow pattern is not identical with that of air moving at high speed past an actual wing.

Consider the first photograph, which corresponds to a plane in level flight. It will be seen that there is relatively little disturbance of the flow below the wing, but because of the shape of the airfoil there is a marked crowding together of the streamlines above it, much as if they were being forced through the throat of a Venturi. Hence the region above the wing is one of increased velocity and reduced pressure, while below the wing the pressure is nearly atmospheric. It is this pressure differential between upper and lower wing surfaces which gives rise to the lift on the wing. The wing is not simply forced up by air blowing against its lower surface.

There is a mistaken impression that the flow around an airplane wing results in an upward "pull" on the upper surface of the wing. Of course this cannot happen. The air presses against all portions of the wing surface, but the reduction below atmospheric pressure at the upper surface usually exceeds the increase above atmospheric pressure at the lower surface.

The second and third photographs show how, as the angle of attack is increased, the streamlines above the wing have to change direction sharply to follow the contour of the wing surface and join smoothly with the streamline flow below the wing. While the slowly moving water in Fig. 17–9 does retain its streamline form even at the large angle of attack in the third photograph, it is much more difficult for the air moving rapidly past an airplane wing to do so. As a consequence, if the angle of attack is too great, the streamline flow in the region above and behind the wing breaks down and a complicated system of whirls and eddies known as *turbulence* is set up. Bernoulli's equation no longer applies; the pressure above the wing rises, the lift on the wing decreases, and the plane stalls.

17–5 Viscosity. Viscosity may be thought of as the internal friction of a fluid. Because of viscosity, a force must be exerted to cause one layer of a fluid to slide past another, or to cause one surface to slide past another if there is a layer of fluid between the surfaces. Both liquids and gases exhibit viscosity, although liquids are much more viscous than gases. In developing the fundamental equations of viscous flow, it will be seen that the problem is very similar to that of the shearing stress and strain in a solid. (See Sections 13–2 and 13–3.)

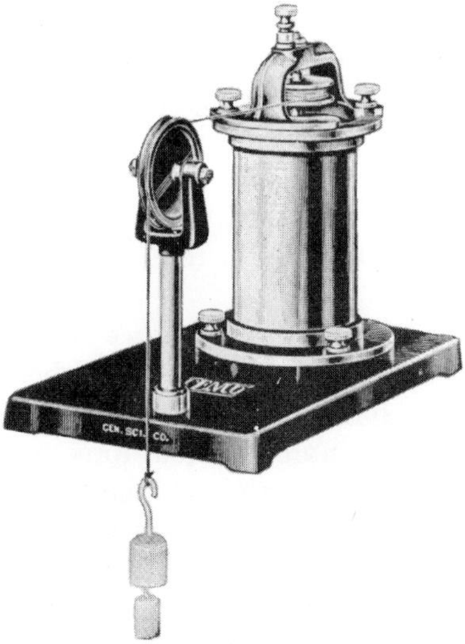

Figure 17–10 illustrates one type of apparatus for measuring the viscosity of a liquid. A cylinder is pivoted on nearly frictionless bearings so as to rotate concentrically within a cylindrical vessel. The liquid whose viscosity is to be measured is poured into the annular space between the cylinders. A torque can be applied to the inner cylinder by the weight-pulley system. When the weight is released, the inner cylinder accelerates momentarily but very quickly comes up to a constant angular velocity and continues to rotate at that velocity so long as the torque acts.

FIG. 17–10. One type of viscosimeter. (Courtesy of Central Scientific Co.)

It is obvious that this velocity will be smaller with a liquid such as glycerin in the annular space than it will be if the liquid is water or kerosene. From a knowledge of the torque, the dimensions of the apparatus, and the angular velocity, the viscosity of the liquid may be computed.

To reduce the problem to its essential terms, imagine that the cylinders are of nearly the same size so that the liquid layer between them is small. A short arc of this layer will then be approximately a straight line. Figure 17–11 shows a portion of the liquid layer between the moving inner

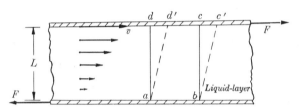

FIG. 17–11. Laminar flow of a viscous liquid.

wall and the stationary outer wall. The liquid in contact with the moving surface is found to have the same velocity as that surface; the liquid adjacent to the stationary inner wall is at rest. The velocity of intermediate layers of the liquid increases uniformly from one wall to the other as shown by the arrows.

Flow of this type is called *laminar*. (A lamina is a thin sheet.) The layers of liquid slide over one another much as do the leaves of a book when it is placed flat on a table and a horizontal force applied to the top cover. As a consequence of this motion, a portion of the liquid which at some instant has the shape $abcd$ will a moment later take the shape $abc'd'$ and will become more and more distorted as the motion continues. In other words, the liquid is in a state of continually increasing shearing strain.

In order to maintain the motion, it is necessary that a force shall be continually exerted to the right on the upper, moving plate, and hence indirectly on the upper liquid surface. This force tends to drag the liquid and the lower plate as well to the right. Therefore an equal force must be exerted toward the left on the lower plate in order to hold it stationary. These forces are lettered F in Fig. 17–11. If A is the area of the liquid over which these forces are applied, the ratio F/A is the shearing stress exerted on the liquid.

When a shearing stress is applied to a solid, the effect of the stress is to produce a certain displacement of the solid, such as dd'. The shearing strain is defined as the ratio of this displacement to the transverse dimension L, and within the elastic limit the shearing stress is proportional to the shearing strain. With a fluid, on the other hand, the shearing strain increases without limit so long as the stress is applied, and the stress is found by experiment to be proportional, not to the shearing strain, but to its *rate of change*. The strain in Fig. 17–11 at the instant when the volume of fluid has the shape $abc'd'$ is dd'/ad, or dd'/L. Since L is constant, the rate of change of strain equals $1/L$ times the rate of change of dd'. But the rate of change of dd' is simply the velocity of point d', or the velocity v of the moving wall. Since shearing stress is proportional to rate of change of shearing strain,

$$\frac{F}{A} \propto \frac{v}{L}, \quad \text{or} \quad \frac{F}{A} = \eta \frac{v}{L},$$

or

$$F = \eta \frac{Av}{L}. \tag{17–10}$$

The proportionality constant, represented by the Greek letter η (eta), is called the *coefficient of viscosity*, or simply the *viscosity*. It is small for liquids which flow readily like kerosene, and larger for liquids like molasses or glycerin.

Equation (17–10) was derived for a special case in which the velocity increased at a uniform rate with increasing height above the lower plate. The general term for the *space* rate of change of velocity in a direction at right angles to the flow is the *velocity gradient*. In this special case it is equal to the ratio v/L. In the general case, the velocity gradient is not uniform and its value at any point is written as dv/dy, where dv is the small difference in velocity between two points separated by a distance dy measured at right angles to the direction of flow. Hence the general form of Eq. (17–10) is

$$F = \eta A \frac{dv}{dy}. \tag{17–11}$$

From Eq. (17–10), the unit of viscosity is that of force times distance divided by area times velocity, or in the cgs system, 1 dyne·sec/cm². A viscosity of 1 dyne·sec/cm² is called a *poise*. Small viscosities are usually expressed in centipoises (1 cp = 10^{-2} poise) or micropoises (1 μp = 10^{-3} poise). Some typical values of viscosity are given in Table VI.

TABLE VI

VISCOSITIES OF LIQUIDS AND GASES

(1 cp $= 10^{-2}$ poise, 1 μp $= 10^{-6}$ poise)

Liquid	t (°C)	η (cp)
Alcohol, ethyl	20	16
Glycerin	20	830
Machine oil:		
Heavy	15	660
Light	15	113
Mercury	20	1.55
Water	20.20	1.0000

Gas	t (°C)	η (μp)
Air	20	181
Argon	23	221
Carbon dioxide	20	148
Helium	23	196
Hydrogen	20	88
Mercury (vapor)	380	654
Neon	15	312
Nitrogen	23	177
Oxygen	15	196

The coefficient of viscosity is markedly dependent on temperature, increasing for gases and decreasing for liquids as the temperature is increased. See Table VII.

TABLE VII

VARIATION OF VISCOSITY WITH TEMPERATURE

Water					
t (°C)	0	20	40	60	80	100
η (cp)	1.792	1.0050	0.6560	0.4688	0.3565	0.2838

Glycerin				
t (°C)	2.8	8.1	14.3	20.3	26.5
η (cp)	4220	2518	1387	830	494

Light Machine Oil		
t (°C)	15.6	37.8	100
η (cp)	114	34.2	4.9

Making use of Eq. (17–10), we can now return to the apparatus described on page 345 and see how viscosity can be measured with its aid. Let the depth of liquid between the cylinders be h, the radius of the inner cylinder R_1, and that of the outer cylinder R_2. The force F is exerted tangent to the surface of the inner cylinder and is distributed over an area

$$A = 2\pi R_1 h.$$

This force produces a torque $\tau = FR_1$, and hence

$$F = \frac{\tau}{R_1}.$$

The velocity v is related to the angular velocity ω by

$$v = \omega R_1.$$

The thickness of the liquid layer is $R_2 - R_1$. Hence, from Eq. (17–10),

$$\frac{\tau}{R_1} = \eta \frac{2\pi R_1 h \times \omega R_1}{R_2 - R_1},$$

$$\eta = \frac{\tau (R_2 - R_1)}{2\pi \omega R_1^3 h}.$$

Since all of the quantities on the right are measurable, the viscosity can be computed.

A common technical method of measuring viscosity makes use of an apparatus called a *viscosimeter*. This consists of a small container in the bottom of which is an orifice of specified dimensions. A specified volume of liquid is poured into the container and the time required for the liquid to run out through the orifice is measured. The viscosity is then computed by an empirical formula (see Problem 25 on page 356).

Viscosities of lubricating oils are commonly expressed on an arbitrary scale established by the Society of Automotive Engineers. An oil whose SAE number is 10 has a viscosity at 130°F between about 160 and 220 centipoise; the viscosity of SAE 20 is between 230 and 300 centipoise, and that of SAE 30 is between 360 and 430 centipoise.

17–6 Stokes' law. When a viscous fluid flows past a sphere with streamline flow or when a sphere moves through a viscous fluid at rest, a friction force is exerted on the sphere. (A force is, of course, experienced by a body of any shape, but only for a sphere is the expression for the force readily calculable.) Analysis which is beyond the scope of this book shows that the friction force is given by

$$F = 6\pi\eta rv,$$

where η is the viscosity of the fluid, r the radius of the sphere, and v the relative velocity of sphere and fluid. This relation was first deduced by Sir George Stokes in 1845 and is called *Stokes' law*. We shall consider it briefly in relation to a sphere falling through a viscous fluid.

If the sphere is released from rest ($v = 0$), the viscous force at the start is zero. The other forces on the sphere are its weight and the buoyant force of the fluid. If ρ is the density of the sphere and ρ_0 the density of the fluid,

$$\text{Weight} = mg = \tfrac{4}{3}\pi r^3 \rho g,$$

$$\text{Buoyant force} = \tfrac{4}{3}\pi r^3 \rho_0 g.$$

Since the net downward force on the sphere is equal to the product of its mass times its acceleration, the initial acceleration is

$$a_0 = \frac{\rho - \rho_0}{\rho}\, g.$$

As a result of this acceleration, the sphere acquires a downward velocity and therefore experiences a retarding force given by Stokes' law. As the velocity increases, the retarding force also increases in direct proportion, and eventually a velocity is reached such that the downward force and the retarding force are equal. The sphere then ceases to accelerate and moves with a constant velocity called its *terminal velocity*. This velocity can be found by setting the downward force equal to the retarding force.

$$\tfrac{4}{3}\pi r^3 (\rho - \rho_0)g = 6\pi\eta rv,$$

or

$$v = \frac{2}{9}\frac{r^2 g}{\eta}(\rho - \rho_0). \qquad (17\text{--}12)$$

The relation above holds provided the velocity is not so great that turbulence sets in. When this occurs the retarding force is much greater than that given by Stokes' law.

EXAMPLE. Find the terminal velocity of a steel ball-bearing 2 mm in radius, falling in a tank of glycerin.

$$\rho_{\text{steel}} = \text{(about) } 8 \text{ gm/cm}^3,$$
$$\rho_{\text{glycerin}} = \text{``} \quad 1.3 \text{ gm/cm}^3,$$
$$\eta_{\text{glycerin}} = \text{``} \quad 8.3 \text{ poise,}$$

$$v = \frac{2}{9} \frac{(0.2)^2 \times 980}{8.3} (8 - 1.3)$$

$$= 7 \text{ cm/sec.}$$

This velocity is attained in a very short distance from the start of the motion. The experiment above is used as one method of measuring viscosity.

17–7 Flow of viscous fluids through tubes. It is evident from the general nature of viscous effects that the velocity of a viscous fluid flowing through a tube will not be the same at all points of a cross section. The walls of the tube exert a backward drag on the outermost layer of fluid, which in turn drags backward on the next layer within it, and so on. As a result, the velocity is a maximum at the center of the tube, decreasing to zero at the walls. The flow is similar to that of a number of telescoping tubes sliding one within another.

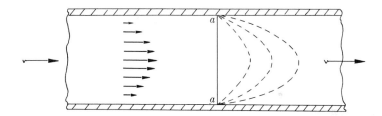

FIG. 17–12. Velocity distribution across a circular tube.

If the cross section of the tube is a circle, the velocity distribution across it will be parabolic, as shown by the arrows in Fig. 17–12. A group of particles which at some instant lie in a transverse plane such as aa will occupy successively the positions shown by the curves at the right of aa. The equation of these curves can be shown to be

$$v = \frac{p_1 - p_2}{4\eta L} (R^2 - r^2),$$

where v is the velocity at a radius r, p_1 and p_2 are the pressures at the ends of the tube, L is its length, and R is its radius.

The discharge rate is

$$Q = \frac{\pi R^4}{8\eta L} (p_1 - p_2), \qquad (17\text{–}13)$$

a relation known as *Poiseuille's law*. It states that the rate of flow of a viscous fluid through a tube is directly proportional to the pressure difference between the ends of the tube, proportional to the 4th power of the radius of the tube, and inversely proportional to the viscosity of the fluid. The viscosity of gases is most conveniently measured by measuring their rate of flow through a capillary tube and computing η from Eq. (17–13).

17–8 Derivation of Poiseuille's law. To derive Poiseuille's law, consider a portion of a tube of radius R and length L, through which is flowing a fluid of viscosity η (Fig. 17–13a). Let the pressures at the ends of the tube be p_1 and p_2. A small cylinder of radius r is in equilibrium

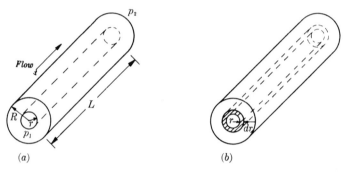

FIGURE 17–13

(moving with constant velocity) under the driving force, due to the pressure difference between its ends and the retarding viscous force at its outer surface. (This derivation is not rigorously correct—one should consider a thin-walled tube of fluid—but it leads to the right answer.) The driving force is

$$(p_1 - p_2)\pi r^2.$$

The viscous force, from Eq. (17–11), is

$$F = -\eta A \frac{dv}{dr}$$

$$= -\eta \times 2\pi r L \times \frac{dv}{dr},$$

where the minus sign is introduced because v decreases as r increases. Equating the forces, we find

$$-\frac{dv}{dr} = (p_1 - p_2)\frac{r}{2\eta L},$$

$$-dv = \frac{p_1 - p_2}{2\eta L} r \, dr.$$

Integration of this equation gives

$$-v = \frac{p_1 - p_2}{4\eta L} r^2 + C.$$

But since $v = 0$ when $r = R$,

$$0 = \frac{p_1 - p_2}{4\eta L} R^2 + C,$$

and therefore

$$v = \frac{p_1 - p_2}{4\eta L}(R^2 - r^2),$$

which is the equation of a parabola.

To find the discharge rate Q, consider the thin-walled element in Fig. 17–13(b). The discharge rate of this element, dQ, is

$$dQ = v \, dA = \frac{p_1 - p_2}{4\eta L}(R^2 - r^2) \times 2\pi r \, dr,$$

and

$$Q = \int dQ = \frac{2\pi(p_1 - p_2)}{4\eta L} \int_0^R (R^2 - r^2) r \, dr,$$

$$Q = \frac{\pi R^4}{8\eta L}(p_1 - p_2),$$

which is Poiseuille's law.

PROBLEMS

17–1. A circular hole 1 inch in diameter is cut in the side of a large standpipe, 20 ft below the water level in the standpipe. Find the velocity of efflux and the discharge rate. Neglect the contraction of the stream lines after emerging from the hole.

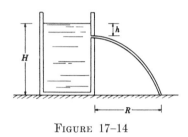

FIGURE 17–14

17–2. Water stands at a depth H in a large open tank whose side walls are vertical. (Fig. 17–14.) A hole is made in one of the walls at a depth h below the water surface. (a) At what distance R from the foot of the wall does the emerging stream of water strike the floor? (b) At what height above the bottom of the tank could a second hole be cut so that the stream emerging from it would have the same range? (c) Find the depth h below the surface which makes R a maximum.

17–3. A tank of large area is filled with water to a height of one foot. A hole of 1 in^2 cross section in the bottom allows water to drain out in a *continuous stream*. (a) What is the rate at which water flows out of the tank, in ft^3/sec? Neglect the convergence of the stream lines. (b) At what distance below the bottom of the tank is the cross-sectional area of the stream equal to one-half the area of the hole?

17–4. A cylindrical tank 1 ft high, of cross section 72 in^2, is initially full of water. How long a time is required for the tank to empty through a hole in the bottom 1 in^2 in area? Neglect the convergence of the stream lines.

17–5. A cylindrical vessel, open at the top, is 20 cm high and 10 cm in diameter. A circular hole whose cross-sectional area is 1 cm^2 is cut in the center of the bottom of the vessel. Water flows into the vessel from a tube above it at the rate of 140 cm^3/sec. (a) How high will the water in the vessel rise? (b) If the flow into the vessel is stopped after the above height has been reached, how long a time is required for the vessel to empty?

17–6. A sealed tank containing sea water to a height of 5 ft also contains air above the water at a gauge pressure of 580 lb/ft^2. Water flows out from a hole at the bottom. The cross-sectional area of the hole is 1.6 in^2. (a) Calculate the efflux velocity of the water. (b) Calculate the reaction force on the tank exerted by the water in the emergent stream.

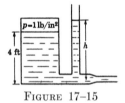

FIGURE 17–15

17–7. Sea water (weight density 64 lb/ft^3) stands to a height of 4 ft in a tank. The tank contains compressed air at a gauge pressure of 1 lb/in^2. The horizontal outlet pipe has cross-sectional areas of 2.88 in^2 and 1.44 in^2 at the larger and smaller sections (Fig. 17–15). (a) What is the discharge rate from the outlet? (b) To what height h does water stand in the open end pipe? (c) If now the tank is punctured at the top and the gauge pressure drops to zero, what will be the height h?

17–8. The depth of sea water (weight density = 64 lb/ft^3) in a closed tank of large cross section is 16 ft. A horizontal pipe line leads from the bottom of the tank, the area of the line tapering from

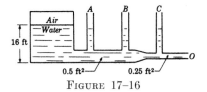

FIGURE 17–16

0.5 ft^2 to 0.25 ft^2 as shown in Fig. 17–16. Three vertical side arms, A, B, C, open at the top, are connected to the horizontal pipe. The gauge pressure of the compressed air in the top of the large tank is 4 lb/in^2. When the pipe is open at point O, what is the height of the water column in side arms A, B, and C? Neglect viscosity and compressibility. Assume that the depth of water in the large tank and the air pressure above it remain constant.

17–9. Water stands at a depth of 4 ft in an enclosed tank whose side walls are vertical. The space above the water surface contains air at a gauge pressure of 120 lb/in^2. The tank rests on a platform 8 ft above the floor. A hole of cross-sectional area 0.5 in^2 is made in one of the side walls just above the bottom of the tank. (a) Where does the stream of water from the hole strike the floor? (b) What is the vertical force exerted on the floor by the stream? (c) What is the horizontal force exerted on the tank? Assume the water level and the pressure in the tank to remain constant.

17–10. A large open tank is filled with water to a depth h_0. The tank has two openings in its bottom, each of cross-sectional area A_0. One of these leads directly to the air. The other is connected to a garden hose having a nozzle with a variable opening of area A_1 at its end. The nozzle is at a distance h_1 below the bottom of the tank. (a) If $A_0 = A_1$, will more or less water flow through the hose than through the other opening? (b) What should be the area A_1 of the nozzle so that there will be the same total flow through each of the holes in the tank?

17–11. The water level in a tank on the top of a building is 100 ft above the ground. The tank supplies water, through pipes of 0.02 ft^2 cross-sectional area, to the various apartments. Each faucet through which the water emerges has an orifice of 0.01 ft^2 effective area. (a) How long will it take to fill a 1 ft^3 pail in an apartment 75 ft above the ground? (b) What is the gauge pressure in a water pipe (not in the faucet) on the ground level when the faucet is closed? (c) What is the gauge pressure in a pipe on the ground level when the faucet is open?

17–12. What gauge pressure is required in the city mains in order that a stream from a fire hose connected to the mains may reach a vertical height of 60 ft? Neglect friction effects.

17–13. A hole of area A_1 is made in the wall of a cylindrical tank of cross-sectional area A_2, at a depth h below the liquid surface. Find the expression for the velocity of efflux from the hole, and show that it reduces to Torricelli's theorem when A_2 is much greater than A_1.

17–14. A pipe line 6 inches in diameter, flowing full of water, has a constriction of diameter 3 inches. If the velocity in the 6-inch portion is 4 ft/sec, find the velocity in the constriction and the discharge rate in ft^3/sec.

17–15. A horizontal pipe of 6 in^2 cross section tapers to a cross section of 2 in^2. If sea water of density 2 slug/ft^3 is flowing with a velocity of 180 ft/min in the large pipe where a pressure gauge reads 10.5 lb/in^2, what is the gauge pressure in the adjoining part of the small pipe? The barometer reads 30 inches of mercury.

17–16. Sea water of weight density 64 lb/ft^3 flows through a horizontal pipe of cross-sectional area 1.44 in^2. At one section the cross-sectional area is 0.72 in^2. The pressure difference between the two sections is 0.048 lb/in^2. How many cubic feet of water will flow out of the pipe in one minute?

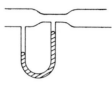

FIGURE 17-17

17-17. The section of pipe line shown in Fig. 17-17 has a cross section of 0.04 ft² at the wider portions and 0.01 ft² at the constriction. One cubic foot of water is discharged from the pipe in 5 sec. (a) Find the velocities at the wide and narrow portions. (b) Find the pressure difference between these portions. (c) Find the difference in height between the mercury columns in the U-tube.

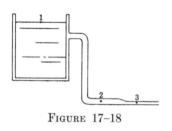

FIGURE 17-18

17-18. Water flows steadily from the reservoir shown in Fig. 17-18. The elevation of point 1 is 40 ft; of points 2 and 3 it is 4 ft. The cross section at point 2 is 0.5 ft² and at point 3 it is 0.25 ft². The area of the reservoir is very large compared with the cross sections of the pipe. (a) Compute the absolute pressure at point 2. (b) Compute the discharge rate in ft³/sec.

17-19. A horizontal glass tube consists of three sections, A, B, and C, in series, each of constant cross section. Water flows through the tube and discharges into the air at the open end of C. Each of the sections A and B has a small hole in the wall. It is found that water is ejected from the hole in A, whereas air bubbles appear in the water at the hole in B. Which section of the

tube has the largest and which the smallest diameter?

17-20. Assume that air is streaming horizontally past an airplane wing such that the velocity is 100 ft/sec over the top surface and 80 ft/sec past the bottom surface. If the wing weighs 600 lb and has an area of 40 ft², what is the net force on the wing?

17-21. Sea water with a weight density of 64 lb/ft³ is siphoned from an open tank through a tube of cross-sectional area 0.5 in². The highest point in the tube is 15 ft above the water surface and the outlet is 9 ft below the water surface. (a) What is the discharge rate in ft³/sec? (b) What is the gauge pressure in lb/in² at the highest point in the tube?

17-22. A fire engine pumps 10,000 lb of water per minute from a lake, and ejects it from a nozzle 17 ft above the lake surface with a velocity of 32 ft/sec. What horsepower output must the engine have, if friction losses are neglected?

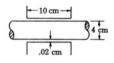

FIGURE 17-19

17-23. A shaft 4 cm in diameter rotates inside a bearing 10 cm long. The clearance between shaft and bearing is 0.02 cm, and the intervening space is filled with oil of viscosity 100 cp. What is the torque required to turn the shaft at a constant angular velocity of 10 rev/sec? (See Fig. 17-19.)

17-24. The following experiment was performed to determine the coefficient of viscosity of a heavy lubricating oil. A tank of oil at 20°C and maintained at a constant gauge pressure of 380 cm of mercury was connected to one end of a glass capillary tube 2 mm in diameter and 1 meter long. The other end of the tube was open to the atmosphere and delivered 30.39 cm^3 of oil in 2 minutes. The barometer read 76 cm of mercury. What was the coefficient of viscosity of the oil?

17-25. The formula for use with the Saybolt viscosimeter is

$$\mu = \rho \left(0.0022\, t - \frac{1.80}{t} \right),$$

where μ is the viscosity in poises, ρ is the density in gm/cm^3, and t is the time in seconds for 60 cm^3 of liquid to run out of the viscosimeter through an orifice 0.483 inch long and 0.0695 inch in diameter. If the time of efflux of 60 cm^3 of an oil of specific gravity 0.92 is 130 sec, find the viscosity of the oil in centipoises. What would be the SAE number of this oil?

17-26. A liquid flows in laminar flow in a vertical tube of radius R under the action of gravity. Show that the velocity of the liquid at a distance r from the axis of the tube is given by

$$v = \frac{\rho g}{4\eta} (R^2 - r^2).$$

What is the discharge rate of the tube?

17-27. A viscous liquid flows in laminar flow under the action of gravity between two vertical plates of large area. (a) If the plates are separated by a distance $2a$, show that the velocity of the liquid at a distance x from a plane halfway between the plates is given by

$$v = \frac{\rho g}{2\eta} (a^2 - x^2).$$

(b) Derive an expression for the volume of liquid flowing per unit time across a horizontal area of width w and breadth $2a$.

17-28. A tank filled with glycerin to a depth of 25 cm has a vertical tube 0.3 cm in radius and 25 cm long connected to its bottom. The lower end of the tube is open to the atmosphere. Consider a cylindrical element of the liquid in the tube, of radius r and 25 cm high. Calculate the force on this element due to the pressure of the glycerin above it, its own weight, and the viscous force exerted on its curved surface. To calculate the velocity of flow at the center of the tube, use the fact that in the steady state the sum of these three forces is zero. The density of glycerin is 1.32 gm/cm^3.

17-29. A water droplet is observed to fall in a gas of viscosity 2×10^{-4} poise with a terminal velocity of 980 cm/sec. What is the radius of the drop? The density of the gas is 10^{-3} gm/cm^3.

17-30. An aluminum sphere 2 mm in radius starts from rest and falls in a tank of glycerin. (a) What is the terminal velocity? (b) What is the acceleration of the sphere when the velocity is one-half of the terminal velocity?

17-31. With what terminal velocity will an air bubble 1 mm in diameter rise in a liquid of viscosity 150 cp and density 0.90 gm/cm^3? What is the terminal velocity of the same bubble in water?

17-32. A wind tunnel is essentially a Venturi meter through which air is drawn by a large fan. The velocity of the air stream in the throat of a certain wind tunnel is 150 ft/sec. Find the pressure in the throat.

SUGGESTED BOOKS FOR COLLATERAL READING

CAJORI, F., *A History of Physics*. Macmillan.

COURANT and ROBBINS, *What is Mathematics?* Macmillan.

GALILEI, GALILEO, *Dialogues Concerning Two New Sciences*. (Henry Crew and Alfonso de Salvio, translators.) Northwestern University Studies.

LENARD, PHILIP, *The Great Men of Science*. (H. S. Hatfield, translator,) Macmillan.

LINDSAY, R. B., *Handbook of Elementary Physics*. Dryden Press.

MACH, ERNST, *The Science of Mechanics*. (T. J. MacCormack, translator.) The Open Court Publishing Co.

MAGIE, W. F., *A Source Book in Physics*. McGraw Hill.

TAYLOR, W. L., *Physics, The Pioneer Science*. Houghton Mifflin.

HEAT

CHAPTER 18

TEMPERATURE—EXPANSION

18–1 Temperature. The temperature of a body is a measure of its relative hotness or coldness. When we touch a body our temperature sense enables us to make a rough estimate of its temperature in somewhat the same way that we can, by muscular effort, make a rough estimate of the magnitude of a force. It is evident, however, that the temperature sense is too limited in its range and not sufficiently precise to be of any value in engineering or scientific work. For the *measurement* of temperature we must make use of some measurable physical property which changes with temperature, just as for the measurement of a force we use some property of a body which changes with the force, as, for example, the length of a coil spring. Any instrument used for the measurement of temperature is called a *thermometer*.

18–2 Thermometers. Some common physical properties which change with temperature are the length of a rod, the volume of a liquid, the electrical resistance of a wire, and the color of a lamp filament; in fact, all these changes are utilized in the construction of various types of thermometers.

We shall consider first the common *liquid-in-glass* thermometer. This instrument, illustrated in Fig. 18–1, consists of a thin-walled glass bulb A,

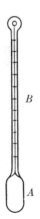

to the top of which is sealed a slender glass capillary tube B. A liquid such as mercury or colored alcohol partially fills the bulb and tube. The upper end of the tube is sealed off, and in most instances the air is removed from the space above the liquid. A scale for measuring the position of the top of the liquid column in the capillary is engraved on tube B, or a separate scale may be mounted behind the tube. As the temperature of the thermometer is increased, the volume of the liquid

Fig. 18–1. Liquid-in-glass thermometer. increases, and the volume of the bulb and capillary increases also. If both expanded alike, the position of the liquid in the capillary would not change, but actually the liquid expands more rapidly than does the bulb. Hence the liquid level rises in the capillary with increasing temperature and falls as the temperature is lowered. This instrument, therefore, utilizes the *difference* between the expansions of the liquid and the glass.

18–3 Temperature scales. In engineering work and in everyday life in this country, the *fahrenheit* temperature scale is used. In scientific work throughout the world, temperatures are expressed on the *centigrade* scale. In defining either scale, two reference temperatures or so-called *fixed points* are chosen, and arbitrary values are assigned to those temperatures, thus fixing the position of the zero point and the size of the temperature unit.

One reference temperature, the *ice point*, is the temperature of a mixture of air-saturated water and ice at a pressure of one atmosphere. The other, the *steam point*, is the boiling temperature of pure water at a pressure of one atmosphere. On the centigrade scale, the ice point is numbered zero and the steam point, 100. On the fahrenheit scale, these temperatures are numbered 32 and 212.

With the aid of a liquid-in-glass thermometer, any other temperature, t, is now defined (on the centigrade scale) as that temperature which produces a relative volume change $t/100$ as great as the change between the ice and steam points. If the capillary is of uniform cross section, this is equivalent to the statement that the temperature is directly proportional to the length of the liquid column above the ice point. In other words, if the position of the top of the liquid is marked on the tube, first when the thermometer is inserted in an ice-water mixture, and second when it is immersed in pure water boiling at atmospheric pressure, the distance between these marks may be divided into 100 equal parts and these divisions numbered from zero to 100. The temperature interval corresponding to each of the divisions is called one centigrade degree, and the numbering may be extended above 100 and below zero. The fahrenheit scale is obtained in a similar way, by dividing the length of the column between ice and steam points into 180 divisions and extending the scale in either direction.

Since the same temperature interval is divided into 100 degrees on the centigrade scale and 180 degrees on the fahrenheit scale, the temperature range corresponding to one centigrade degree is $\frac{180}{100}$, or $\frac{9}{5}$ as great as that corresponding to one fahrenheit degree.

The zero point on the fahrenheit scale is obviously 32 fahrenheit degrees below the ice point. Temperatures below the zero of either scale are considered negative. The relation between the scales is best kept in mind by a diagram such as that of Fig. 18–2.

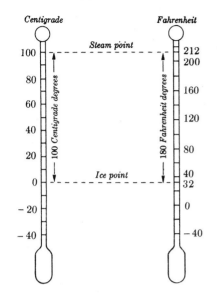

FIG. 18–2. Relation between centigrade and fahrenheit scales.

NOTE. Suppose the temperature of a beaker of water is raised from 20°C to 30°C, through a temperature interval of 10 centigrade degrees. It is desirable to distinguish between such a temperature interval and the actual temperature of 10 degrees above the centigrade zero. Hence we shall use the phrase "10 degrees centigrade," or "10°C," when referring to an *actual temperature*, and "10 centigrade degrees," or "10 C°" to mean a temperature *interval*. Thus there is an interval of 10 centigrade degrees between 20 degrees centigrade and 30 degrees centigrade.

The following process of reasoning may be used to convert a temperature expressed on one scale to its value on the other scale. A temperature of 30°C, for instance, means that there is an interval of 30 centigrade degrees between this temperature and the ice point. Since 1 C° = $\frac{9}{5}$ F° (*not* 1°C = $\frac{9}{5}$°F), 30 C° = 30 × $\frac{9}{5}$ = 54 F°. Hence this temperature lies 54 fahrenheit degrees above the ice point. Since the temperature of the ice point is 32°F, the temperature corresponding to 30°C is 54 + 32, or 86°F. It will be left as an exercise to show that the reasoning above leads to the following relations:

$$F = \tfrac{9}{5}C + 32, \qquad C = \tfrac{5}{9}(F - 32),$$

where F and C refer to the same temperature on the fahrenheit and centigrade scales respectively. What is the relation between a given temperature *interval* on the fahrenheit and centigrade scales?

There is, of course, no reason why the numbering of either scale cannot be extended indefinitely both above and below zero. We shall see later however, that both theory and experiment show that there is a limit to the *lowest* temperature which can ever be attained, although there is no theoretical limit to the highest possible temperature. This lowest attainable temperature is known as *absolute* zero, and its location is at −273.2°C or, in round numbers, −273°C. For some purposes, it is found convenient to use a temperature scale whose zero point is at absolute zero. Temperatures on this scale are called absolute temperatures, and both fahrenheit and centigrade absolute scales are used. The centigrade absolute is also called the Kelvin scale, in honor of Lord Kelvin who first suggested its use. Temperatures on the Kelvin scale are numerically 273 degrees larger than those on the centigrade scale, so the temperature of the ice point is 273°K and that of the steam point is 373°K.

The temperature of absolute zero on the fahrenheit scale is —460°F, and temperatures on the fahrenheit absolute scale are numerically 460° larger than on the fahrenheit scale. The temperature of the ice point is thus 492° fahrenheit absolute, and that of the steam point is 672° fahrenheit absolute.

18-4 Other methods of thermometry. Mercury freezes at —40°C, and its vapor pressure (see Section 23-3) becomes unduly high at temperatures much above 360°C. Hence the mercury-in-glass thermometer is limited to this temperature range. The lower range of liquid-in-glass thermometers can be extended by the use of liquids such as alcohol or pentane, which freeze at —130°C and —200°C respectively.

The *resistance thermometer* makes use of the fact that the electrical resistance of metals increases with increasing temperature. The thermometer itself consists of a fine wire, usually of platinum, wound on a mica frame and enclosed in a thin-walled silver tube for protection. Copper wires lead from the thermometer unit to a resistance measuring device which may be located at any convenient point. Since resistance may be measured with a high degree of precision, the resistance thermometer is one of the most precise instruments for the measurement of temperature, a precision of 0.001°C being attainable. The range of a platinum resistance thermometer is from the lowest attainable temperature to 1760°C, the melting point of platinum.

The *thermocouple* consists of an electrical circuit such as that shown in Fig. 18-3(a). When wires of any two unlike metals are joined so as to form a complete circuit, it is found that an electromotive force exists in the circuit whenever the junctions A and B are at different temperatures.* The emf, for any given pair of metals, depends on the difference in temperature between the junctions. The thermocouple may be used as a thermometer by placing one junction

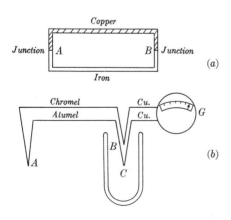

FIG. 18-3. Thermocouple circuit.

*A text on electricity should be consulted for a more complete discussion of thermal emf's.

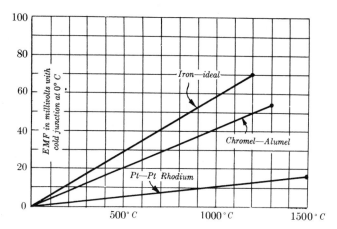

FIG. 18–4. Emf's of some common thermocouples.

in contact with the body whose temperature is to be measured, keeping the other junction at some known temperature (usually 0°C), and measuring the emf. The usual thermocouple circuit is shown in Fig. 18–3(b). Junctions B and C are kept at 0°C by ice and water in a Dewar flask, and junction A is placed in contact with the body whose temperature is to be measured. The emf is read on galvanometer G. Figure 18–4 shows the emf's developed by some of the pairs of metals commonly used in thermocouples, for various temperatures of the "hot junction," when the "cold junction" is at 0°C.

The *constant-volume gas thermometer*, shown in Fig. 18–5, makes use of the pressure changes of a gas kept at constant volume. The gas, usually hydrogen or helium, is contained in bulb C, and the pressure exerted by it can be measured by the open-tube mercury manometer. As the temperature of the gas is increased, it expands, forcing the mercury down in tube B and up in tube A. A and B are connected by the flexible rubber tube D, and by raising A the mercury level in B may be brought back to the reference mark E. The gas is thus kept at constant volume.

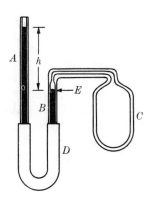

FIG. 18–5. Constant-volume gas thermometer.

Any type of thermometer, such as a liquid-in-glass thermometer, a resistance thermometer, a thermocouple, or a gas thermometer may be used to establish a scale of temperature. That is, $t°C$ is defined for any one thermometer as that temperature which produces $t/100$ as great a change in the physical property used by that particular thermometer, as the change occurring between 0°C and 100°C. The temperature scales of different kinds of thermometers do not agree. That is, liquid-in-glass thermometers filled with different liquids do not agree among themselves except at the fixed points of 0°C and 100°C, nor do any of the liquid-in-glass scales agree with the resistance thermometer scales, and so on. The least variation is found among gas thermometers using different kinds of gas, and by applying certain corrections which are beyond the scope of this book, it is possible to bring all gas thermometers into agreement. Furthermore, it can be shown that such corrected temperatures agree with the Kelvin temperature scale. The differences between the gas thermometer scale and other scales is not large and for many purposes can be neglected. Some typical values are given in Table VIII.

TABLE VIII

COMPARISON OF CONSTANT-VOLUME HYDROGEN THERMOMETER
WITH OTHER THERMOMETERS

Constant-volume hydrogen thermometer	Mercury-in-glass thermometer	Platinum resistance thermometer	Platinum, pt-rhodium thermocouple
0°C	0°C	0°C	0°C
20	20.091	20.240	20.150
40	40.111	40.360	40.297
60	60.086	60.360	60.293
80	80.041	80.240	80.147
100	100	100	100

The *optical pyrometer*, illustrated in Fig. 18–6, consists essentially of a telescope, in the tube of which is mounted a filter A of red glass and a small electric lamp bulb B. When the pyrometer is directed toward a furnace, an observer looking through the telescope sees the dark lamp filament against the bright background of the furnace. The lamp filament is connected to a battery C and a rheostat D. By turning the rheostat knob, the current in the filament, and hence its brightness, may be gradually increased until the brightness of the filament just matches the brightness of the background. From previous calibration

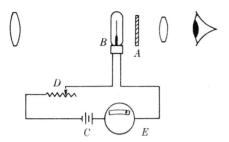

FIG. 18-6. Principle of the optical pyrometer.

of the instrument at known temperatures, the scale of the ammeter E in the circuit may be marked to read the unknown temperature directly. Since no part of the instrument needs to come into contact with the hot body, the optical pyrometer may be used at temperatures above the melting points of resistance thermometers or of thermocouples.

The melting and boiling points of a large number of substances have been carefully measured and tabulated, and these temperatures may now be used in the calibration of any type of thermometer. A number of such temperatures are listed in Table IX.

TABLE IX

TABLE OF FIXED POINTS

Hydrogen boiling point	−252.78°C
Nitrogen boiling point	−195.81
Mercury freezing point	−38.87
Ice point	0.00
Steam point	100.00
Sulphur boiling point	444.60
Silver melting point	960.5
Gold melting point	1063.0

18-5 Linear expansion. With a few exceptions, the dimensions of all substances increase as the temperature of the substance is increased. If a given specimen is in the form of a rod or cable, one is usually interested in its change of *length* with changes in temperature. (The change in cross section is so small it may be neglected.) Figure 18-7 represents a rod whose length is L_0 at some reference temperature t_0, and whose length is L at some higher temperature t. The difference $L - L_0 = \Delta L$ is the amount the rod has expanded on heating. It is found experimentally that the increase in length, ΔL, is proportional to the original length L_0, and very nearly proportional to the increase in temperature, $t - t_0$ or Δt. That is,

$$\Delta L \propto L_0 \, \Delta t,$$

or

$$\Delta L = \alpha L_0 \, \Delta t, \qquad (18\text{-}1)$$

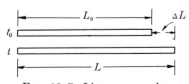

FIG. 18-7. Linear expansion.

where α is a proportionality constant, different for different materials, and is called the *coefficient of linear expansion*.

Equation (18–1) may be solved for α and written

$$\alpha = \frac{\Delta L}{L_0} \frac{1}{\Delta t}. \tag{18–2}$$

The coefficient of linear expansion of a substance may therefore be described as the fractional change in length per degree rise in temperature. Another useful relation is obtained by replacing ΔL by $L - L_0$ and solving for L.

$$L = L_0(1 + \alpha \, \Delta t). \tag{18–3}$$

Since L_0, L, and ΔL are all expressed in the same unit, the units of α are "reciprocal degrees" (centigrade or fahrenheit). Thus the coefficient of linear expansion of copper is written

$$\alpha = 14 \times 10^{-6} \text{ per centigrade degree,}$$

or

$$\alpha = 14 \times 10^{-6} \, (°C^{-1}).$$

This means that a copper rod one centimeter long at 0°C, increases in length by 0.000014 cm when heated to 1°C. A rod one foot long at 0°C increases by 0.000014 ft, and so on.

Since the fahrenheit degree is only $\frac{5}{9}$ as large as the centigrade degree, coefficients of expansion per fahrenheit degree are $\frac{5}{9}$ as large as their values on the centigrade scale.

EXAMPLE. An iron steam pipe is 200 ft long at 0°C. What will be its increase in length when heated to 100°C? $\alpha = 10 \times 10^{-6}$ per centigrade degree.

$L_0 = 200$ ft, $\alpha = 10 \times 10^{-6}$ per C°, $t = 100°C$, $t_0 = 0°C$.

$$\begin{aligned} \text{Increase in length} = \Delta L &= \alpha L_0 \, \Delta t \\ &= (10 \times 10^{-6})(200)(100) \\ &= 0.20 \text{ ft.} \end{aligned}$$

The coefficient of expansion of a substance in the form of a rod is measured by making two fine lines on the rod near its ends, and measuring the displacement of each line with a measuring microscope while the temperature of the rod is changed by a measured amount.

The *bimetallic* element is a device which has come into wide use in recent years, both as a thermometer and as a part of many thermostatic controls. It consists of two thin flat strips of different metals, welded or riveted together as in Fig. 18–8(a). If metal A has a larger coefficient of expansion than metal B, the compound strip, if originally straight, bends into a curve when heated, as shown in Fig. 18–8(b). The transverse motion of the end of the strip is very much larger than the increase in length of either metal.

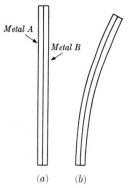

Metal A

Metal B

(a)   (b)

FIG. 18–8. The bimetallic element.

When used as a thermostat, one end of the strip is fixed and the motion of the other end is made to open or close an electrical control circuit. The common oven thermometer consists of a bimetallic strip coiled in a helix. With changes in temperature the helix winds up or unwinds, and this motion is transmitted to a pivoted pointer which moves over a calibrated scale. Because of lost motion and friction, such thermometers are not precision instruments.

If both strips are of equal thickness, d, it can be shown that a straight bimetallic strip will bend into a circle of radius

$$R = \frac{d}{\Delta t(\alpha_2 - \alpha_1)},$$

where Δt is the temperature interval above or below the temperature at which the strip is straight, and α_1 and α_2 are the two coefficients of linear expansion.

TABLE X

COEFFICIENTS OF LINEAR EXPANSION

Substance	α (°C^{-1})
Aluminum	24×10^{-6}
Brass	20
Copper	14
Glass	4–9
Steel	12
Invar	0.9
Quartz (fused)	0.4
Zinc	26

TABLE XI

COEFFICIENTS OF CUBICAL EXPANSION

Substance	β (C°)$^{-1}$
Alcohol, ethyl	0.745×10^{-3}
Carbon disulphide	1.140
Glycerine	0.485
Mercury	0.182
Petroleum	0.899

18–6 Surface and volume expansion. When a plate or sheet of material is heated, both the length and breadth of the plate increase. Consider a rectangular plate whose length and breadth at temperature t_0 are L_0 and b_0 respectively. When heated to a temperature t, these dimensions become

$$L = L_0(1 + \alpha\,\Delta t),$$
$$b = b_0(1 + \alpha\,\Delta t).$$

The original area of the plate was

$$A_0 = L_0 b_0,$$

and the area after heating is

$$A = Lb = L_0 b_0 (1 + \alpha\,\Delta t)(1 + \alpha\,\Delta t)$$
$$= A_0[1 + 2\alpha\,\Delta t + (\alpha\,\Delta t)^2].$$

But since α is a small quantity, α^2 will be extremely small and the term $(\alpha\,\Delta t)^2$ may be neglected. Hence

$$A = A_0(1 + 2\alpha\,\Delta t).$$

If we now define a *surface* coefficient of expansion γ so that

$$A = A_0(1 + \gamma\,\Delta t), \tag{18–4}$$

it follows that

$$\gamma = 2\alpha,$$

and the coefficient of surface expansion is twice the coefficient of linear expansion. Although derived for the special case of a rectangular plate, the result holds for a plate of any shape whatever.

If the plate contains a hole, the area *of the hole* expands at the same rate as does the surrounding material. This remains true even if the hole becomes so large that the "plate" is reduced to nothing but a rim around the hole. Thus the area of the "hole" enclosed by a steel wagon tire expands at the same rate as would a disk of this size, if constructed of the same kind of steel as is the rim.

By considering a solid block of material in the form of a rectangular parallelepiped whose dimensions at t_0 are L_0, b_0, and c_0, it is easy to show by the same type of reasoning that

$$V = V_0(1 + 3\alpha \, \Delta t)$$

$$= V_0(1 + \beta \, \Delta t), \tag{18–5}$$

where V is the volume at the temperature t, V_0 is the volume at t_0, and $\beta \, (=3\alpha)$ is the *volume* coefficient or *cubical* coefficient of expansion. This equation holds regardless of the shape of the body.

It is also true that the volume enclosed by a solid, such as the volume of a tank, a flask, or thermometer bulb, expands at the same rate as would a solid body of the same material as that of which the walls are composed.

Equation (18–5) may also be used to compute the expansion of a liquid. The linear and surface coefficients of expansion of a liquid are of little significance.

––––––––––

EXAMPLE. The volume of the bulb of a mercury thermometer at 0°C is V_0, and the cross section of the capillary is A_0. The linear coefficient of expansion of the glass is α_G per C° and the cubical coefficient of expansion of mercury is β_M per C°. If the mercury just fills the bulb at 0°C, what is the length of the mercury column in the capillary at a temperature of t°C?

The volume of the bulb at a temperature t is

$$V = V_0(1 + \beta_G t),$$

where $\beta_G = 3\alpha_G$ is the cubical coefficient of expansion of the glass.

The volume of the mercury at a temperature t is

$$V_M = V_0(1 + \beta_M t).$$

The volume of mercury that has been expelled from the bulb is the difference between these, or

$$V_0(1 + \beta_M t) - V_0(1 + \beta_G t) = V_0 t(\beta_M - \beta_G).$$

This volume is also equal to the length l of the mercury column multiplied by the cross section A of the capillary, where

$$A = A_0(1 + 2\alpha_G t).$$

Table X shows that the linear coefficient of expansion of glass is of the order of 5×10^{-6} per C°. Hence even if t is as great as 300°C the term $2\alpha_G t$ is only 0.003. It may therefore be neglected in comparison with unity, which is equivalent to considering the cross section of the capillary constant. Then

and
$$lA_0 = V_0 t(\beta_M - \beta_G),$$

$$l = \frac{V_0}{A_0} (\beta_M - \beta_G)t.$$

The length of the mercury column is therefore proportional to the temperature and to the difference between the cubical coefficients of expansion of mercury and the glass of which the thermometer is constructed.

Water in the temperature range from 0°C to 4°C *decreases* in volume with increasing temperature, contrary to the behavior of most substances. That is, between 0°C and 4°C the coefficient of expansion of water is *negative*. Above 4°C, water expands when heated. Since the volume of a given mass of water is smaller at 4°C than at any other temperature, the density of water is a maximum at 4°C. This behavior of water is the reason why lakes and ponds freeze first at their upper surface, an effect which will be explained in Chapter 20. Table XII illustrates this anomalous expansion of water.

TABLE XII

DENSITY AND VOLUME OF WATER

t°C	Density (gm/cm^3)	Volume of 1 gm, in cm^3
0	0.99987	1.00013
2	0.99997	1.00003
4	1.00000	1.00000
6	0.99997	1.00003
10	0.99973	1.00027
20	0.99823	1.00177
50	0.98807	1.01207
75	0.97489	1.02576
100	0.95838	1.04343

18–7 Thermal stresses. If the ends of a rod are rigidly fixed so as to prevent expansion or contraction and the temperature of the rod is changed, tensile or compressive stresses, called *thermal stresses*, will be set up in the rod. These stresses may become very large, sufficiently so to stress the rod beyond its elastic limit or even beyond its breaking strength. Hence in the design of any structure which is subject to changes in temperature, some provision must, in general, be made for expansion. In a long steam pipe this is accomplished by the insertion of expansion joints or a section of pipe in the form of a U. In bridges, one end may be rigidly fastened to its abutment while the other rests on rollers.

It is a simple matter to compute the thermal stress set up in a rod which is not free to expand or contract. Suppose that a rod at a temperature t has its ends rigidly fastened, and that while they are thus held the temperature is reduced to a lower value, t_0.

The fractional change in length if the rod were free to contract would be

$$\frac{\Delta L}{L_0} = \alpha(t - t_0) = \alpha \, \Delta t. \tag{18–6}$$

Since the rod is not free to contract, the tension must increase by a sufficient amount to produce the same fractional change in length. But from the general definition of Young's modulus (see page 250),

$$Y = \frac{\Delta F/A}{\Delta L/L_0},$$

and hence

$$\Delta F = AY \frac{\Delta L}{L_0}.$$

Introducing the expression for $\Delta L/L_0$ from Eq. (18–6), we have

$$\Delta F = AY\alpha \, \Delta t, \tag{18–7}$$

which gives the increase in tension. The increase in *stress* is

$$\frac{\Delta F}{A} = Y\alpha \, \Delta t. \tag{18–8}$$

PROBLEMS

18–1. (a) What is the melting point of silver in degrees fahrenheit? (See Table IX.) (b) What is the coefficient of linear expansion of copper in $(F°)^{-1}$? (c) What is the coefficient of volume expansion of copper in $(C°)^{-1}$? (d) At what temperature do the fahrenheit and centigrade scales coincide?

18–2. One steel meter bar is correct at 0°C, another at 25°C. What is the difference between their lengths at 20°C?

18–3. The length of Technology Bridge is about 2000 ft. Find the difference between its length on a winter day when the temperature is −20°F, and a summer day when the temperature is 100°F. Use the coefficient of expansion of steel.

18–4. A surveyor's 100-ft steel tape is correct at a temperature of 65°F. The distance between two points, as measured by this tape on a day when the temperature is 95°F, is 86.57 ft. What is the true distance between the points?

18–5. To ensure a tight fit, the aluminum rivets used in airplane construction are made slightly larger than the rivet holes and cooled by "dry ice" (solid CO_2) before being driven. If the diameter of a hole is 0.2500 inch, what should be the diameter of a rivet at 20°C if its diameter is to equal that of the hole when the rivet is cooled to −78°C, the temperature of dry ice?

18–6. (a) A steel ring of 3.000 inches inside diameter at 20°C is to be heated and slipped over a brass shaft measuring 3.002 inches in diameter at 20°C. To what temperature should the ring be heated? (b) If the ring and shaft together are cooled by some means such as liquid air, at what temperature will the ring just slip off the shaft?

18–7. A wire 60 cm long is bent into a circular ring, having a gap of 1.0 cm. The temperature of the wire is increased uniformly by 100°C. At the new temperature the gap is found to be 1.002 cm. What is the linear temperature coefficient of expansion of the wire? Assume no stresses within the wire before or after heating.

18–8. A glass flask whose volume is exactly 1000 cm^3 at 0°C is filled level full of mercury at this temperature. When flask and mercury are heated to 100°C, 15.2 cm^3 of mercury overflow. If the cubical coefficient of expansion of mercury is 0.000182 per centigrade degree, compute the linear coefficient of expansion of the glass.

18–9. A closed glass vessel is partially filled with mercury and evacuated. It is found that upon heating the whole, the ullage remains constant. What fraction of the whole volume did the mercury originally occupy? The cubical coefficient of expansion of the glass is 2.5×10^{-5} per C°.

18–10. A simple pendulum is constructed of a fine steel wire supporting a small brass ball. Its period is measured at a point on the earth's surface where $g = 980.00$ cm/sec^2, and where the temperature is 27°C. The pendulum is then taken to an elevation of 8 km above the earth's surface. At what temperature will its period equal its original period? (Approximate finite changes by differentials.)

18–11. A clock whose pendulum makes one vibration in 2 sec is correct at 25°C. The pendulum shaft is of steel and its moment of inertia may be neglected compared with that of the bob. (a) What is the fractional change in length of the shaft when it is cooled to 15°C? (b) How many seconds per day will the clock gain or lose at 15°C?

18–12. A watch with a balance wheel in the form of an annular brass ring (neglect the effect of the spokes) keeps correct time at a temperature of 30°C. By how many seconds per day will it be in error at a temperature of 15°C? Will it gain or lose?

18–13. A glass flask, when filled at a temperature of 20°C, holds 680 gm of mercury. How much mercury overflows when the whole is heated to 100°C? The linear coefficient of expansion of the glass is 8×10^{-6} per C°.

18–14. A slender steel rod oscillates as a physical pendulum about a horizontal axis through one end. If the rod is 8 ft long at 30°C, compute the change in its period when the temperature is decreased to 0°C.

18–15. A steel rod 1.5 cm² in cross section is 70 cm long at 20°C. If it is heated to 520°C and cooled to 20°C without being allowed to contract, compute the stress in the rod.

18–16. The cross section of a steel rod is 1.5 in². What is the least force that will prevent it from contracting while cooling from 520°C to 20°C?

FIGURE 18–9

18–17. A heavy brass bar has projections at its ends, as in Fig. 18–9. Two fine steel wires fastened between the projections are just taut (zero tension), when whole system is at 0°C. What is the tensile stress in the steel wires when the temperature of the system is raised to 300°C? Make any simplifying assumptions you think are justified, but state what they are.

18–18. What hydrostatic pressure is necessary to prevent a copper block from expanding when its temperature is increased from 20°C to 30°C?

18–19. A steel bomb is filled with water at 10°C. If the whole is heated to 75°C and no water is allowed to escape, compute the increase in pressure in the bomb. Assume the bomb to be sufficiently strong so that it is not stretched by the increased pressure.

18–20. Steel railroad rails 60 ft long are laid on a winter day when the temperature is 20°F. (a) How much space must be left between rails if they are to just touch on a summer day when the temperature is 110°F? (b) If the rails were originally laid in contact, what would be the stress in them on a summer day when the temperature is 110°F?

18–21. A liquid is enclosed in a metal cylinder provided with a piston of the same metal. The system is originally at atmospheric pressure and at a temperature of 80°C. The piston is forced down until the pressure on the liquid is increased by 100 atm, and it is then clamped in this position. Find the new temperature at which the pressure of the liquid is again 1 atmosphere. Assume that the cylinder is sufficiently strong so that its volume is not altered by changes in pressure, but only by changes in temperature. Compressibility of liquid $(\kappa) = 50 \times 10^{-6} \text{ atm}^{-1}$. Cubical coefficient of expansion of liquid $(\beta) = 5.3 \times 10^{-4} (°\text{C})^{-1}$. Linear coefficient of expansion of metal $(\alpha) = 10 \times 10^{-6} (°\text{C})^{-1}$.

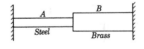

FIGURE 18–10

18–22. A steel bar A, whose cross section is 1 in², and a brass bar B, whose cross section is 2 in², are each exactly 20 cm long at 0°C. The bars are placed end to end between two supports 40 cm apart. When the temperature is increased to 200°C, find the stress in each bar, if the distance between the supports does not change (see Fig. 18–10).

18–23. A steel rod of length 40 cm and a copper rod of length 36 cm, both of the same diameter, are placed end to end between two rigid supports, with no initial stress in the rods. The temperature of the rods is now raised 50 C°. What is the stress in either rod?

CHAPTER 19

QUANTITY OF HEAT

19-1 Heat, a form of energy. Heat was formerly thought to be an invisible weightless fluid called *caloric*, which was produced when a substance burned and which could be transmitted by conduction from one body to another. The abandonment of the caloric theory was a part of the general development of physics during the 18th and 19th centuries. The two men who were probably chiefly responsible for the views we hold today were Count Rumford (1753–1814) (a native of Woburn, Mass.) and Sir James Prescott Joule.

Rumford was engaged in supervising the boring of cannon for the government of Bavaria. To prevent overheating, the bore of the cannon was kept filled with water, and as this boiled away during the boring process the supply had to be continually replenished. It was admitted that caloric had to be supplied to water in order to boil it, and the continual production of caloric was explained by the hypothesis that when matter was more finely subdivided (as in the process of boring) its capacity for retaining caloric grew smaller and the caloric thus released was what caused the water to boil.

Rumford noted, however, that the cooling water continued to boil away even when his boring tools became so dull that they were no longer cutting. That is, even a dull boring tool was apparently an inexhaustible supply of caloric *so long as mechanical work was being done to rotate the tool.*

Now one of the features which justifies our acceptance of many abstract ideas in physics is that they obey a "conservation principle." Here was a process in which *two* quantities failed to be conserved. Mechanical energy was not conserved since work was continually being expended, and caloric was not conserved since it was continually being created. Although Rumford did not express his ideas in just this way, he saw the opportunity to eliminate two cases of nonconservation and at the same time to extend the principle of the conservation of energy as it was then understood. He asserted that what had formerly been thought a separate entity, namely caloric, was in reality merely energy in another form. The process was not the continual disappearance of one thing and the appearance of another, but merely the transformation of energy from one form to another. As we would say today, mechanical energy was continually being transformed into heat, the process being one example of the conservation of energy.

Rumford made some measurements of the quantities of work done and of cooling water boiled away, but his experiments were not of great precision. When Joule, in the period from 1843 to 1878, showed that whenever a given quantity of mechanical energy was converted to heat the *same* quantity of heat was always developed, the equivalence of heat and work as two forms of energy was definitely established.

There are, of course, processes for whose explanation the caloric theory is entirely satisfactory. When heat flows from one body to another by conduction, or when substances at different temperatures are mixed in a calorimeter, heat is conserved, and for such processes the caloric theory would serve perfectly well.

19–2 Quantity of heat. Heat, like mechanical energy, is an intangible thing and a unit of heat is not something that can be preserved in a Standards laboratory. The quantity of heat involved in a process is measured by some change which accompanies the process, and a unit of heat is defined as the heat necessary to produce some standard, agreed-on change. Three such units are in common use, the kilogram·calorie, the gram·calorie, and the British thermal unit (Btu).

One kilogram·calorie is the quantity of heat which must be supplied to one kilogram of water to raise its temperature through one centigrade degree.

One gram·calorie is the quantity of heat which must be supplied to one gram of water to raise its temperature through one centigrade degree.

Evidently, 1 kilogram·calorie = 1000 gram·calories.

One Btu is the quantity of heat which must be supplied to one pound of water to raise its temperature through one fahrenheit degree.*

Since 454 gm = 1 lb, and since $1\,F° = \frac{5}{9}\,C°$, the Btu may be defined as the quantity of heat which must be supplied to 454 gm (0.454 kgm) of water to raise its temperature through $\frac{5}{9}\,C°$, which is $454 \times \frac{5}{9} = 252$ gm·cal or 0.252 kgm·cal. Hence

$$1\ \text{Btu} = 252\ \text{gm·cal} = 0.252\ \text{kgm·cal}.$$

---

*In this part of the subject we shall depart from the English gravitational system of units which we used throughout mechanics, and adopt as a mass unit the *mass of the standard pound*. This unit is also called one pound, and is equal to a mass of 454 grams, or (about) $\frac{1}{32}$ slug. Also, for simplicity, we shall confine metric units chiefly to the cgs system.

The gram-calorie is much more widely used in physics and chemistry than is the kilogram-calorie, and from now on, unless stated otherwise, we shall use the term calorie to mean gram-calorie.

The heat units here defined vary somewhat with the location of the degree, i.e., whether it is from 0° to 1°, 47° to 48°, etc. It is generally agreed to use the temperature interval from 14.5°C to 15.5°C (the "15° calorie"), and in English units to use the temperature interval from 63°F to 64°F. For most purposes this variation is small enough to be neglected.

It is essential that the distinction between "quantity of heat" and "temperature" shall be clearly understood. The terms are commonly misused in everyday life. Suppose that two pans, one containing a small and the other a large amount of water, are placed over identical gas burners and heated for the same length of time. It is obvious that at the end of this time the temperature of the small amount of water will have risen higher than that of the large amount. In this instance, equal quantities of heat have been supplied to each pan of water, but the increases in temperature are not equal.

On the other hand, suppose the two pans are both initially at a temperature of 60°F and that both are to be heated to 212°F. It is evident that more heat must be supplied to the pan containing the larger amount of water. The temperature change is the same for both but the quantities of heat supplied are very different.

We shall represent a quantity of heat by the letter Q.

19–3 The mechanical equivalent of heat. Energy in mechanical form is usually expressed in ergs, joules, or ft·lb; energy in the form of heat is expressed in calories or Btu. The relative magnitudes of the "heat units" and the "mechanical units" can be found by an experiment in which a measured quantity of mechanical energy is completely converted into a measured quantity of heat. The first accurate experiments were performed by Joule, using an apparatus in which falling weights rotated a set of paddles in a container of water. The energy transformed was computed in mechanical units from a knowledge of the weights and their height of fall, and in heat units from a measurement of the mass of water and its rise in temperature. In more recent methods, which are also more precise, electrical energy is converted to heat in a resistance wire immersed in water. The best results to date give:

$$778 \text{ ft·lb} = 1 \text{ Btu,}$$
$$4.186 \text{ joules} = 1 \text{ gm·cal,}$$
$$4186 \text{ joules} = 1 \text{ kgm·cal.}$$

That is, 778 ft·lb of mechanical energy, when converted to heat, will raise the temperature of 1 lb of water through 1 F°, etc.

These relations are often expressed by the statement that *the mechanical equivalent of heat* is 4.186 joules/gm·cal, or 778 ft·lb/Btu. The phraseology is a carry-over from the early days when the equivalence of mechanical energy and heat was being established.

The precise value of the mechanical equivalent of heat depends on the particular temperature interval used in the definition of the calorie or Btu. To avoid this confusion, an international commission has agreed to *define* 1 kgm·cal as *exactly* $\frac{1}{860}$ kilowatt·hour. Then by definition, 1 gm·cal = 4.18605 joules and 1 Btu = 778.26 ft·lb. It follows that 1 Btu = 251.996 gm·cal.

19–4 Heat capacity. Specific heat. Materials differ from one another in the quantity of heat required to produce a given elevation of temperature in a given mass. Suppose that a quantity of heat Q is supplied to a given body, resulting in a temperature rise Δt. The ratio of the heat supplied to the corresponding temperature rise is called the *heat capacity* of the body.

$$\text{Heat capacity} = \frac{Q}{\Delta t}. \tag{19–1}$$

Heat capacities are ordinarily expressed in calories per centigrade degree, or Btu per fahrenheit degree. If we set $\Delta t = 1$ degree in Eq. (19–1), it will be seen that the heat capacity of a body is numerically equal to the quantity of heat which must be supplied to it to increase its temperature by one degree.

To obtain a figure which is characteristic of the material of which a body is composed, the *specific heat capacity* of a material is defined as the *heat capacity per unit mass* of a body composed of the material. We shall represent specific heat capacity by the letter c.

$$c = \frac{\text{heat capacity}}{\text{mass}} = \frac{Q/\Delta t}{m} = \frac{Q}{m \, \Delta t}. \tag{19–2}$$

Specific heat capacity is expressed in calories per gram·centigrade degree, or Btu per pound·fahrenheit degree.

The specific heat capacity of a material is numerically equal to the quantity of heat which must be supplied to *unit mass* of the material to increase its temperature through 1 degree. The specific heat capacities of a few common materials are listed in Table XIII.

It follows from Eq. (19–2) that the heat which must be supplied to a body of mass m, whose specific heat capacity is c, to increase its temperature through an interval Δt, is

$$Q = mc\,\Delta t = mc(t_2 - t_1). \tag{19–3}$$

The *specific heat* of a material is defined as the ratio of its specific heat capacity to the specific heat capacity of water. Reference to the definitions of the heat units on page 377 will show that the specific heat capacity of water is 1 cal/gm·C°, or 1 Btu/lb·F°. Hence the specific heat of a material is numerically equal to its specific heat capacity, but from its definition as a ratio it is a pure number. For example, the specific heat capacity of copper is 0.093 cal/gm·C°, while the specific heat of copper is 0.093. This distinction between specific heat capacity and specific heat is not always rigidly adhered to, and the term "specific heat" is often used for the quantity we have defined as "specific heat capacity."

TABLE XIII

Substance	Specific heat	Temperature interval
Aluminum	0.217	17–100°C
Brass	0.094	15–100
Copper	0.093	15–100
Glass	0.199	20–100
Ice	0.55	10–0
Iron	0.113	18–100
Lead	0.031	20–100
Mercury	0.033	0–100
Silver	0.056	15–100

Since a specific heat is a pure number it has the same numerical value in all systems of units, and since specific heats and specific heat capacities are numerically equal, it follows that specific heat capacities are also equal in all systems. Referring to the example above, the specific heat capacity of copper is also 0.093 Btu/lb·F°.

It is easy to see from the definitions above that the heat capacity of a body is the product of its mass and its specific heat capacity.

Strictly speaking, Eq. (19–2) defines the *average* specific heat capacity over the temperature range Δt. It is found, however, that the quantity of heat required to raise the temperature of a material through a small interval varies with the location of the interval in the temperature scale. The *true* specific heat capacity of a material at any temperature is defined from Eq. (19–2) by considering an infinitesimal temperature rise dt, and letting dQ be the heat required to produce this rise. We then have

$$\text{True specific heat } c = \frac{1}{m}\frac{dQ}{dt},$$

$$dQ = mc\,dt,$$

$$Q = m\int_{t_1}^{t_2} c\,dt.$$

In general, c is a function of the temperature and must be so expressed to carry out the integration above.

At ordinary temperatures, and over temperature intervals which are not too great, specific heats may be considered constant. At extremely low temperatures, approaching absolute zero, all specific heats decrease and for certain substances approach zero.

It should be pointed out that the significance of the word "capacity" in "heat capacity" is not the same as when one speaks of the "capacity" of a bucket. The bucket can hold just so much water and no more, while heat can be added to a body indefinitely with, of course, a corresponding rise in temperature.

For some purposes, particularly in dealing with gases, it is more convenient to express specific heat capacities on the basis of one gram·atomic weight rather than one gram. It was first noted in 1819 by Dulong and Petit that the specific heat capacities of the metals, expressed in this way, were all very nearly equal to 6 cal/gm·atomic wt·C°. This fact is known as the *Dulong and Petit law*.

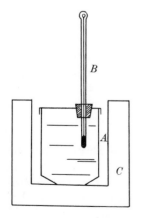

FIG. 19–1. The water calorimeter.

19–5 Calorimetry. The term calorimetry relates to the measurement of quantities of heat. Two types of calorimeter, the *water calorimeter* and the *continuous-flow calorimeter*, will be described.

The water calorimeter in its simple form consists of a thin-walled metal can A (Fig. 19–1) whose capacity is about two liters, and whose outer surface is nickel-plated to minimize loss of heat by radiation (see Chapter 20). The can contains a measured quantity of water, and is provided with a cover through which passes thermometer B. Heat losses are further reduced by surrounding the can with the heat-insulating jacket C. If the thermometer is read before and after an unknown quantity of heat Q is introduced into the calorimeter, Q may be found from the measured rise in temperature.

The water calorimeter may be used to measure specific heats as follows: a sample of the material whose specific heat is desired is heated in a furnace or steam bath to a known temperature, say t_s. Let the mass of the sample be m_s and its specific heat c_s.

The water in the calorimeter is thoroughly stirred and its temperature is measured. The sample is then quickly transferred to the calorimeter, the water is again thoroughly stirred, and the new temperature of the water is recorded. Let t_1 and t_2 be the initial and final temperatures of the water, m_w the mass of the water, m_c the mass of the calorimeter can, and c_c its specific heat.

If no heat is lost from the calorimeter during the experiment, the heat given up by the sample in cooling from t_s to t_2 must equal the heat gained by the water and the calorimeter can. Hence

$$m_s c_s(t_s - t_2) = m_w \times 1(t_2 - t_1) + m_c c_c(t_2 - t_1)$$

$$= (m_w + m_c c_c)(t_2 - t_1),$$

and c_s may be found, since the other factors are known.

The effect of the heat capacity of the calorimeter, $m_c c_c$, is evidently equivalent to increasing the mass of the water by an amount $m_c c_c$ and using a calorimeter of zero heat capacity. The product $m_c c_c$ is called the *water equivalent* of the calorimeter.

Actually the calorimeter will gain (or lose) heat from its surroundings during an experiment unless special precautions are taken. One way of minimizing the heat transfer is to start with the calorimeter somewhat cooler than its surroundings and finish with its temperature the same amount higher than the surroundings. Then the heat gained during the first part of the experiment offsets the heat lost in the latter part. Another method (the so-called "adiabatic jacket") is to heat the jacket by an electric heating coil so that its temperature rises at the same rate as does that of the calorimeter. If both temperatures are always equal there will be no gain or loss of heat.

It should be noted that this method of measuring specific heat gives only the *average* specific heat over the temperature range from t_s to t_2. Much more elaborate apparatus is required to measure the true specific heat at any desired temperature.

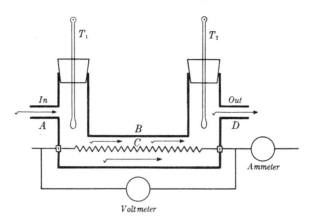

Fig. 19–2. A continuous-flow calorimeter.

The *continuous-flow* calorimeter, as used to measure the mechanical equivalent of heat, is illustrated in Fig. 19–2. A continuous stream of water enters the apparatus at A, flows through tube B around the resistance wire C, and leaves at D. Thermometers T_1 and T_2 read the temperatures t_1 and t_2 at inlet and outlet, and the electrical power expended is measured by the ammeter and voltmeter.

To use the calorimeter, the water is started flowing and the heating current is turned on. Thermometers T_1 and T_2 are read at intervals of, say, one minute, and their temperatures recorded. After sufficient time has elapsed, both thermometer readings become constant. Of course the temperature t_2 at the outlet is higher than the temperature t_1 at the inlet. When this steady state has been reached the apparatus itself is absorbing no heat, since its temperature remains constant. Heat is therefore being carried away by the flowing water at exactly the same rate as it is developed by the heating coil.

If then the mass of water passing through the calorimeter in a certain time is found, usually by catching the water in a beaker placed below the outlet, the quantity of heat developed can be computed from the rise in temperature of this mass of water. The energy input in the same time can be found from the ammeter and voltmeter readings.

A modified form of continuous-flow calorimeter is used to measure the heat of combustion of gas, the flowing water being heated by a gas flame instead of an electrical heater.

19–6 Heat of combustion. The heat of combustion of a substance is the quantity of heat liberated per unit mass, or per unit volume, when the substance is completely burned. Heats of combustion of solid and

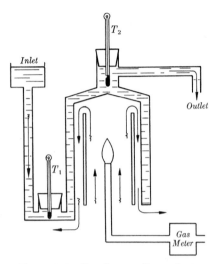

FIG. 19–3. Continuous-flow calorimeter used to measure the heat of combustion of gaseous fuel.

liquid fuels are usually expressed in Btu/lb or in cal/gm. The heat of combustion of gases is commonly expressed in Btu/ft^3. Some values are given in Table XIV.

Heats of combustion of solid and liquid fuels are measured with a *bomb calorimeter*. A measured mass of the fuel is inserted in a strong steel bomb which is filled with oxygen under pressure to ensure complete combustion. The bomb is placed in a water calorimeter and the fuel ignited by sending a momentary electric current through a fine heater wire. From the measured temperature rise, the mass of water, and the water equivalent of calorimeter and bomb, the heat of combustion can be computed.

The heat of combustion of gaseous fuels is usually measured with the type of continuous-flow calorimeter illustrated in Fig. 19–3.

TABLE XIV

HEATS OF COMBUSTION

Coal gas	600 Btu/ft^3
Natural gas	1000–2500 Btu/ft^3
Coal	11,000–14,000 Btu/lb
Ethyl alcohol	14,000 Btu/lb
Fuel oil	20,000 Btu/lb

19–7 Internal energy. A body may be warmed either by placing it in contact with a second body at a higher temperature, or by doing mechanical work on the body. For example, the air in a bicycle pump becomes hotter when the piston is pushed down, although it could also be heated by placing it in a furnace.

If one were given a sample of hot air, it would be impossible to tell by any tests whether it had been heated by compression or by heat flow from a hotter body. This raises the question as to whether one is justified in speaking of the "heat in a body," since the present state of the body may have been brought about either by adding heat to it or by doing work on it. We shall show in Chapter 22 that the proper term to use is "internal energy," and that the expression "heat energy of a body" is meaningless.

From the atomic point of view, the internal energy of a body is the sum total of the kinetic and potential energies of its atoms, apart from any kinetic or potential energy of the body as a whole. Not enough is known at present about the atomic structure of matter to be able to express internal energies wholly in terms of an atomic model, but we shall show in Chapter 25, in connection with the atomic model of a gas, how one goes about the problem. To a first approximation, the internal energy of a gas at low pressure may be identified with the aggregate kinetic energy of its atoms.

Even though the details of the atomic picture of matter are not fully understood, we do have definite evidence that atomic energies and velocities, whether in a solid, liquid, or gas, increase with increasing temperature. Such statements as "the heat in a body is the energy of motion of its atoms" should, however, be avoided.

Problems

19–1. How many cubic feet of a coal gas must be burned to heat 40 gallons of water from 50°F to 150°F, assuming 25% stack loss? (There are 7.5 gallons in a cubic foot.)

19–2. A certain Diesel engine consumes 20 lb of fuel oil per hour. The heat of combustion of the oil is 20,000 Btu/lb. If the over-all efficiency of the engine is 30%, (a) how many Btu/hr are converted into mechanical work? (b) How many Btu are wasted? (c) What horsepower does the engine develop?

19–3. An automobile weighing 2000 lb is traveling at 100 ft/sec. How many Btu are developed in the brakes when it is brought to rest?

19–4. The electric power input to a certain electric motor is 0.5 kw and the mechanical power output is 0.54 hp. (a) What is the efficiency of the motor? (b) How many Btu are developed in the motor in one hour of operation?

19–5. 400 gm of water are contained in a copper vessel of mass 200 gm. The water is heated by a friction device which dissipates mechanical energy, and it is observed that the temperature of the system rises at the rate of 3C° per minute. Neglect heat losses to the surroundings. What power in watts is dissipated in the water?

19–6. How long could a 2000-hp motor be operated on the heat energy liberated by one cubic mile of ocean water when the temperature of the water is lowered by 1C°, if all of this heat were converted to mechanical energy? Why do we not utilize this tremendous reservoir of energy?

19–7. A lead bullet of mass 5 gm, traveling with a kinetic energy of 12.6 joules, strikes a target and is brought to rest. What would be the rise in temperature of the bullet if none of the heat developed were lost to the surroundings?

19–8. Compute from Table XIII the heat capacities of one gram atomic weight of Al, Cu, Pb, Hg, and Ag, and compare with the values predicted by the Dulong and Petit law.

19–9. Compare the heat capacities of equal *volumes* of water, copper, and lead.

19–10. An aluminum can of mass 500 gm contains 117.5 gm of water at a temperature of 20°C. A 200-gm block of iron at 75°C is dropped into the can. (a) Find the final temperature, assuming no heat loss to the surroundings. (b) What is the water equivalent of the calorimeter?

19–11. A billet of iron weighing 30 lb is taken from an annealing furnace and quenched in a tank containing 100 lb of oil at a temperature of 72°F. The temperature of the oil increases to 116°F. The specific heat of the oil is 0.45. Neglect the heat capacity of the tank and heat losses to the surroundings. Find the temperature of the annealing furnace.

19–12. A casting weighing 100 lb is taken from an annealing furnace where its temperature was 900°F and plunged into a tank containing 800 lb of oil at a temperature of 80°F. The final temperature is 100°F, and the specific heat of the oil is 0.5. What was the specific heat of the casting? Neglect the heat capacity of the tank itself and any heat losses.

19–13. The specific heat capacity c of a substance is given by the empirical equation $c = a + bt^2$, where a and b are constants and t is the centigrade temperature. (a) Compute the heat required to raise the temperature of a mass m of the substance from 0°C to t°C. (b) What is the mean specific heat of the substance in the temperature range between 0°C and t°C? (c) Compare this with the true specific heat at a temperature midway between 0°C and t°C.

19–14. At very low temperatures, in the neighborhood of absolute zero, the specific heat of solids is given by the

Debye equation, $c = kT^3$, where T is the absolute or Kelvin temperature and k is a constant, different for different materials. (a) Compute the heat required to raise the temperature of a mass m of a solid from 0°K to 10°K. (b) Compute the mean specific heat in the temperature range between 0°K and 10°K. (c) Compute the true specific heat at a temperature of 10°K.

19–15. A steel cylinder of cross-sectional area 0.1 ft² contains 0.4 ft³ of glycerin. The cylinder is equipped with a tightly fitting piston which supports a load of 6000 lb. The cylinder is heated from 60°F to 160°F. Neglect the expansion of the steel cylinder. Find (a) the increase in volume of the glycerin, (b) the mechanical work done against the 6000-lb force by the glycerin, (c) the amount of heat added to the glycerin (specific heat of glycerin = 0.58), (d) the change in internal energy of the glycerin (i.e., the mechanical equivalent of the heat added minus the work done).

CHAPTER 20

TRANSFER OF HEAT

20–1 Conduction. If one end of a metal rod is placed in a flame while the other is held in the hand, that part of the rod one is holding will be felt to become hotter and hotter, although it was not itself in direct contact with the flame. Heat is said to reach the cooler end of the rod by *conduction* along or through the material of the rod. The atoms at the hot end of the rod increase the violence of their vibration as the temperature of the hot end increases. Then, as they collide with their more slowly moving neighbors further out on the rod, some of their energy of motion is shared with these neighbors and they in turn pass it along to those further out from the flame. Hence energy of thermal motion is passed along from one atom to the next, while each individual atom remains at its original position.

It is well known that metals are good conductors of electricity and also good conductors of heat. The ability of a metal to conduct an electric current is due to the fact that there are within it so-called "free" electrons, that is, electrons that have become detached from their parent atoms. The free electrons also play a part in the conduction of heat, and the reason metals are such good heat conductors is that the free electrons, as well as the atoms, share in the process of handing on thermal energy from the hotter to the cooler portions of the metal.

Conduction of heat can take place in a body only when different parts of the body are at different temperatures, and the direction of heat flow is always from points of higher to points of lower temperature. The phenomenon of heat flow is sometimes made the basis for the definition of temperature equality or inequality. That is, if heat flows from one body to another when the two are in contact, the temperature of the first, by definition, is higher than that of the second. If there is no heat flow, the temperatures are equal.

Figure 20–1 represents a slab of material of cross section A and thickness L. Let the whole of the left face of the slab be kept at a temperature t_2, and the whole of the right face at a lower temperature t_1. The direction of the heat current is then from left to right through the slab.

After the faces of the slab have been kept at the temperatures t_1 and t_2 for a sufficient length of time, the temperature at points within the slab is found to decrease uniformly with distance from the hot to the cold face. At each point, however, the temperature remains constant with time.

The slab is said to be in a "steady state." (Nonsteady state problems in heat conduction involve mathematical methods beyond the scope of this book.)

It is found by experiment that the rate of flow of heat through the slab in the steady state is proportional to the area A, proportional to the temperature difference $(t_2 - t_1)$, and inversely proportional to the thickness L. Let H represent the rate of heat flow. Then

$$H \propto \frac{A(t_2 - t_1)}{L}.$$

This proportion may be converted to an equation on multiplication by a constant K whose numerical value depends on the material of the slab. The quantity K is called the *coefficient of thermal conductivity* or simply the thermal conductivity of the material.

$$H = \frac{KA(t_2 - t_1)}{L}. \tag{20–1}$$

Equation (20–1) may also be used to compute the rate of heat flow along a rod whose side walls are thermally insulated. See Fig. 20–2, where the letters have the same meaning as in Fig. 20–1.

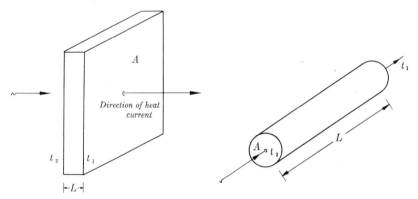

Fig. 20–1. Conduction of heat through a slab.

Fig. 20–2. Conduction of heat along a rod.

In some circumstances, either because of nonsteady conditions or because of the geometry of the conductor, the temperature in a body through which heat is flowing does not decrease uniformly along the direction of heat flow as it did in the slab of Fig. 20–1. We can then consider a thin slab of thickness dx between whose faces the temperature difference is dt, and Eq. (20–1) becomes

$$H = -KA\frac{dt}{dx}.$$ (20–2)

The minus sign is introduced since, if the temperature increases from left to right, the direction of the heat current is from right to left. Equation (20–2) is the general equation of heat conduction. The ratio dt/dx is called the *temperature gradient*. Equation (20–1) evidently relates to a special case in which the temperature gradient is constant and equal to $(t_2 - t_1)/L$.

The cgs unit of rate of heat flow, or heat current, is one calorie per second. Temperatures are expressed on the centigrade scale. The units of A and L are obvious. The thermal conductivities of commercial insulating materials such as cork or rock wool are usually stated in a "hybrid" system in which areas are in square feet, temperatures in fahrenheit degrees, thicknesses in inches, and the rate of flow of heat in Btu per hour. In any system of units, H and K are numerically equal when $A = $ one unit of area, $t_2 - t_1 = $ one degree, and $L = $ one unit of length. Thus in the commercial system, the thermal conductivity of a material is numerically equal to the number of Btu that flow in one hour through a slab one inch thick and one square foot in cross section, when the temperature difference between the faces of the slab is one fahrenheit degree.

It is evident from Eq. (20–1) that the larger the thermal conductivity K, the larger the heat current, other factors being equal. A material for which K is large is therefore a good heat conductor, while if K is small, the material is a poor conductor or a good insulator. There is no such thing as a "perfect heat conductor" ($K = \infty$) or a "perfect heat insulator" ($K = 0$). However, it will be seen from Table XV, which lists some representative values of thermal conductivity, that the metals as a group have much greater thermal conductivities than the nonmetals.

20–2 Heat flow through a compound wall. Figure 20–3 illustrates a so-called compound wall, constructed of two materials having different thicknesses and different thermal conductivities. The heat current through section 1 is

$$H = \frac{K_1 A(t_2 - t_x)}{L_1},$$ (20–3)

and through section 2 it is

$$H = \frac{K_2 A(t_x - t_1)}{L_2}.$$ (20–4)

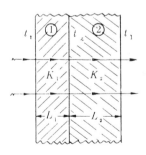

FIG. 20–3. Heat flow through a compound wall.

TABLE XV

THERMAL CONDUCTIVITY

	K (cal·cm/sec·cm^2·C°)	K (Btu·in/hr·ft^2·F°)
Metals:		
Aluminum	0.49	
Brass	0.26	
Copper	0.92	
Lead	0.083	
Mercury	0.020	
Silver	0.97	
Steel	0.12	
Various Solids: (Representative values)		
Firebrick	0.0025	8
Insulating brick	0.00035	1
Red brick	0.0015	4
Concrete	0.002	6
Cork	0.0001	0.3
Felt	0.0001	0.3
Glass	0.002	6
Ice	0.004	12
Rock wool	0.0001	0.3
Wood	0.0003–0.0001	9–3
Gases:		
Air	0.000057	
Argon	0.000039	
Helium	0.00034	
Hydrogen	0.00033	
Oxygen	0.000056	

In the steady state these currents must be equal. (Why?) Hence

$$\frac{K_1 A (t_2 - t_x)}{L_1} = \frac{K_2 A (t_x - t_1)}{L_2}.$$

Solving for t_x and substituting in Eq. (20–3) or (20–4) we find

$$H = \frac{A(t_2 - t_1)}{L_1/K_1 + L_2/K_2}.$$

In general, for any number of sections in series,

$$H = \frac{A(t_2 - t_1)}{\Sigma L/K}. \tag{20–5}$$

20–3 Heat flow through a cylindrical pipe cover. Figure 20–4 represents a section through the cylindrical heat insulator surrounding a steam pipe. The inner and outer radii of the insulator are a and b respectively, t_a and t_b are the temperatures at its inner and outer surfaces, and K is its thermal conductivity. Consider the thin cylindrical shell shown shaded. From Eq. (20–2) the heat current through the shell is

$$H = -2\pi LKr \frac{dt}{dr},$$

where L is the length of the shell.

(Question: Should one write dH instead of H?) This equation may now be integrated to express the heat current in terms of inside and outside temperatures.

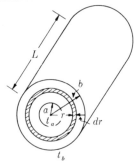

$$H \frac{dr}{r} = -2\pi LK \, dt,$$

$$H \int_a^b \frac{dr}{r} = -2\pi LK \int_{t_a}^{t_b} dt,$$

$$H \ln \frac{b}{a} = 2\pi LK(t_a - t_b),$$

$$H = \frac{2\pi LK(t_a - t_b)}{\ln b/a}. \quad (20\text{–}6)$$

FIG. 20–4. Radial heat flow in a cylinder.

NOTE. "Hybrid" units of K such as those in Table XV cannot be used in this equation. For example, K must represent the number of Btu that flow in unit time through a slab one foot square and one *foot* thick, with 1F° temperature difference between its faces.

20–4 Convection. The term *convection* is applied to the transfer of heat from one place to another by the actual motion of hot material. The hot-air furnace and the hot-water heating system are examples. If the heated material is forced to move by a blower or pump, the process is called *forced* convection; if the material flows due to differences in density, the process is called *natural* or *free* convection. To understand the latter, consider a U-tube as illustrated in Fig. 20–5.

In (a), the water is at the same temperature in both arms of the U and hence stands at the same level in each. In (b), the right side of the U has been heated. The water in this side expands and therefore, being of smaller density, a longer column is needed to balance the pressure produced by the cold water in the left column. The stopcock may now be opened and water will flow from the top of the warmer column into the colder column. This increases the pressure at the bottom of the U produced by the cold column, and decreases the pressure at this point due to the hot column. Hence at the bottom of the U, water is forced from the cold to the hot side. If heat is continually applied to the hot side and removed from the cold side, the circulation continues of itself. The net result is a continual transfer of heat from the hot to the cold side of the column. In the common household hot-water heating system, the "cold" side corresponds to the radiators and the "hot" side to the furnace.

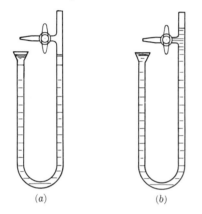

(a)　　　　　　　　(b)

Fig. 20–5. Convection is brought about by differences in density.

The anomalous expansion of water which was mentioned in Chapter 18 has an important effect on the way in which lakes and ponds freeze in winter. Consider a pond at a temperature of, say, 20°C throughout, and suppose the air temperature at its surface falls to −10°C. The water at the surface becomes cooled to, say, 19°C. It therefore contracts, becomes more dense than the warmer water below it, and sinks in this less dense water, its place being taken by water at 20°C. The sinking of the cooled water causes a mixing process, which continues until all of the water has been cooled to 4°C. Now, however, when the surface water cools to 3°C, it expands, is less dense than the water below it and hence floats on the surface. Convection and mixing then cease, and the remainder of the water can lose heat only by *conduction*. Since water is an extremely poor heat conductor cooling takes place very slowly after 4°C is reached, with the result that the pond freezes first at its surface. Then, since the density of ice is even smaller than that of water at 0°C, the ice floats on the water below it, and further freezing can result only from heat flow upward by conduction.

20–5 Radiation. When one's hand is placed in direct contact with the surface of a hot-water or steam radiator, heat reaches the hand by *conduction* through the radiator walls. If the hand is held above the radiator but not in contact with it, heat reaches the hand by way of the upward-moving *convection* currents of warm air. If the hand is held at one side of the radiator it still becomes warm, even though conduction through the air is negligible and the hand is not in the path of the convection currents. Energy now reaches the hand by *radiation.*

The term radiation refers to the continual emission of energy from the surface of all bodies. This energy is called *radiant energy* and is in the form of electromagnetic waves which are identical in nature with light waves, radio waves, or X-rays, differing from these only in wave length. These waves travel with the velocity of light and are transmitted through a vacuum as well as through air (better, in fact, since they are absorbed by air to some extent). When they fall on a body which is not transparent to them, such as the surface of one's hand or the walls of the room, they are absorbed and their energy converted to heat.

The radiant energy emitted by a surface, per unit time and per unit area, depends on the nature of the surface and on its temperature. At low temperatures the rate of radiation is small and the radiant energy is chiefly of relatively long wave length. As the temperature is increased, the rate of radiation increases very rapidly, in proportion to the 4th power of the absolute temperature. At the same time, the relative amount of radiant energy of short wave length increases. For example, a copper block at a temperature of 100°C (373°K) radiates about 300,000 ergs/sec or 0.03 watts from each square centimeter of its surface. At a temperature of 500°C (773°K) it radiates about 0.54 watts from each square centimeter, and at 1000°C (1273°K) it radiates about 4 watts per square centimeter. This rate is 130 times as great as that at a temperature of 100°C.

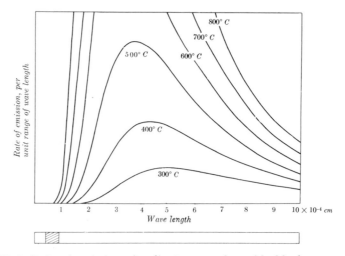

Fig. 20–6. Rate of emission of radiant energy by a blackbody, per unit of wave length, as a function of wave length. Shaded area indicates visible spectrum.

At each of these temperatures the radiant energy emitted is a mixture of waves of different wave lengths. At a temperature of 300°C the most intense of these waves has a wave length of about 5×10^{-4} cm; for wave lengths either greater or less than this value the intensity decreases as shown by the curve in Fig. 20–6. The corresponding distribution of energy at higher temperatures is also shown in the figure. The area between each curve and the horizontal axis represents the total rate of radiation at that temperature. It is evident that this rate increases rapidly with increasing temperature, and also that the wave length of the most intense wave shifts toward the left, or toward shorter wave lengths, with increasing temperature.

At a temperature of 300°C, practically all of the radiant energy emitted by a body is carried by waves longer than those corresponding to red light. Such waves are called *infrared*, meaning "beyond the red." At a temperature of 800°C a body emits enough visible radiant energy to be self-luminous and appears "red-hot." By far the larger part of the energy emitted, however, is still carried by infrared waves. At 3000°C, which is about the temperature of an incandescent lamp filament, the radiant energy contains enough of the shorter wave lengths so that the body appears nearly "white-hot."

20–6 Stefan's law. Experimental measurements of the rate of emission of radiant energy from the surface of a body were made by John Tyndall (1820–1893) and on the basis of these Josef Stefan (1835–1893), in 1879, concluded that the rate of emission could be expressed by the relation

$$W = e\sigma T^4, \tag{20–7}$$

which is *Stefan's law*. W is the rate of emission of radiant energy per unit area and is expressed in ergs per second per square centimeter in the cgs system, and in watts per square meter in the mks system. The constant σ has a numerical value of 5.672×10^{-5} in cgs units and 5.672×10^{-8} in mks units. T is the Kelvin temperature of the surface and e is a quantity called the emissivity of the surface. The emissivity lies between zero and unity, depending on the nature of the surface. The emissivity of copper, for example, is about 0.3. (Strictly speaking, the emissivity varies somewhat with temperature even for the same surface.) In general, the emissivity is larger for rough and smaller for smooth, polished surfaces.

It may be wondered why it is, if the surfaces of all bodies are continually emitting radiant energy, that all bodies do not eventually radiate away all of their internal energy and cool down to a temperature of absolute zero (where $W = 0$ by Eq. 20–7). The answer is that they would do so if energy were not supplied to them in some way. In the case of a Sunbowl heater element or the filament of an electric lamp, energy is supplied electrically to make up for the energy radiated. As soon as this energy supply is cut off, these bodies do, in fact, cool down very quickly to room temperature. The reason that they do not cool further is that their surroundings (the walls, and other objects in the room) are also radiating, and some of this radiant energy is intercepted, absorbed, and converted into heat. The same thing is true of all other objects in the room—each is both emitting and absorbing radiant energy simultaneously. If any object is hotter than its surroundings, its rate of emission will exceed its rate of absorption. There will thus be a net loss of energy and the body will cool down unless heated by some other method. If a body is at a lower temperature than its surroundings, its rate of absorption will be larger than its rate of emission and its temperature will rise. When the body is at the same temperature as its surroundings the two rates become equal, there is no net gain or loss of energy, and no change in temperature.

If a small body of emissivity e is completely surrounded by walls at a temperature T, the rate of *absorption* of radiant energy by the body is

$$W = e\sigma T^4.$$

Hence for such a body at a temperature T_1, surrounded by walls at a temperature T_2, the *net* rate of loss (or gain) of energy by radiation is

$$W_{\text{net}} = e\sigma T_1^4 - e\sigma T_2^4$$

$$= e\sigma(T_1^4 - T_2^4). \tag{20-8}$$

20–7 The ideal radiator. Imagine that the walls of the enclosure in Fig. 20–7 are kept at the temperature T_2 and a number of different bodies

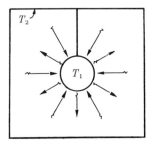

FIG. 20–7. In thermal equilibrium, the rate of emission of radiant energy equals the rate of absorption. Hence a good absorber is a good emitter.

having different emissivities are suspended one after another within the enclosure. Regardless of their temperature when they were inserted, it will be found that eventually each comes to the same temperature as that of the walls, T_2, even if the enclosure is evacuated. If the bodies are small compared to the size of the enclosure, radiant energy from the walls strikes the surface of each body at the same rate. Of this energy, a part is reflected and the remainder absorbed. In the absence of any other process, the energy absorbed would raise the temperature of the absorbing body, but since the temperature is observed *not* to change, each body must *emit* radiant energy at the same rate as it *absorbs* it. Hence a good absorber is a good emitter, and a poor absorber is a poor emitter. But since each body must either absorb or reflect the radiant energy reaching it, a poor absorber must also be a good reflector. Hence a *good reflector* is a *poor emitter*.

This is the reason for silvering the walls of vacuum ("Thermos") bottles. A vacuum bottle is constructed with double glass walls, the space between the walls being evacuated so that heat flow by conduction and convection is practically eliminated. To reduce the radiant emission to as low a value as possible, the walls are covered with a coating of silver which is highly reflecting and hence is a very poor emitter.

Since a good absorber is a good emitter, the *best* emitter will be that surface which is the best absorber. But no surface can absorb more than all of the radiant energy which strikes it. Any surface which does absorb all of the incident energy will be the best emitting surface possible. Such a surface would reflect no radiant energy, and hence would appear black in color (provided its temperature is not so high that it is self-luminous). It is called an ideally black surface, and a body having such a surface is called an ideal blackbody, an ideal radiator, or simply a blackbody.

No actual surface is ideally black, the closest approach being lampblack, which reflects only about 1%. Blackbody conditions can be closely realized, however, by a small opening in the walls of a closed container. Radiant energy entering the opening is in part absorbed by the interior walls. Of the part reflected, only a very little escapes through the opening, the remainder being eventually absorbed by the walls. Hence the *opening* behaves like an ideal absorber.

Conversely, the radiant energy emitted by the walls or by any body within the enclosure, and escaping through the opening, will, if the walls are at uniform temperature, be of the same nature as that emitted by an ideal radiator. This fact is of importance when using an optical pyrometer, described on page 366. The readings of such an instrument are correct only when it is sighted on a blackbody. If used to measure the temperature of a red-hot ingot of iron in the open, its readings will be too low, since iron is a poorer emitter than a blackbody. If, however, the pyrometer is sighted on the iron while still in the furnace, where it is surrounded by walls at the same temperature, "blackbody conditions" are fulfilled and the reading will be correct. The failure of the iron to emit as effectively as a blackbody is just compensated by the radiant energy which it reflects.

The emissivity e of an ideally black surface is equal to unity. For all real surfaces, it is a fraction, less than one.

Problems

20-1. A slab of a thermal insulator is 100 cm^2 in cross section and 2 cm thick. Its thermal conductivity is 2×10^{-4} cal·sec·cm·C°. If the temperature difference between opposite faces is 100°C, how many calories flow through the slab in one day?

20-2. A long rod, insulated to prevent heat loses has one end immersed in boiling water (at atmospheric pressure) and the other end in a water-ice mixture. The rod consists of 100 cm of copper (one end in steam) and a length, L_2, of steel (one end in ice). Both rods are of cross-sectional area 5 cm^2. The temperature of the copper-iron junction is 60°C, after a steady state has been set up. (a) How many calories per second flow from the steam bath to the ice-water mixture? (b) How long is L_2?

20-3. A compound bar 2 meters long is constructed of a solid steel core 1 cm in diameter surrounded by a copper casing whose outside diameter is 2 cm. The outer surface of the bar is thermally insulated and one end is maintained at 100°C, the other at 0°C. Find the total heat current in the bar. What fraction is carried by each material?

20-4. Heat flows radially outward through a cylindrical insulator of outside radius R_2 surrounding a steam pipe of outside radius R_1. The temperature of the inner surface of the insulator is t_1, that of the outer surface is t_2. At what radial distance from the center

of the pipe is the temperature just halfway between t_1 and t_2?

20-5. A rod is initially at a uniform temperature of 0°C throughout. One end is kept at 0°C and the other is brought into contact with a steam bath at 100°C. The surface of the rod is insulated so that heat can flow only lengthwise along the rod. The cross-sectional area of the rod is 2 cm^2, its length is 100 cm, its thermal conductivity is 0.8 cal·cm/sec·cm^2·C°, its density is 10 gm/cam^3, and its specific heat capacity is 0.10 cal/gm·C°. Consider a short cylindrical element of the rod 1 cm in length, (a) if the temperature gradient at one end of this element is 200 C°/cm, how many calories flow across this end per second? (b) If the average temperature of the element is increasing at the rate of 5 C°/sec, what is the temperature gradient at the other end of the element?

20-6. A copper sphere of mass 4700 gm and radius 5 cm is covered with an insulating layer of thickness 5 cm (outer radius 10 cm). The thermal conductivity of the covering is 0.002 cal·cm/sec·cm^2·C°, and the outer surface is maintained at a temperature of 20°C. The specific heat of copper is 0.093. (a) When the copper is at a temperature of 100°C, what is the heat flow through the insulating covering? (b) Approximately how long will it take for the copper to cool down from 100°C to 99°C?

20–7. (a) What would be the difference in height between the columns in the U-tube in Fig. 20–5 if the liquid is water and one arm is at 4°C while the other is at 80°C? (b) What is the difference between the pressures at the foot of two columns of water each 10 meters high if the temperature of one is 4°C and that of the other is 80°C?

20–8. The walls of a furnace are constructed of firebrick 6 inches thick. The thermal conductivity of the brick is such that 3.6 Btu flow in 1 hr through a section of area 1 ft^2 and thickness 1 inch when the temperature difference between the faces is 1 F°. The temperature of the inner wall is 1000°F and that of the outer wall is 100°F. (a) What thickness of insulating brick whose thermal conductivity is $\frac{1}{3}$ that of the firebrick should be laid outside the latter to reduce the heat loss to 50% if its former value? (b) How many Btu flow in 1 hr through each square foot of the insulated wall? (c) What is the temperature of the interface between firebrick and insulating brick? Assume that inside and outside temperatures are unaffected by the addition of the insulating brick.

20–9. Rods of copper, brass, and steel are welded together to form a Y-shaped figure. The cross-sectional area of each rod is 2 cm^2. The end of the copper rod is maintained at 100°C and the ends of the brass and steel rods at 0°C. Assume there is no heat loss from the surfaces of the rods. The lengths of the rods are: copper, 46 cm; brass, 13 cm; steel, 12 cm. (a) What is the temperature of the junction point? (b) What is the heat current in the copper rod?

20–10. A container of wall area 5000 cm^2 and thickness 2 cm is filled with water in which there is a stirrer. The outer surface of the walls is kept at a constant temperature of 0°C. The thermal conductivity of the walls is

0.000478 cal·cm/sec·cm^2·C°, and the effect of edges and corners can be neglected. The power required to run the stirrer at an angular velocity of 1800 rpm is found to be 100 watts. What will be the final steady-state temperature of the water in the container? Assume that the stirrer keeps the entire mass of water at a uniform temperature.

20–11. A copper bar 20 cm long and of cross section 2 cm^2 is surrounded by an insulating jacket 0.4 cm thick of material having a coefficient of thermal conductivity 0.01 cal/sec·cm C°. One end of the copper bar is maintained at 0°C and the other end at 100°C. Assume the temperature gradient along the bar to be linear. (a) Set up an expression for the heat lost per second through an element of the jacket of length dx at a distance x from the 0°C end. (b) Integrate the above expression to obtain the total heat lost per second through the jacket.

20–12. A cubical box 2 ft on an edge is constructed of insulating wallboard of thermal conductivity 0.4 Btu·in/hr·ft^2·F°, and of thickness $\frac{1}{2}$ inch. Mounted within the box is an electric heater developing 600 watts. Find the temperature difference between inner and outer surfaces of the walls of the box after the steady state has been reached.

20–13. The thermal conductivity of all substances varies to some extent with temperature. Suppose that the thermal conductivity of a certain material is given by the equation $K = K_0 (1 + at)$, where a is a constant and t is the centigrade temperature. (a) Derive the equation for the heat current through a flat slab of this material of area A, thickness L, when the centigrade temperatures of its opposite faces are t_1 and t_2. (b) Find the temperature at a plane midway between the faces of the slab if $t_2 = 100°C$, $t_1 = 0°C$, $a = 0.02(°C)^{-1}$.

CHAPTER 21

CHANGE OF PHASE

21–1 Change of phase. The term *phase* as used here relates to the fact that matter exists either as a solid, liquid or gas. Thus the chemical substance H_2O exists in the *solid phase* as ice, in the *liquid phase* as water, and in the *gaseous phase* as steam. Provided they do not decompose at high temperatures, all substances can exist in any of the three phases under the proper conditions of temperature and pressure. Transitions from one phase to another are accompanied by the absorption or liberation of heat and usually by a change in volume.

As an illustration, suppose that ice is taken from a refrigerator where its temperature was, say, $-25°C$. Let the ice be crushed quickly, placed in a container, and a thermometer inserted in the mass. Imagine the container to be surrounded by a heating coil which supplies heat to the ice at a uniform rate, and suppose that no other heat reaches the ice. The temperature of the ice would be observed to increase steadily as shown by the portion of the graph (Fig. 21–1) from a to b, or until the temperature has risen to $0°C$. As soon as this temperature is reached, some liquid water will be observed in the container. In other words, the ice begins to melt. The melting process is a *change of phase*, from the solid phase to the liquid phase. The thermometer, however, will show *no increase in temperature*, and even though heat is being supplied at the same rate as

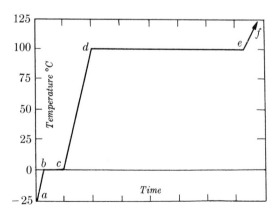

Fig. 21–1. The temperature remains constant during each change of phase.

before, the temperature will remain at 0°C until all of the ice is melted (point *c*, Fig. 21–1). (The ice and water mixture must be kept thoroughly stirred, otherwise the temperature of that part of the water closest to the heater will rise above 0°C.)

As soon as the last of the ice has melted, the temperature begins to rise again at a uniform rate, from *c* to *d*, Fig. 21–1, although this rate will be slower than that from *a* to *b* because the specific heat of water is greater than that of ice. When a temperature of 100°C is reached, (point *d*) bubbles of steam (gaseous water or water vapor) start to escape from the liquid surface, or the water begins to boil. The temperature remains constant at 100°C until all of the water has boiled away. Another change of phase has therefore taken place, from the liquid phase to the gaseous phase.

If all of the water vapor had been trapped and not allowed to diffuse away (a very large container would be needed), the heating process could be continued as from *e* to *f*. The gas would now be called "superheated steam."

Although water was chosen as an example in the process just described, the same type of curve as in Fig. 21–1 is obtained for many other substances. Some, of course, decompose before reaching a melting or boiling point, and others, such as glass or pitch, do not change state at a definite temperature but become gradually softer as their temperature is raised. Crystalline substances, such as ice or a metal, melt at a definite temperature. Glass and pitch are actually supercooled liquids of very high viscosity.

The temperature at which a crystalline solid melts when heat is supplied to it is called its *melting point*, and the temperature at which a liquid boils when heat is supplied to it is called its *boiling point*. The quantity of heat per unit mass that must be supplied to a material at its melting point to convert it completely to a liquid at the same temperature is called the *heat of fusion* of the material. The quantity of heat per unit mass that must be supplied to a material at its boiling point to convert it completely to a gas at the same temperature is called the *heat of vaporization* of the material. Heats of fusion and vaporization are expressed in calories per gram or Btu per pound. Thus the heat of fusion of ice is about 80 cal/gm or 144 Btu/lb. The heat of vaporization of water (at 100°C) is 539 cal/gm or 970 Btu/lb. Some heats of fusion and vaporization are listed in Table XVI.

When heat is removed from a gas its temperature falls, and at the same temperature at which it boiled, it returns to the liquid phase, or *condenses*. In so doing it gives up to its surroundings the same quantity of heat which was required to vaporize it. The heat so given up, per unit mass, is called the *heat of condensation* and is equal to the heat of vaporization. Similarly, a liquid returns to the solid phase, or *freezes*, when cooled to the temperature at which it melted, and gives up heat called *heat of solidification* exactly equal to the heat of fusion. Thus the melting point and the freezing point are at the same temperature, and the boiling point and condensation point are at the same temperature.

Whether a substance, at its melting point, is freezing or melting depends on whether heat is being supplied or removed. That is, if heat is supplied to a beaker containing both ice and water at 0°C some of the ice will melt; if heat is removed, some of the water will freeze; the temperature in either case remains at 0°C as long as both ice and water are present. If heat is *neither supplied nor removed*, no change at all takes place and the relative amounts of ice and water, and the temperature, all remain constant.

This furnishes, then, another point of view which may be taken regarding the melting point. That is, the melting (or freezing) point of a substance is *that temperature at which both the liquid and solid phases can exist together*. At any higher temperature, the substance can only be a liquid; at any lower temperature, it can only be a solid.

The general term *heat of transformation* is applied both to heats of fusion and heats of vaporization, and both are designated by the letter L. Since L represents the heat absorbed or liberated in the change of phase of unit mass, the heat Q absorbed or liberated in the change of phase of a mass m is

$$Q = mL. \qquad (21\text{--}1)$$

The household steam-heating system makes use of a boiling-condensing process to transfer heat from the furnace to the radiators. Each pound of water which is turned to steam in the furnace absorbs 970 Btu (the heat of vaporization of water) from the furnace, and gives up 970 Btu when it condenses in the radiators. (This figure is correct if the steam pressure is one atmosphere. It will be slightly smaller at higher pressures.) Thus the steam-heating system does not need to circulate as much water as a hot-water heating system. If water leaves a hot-water furnace at 140°F and returns at 100°F, dropping 40 F°, about 24 lb of water must circulate to carry the same heat as is carried in the form of heat of vaporization by one pound of steam.

Under the proper conditions of temperature and pressure, a substance can change directly from the solid to the gaseous phase without passing through the liquid phase. The transfer from solid to vapor is called *sublimation*, and the solid is said to *sublime*. "Dry ice" (solid carbon dioxide) sublimes at atmospheric pressure. Liquid carbon dioxide can not exist at a pressure lower than about 73 lb/in^2.

Heat is absorbed in the process of sublimation and liberated in the reverse process. The quantity of heat per unit mass is called the *heat of sublimation*.

TABLE XVI

Substance	Melting point		Heat of fusion		Boiling point (1 atm)		Heat of vaporization	
	°C	°F	$\dfrac{cal}{gm}$	$\dfrac{Btu}{lb}$	°C	°F	$\dfrac{cal}{gm}$	$\dfrac{Btu}{lb}$
Lead	327	621	5.86	10.59				
Mercury	−39	−38	2.82	5.08	357	675	65	117
Nitrogen	−210	−346	6.09	10.95	−196	321	48	87
Oxygen	−219	−363	3.30	5.95	−183	297	51	92
Platinum	1775	3232	27.2	49.0				
Silver	961	1762	21.1	38.0				
Sulphur	119	246	13.2	23.8	444	831		
Water	0	32	79.7	144	100	212	539	970
Ethyl alcohol	−114	−174	24.9	448	78	172	204	368
Sulphur dioxide					138	280	50	90

21–2 Work done in a volume change. When a boiling liquid is open to the atmosphere the vapor escapes into the air as fast as it is formed. The process could equally well be carried out, however, in a closed cylinder as shown in Fig. 21–2. The piston is supposed to make a tight, frictionless fit in the cylinder and to be of negligible weight, so that the pressure exerted by it on the contents of the cylinder is equal to the atmospheric pressure, p.

In Fig. 21–2(a), the cylinder contains a mass m of a liquid at its boiling point, with the piston resting directly on the liquid surface. In Fig. 21–2(b), heat is being supplied to the liquid, which is boiling at constant temperature. The vapor escaping from the liquid is forcing the piston up against atmospheric pressure. In Fig. 21–2(c), the liquid has been completely boiled away, and the cylinder contains a mass m of the substance in the form of a vapor. By definition, the quantity of heat which has been supplied during the process is the product of the mass and heat of vaporization, mL.

The principle of the conservation of energy, in mechanics, states that when work is done on a body (that is, when energy is supplied to the

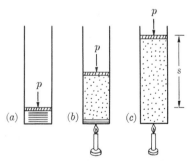

FIG. 21–2. External work is done against atmospheric pressure during the change of phase.

body) all of the work must be accounted for in one form or another. The same principle must apply to our present example, and we ask, in what form or forms can we account for the heat mL which has been supplied to the liquid? Since the same amount of heat can be recovered when the vapor condenses, it seems at first as if this energy were all stored up in the vapor in some form analogous to potential energy in mechanics. However, a study of the boiling process as illustrated in Fig. 21–2 shows that mechanical work has been done in the process, since the vapor has exerted a certain force, say F, against the lower surface of the piston, and the piston has moved up a distance s. The quantity of work, W, which has been done is therefore

$$W = F \times s,$$

and since this work can only have come from the heat mL which was supplied, it is evident that not all of the heat has been "absorbed" by the vapor. The work W is called the *external work* or the external part of the heat of vaporization. The remainder of the heat is said to increase the *internal energy* of the substance. We shall represent the internal energy of a body by the letter U. If U_V is the internal energy of the vapor and U_L that of the liquid, the increase in internal energy is $U_V - U_L$.

Setting up our balance sheet of energy, we have

$$mL = (U_V - U_L) + W. \tag{21–2}$$

That is, the heat supplied equals the increase in internal energy plus the external work.

The work W can be expressed in a more convenient form. Let the area of the piston be A. The force F is then

$$F = pA,$$

and the work is

$$W = Fs = pAs.$$

But As equals the increase in volume, which can be written $V_V - V_L$, where V_V is the volume of the vapor and V_L the volume of the liquid. Hence

$$W = p(V_V - V_L),$$

and Eq. (21–2) becomes

$$mL = (U_V - U_L) + p(V_V - V_L). \tag{21–3}$$

Careful attention must be paid to units when using this equation. Consider the term $p(V_V - V_L)$. In the English system, p must be expressed in pounds/square *foot* and V in cubic feet. Then, since

$$\frac{\text{lb}}{\text{ft}^2} \times \text{ft}^3 = \text{ft·lb} ,$$

this term is in ft·lb. (Note that most pressure gauges read in lb/in^2, and such readings must be multiplied by 144 to obtain lb/ft^2.) Since heats of vaporization are expressed in Btu/lb, either mL must be converted to ft·lb or W to Btu before using Eq. (21–3).

In the cgs system, where p is in dynes/cm^2 (1 atm $= 1.013 \times 10^6$ $\text{dynes/cm}^2 = 14.7 \text{ lb/in}^2$) and V in cm^3, the work is in

$$\frac{\text{dynes}}{\text{cm}^2} \times \text{cm}^3 = \text{dyne·cm or ergs.}$$

EXAMPLE. One gram of water (1 cm^3) becomes 1671 cm^3 of steam when boiled at a pressure of 1 atm. The heat of vaporization at this pressure is 539 cal/gm. Compute the external work and the increase in internal energy.

External work $= p(V_V - V_L)$
$= 1.013 \times 10^6 (1671 - 1)$
$= 1.695 \times 10^9 \text{ ergs}$
$= 169.5 \text{ joules}$
$= 41 \text{ calories.}$

From Eq. (21–2)

$$U_V - U_L = mL - W = 539 - 41$$
$$= 498 \text{ calories.}$$

Hence the external work, or the external part of the heat of vaporization, equals 41 calories, and the increase in internal energy, or the internal part of the heat of vaporization, is 498 calories.

Although the expression for external work, $p(V_V - V_L)$, was derived for the special case of boiling under constant pressure, it can be used for any process in which a substance expands from a volume V_1 to a volume V_2 under a constant pressure p. In general

$$W = p(V_2 - V_1) \quad (p \text{ constant}). \tag{21–4}$$

If the pressure varies during the expansion, the work done in a small volume increase dV is

$$dW = p \, dV, \tag{21–5}$$

and if the volume increases from V_1 to V_2,

$$W = \int_{V_1}^{V_2} p \, dV. \tag{21–6}$$

Of course, p must be known in terms of V in order to carry out the integration. Note that if p is constant, the integral becomes simply $p(V_2 - V_1)$ which is the same as Eq. (21–4).

Graphically, the work done in an expansion is represented by the *area* under a graph in which p is plotted vertically and V horizontally. This is evident from the integral of Eq. (21–6), or from the fact that area on such a graph represents pressure times volume, which has the dimensions of work. See Fig. 21–3.

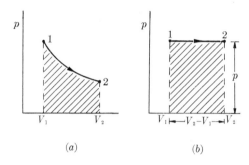

(a)   (b)

FIG. 21–3. The area under a curve in the p–V plane represents external work.

If the volume of a substance increases during a process, as in Fig. 21–3, work is done *by* the substance. If the volume decreases, work is done *on* the substance. Equation (21–6) applies in either case. If V_2 is less than V_1, which it will be in a compression process, then W has a minus sign. That is, if W is positive, work is done *by* the substance; if W is negative, work is done *on* the substance.

21–3 Effect of dissolved substances on freezing and boiling points. The freezing point of a liquid is lowered when some other substance is dissolved in the liquid. A common example is the use of an "anti-freeze" to lower the freezing point of the water in the cooling system of an automobile engine.

The freezing point of a saturated solution of common salt in water is about $-20°C$. To understand why a mixture of ice and salt may be used as a freezing mixture, let us make use of the definition of the freezing point as the only temperature at which the liquid and solid states can exist in equilibrium. When a concentrated salt solution is cooled, it freezes at $-20°C$, and crystals of ice (pure H_2O) separate from the solution. In other words, ice crystals and a salt solution can exist in equilibrium only at $-20°C$, just as ice crystals and pure water can exist together only at $0°C$.

When ice at 0°C is mixed with a salt solution at 20°C, some of the ice melts, abstracting its heat of fusion from the solution until the temperature falls to 0°C. But ice and salt solution cannot remain in equilibrium at 0°C, so that the ice continues to melt. Heat is now supplied both by the ice and the solution, and both cool down until the equilibrium temperature of −20°C is reached. If no heat is supplied from outside, the mixture remains unchanged at this temperature. If the mixture is brought in contact with a warmer body, say an ice cream mixture at 20°C, heat flows from the ice cream mixture to the cold salt solution, melting more of the ice but producing no rise in temperature so long as any ice remains. The flow of heat from the ice cream mixture lowers its temperature to its freezing point (which will be below 0°C since it is itself a solution). Further loss of heat to the ice-salt mixture causes the ice cream to freeze.

The boiling point of a liquid is also affected by dissolved substances, but may be either increased or decreased. Thus the boiling point of a water-alcohol solution is *lower* than that of pure water, while the boiling point of a water-salt solution is *higher* than that of pure water.

Both boiling and freezing points are affected by the external pressure, a matter which will be taken up in Chapter 23.

21–4 Measurement of heats of fusion and vaporization. The method of mixtures is used to measure heats of fusion and of vaporization. For example, the heat of fusion of ice may be found by dropping a weighed sample of ice at 0°C into a calorimeter containing a measured amount of water, and observing the temperature of the water before and after the addition of the ice. Let a mass m_i of ice at 0°C be dropped in a calorimeter containing a mass m_w of water, lowering the temperature of the water from t_1 to t_2. We shall assume that all of the ice melts, and neglect the heat capacity of the calorimeter. Then if L represents the heat of fusion of the ice, the ice absorbs a quantity of heat m_iL on melting (this converts it to water at 0°C) and a further quantity of heat m_it_2 on warming up to the final temperature t_2. The water in the calorimeter gives up a quantity of heat $m_w(t_1 - t_2)$. Hence,

$$m_w(t_1 - t_2) = m_i(L + t_2) ,$$

so that L can be found if the other quantities are known.

The heat of condensation (= heat of vaporization) of steam can be measured in a similar way, by allowing steam from a boiler to condense within a calorimeter. The condensation usually takes place in a coiled tube immersed in the calorimeter, so that the amount of steam condensing may be found by weighing the coil before and after the experiment. We have:

Heat given up by condensing steam $= m_s L$.

Heat given up by condensed steam (water at 100°C) cooling to $t_2 = m_s(100 - t_2)$.

Heat absorbed by calorimeter $= m_w(t_2 - t_1)$.

The heat of vaporization may be found by equating heat loss to heat gain.

Problems

21-1. An open vessel contains 500 gm of ice at -20°C. The heat capacity of the container may be neglected. Heat is supplied to the vessel at the constant rate of 1000 cal/min for 100 min. Plot a curve showing the elapsed time as abscissa and the temperature as ordinate.

21-2. A beaker whose heat capacity is negligible contains 500 gm of water at a temperature of 80°C. How many grams of ice at a temperature of -20°C must be dropped in the water so that the final temperature of the system will be 50°C?

21-3. A copper calorimeter of mass 100 gm contains 150 gm of water and 8 gm of ice in thermal equilibrium at atmospheric pressure. 100 gm of lead at a temperature of 200°C are dropped into the calorimeter. Find the final temperature, if no heat is lost to the surroundings.

21-4. 500 gm of ice at -16°C are dropped into a calorimeter containing 1000 gm of water at 20°C. The calorimeter can is of copper and has a mass of 278 gm. Compute the final temperature of the system, assuming no heat losses.

21-5. A calorimeter contains 500 gm of water and 300 gm of ice, all at a temperature of 0°C. A block of metal of

mass 1000 gm is taken from a furnace where its temperature was 240°C and is dropped quickly into the calorimeter. As a result, all of the ice is just melted. What would the final temperature of the system have been if the mass of the block had been twice as great? Neglect heat loss from the calorimeter, and the heat capacity of the calorimeter.

21-6. A tube leads to a calorimeter from a flask in which water is boiling under atmospheric pressure. The mass of the calorimeter is 150 gm, its water equivalent is 15 gm, and it contains originally 340 gm of water at 15°C. Steam is allowed to condense in the calorimeter until its temperature increases to 71°C, after which the total mass of calorimeter and contents is found to be 525 gm. Compute the heat of condensation of steam from these data.

21-7. A copper calorimeter can, having a water equivalent of 30 gm, contains 50 gm of ice. The system is initially at 0°C. 12 gm of steam at 100°C and 1 atm pressure are run into the calorimeter. What is the final temperature of the calorimeter and its contents?

21-8. A vessel whose walls are thermally insulated contains 2100 gm of water and 200 gm of ice, all at a temperature of 0°C. The outlet of a tube leading from a boiler, in which water is boiling at atmospheric pressure, is inserted in the water. How many grams of steam must condense to raise the temperature of the system to 20°C? Neglect the heat capacity of the container.

21-9. A 2-kgm iron block is taken from a furnace where its temperature was 650°C and placed on a large block of ice at 0°C. Assuming that all of the heat given up by the iron is used to melt the ice, how much ice is melted?

21-10. A copper bar 15 cm long and 6 cm^2 in cross section is placed with one end in a steam bath and the other in a mixture of ice and water, both at atmospheric pressure. The sides of the bar are thermally insulated. After the steady state has been established, (a) how much ice melts in 2 min? (b) How much steam condenses in the same time?

21-11. A boiler with a steel bottom 1.5 cm thick rests on a hot stove. The area of the bottom of the boiler is 1500 cm^2. The water inside the boiler is at 100°C and 750 gm are evaporated every 5 minutes. Find the temperature of the lower surface of the boiler, which is in contact with the stove.

21-12. An icebox, having wall area of 2 m^2 and thickness 5 cm, is constructed of insulating material having a thermal conductivity of 10^{-4} cal/sec·cm·C°. The outside temperature is 20°C, and the inside of the box is to be maintained at 5°C by ice. The melted ice leaves the box at a temperature of 15°C. If ice costs one cent per kgm, what will it cost to run the icebox for one hour?

21-13. In a household hot-water heating system, water is delivered to the radiators at 140°F and leaves at 100°F. The system is to be replaced by a steam system in which steam at atmospheric pressure condenses in the radiators, the condensed steam leaving the radiators at 180°F. How many pounds of steam will supply the same heat as was supplied by 1 lb of hot water in the first installation?

21-14. A "solar house" has storage facilities for 4 million Btu. Compare the space requirements for this storage on the assumption (a) that the heat is stored in water heated from a minimum temperature of 80°F to a maximum of 120°F, and (b) that the heat is stored in Glauber salt (Na$_2$ SO$_4$·10 H$_2$O) heated in the same temperature range.

Properties of Glauber salt

Specific heat (solid)	0.46
Specific heat (liquid)	0.68
Specific gravity	1.6
Melting point	90°F
Heat of fusion	104 Btu/lb

21-15. 1 lb of water when boiled at 212°F and atmospheric pressure becomes 26.8 ft^3 of steam. (a) Compute the external work in ft·lb. (b) Compute the increase in internal energy in Btu.

21-16. An aluminum canteen whose mass is 500 gm contains 750 gm of water and 100 gm of ice. The canteen is dropped from an airplane to the ground. After landing, the temperature of the canteen is found to be 25°C. Assuming that no energy is given to the ground in the impact, what was the velocity of the canteen just before it landed?

21-17. If no heat is lost to the surroundings, what must be the initial velocity of a lead bullet at a temperature of 25°C, so that the heat developed when it is brought to rest shall be just sufficient to melt it?

CHAPTER 22

PROPERTIES OF GASES—THE IDEAL GAS

22–1 Boyle's law. This chapter and the next will be devoted to a study of the properties of gases and to some important principles which are best illustrated with the help of a gas. We know that the atoms of a gas are much more widely separated than are those of a liquid or solid, hence the forces between atoms are of less consequence and the behavior of a gas is governed by simpler laws than those applying to liquids and solids. Discussion of the atomic model of a gas will be postponed until Chapter 25, however, and for the present we shall be concerned only with quantities that are directly measurable.

Robert Boyle, in 1660, reported on one of the first quantitative experiments relating to gaseous behavior. He found that if the temperature of a fixed mass of gas was held constant while its volume was varied over wide limits, the pressure exerted by the gas varied also, and in such a way that the product of pressure and volume remained essentially constant Mathematically,

$$pV = \text{constant (at constant temperature and for a fixed mass of gas).} \qquad (22\text{--}1)$$

This relation is *Boyle's law.* (p represents the *absolute* pressure.)

If the subscripts 1 and 2 refer to two different states of the gas at the same temperature, Boyle's law may also be written

$$p_1 V_1 = p_2 V_2 \text{ (states 1 and 2 at the same temperature).} \qquad (22\text{--}2)$$

The pV product, while nearly constant at a given temperature, does vary somewhat with the pressure. We therefore find it convenient to postulate an imaginary substance called an *ideal gas* which *by definition* obeys Boyle's law exactly at all pressures. Real gases at low pressure closely approximate an ideal gas.

The relation between the pressure and volume of an ideal gas at constant temperature is shown by the curves of Fig. 22–1, where p is plotted vertically and V horizontally. The axes are said to define a "p-V plane." The curves are equilateral hyperbolas, asymptotic to the p and V axes.

410

Each curve corresponds to a different temperature. That is, while $pV =$ *a constant* at any one temperature, the constant is larger the higher the temperature.

It should be pointed out that in order to be able to speak of *the* pressure or *the* temperature of a gas (or of any other body) it is necessary that all parts be at the *same* pressure or temperature. When its pressure and temperature are the same throughout, the gas is said to be in *equilibrium*. For example, a gas in a tank which is being heated by a blowtorch directed against one side is not in equilibrium, and it is meaningless to speak of *the* temperature of the gas. Only when the gas has been left to itself for a long enough time for its temperature to be the same throughout can we associate any one temperature with it.

Each of the curves in Fig. 22–1 can be considered to represent a *process* through which the gas is carried, for example, a process in which the gas is compressed from a large to a small volume in a bicycle pump. We shall show later that in order to keep the temperature constant in such a compression it would be necessary to remove heat from the gas, and therefore the process would have to be carried out slowly in order to keep the temperature the same throughout the gas. Such a slow compression carries the gas through a series of states, each of which is very nearly an equilibrium state, and it is called a *quasi-static*, or a "nearly static" process. Unless stated otherwise, we shall assume in what follows that all processes are to be carried out in this way, so that at each stage of the process the gas is, for all practical purposes, in an equilibrium state.

Any process in which the temperature remains constant is called *isothermal*, and the curves in Fig. 22–1 are the isothermal curves, or, more briefly, the *isotherms* of an ideal gas.

It is evident that work will be done *by* a gas in an isothermal expansion, and work will be done *on* a gas when it is compressed. The work done may be computed from the relation

$$W = \int_{V_1}^{V_2} p \, dV.$$

Let us write Boyle's law as

$$pV = p_1 V_1 = p_2 V_2 = C.$$

Then $p = C/V$, and

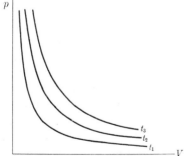

FIG. 22–1. The product of the pressure and volume of an ideal gas is constant at constant temperature.

$$W = \int_{V_1}^{V_2} C \frac{dV}{V} = C \ln \frac{V_2}{V_1},$$

and since C is equal to the constant pV product at all stages of the process, we may write

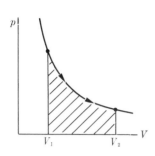

$$W = p_1 V_1 \ln \frac{V_2}{V_1} = p_2 V_2 \ln \frac{V_2}{V_1} .$$

Also, since

$$\frac{V_2}{V_1} = \frac{p_1}{p_2} ,$$

$$W = p_1 V_1 \ln \frac{p_1}{p_2} , \text{ etc.}$$

Fig. 22–2. Work done in the iso-thermal expansion of an ideal gas.

Graphically, the work done in an isothermal expansion is represented by the area under an isothermal curve in the p-V plane, shown shaded in Fig. 22–2. If the process is an expansion as in the figure, the ratio $V_2/V_1 > 1$, its natural logarithm is positive, and W is positive. If the process is a compression, the ratio V_2/V_1 is a fraction whose logarithm is negative. Hence W is negative when work is done on a body.

The units of W are the same as those of the pV product, since the logarithm is a pure number. If p is in lb/ft^2 and V in ft^3, W is in ft·lb. If p is in dynes/cm^2 and V in cm^3, W is in dyne·cm or ergs. In the mks system W is in joules.

22–2 Gay-Lussac's law. The first accurate statement of the law connecting the volume changes of a gas with changes in its temperature was published by Joseph Louis Gay-Lussac in 1802. Earlier work on the subject had been carried on by many other investigators, among them being Jacques A. C. Charles, whose name is often associated with Gay-Lussac's in connection with the law.

Gay-Lussac measured what we would now call the cubical coefficient of expansion of a number of different gases, and was apparently the first to recognize that when making such measurements with gases it is essential that the pressure be kept constant. If this is not done, the volume changes due to changes in pressure will obscure those due to temperature alone. The quantity measured was therefore the cubical coefficient of expansion *at constant pressure*. The experimental results can be expressed by the relation

$$V = V_0[1 + \beta(t - t_0)], \tag{22–3}$$

where V_0 is the volume at some reference temperature t_0 and V is the volume at temperature t. The coefficient β is the cubical coefficient of expansion, expressed in reciprocal degrees. Its numerical value depends on the size of the degree (i.e., whether fahrenheit or centigrade) and on the reference temperature t_0. If, as is usually the case, the reference temperature is taken as 0°C, Eq. (22–3) becomes

$$V = V_0(1 + \beta_0 t). \tag{22–4}$$

The symbol β_0 indicates that the reference temperature is 0°C.

The first point of interest is that the volume is a *linear* function of the temperature. The second, which is more surprising, is that the coefficient of expansion, β_0, has *very nearly the same value for all gases.* (Compare with the coefficients of expansion of liquids and solids, which differ widely for different materials.) By measuring β_0 for a number of gases in a series of measurements at different pressures, it is found that the lower the pressure the more closely do the values of β_0 agree for different gases. Extrapolating such a series of measurements to zero pressure gives the following value, common to all gases:

$$\beta_0 = 0.003660 \text{ per centigrade degree.}$$

We accordingly extend our definition of an ideal gas, and state that, in addition to obeying Boyle's law at all pressures, it obeys Gay-Lussac's law with $\beta_0 = 0.003660$ per centigrade degree.

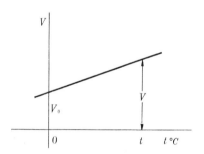

FIG. 22–3. At constant pressure the volume of an ideal gas is a linear function of the temperature.

The decimal 0.003660 is very nearly equal to 1/273, and the experimental facts above are often stated: the volume of a fixed mass of any gas kept at constant pressure increases by 1/273 of its value at 0°C for each centigrade degree increase in temperature. The linear relation between volume and temperature, at constant pressure, is illustrated in Fig. 22–3.

22–3 The equation of state of an ideal gas. We may now combine the laws of Boyle and Gay-Lussac to obtain a single equation connecting the pressure, volume, and temperature of an ideal gas. The analysis is best understood with the help of a diagram in the p-V plane. The coordinates of point 1 in Fig. 22–4 represent the pressure and volume of a certain mass of an ideal gas at a pressure $p_0 = 1$ atm and a temperature $t_0 = 0°C$. Point 2 is any other point at which the pressure is p, the volume V, and the temperature $t°C$. Consider a process shown by the heavy lines, in which the gas is first expanded from its initial state to another (point 3) at the same pressure but at a temperature t; and second, compressed isothermally to a point 2.

Since points 1 and 3 are at the same pressure, it follows from Gay-Lussac's law that

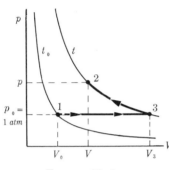

FIGURE 22–4

$$V_3 = V_0(1 + \beta_0 t). \quad (22\text{–}5)$$

Since points 3 and 2 are at the same temperature, it follows from Boyle's law that

$$pV = p_0 V_3. \quad (22\text{–}6)$$

When the expression for V_3 from Eq. (22–5) is substituted in (22–6), we obtain

$$pV = p_0 V_0(1 + \beta_0 t),$$

which can be written

$$pV = p_0 V_0 \beta_0 \left(t + \frac{1}{\beta_0}\right). \quad (22\text{–}7)$$

The term $p_0 V_0 \beta_0$ may be evaluated as follows: the pressure p_0 is 1 atm $= 1.013 \times 10^6$ dynes/cm^2. The temperature $t_0 = 0°C$. The volume V_0 is therefore the volume occupied by the gas at 1 atm and 0°C ("standard conditions"). It is well known that at standard conditions one gram-mole of all the common gases occupies a volume of 22.415 liters or 22,415 cm^3 (approximately 22,400 cm^3). Hence if the sample comprises n gram-moles, its volume $V_0 = n \times 22.4$ liters $= n \times 22,400$ cm^3. Therefore in cgs units

$$p_0 V_0 \beta_0 = 1.013 \times 10^6 \times n \times 22,400 \times 0.00366$$

$$= n \times 8.31 \times 10^7 \text{ dyne·cm/°C or ergs/°C},$$

and in a common hybrid system in which the pressure unit is 1 atm and the volume unit is 1 liter,

$$p_0 V_0 \beta_0 = 1 \times n \times 22.4 \times 0.00366$$

$$= n \times 0.08207 \text{ liter·atm/°C}.$$

The coefficients 8.31×10^7 or 0.08207 are the same for all gases; their numerical value is determined by the choice of units. Both are represented by the letter R, referred to as the *universal gas constant*.

$$R = 8.31 \times 10^7 \text{ ergs/mole·°C,}$$
$$R = 0.08207 \text{ liter·atm/mole·°C.}$$

$$(22\text{–}8)$$

Consider next the term in parentheses in Eq. (22–7), $(t + 1/\beta_0)$. Since β_0 is in reciprocal degrees centigrade, $1/\beta_0$ is in centigrade degrees and represents a temperature. The addition of this temperature to the centigrade temperature t is equivalent to expressing temperatures on a new scale whose zero point is lower than the centigrade zero by $1/\beta_0$ degrees, but in which the unit temperature interval is the same as on the centigrade scale. Temperatures on this scale are called *centigrade absolute* or *Kelvin* temperatures and will be represented by T.

$$T = t + \frac{1}{\beta_0} = t + \frac{1}{0.00366} = t + 273.2. \qquad (22\text{–}9)$$

Equation (22–7) may now be written

$$\boxed{pV = nRT,} \qquad (22\text{–}10)$$

where if p is in atmospheres, V in liters, n in gram·moles and T in degrees Kelvin, $R = 0.08207$ lit·atm/mole·°K, and if p is in dynes/cm², V in cm³, n in gram·moles and T in degrees Kelvin, $R = 8.31 \times 10^7$ ergs/mole·°K. Equation (22–10) is known as *the equation of state of an ideal gas*.

Notice that nothing in the foregoing discussion precludes the existence of temperatures lower than zero on the Kelvin scale. It is true, from Eq. (22–10), that the volume of an ideal gas kept at constant pressure would become zero at the Kelvin zero of temperature, or the pressure of an ideal gas kept at constant volume would become zero at this temperature. These consequences alone do not imply that still lower temperatures would be unattainable. The true significance of absolute zero as the lower limit of attainable temperatures can only be brought out in the light of reasoning based on the second law of thermodynamics. We shall consider this briefly in Section 24–9, where we give an alternate definition of the Kelvin temperature scale.

The number of moles n in a sample of gas equals the mass m of the gas divided by its molecular weight. If the latter is represented by M, then

$$n = \frac{m}{M}.$$

Hence we may write

$$pV = m \frac{R}{M} T.$$

By definition, the density of the gas, ρ, is

$$\rho = \frac{m}{V},$$

and alternate forms of Eq. (22–10) are

$$p = \rho \frac{R}{M} T, \qquad \text{or} \qquad \frac{p}{\rho} = \frac{RT}{M}, \qquad \text{or} \qquad \rho = \frac{pM}{RT}.$$

The density of a gas is seen to depend on its pressure and temperature as well as on its molecular weight. Hence in tabulating gas densities the pressure and temperature must be specified. The densities of a few common gases are listed in Table XVII.

The volume per unit mass of a substance is called its *specific volume* and will be represented by v.

$$v = \frac{V}{m}.$$

The specific volume is simply the reciprocal of the density, m/V. Another form of Eq. (22–10) is therefore

$$pv = \frac{R}{M} T. \qquad (22\text{–}11)$$

The molal specific volume is defined as the volume per mole, V/n. This is also represented by v and in terms of it Eq. (22–10) becomes

$$pv = RT. \qquad (22\text{–}12)$$

(It is not worth while to attempt to memorize these various forms of the ideal gas equation. Equation (22–10) is sufficient.)

A more familiar form of the ideal gas equation may be obtained by writing Eq. (22–10) in the form

$$\frac{pV}{T} = nR. \qquad (22\text{–}13)$$

If a fixed mass of gas is carried through any sort of process, the right side of Eq. (22–13) will have the same value at all stages of the process. Hence if subscripts 1 and 2 refer to any two states,

$$\frac{p_1 V_1}{T_1} = \frac{p_2 V_2}{T_2}. \qquad (22\text{–}14)$$

TABLE XVII

DENSITIES OF GASES

(gm/cm^3 at 1 atm and 0°C)

Air	1.2929×10^3
Argon	1.7832
Carbon dioxide	1.9769
Helium	0.1785
Hydrogen	0.0899
Nitrogen	1.2506
Oxygen	1.4290

EXAMPLES. (1) How many kilograms of O_2 are contained in a tank whose volume is 2 ft^3 when the gauge pressure is 2000 lb/in^2 and the temperature is 27°C? Assume the ideal gas laws to hold.

$$2 \text{ ft}^3 = 2 \times 28.3 = 56.6 \text{ liters.}$$

$$p_{abs} = 2015 \text{ lb/in}^2 = \frac{2015}{14.7} = 137 \text{ atm.}$$

$$T = t + 273 = 300°K.$$

Hence

$$n = \frac{pV}{RT} = \frac{137 \times 56.6}{.082 \times 300}$$

$$= 315 \text{ moles}$$

$$= 315 \times 32 = 10,100 \text{ gm}$$

$$= 10.1 \text{ kgm.}$$

(2) What volume would be occupied by this gas if it were allowed to expand to atmospheric pressure at a temperature of 50°C?

Since we are dealing with a fixed mass of gas, we may write

$$\frac{p_1 V_1}{T_1} = \frac{p_2 V_2}{T_2},$$

$$V_2 = V_1 \frac{p_1}{p_2} \frac{T_2}{T_1}$$

$$= 2 \times \frac{137}{1} \times \frac{323}{300}$$

$$= 295 \text{ ft}^3.$$

22–4 Internal energy of a gas. As a part of his program of measuring the mechanical equivalent of heat, Joule wished to show that whenever mechanical energy is converted to heat the same quantity of heat is developed by a given quantity of mechanical energy *whatever the mechanism by which the conversion is brought about.* One process in which heat may be produced by the expenditure of mechanical energy is the compression of a gas, when, as is well known, the temperature of the gas rises unless heat is withdrawn. If the gas is contained in a cylinder immersed in a water bath, the heat evolved can be measured; and since the work done is known, the mechanical equivalent of heat can be computed, provided all of the work is converted into heat and none into any other form. This was one of the methods used by Joule in the course of his experiments.

It is conceivable, however, that some of the work done in compressing a gas may go into increasing the internal energy of the gas and hence may not appear as heat. A separate investigation was therefore necessary to find how the internal energy of a gas depended on its temperature and volume. This was done by Joule in a classic experiment which bears his name.

Let us first formulate the principle of the conservation of energy as follows. On page 404, in connection with heats of transformation, we used the following relation:

$$mL = (U_V - U_L) + W,$$

which expressed the fact that the heat of transformation, mL, was accounted for in part by an increase of internal energy, $(U_V - U_L)$, and in part by mechanical work W performed during the change of phase. The same conservation principle applies to any process in which heat is supplied to a system—the heat must be accounted for either by an increase in internal energy or by work done by the system. Hence, if the heat supplied in any process is represented by Q, we may write

$$Q = (U_2 - U_1) + W \qquad (22\text{--}15)$$

or, if the process is an infinitesimal one,

$$dQ = dU + dW. \qquad (22\text{--}16)$$

(Q, U, and W must all be expressed in the same unit.)

This relation, the conservation of energy principle including heat as a form of energy, is called the *first law of thermodynamics.*

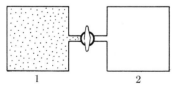

FIG. 22–5. Free expansion of a gas into an evacuated container.

Let us return now to the Joule experiment, illustrated schematically in Fig. 22–5. Compressed gas is forced into container 1, container 2 is evacuated, and the system is allowed to come to a uniform temperature.

The stopcock connecting the containers is then opened, and gas rushes into the evacuated container until the pressures become equalized. The gas remaining in container 1 is cooled by the expansion, while that which flows into container 2 becomes warmer. If both containers are immersed in a tank of water which is kept stirred, it is found that no net temperature change of the water can be observed—that is, the cooling of one part just offsets the heating of the other.

Since the temperature of the water does not change, no heat has left it. Hence no heat is supplied to the gas, and $Q = 0$. Furthermore, the external work done by the gas is also zero. This seems puzzling at first, since the volume of the gas increases, and the gas remaining in container 1 does work on the gas which first rushes into container 2. However, this is merely work done by one part of the gas on another part, or a transfer of energy within the gas itself, and does not represent work done by the gas as a whole on its surroundings. Therefore $W = 0$, and from the first law,

$$0 = U_2 - U_1 + 0,$$

or

$$U_2 = U_1.$$

That is, the internal energy of the gas is the same whether it occupies a large or small volume, *so long as its temperature does not change*. This fact is usually stated: *the internal energy of a gas is a function of its temperature only, and does not depend on its volume*.

More precise experiments in recent years have proved that real gases actually show a small temperature drop in a Joule expansion. We extend our definition of an ideal gas, and say that for such a gas the temperature change in a Joule expansion is zero, and that the internal energy of an ideal gas depends only on its temperature.

The results of this experiment showed Joule that he could safely assume that all of the work done in compressing a gas *isothermally* was converted into heat. Of course, this is strictly true for an ideal gas, by definition.

22–5 Specific heats of a gas. The true specific heat capacity of a substance was defined in Chapter 19 by the equation

$$c = \frac{dQ}{m \, dT}.$$

When dealing with gases, it is more convenient to use the *molal* specific heat, equal numerically to the heat required to increase the temperature of one mole by one degree. The molal specific heat is represented by C, and its definition is

$$C = \frac{dQ}{n \, dT},$$

where n is the number of moles.

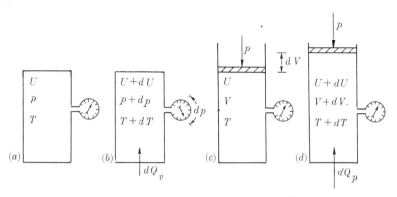

Fig. 22-6. Since external work is done when a gas is heated at constant pressure, $C_p > C_v$.

Let us apply this relation to the process of raising the temperature of a gas. Suppose we have two cylinders, as in Fig. 22-6, (a) and (c), containing equal masses of an ideal gas at the same initial pressure, volume, and temperature. The volume of cylinder (a) is kept constant while the other is provided with a freely moving piston so the gas in it can expand at constant pressure. Let sufficient heat be supplied to each cylinder to increase the temperature of each by the same amount. Then from the first law,

$$dQ_v = dU_v + dW_v,$$
$$dQ_p = dU_p + dW_p,$$

(22-17)

where the subscripts p and v distinguish corresponding quantities in the constant-pressure and constant-volume processes.

That part of the heat which goes into increasing internal energy is the same in each cylinder, since by definition the internal energy of an ideal gas depends only on its temperature, and the temperature rise is the same in each cylinder. Hence

$$dU_v = dU_p = \text{(say)} \ dU.$$

The mechanical work done, however, is not the same for both cylinders. It is zero for the one kept at constant volume and equal to $p \, dV$ for the one at constant pressure. That is,

$$dW_v = 0, \qquad dW_p = p \, dV.$$

Equations (22-17) then become

$$dQ_v = dU, \qquad dQ_p = dU + p \, dV,$$

and the heat that must be supplied in the constant pressure process, during which the gas expands and does work, is greater than the heat supplied in the constant-volume process, during which the gas does no work.

From the ideal gas law, for a process at constant pressure,

$$p \, dV = nR \, dT.$$

Hence

$$dQ_p = dU + nR \, dT.$$

Finally, from the definition of molal specific heat,

$$C_v = \frac{dQ_v}{n \, dT} = \frac{dU}{n \, dT}, \tag{22-18}$$

$$C_p = \frac{dQ_p}{n \, dT} = \frac{dU}{n \, dT} + \frac{nR \, dT}{n \, dT}, \tag{22-19}$$

or

$$\boxed{C_p = C_v + R.} \tag{22-20}$$

That is, the molal specific heat of an ideal gas, at constant pressure, is greater than its specific heat at constant volume by the gas constant R. Of course R must be expressed in the same units as C_v and C_p, usually cal/mol·C°. Since $R = 8.31$ joules/mol·C° and 4.19 joules $= 1$ cal,

$$R = \frac{8.31}{4.19} = 1.98 \text{ cal/mol·C°},$$

or very nearly 2 cal/mol·C°.

Equation (22–20) is strictly true for an ideal gas and very nearly true for real gases.

It should be pointed out that there are many ways in which a gas can be heated other than at constant volume or constant pressure. For example, in Fig. 22–6, the piston could have been forcibly pushed down or pulled up while heat was supplied to the gas, so that temperature, volume, and pressure all changed during the heating process. For each way in which the volume is caused to vary, the external work will be different, hence the specific heat will be different. Since there are an infinite number of ways in which the volume can be varied, a gas has an infinite number of specific heats, which range from $-\infty$ to $+\infty$. However, since most actual processes are carried out either in a closed container (volume constant) or in a container open to the atmosphere (pressure constant), C_v and C_p are the most useful of the specific heats.

Solids and liquids also expand when heated, if free to do so, and hence perform work. The coefficients of volume expansion of solids and liquids are, however, so much smaller than those of gases that the external work is small. The internal energy of a solid or liquid *does* depend on its volume as well as its temperature and this must be considered when evaluating the difference between specific heats of solids or liquids. It turns out that here also $C_p > C_v$, but the difference is small and is not expressible as simply as that for a gas. Because of the large stresses set up when solids or liquids are heated and *not* allowed to expand, most heating processes involving them take place at constant pressure and hence C_p is the quantity usually measured for a solid or liquid.

TABLE XVIII

SPECIFIC HEATS OF GASES AT ROOM TEMPERATURE

Gas	c_p(cal/gm·°C)	C_p(cal/mole·°C)	γ
He	1.25	5.00	1.660
A	0.1253	5.01	1.668
H_2	3.389	6.778	1.410
N_2	0.2477	6.95	1.404
O_2	0.2178	6.98	1.401
Air	0.240	6.96	1.40
CO_2	0.1989	8.76	1.304
NH_3	0.5232	8.90	1.310
H_2O	0.465	8.37	1.27

A part of the definition of an ideal gas is that its internal energy is a function of its temperature only. The precise form of the function is given by Eq. (22–18), namely

$$dU = nC_v \, dT, \qquad (22\text{–}18)$$

or in a finite process

$$U_2 - U_1 = nC_v(T_2 - T_1). \qquad (22\text{–}21)$$

As with all other forms of energy, only *differences* in internal energy can be computed or measured.

The significance of Eq. (22–21) is that the change in internal energy of n moles of an ideal gas in *any* process equals the product of the number of moles times the temperature change times the molal specific heat *at constant volume*, even though the process is not at constant volume. This is puzzling and calls for further explanation.

Suppose we wish to find the internal energy change of an ideal gas which is carried along the line 1–2 from the state represented by point 1 in Fig. 22–7 to the state represented by 2, a process in which neither volume, pressure, nor temperature is constant. If the gas were to be brought from the same initial state (1) to the same final state (2) by some other process (such as 1–3–2) its end state would be indistinguishable from that at the end of the process 1–2. Hence its internal energy change between states 1 and 2 is the same for all processes connecting these points. Now, no matter where points 1 and 2 may be located in the p-V plane, it is always possible to go from one to the other by a path such as 1–3–2, that is, a constant-volume process combined with an isothermal process.

The internal energy change along path 1–3 is obviously

$$U_3 - U_1 = nC_v(T_2 - T_1),$$

since this part of the path is at constant volume.

The internal energy change along path 3–2 is *zero*, since this path is at constant temperature. That is,

$$U_3 = U_2.$$

Hence
$$U_2 - U_1 = nC_v(T_2 - T_1),$$

and since this is the same for *all* paths between 1 and 2 it is true for the process 1–2 also.

22–6 Internal energy and heat. In the preceding section we justified the application of Eq. (22–21) to *any* process by the argument that the end state of the gas, if brought by any path from point 1 to point 2 in Fig. 22–7, was indistinguishable from its state when brought from 1 to 2 by any other path. In other words, by an appeal to reason alone, we concluded that the internal energy change between two states was independent of the path connecting them. Physics, however, is an experimental science and the conservation of energy is an experimental law.

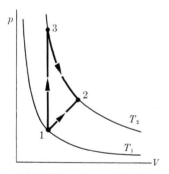

FIG. 22–7. The change in internal energy is the same for all paths between the same two end points.

Let us see precisely what are the experimental facts expressed by the first law of thermodynamics (Eq. 22–15), which we may write

$$Q - W = U_2 - U_1. \quad (22\text{–}15)$$

This equation differs from many physical relations in that it does not describe an equality which is found to hold between quantities which can be measured. That is, while the heat Q supplied to a system can be measured with a calorimeter and the work W done by the system can be found from mechanical measurements, *there are no instruments for measuring internal energy.* Equation (22–15) does *not* imply that when we measure Q, W, U_2, and U_1 we find the relation above to hold true.

The true significance of the first law is that when we measure Q and W for any number of paths between the same two end points, we find that while the heat added may vary from one path to another, and the work done may vary also, the difference $Q - W$ is found by experiment to have the same value along all paths. This difference then *defines* the internal energy change $U_2 - U_1$. While Eq. (22–15) is an experimental law, the experimental part is not that $Q - W = U_2 - U_1$ (this is true by definition) but that $Q - W$ has the same value for all paths between the same two end points.

It follows that if some arbitrary value is assigned to the internal energy in some standard reference state, its value in any other state is uniquely defined, since $Q - W$ is the same for all processes connecting the states.

We can now see why the expression "the heat in a body" is meaningless. Suppose we assigned an arbitrary value to "the heat in a body" in some standard reference state. The "heat in the body" in some other state would then equal the "heat" in the reference state plus the heat added when the body is carried to the second state. But the heat added depends entirely on the path by which we go from one state to the other, and since there are an infinite number of paths which might be followed, there are an infinite number of values which might equally well be assigned to the "heat in the body" in the second state. Since it is not possible to assign any one value to the "heat in the body" we conclude that this concept is meaningless, or at any rate useless.

22–7 Adiabatic processes. Any process in which there is no flow of heat into or out of a system is called *adiabatic*. To perform a truly adiabatic process it would be necessary that the system be surrounded by a perfect heat insulator, or that the surroundings of the system be kept always at the same temperature as the system. However, if a process such as the compression or expansion of a gas is carried out very rapidly, it will be nearly adiabatic, since the flow of heat into or out of the system is slow even under favorable conditions. Thus the compression stroke of a gasoline or Diesel engine is approximately adiabatic.

Note that external work may be done on or by a system in an adiabatic process, and that the temperature usually changes in such a process.

Let an ideal gas undergo an infinitesimal adiabatic process. Then $dQ = 0$, $dU = nC_v\, dT$, $dW = p\, dV$, and from the first law,

$$nC_v\, dT = -p\, dV. \qquad (22\text{–}22)$$

From the equation of state,

$$p\, dV + V\, dp = nR\, dT. \qquad (22\text{–}23)$$

Eliminating dT between Eqs. (22–22) and (22–23), and making use of the fact that $C_p = C_v + R$, we obtain the relation

$$\frac{dp}{p} + \frac{C_p}{C_v}\frac{dV}{V} = 0,$$

or, if C_p/C_v is denoted by γ,

$$\frac{dp}{p} + \gamma\frac{dV}{V} = 0. \tag{22–24}$$

To obtain the relation between p and V in a finite adiabatic change we may integrate Eq. (22–24). This gives

$$\ln p + \gamma \ln V = \ln (\text{constant}),$$

or

$$pV^\gamma = \text{constant}.$$

If subscripts 1 and 2 refer to any two points of the process,

$$p_1V_1^\gamma = p_2V_2^\gamma. \tag{22–25}$$

It is left as a problem to show that by combining Eq. (22–25) with the equation of state (Eq. 22–10) we obtain the alternate forms

$$T_1V_1^{\gamma-1} = T_2V_2^{\gamma-1}, \tag{22–26}$$

$$T_1p_1^{1-\gamma/\gamma} = T_2p_2^{1-\gamma/\gamma}. \tag{22–27}$$

Values of the specific heat ratio, $C_p/C_v = \gamma$, are listed in Table XVIII for some common gases.

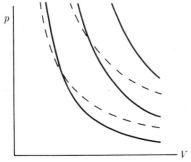

An adiabatic expansion or compression may be represented graphically by a plot of Eq. (22–25) as in Fig. 22–8, in which a number of isothermal curves are shown for comparison. The adiabatic curves, at any point, have a somewhat steeper slope than the isothermal curve passing through the same point. That is, as one follows along an adiabatic from right to left (compression process) the curve continually cuts across isotherms of higher and higher temperatures, in agreement with the fact that the temperature continually increases in an adiabatic compression.

Fig. 22–8. Adiabatic curves (full lines) vs. isothermal curves (dashed lines).

EXAMPLE. The compression ratio of a Diesel engine, V_1/V_2, is about 15. If the cylinder contains air at 15 lb/in^2 (absolute) and 60°F ($= 520$°F abs) at the start of the compression stroke, compute the pressure and temperature at the end of this stroke. Assume that air behaves like an ideal gas and that the compression is adiabatic. The value of γ for air is 1.40.

From Eq. (22–25),

$$p_2 = p_1 \left(\frac{V_1}{V_2}\right)^{\gamma},$$

or

$$\log p_2 = \log p_1 + \gamma \log \left(\frac{V_1}{V_2}\right)$$

$$= \log 15 + 1.4 \log 15$$

$$= 1.176 + 1.646 = 2.822,$$

$$\therefore p_2 = 663 \text{ lb/in}^2.$$

The temperature may now be found from Eq. (22–26) or (22–27) or by the ideal gas law. Thus, from Eq. (22–26), we have

$$T_2 = T_1 \left(\frac{V_1}{V_2}\right)^{\gamma - 1},$$

$$\log T_2 = \log T_1 + (\gamma - 1) \log \left(\frac{V_1}{V_2}\right)$$

$$= \log 520 + (1.4 - 1) \log 15$$

$$= 2.716 + 0.470 = 3.186,$$

$$T_2 = 1535 \text{°F abs}$$

$$= 1075 \text{°F}.$$

Or,

$$T_2 = T_1 \times \frac{p_2 V_2}{p_1 V_1}$$

$$= 520 \times \frac{663 \times 1}{15 \times 15}$$

$$= 1535 \text{°F abs}.$$

The work done by an ideal gas in an adiabatic expansion is computed as follows. We have, from Eq. (22–22),

$$W = \int_{V_1}^{V_2} p \, dV = \int_{T_1}^{T_2} -nC_v \, dT. \qquad (22\text{–}28)$$

Hence the work may be found from either integral. Consider first $\int p \, dV$. Since $pV^{\gamma} = p_1 V_1^{\gamma} = p_2 V_2^{\gamma} = $ a constant, C, we may write

$$W = \int_{V_1}^{V_2} p\, dV = C \int_{V_1}^{V_2} \frac{dV}{V^\gamma}$$

$$= \frac{1}{1-\gamma} [CV_2^{1-\gamma} - CV_1^{1-\gamma}].$$

In the first term let $C = p_2 V_2^\gamma$ and in the second term let $C = p_1 V_1^\gamma$. Then

$$W = \frac{p_2 V_2 - p_1 V_1}{1 - \gamma}. \tag{22-29}$$

This expresses the work in terms of initial and final pressures and volumes. If the initial and final temperatures are known, we may return to Eq. (22–28) and write

$$W = \int_{T_1}^{T_2} -nC_v\, dT = nC_v(T_1 - T_2).$$

EXAMPLE. Compute the work done in the compression stroke of the Diesel engine described in the previous example. Let the initial volume be 60 in³.
From Eq. (22–29),

$$W = \frac{p_2 V_2 - p_1 V_1}{1 - \gamma}.$$

The pressures must be expressed in pounds per square *foot* and the volumes in cubic *feet*.

$$W = \frac{(663 \times 144)(4/1728) - (15 \times 144)(60/1728)}{1 - 1.40} = \frac{146}{-0.40} = -365 \text{ ft·lb.}$$

The work is negative since, according to our convention of signs, W is negative when work is done *on* a system.

22–8 Compressibilities of a gas. The compressibility of a substance is defined (see page 250) as the fractional decrease in volume per unit increase in pressure:

$$k = -\frac{1}{V}\frac{dV}{dp}.$$

It is evident, however, that if any temperature change occurs while the pressure is increased, the volume will change for that reason also. In other words, the compressibility of a substance, like its specific heat, may have any number of values, depending on the conditions which hold during the compression process. The *isothermal* compressibility and the *adiabatic* compressibility are the most useful. They can readily be derived for an ideal gas.

If the temperature is constant, we obtain from Boyle's law

$$pV = C,$$

$$\frac{dV}{dp} = -\frac{C}{p^2},$$

$$k_{is} = -\frac{1}{V}\frac{dV}{dp} = -\frac{p}{C}\left(-\frac{C}{p^2}\right),$$

$$k_{is} = \frac{1}{p}, \tag{22-30}$$

and the isothermal compressibility, k_{is}, equals the reciprocal of the pressure.

If the process is adiabatic,

$$pV^\gamma = C,$$

$$\frac{dV}{dp} = -\left(\frac{C^{1/\gamma}p^{-(\gamma+1/\gamma)}}{\gamma}\right),$$

$$k_{ad} = \frac{1}{\gamma p}. \tag{22-31}$$

The adiabatic compressibility is also inversely proportional to the pressure but is smaller than the isothermal compressibility since $\gamma > 1$.

A gas is therefore more readily compressed at low than at high pressure. This explains, in part, the use of pneumatic rather than solid tires on motor vehicles. The air in a pneumatic tire is like a spring with a variable force constant; it yields readily at first but less and less readily the more it is compressed. No combination of springs can behave in this way; the force constant of a spring is independent of the load.

We shall see in a later chapter that the adiabatic compressibility must be used to compute the velocity of sound waves.

Problems

22–1. A tank contains 1.5 ft³ of nitrogen at an absolute pressure of 20 lb/in² and a temperature of 40°F. What will be the pressure if the volume is increased to 15 ft³ and the temperature raised to 440°F?

22–2. A liter of helium under a pressure of 2 atm and at a temperature of 27°C is heated until both pressure and volume are doubled. (a) What is the final temperature? (b) How many grams of helium are there?

22–3. A flask of volume 2 liters, provided with a stopcock, contains oxygen at 300°K and atmospheric pressure. The system is heated to a temperature of 400°K, with the stopcock open to the atmosphere. The stopcock is then closed and the flask cooled to its original temperature. (a) What is the final pressure of the oxygen in the flask? (b) How many grams of oxygen remain in the flask?

22–4. A bubble of air having a radius of 1 cm is formed at the bottom of a lake 68 ft deep, where the temperature is 4°C, and rises to the top, where the temperature is 27°C. Neglect surface tension. What is the radius of the bubble as it reaches the water surface (a) if it is continually at the same temperature as the surrounding water, and (b) if there is no heat transfer between the bubble and the water?

22–5. The submarine Squalus sank at a point where the depth of water was 240 ft. (a) If a diving bell in the form of a circular cylinder 8 ft high, open at the bottom and closed at the top, is lowered to this depth, to what height will the water rise within it when it reaches the bottom? (b) The temperature at the surface is 27°C and at the bottom it is 7°C. At what gauge pressure must compressed air be supplied to the bell while on the bottom to expel all the water from it? The density of sea water may be taken as 2 slugs/ft³.

22–6. The cylinder of a pump compressing air from atmospheric pressure

into a very large tank at 60 lb/in² gauge pressure is 10 inches long. (a) At what position in the stroke will air begin to enter the tank? Assume the compression to be adiabatic. (b) If the air is taken into the pump at 27°C, what is the temperature of the compressed air?

22–7. A capillary tube 1 m long, of inside diameter 1 mm, is closed at its upper end. The lower end is dipped just below the surface of water in a large container. (a) What is the height of the meniscus in the tube? (b) How far must the lower end of the tube be below the surface of the surrounding liquid so that the water level is the same both inside and outside the tube?

22–8. (a) Derive from the equation of state an equation for the density of an ideal gas in terms of the pressure, temperature, and appropriate constants. (b) By considering the forces acting on an infinitesimal volume of air of height dh, derive an equation, in terms of the density, for the rate of change of pressure with height in the earth's atmosphere. (c) Combine these two results and integrate to get the equation expressing the pressure variation with altitude. Assume temperature and gravitational field intensity independent of height.

22–9. 2 moles of oxygen are initially at a temperature of 27°C and volume 20 liters. The gas is expanded first at constant pressure until the volume has doubled, and then adiabatically until the temperature returns to the original value. (a) What is the total increase in internal energy in calories? (b) What is the total heat added in calories? (c) What is the total work done by the gas in joules? (d) What is the final volume?

22–10. Derive Eqs. (22–26) and (22–27) from Eq. (22–25).

22–11. Air at a gauge pressure of 300 lb/in² is used to drive an air engine which exhausts at a gauge pressure of 15 lb/in². What must be the temperature of the compressed air in order that there may be no possibility of frost forming in the exhaust ports of the engine? Assume the expansion to be adiabatic. Note: frost frequently forms in the exhaust ports of an air-driven engine. This occurs when the moist air is cooled below 0°C by the expansion which takes place in the engine.

22–12. In a certain process, 500 cal of heat are supplied to a system and at the same time 100 joules of mechanical work are done on the system. What is the increase in its internal energy?

FIGURE 22–9

22–13. The cylinder in Fig. 22–9 has a total volume of 4 liters and contains 0.2 mole of an ideal gas at a temperature of 300°K. γ for the gas is 1.5. The cylinder is thermally insulated from the surroundings, and equipped with a tightly fitting, frictionless piston. Initially the gas occupies a volume of only 1 liter, and the remaining volume is completely evacuated. The gas is allowed to expand until it occupies the entire volume of the cylinder. (a) If the expansion were performed by slowly raising the piston, calculate the final temperature and pressure and find the work done, heat absorbed, and change in internal energy. (b) If the expansion were performed by keeping the piston in its original position and opening a small valve, find the final temperature and pressure, and the work done, heat absorbed, and change in internal energy after equilibrium has been reached.

22–14. (a) Why does $dU = nC_v\, dT$ for a constant volume process whereas $dU \neq nC_p\, dT$ for a constant pressure process? (b) Why does $dU = nC_v\, dT$ for an ideal gas, regardless of whether the process is adiabatic, isothermal,

constant volume, constant pressure, etc.?

22–15. 200 Btu are supplied to a system in a certain process, and at the same time the system expands against a constant external pressure of 100 lb/in². The internal energy of the system is the same at the beginning and at the end of the process. Find the increase in volume of the system.

22–16. A mole of an ideal gas at 27°C is placed in a container covered by a piston which maintains atmospheric pressure on the gas. The gas is heated until its temperature rises to 127°C. (a) Draw a pV diagram showing the process. (b) How much mechanical work is done in this process? (c) On what is this work done? (d) What is the change in the internal energy of the gas? (e) How much heat was supplied to the gas? (f) How much mechanical work would have been done if the external pressure had been $\frac{1}{2}$ atmospheric instead of atmospheric?

22–17. 2 moles of an ideal gas for which $C_v = 3$ cal/mole·deg are carried around the cycle abc in Fig. 22–10. Process bc is an isothermal compression. Compute the work done by the gas in each part of the cycle, the heat added to the gas, and the change in its internal energy.

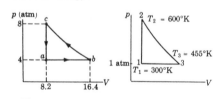

FIGURE 22–10   FIGURE 22–11

22–18. A heat engine carries 0.1 mole of an ideal gas around the cycle shown in the pV diagram in Fig. 22–11. Process 1–2 is at constant volume, 2–3 is adiabatic, and 3–1 is at a constant pressure of 1 atm. γ for this gas is $\frac{5}{3}$. (a) Find the pressure and volume at points 1, 2, and 3. (b) Find the net work done by the gas during the cycle.

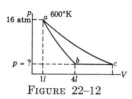

FIGURE 22-12

22-19. Fig. 22-12 shows three processes for an ideal gas. The temperature at point a is 600°K, the pressure is 16 atm, and the volume is 1 liter. The volume at point b is 4 liters. One of the processes ab and ac is isothermal and the other is adiabatic. The ratio of specific heats for the gas is 1.50. (a) Which of the processes ab or ac is isothermal and which is adiabatic? How do you know? (b) Compute the pressure at points b and c. (c) Compute the temperatures at b and c. (d) Compute the volume at point c.

22-20. A cylinder containing 10 gm of gas is compressed from a volume of 500 cm³ to a volume of 100 cm³. During the process, 100 cal of heat are removed from the gas, and at the end of the process the temperature of the gas has increased by 5°C. Compute the specific heat of the gas in this particular process.

22-21. What is the isothermal compressibility of 1 mole of helium at 200°C in a cylinder whose volume is 1 liter? Treat helium as an ideal gas.

FIGURE 22-13

22-22. One-tenth mole of an ideal gas is contained in the lower part of a cylinder beneath a piston of area 50 cm², as shown in Fig. 22-13. The specific heat of the gas at constant volume is 5 cal/mole·C°. The piston has a negligible mass, but carries a weight of mass 100 kgm. The region above the piston is evacuated. The initial temperature of the gas is 0°C and the piston is initially at a distance h from the bottom of the cylinder. If the gas is heated until the weight has

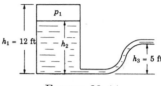

FIGURE 22-14

risen 10 cm, (a) find the original height h, (b) find the final temperature. (c) What was the increase in internal energy of the gas in calories? (d) How many calories of heat were supplied to the gas?

22-23. A large tank of water has a hose connected to it, as shown in Fig. 22-14. The tank is sealed at the top and has compressed air between the water surface and the top. When the water height, h_2 is 10 ft, the gauge pressure, p_1, is 15 lb/in². Assume that the air above the water surface expands isothermally and take the weight density of water to be 64 lb/ft³. (a) What is the velocity of flow out of the hose when $h_2 = 10$ ft? (b) What is the velocity of flow of the hose when h_2 has decreased to 8 ft?

CHAPTER 23

REAL GASES

23–1 Liquefaction of gases. An ideal gas, if compressed isothermally, remains a gas no matter how great a pressure is applied to it; the volume decreases continually with increasing pressure according to Boyle's Law. All real gases, however, become liquids when the pressure is increased sufficiently (provided the temperature is below the critical temperature of the gas, see page 433).

The difference between the behavior of an ideal and a real gas in an isothermal compression is illustrated in Fig. 23–1(a) and (b). Imagine the gases to be contained within two similar cylinders each provided with a piston and a pressure gauge. The volume at each stage of the compression process is proportional to the distance of the piston from the closed end of the cylinder, and the pressure at each stage may be read from the gauge. Several stages of the compression are illustrated in each

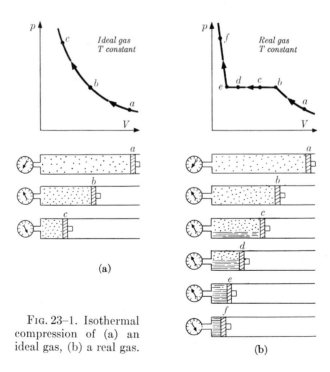

(a)

(b)

Fig. 23–1. Isothermal compression of (a) an ideal gas, (b) a real gas.

432

part of the figure, and corresponding pressures and volumes are plotted in the p-V diagrams. Each gas is initially at the same pressure and volume (point a), and the temperature of each is held constant throughout by the removal of heat.

As the piston in (a) is forced to the left, the reading of the pressure gauge rises steadily, and the relation between p and V is the familiar Boyle's law. As the volume of the real gas is decreased in Fig. 23–1(b), the pressure rises at first along the curve a-b in a manner not very different from that of the ideal gas. When point b is reached, however, a sharp break occurs in the curve, and the volume continues to decrease *without further increase of pressure*. At the same time, drops of liquid begin to appear on the walls of the cylinder. In other words, the process of *liquefaction* or *condensation* begins at point b.

As the volume is further reduced, from b to c and d, the quantity of liquid in the cylinder increases and the quantity of gas decreases. At point e, all of the substance has been converted into the liquid phase. Of course, during the transition from b to e, it was necessary to remove from the substance its heat of condensation.

Since liquids are nearly incompressible, a very large pressure increase is necessary to reduce the volume below that at point e. That is, the curve rises nearly vertically from e to f, having a very small inclination toward the left.

As an example, if cylinder (b) had initially contained one gram of steam at 100°C, at a pressure of about $\frac{1}{2}$ atm and a volume of about 3000 cm^3 (point a), condensation would begin when the pressure had increased to 1 atm and the volume decreased to 1670 cm^3 (point b). The pressure would remain constant at 1 atm from b to e, while the volume decreased from 1670 cm^3 to 1 cm^3. 539 cal would have to be removed during the condensation process. To produce a further decrease in volume of 0.001 cm^3, it would be necessary to increase the pressure to about 20 atm (point f).

If the experiment illustrated in Fig. 23–1(b) is repeated, starting at higher and higher temperatures, it is found that greater and greater pressures must be exerted on the gas before condensation begins. Figure 23–2 illustrates the curves obtained. It will be seen that point b moves toward the left and point e toward the right until at the particular temperature lettered T_c in Fig. 23–2 the two points coincide. Above this temperature there is no straight horizontal portion to the curve. In other words, there is no stage in the compression process at which the substance separates into two distinct portions, one of which is a gas while the other is a liquid. The temperature T_c is called the *critical* temperature, and it is now evident why a gas must first be cooled below its critical temperature before it can be liquefied by compression.

It is customary to refer to a gas below its critical temperature as a *vapor*, although this distinction is not rigidly adhered to. Critical temperatures of some common gases are given in Table XIX.

TABLE XIX

CRITICAL TEMPERATURES AND PRESSURES

Substance	Critical temperature (°C)	Critical pressure (atm)	Critical volume (cm³/gm)
Ammonia	132	112	4.25
Argon	−122	48	1.88
Carbon dioxide	31	73.0	2.17
Helium	−268	2.26	14.4
Hydrogen	−240	12.8	32.3
Oxygen	−119	49.7	2.33
Sulphur dioxide	157	77.7	1.92
Water	374	218	3.14

The dotted line in Fig. 23–2 divides the p-V plane into three regions. At all values of p, V, and T underneath this line, the substance is partly in the liquid and partly in the vapor phase. At the right of the curve it is a vapor or gas, at the left of the curve it is a liquid.

An examination of Fig. 23–2 shows that at any given temperature below the critical temperature there is one pressure and one pressure only at which the substance can exist in either the liquid or the vapor phase, or in both phases simultaneously. This is the pressure corresponding to the horizontal portion of the isothermal curve for that particular temperature. If the pressure is any higher than that at the horizontal portion, the substance can only be a liquid. If the pressure is any lower, it can only be a vapor. Precisely at this pressure both liquid and vapor can exist together. The pressure at which a liquid and its vapor can exist in equilibrium, at any given temperature, is called the *vapor pressure* at that temperature. A vapor whose pressure and temper-

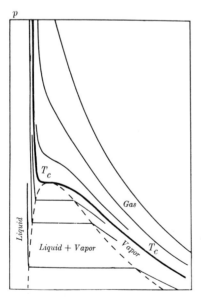

FIG. 23–2. Isotherms of a real gas.

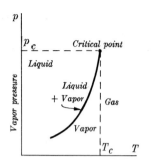

FIG. 23-3. Vapor pressure *vs.* temperature.

ature are those corresponding to the horizontal portion of any of the curves in Fig. 23-2 is called a *saturated vapor*. A saturated vapor may also be defined as one which is in equilibrium with its liquid.

From the straight portions of the curves of Fig. 23-2, we can read off a series of vapor pressures and their corresponding temperatures. If these are plotted as in Fig. 23-3, we obtain the *vapor pressure curve* of the substance. All vapor pressure curves are similar in form to that of Fig. 23-3, rising with a continually increasing slope and ending at the critical point. An abridged table of the vapor pressure of water is given in Table XX on page 437.

Reference to Table XIX will show that the critical temperatures of carbon dioxide, ammonia, and sulphur dioxide are higher than "room temperature." Hence these gases can be liquefied at room temperature without precooling, simply by increasing the pressure. Oxygen, nitrogen, or hydrogen, however, must evidently be precooled below room temperature before they can be liquefied. The Linde process for producing liquid air (or liquid oxygen or nitrogen) will be described briefly. See Figure 23-4.

Compressor *A* maintains a continuous circulation of air as shown by the arrows. At *B*, the air leaving the compressor is at high pressure and high temperature. It enters cooling coils *C*, where it is cooled by air or water, and, still at high pressure, escapes through the small orifice or nozzle *D*, performing what is known as a *throttling* process. If air were an ideal gas, no temperature change would result from a throttling process. Real gases, however, undergo marked temperature changes in such a process, and if not too hot to begin with, are cooled in passing through the orifice. The pressure in *E* and *F* is kept low by the pump, and the cooled air passes up through *E* and *F* and repeats the cycle. The cooled air in *E*, circulating around the incoming air in *D*, cools it still further, hence an even lower temperature is reached by the air escaping from *D*, until eventually the temperature falls sufficiently so that some of the air liquefies as it leaves the nozzle. The liquid air collects at *G*, where it may be drawn off.

23-2 Effect of pressure on boiling and freezing points. A pan of water exposed to the air of the room will evaporate, whatever its temperature, provided only that there is opportunity for the vapor to diffuse away, or

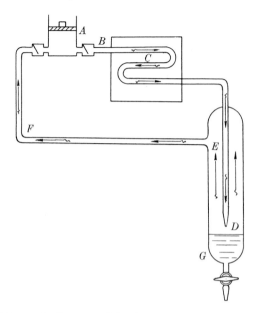

FIG. 23–4. Schematic diagram of Linde process for producing liquid air.

to be removed in some manner from above the water surface. If the temperature of the water is increased to 100°C (at normal atmospheric pressure) the nature of the evaporation process changes completely. Vapor is formed not only at the liquid surface but throughout the entire volume of the liquid, which becomes violently agitated by the bubbles of vapor which rise through it and break at the surface. What distinguishes this violent process of *boiling* from the slow evaporation which goes on at temperatures below the boiling point?

It will be recalled that every liquid has a certain *vapor pressure* which depends on the temperature of the liquid. If, keeping the temperature constant, an attempt is made to increase the pressure *above* the vapor pressure, the vapor immediately condenses. The water in a vessel open to the atmosphere is subjected to atmospheric pressure. Say that the temperature of the water is 80°C. Its vapor pressure, from Table XX, is then 355 mm of mercury or 6.87 lb/in². Hence if a small bubble of vapor should chance to form within the liquid, where it is subjected to a pressure of 760 mm or 14.7 lb/in², it would immediately collapse under the pressure and condense.

Suppose now the temperature of the liquid is increased to 100°C. At this temperature the vapor pressure is 760 mm or 14.7 lb/in². Hence bubbles of vapor can form at this temperature, and if the temperature were to increase only slightly above 100°C, the entire mass of water would change to the vapor phase *if its heat of vaporization could be supplied to it.* What actually happens is that the water does so change as fast as heat is supplied. As long as any liquid water remains, the temperature cannot rise above 100°C and all of the heat supplied is used to produce a *change of phase* rather than an *increase in temperature.*

If the external pressure is suddenly increased above 14.7 lb/in^2, boiling immediately ceases, since the pressure is higher than the vapor pressure of water at 100°C. Assuming that heat is still supplied to the water, its temperature increases until the vapor pressure equals the applied pressure, when boiling commences again.

It is evident that under an external pressure less than atmospheric, boiling will take place at a temperature below 100°C. From Table XX, if the pressure is reduced to 0.339 lb/in^2, water will boil at room temperature (20°C).

> *The boiling point of a liquid is that temperature at which the vapor pressure of the liquid is equal to the external pressure.*

Freezing points as well as boiling points are affected by external pressure. The freezing point of a substance like water, which expands on solidifying, is *lowered* by an increase in pressure. The reverse is true for substances which contract on solidifying. The change in the freezing-point temperature is much smaller than is that of the boiling point, an increase of one atmosphere lowering the freezing point of water by only about 0.007°C.

TABLE XX

VAPOR PRESSURE OF WATER (ABSOLUTE)

t (°C)	Vapor pressure		t (°F)
	mm of mercury	lb/in^2	
0	4.58	0.0886	32
5	6.51	0.126	41
10	8.94	0.173	50
15	12.67	0.245	59
20	17.5	0.339	68
40	55.1	1.07	104
60	149	2.89	140
80	355	6.87	176
100	760	14.7	212
120	1490	28.8	248
140	2710	52.4	284
160	4630	89.6	320
180	7510	145	356
200	11650	225	392
220	17390	336	428

The lowering of the freezing point of water (or the melting point of ice) can be demonstrated by passing a loop of fine wire over a block of ice and hanging a weight of a few pounds from the loop. The high pressure directly under the wire lowers the *melting point* (*not* the *temperature*) below 0°C. Hence, if the ice is at 0°, it is warmer than its melting point and therefore melts. The water thus formed is squeezed out from under the wire, the pressure on it is relieved, and it immediately refreezes. The wire meanwhile sinks further and further into the block, eventually cutting its way completely through but leaving a solid block of ice behind it. This phenomenon is known as *regelation* (refreezing).

23–3 The Clausius-Clapeyron equation. The vapor pressure curve of Fig. 23–3 can be considered to represent the pressure and temperature at which a change of phase from liquid to vapor will occur, or the pressure and temperature at which both phases can remain in equilibrium with each

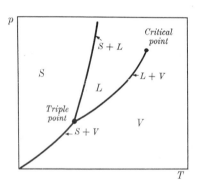

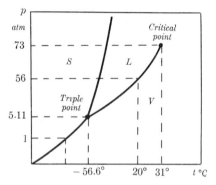

FIG. 23–5. Pressure-temperature diagram.

FIG. 23–6. Pressure-temperature diagram of CO_2. (Not to a uniform scale.)

other. There exist similar curves representing the pressure and temperature at which solid and liquid can be in equilibrium, and at which solid and vapor can be in equilibrium. These curves are shown in Fig. 23–5 and they intersect at a common point, called the *triple point*, at which all three phases can exist simultaneously.

The three curves in Fig. 23–5 divide the p-T plane into three regions. At any point (i.e., any pair of values of p and T) within one of these regions, the substance can exist in one phase only: solid, liquid, or vapor. Along each line, two phases can exist together, while only at the triple point can all three phases coexist.

For example, the triple-point temperature of CO_2 is $-56.6°C$ and the triple-point pressure is 5.11 atm. It is evident from Fig. 23–6 (not to a uniform scale) that at atmospheric pressure CO_2 can exist only as a solid or a vapor. Hence solid CO_2 (dry ice) transforms directly to CO_2 vapor when open to the atmosphere, without passing through the liquid state. This direct transition from solid to vapor is called *sublimation*. Liquid CO_2 can exist only at a pressure greater than 5.11 atm. The steel tanks in which CO_2 is commonly stored contain liquid and vapor. The pressure in these tanks is the vapor pressure of CO_2 at the temperature of the tank. If the latter is 20°C the vapor pressure is about 56 atm or 830 lb/in².

The curve separating the liquid and vapor regions is the vapor pressure or boiling-point curve, that separating the solid and liquid regions is the freezing-point curve, and that separating the solid and vapor regions is the sublimation-point curve. The equations of these curves are rather complex and will not be given. There is, however, a relatively simple relation known as the Clausius-Clapeyron equation which gives the *slope* of each curve at any point. This equation is

$$\frac{dp}{dT} = \frac{L}{T(v_2 - v_1)}, \tag{23–1}$$

where dp/dT is the slope of the curve, T the Kelvin temperature, L the heat of transformation appropriate to that curve (i.e., heat of vaporization, heat of fusion, or heat of sublimation), and $v_2 - v_1$ is the change

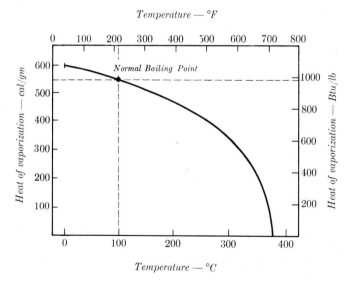

Fig. 23–7. Heat of vaporization of water as a function of temperature. The heat of vaporization becomes zero at the critical temperature of 705°F = 374°C.

in specific volume which takes place in the corresponding change of state ($v_{\text{vapor}} - v_{\text{liquid}}$, $v_{\text{liquid}} - v_{\text{solid}}$, or $v_{\text{vapor}} - v_{\text{solid}}$).

Some interesting conclusions can be drawn from an inspection of the Clausius-Clapeyron equation. At the critical point the volumes of a unit mass of vapor and liquid are equal, and $v_v - v_l = 0$. The slope of the vapor pressure curve remains finite at this temperature, and, as a consequence, the heat of vaporization at the critical point is zero. This would also be expected from the fact that the properties of the liquid and vapor are identical at the critical point. It follows that the heat of vaporization of a liquid is not a constant but decreases with increasing temperature, becoming zero at the critical point. (See Fig. 23–7.)

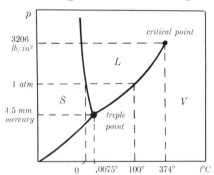

FIG. 23–8. Pressure-temperature diagram of H_2O. (Not to a uniform scale.)

While most substances increase in volume in the change of phase from solid to liquid, there are a few for which the reverse is true. Water, of course, is one of the latter. It follows that when the Clausius-Clapeyron equation is applied to the melting of ice, the term dp/dT is negative, since L and T are both positive but $v_{\text{water}} - v_{\text{ice}}$ is negative. The freezing-point curve of water therefore has a negative slope (Fig. 23–8) or, in other words, an *increase* of pressure produces a *lowering* of the freezing point of water. This effect was mentioned earlier in connection with regelation.

EXAMPLE. At a pressure of 760 mm of mercury, water boils at a temperature of 100°C. The heat of vaporization of water at this temperature is 539 cal/gm, and the volume of one gram of saturated water vapor is 1671 cm³. Compute the boiling temperature of water at a pressure of 770 mm of mercury.

Let us use cgs units. Then

$$L = 539 \text{ cal/gm} = 226 \times 10^8 \text{ ergs/gm},$$

and since v_1 (equals 1 cm³/gm) is negligible,

$$\frac{dp}{dT} = \frac{226 \times 10^8}{373 \times 1671} = 3.61 \times 10^4 \text{ dynes/cm}^2, \text{ per C}°.$$

That is, at 100°C, the vapor pressure of water increases at the rate of 3.61×10^4 dynes/cm² per centigrade degree increase in temperature. If this rate is assumed constant over a small change of temperature, we may replace dp/dT by $\Delta p/\Delta T$, and

$$\frac{\Delta p}{\Delta T} = 3.61 \times 10^4.$$

But $\Delta p = 770 - 760 = 10$ mm mercury $= 1.33 \times 10^4$ dynes/cm^2. Hence

$$\Delta T = \frac{1.33 \times 10^4}{3.61 \times 10^4} = 0.37 \ \mathrm{C}°,$$

and the new boiling temperature is

$$100 + 0.37 = 100.37°\mathrm{C}.$$

The observed boiling point is 100.38°C.

23–4 Humidity. Atmospheric air is a mixture of gases consisting of about 80% nitrogen, 18% oxygen, and small amounts of carbon dioxide, water vapor, and other gases. The mass of water vapor per unit volume is called the *absolute humidity*. The total pressure exerted by the atmosphere is the sum of the pressures exerted by its component gases. These pressures are called the *partial pressures* of the components. It is found that the partial pressure of each of the component gases of a gas mixture is very nearly the same as would be the actual pressure of that component alone if it occupied the same volume as does the mixture, a fact known as *Dalton's law*. That is, each of the gases of a gas mixture behaves independently of the others. The partial pressure of water vapor in the atmosphere is ordinarily a few millimeters of mercury.

It should be evident that the partial pressure of water vapor at any given air temperature can never exceed the vapor pressure of water at that particular temperature. Thus at 10°C, from Table **XX**, the partial pressure cannot exceed 8.94 mm, or at 15°C it cannot exceed 12.67 mm. If the concentration of water vapor, or the absolute humidity, is such that the partial pressure equals the vapor pressure, the vapor is said to be *saturated*. If the partial pressure is less than the vapor pressure, the vapor is *unsaturated*. The ratio of the partial pressure to the vapor pressure at the same temperature is called the *relative humidity*, and is usually expressed as a percentage.

$$\text{Relative humidity (\%)} = 100 \times \frac{\text{partial pressure of water vapor}}{\text{vapor pressure at same temperature}}.$$

The relative humidity is 100% if the vapor is saturated and zero if no water vapor at all is present.

EXAMPLE. The partial pressure of water vapor in the atmosphere is 10 mm and the temperature is 20°C. Find the relative humidity.

From Table **XX**, the vapor pressure at 20°C is 17.5 mm. Hence,

$$\text{relative humidity} = \frac{10}{17.5} \times 100 = 57\%.$$

Since the water vapor in the atmosphere is saturated when its partial pressure equals the vapor pressure at the air temperature, saturation can be brought about either by increasing the water vapor content or by lowering the temperature. For example, let the partial pressure of water vapor be 10 mm when the air temperature is 20°C, as in the preceding example. Saturation or 100% relative humidity could be attained either by introducing enough more water vapor (keeping the temperature constant) to increase the partial pressure to 17.5 mm, *or by lowering the temperature* to 11.4°C, at which, by interpolation from Table XX, the vapor pressure is 10 mm.

If the temperature were to be lowered *below* 11.4°C, the vapor pressure would be less than 10 mm. The partial pressure would then be higher than the vapor pressure and enough vapor would condense to reduce the partial pressure to the vapor pressure at the lower temperature. It is this process which brings about the formation of clouds, fog, and rain. The phenomenon is also of frequent occurrence at night when the earth's surface becomes cooled by radiation. The condensed moisture is called *dew*. If the vapor pressure is so low that the temperature must fall below 0°C before saturation exists, the vapor condenses into ice crystals in the form of frost.

The temperature at which the water vapor in a given sample of air becomes saturated is called the *dew point*. Measuring the temperature of the dew point is the most accurate method of determining relative humidity. The usual method is to cool a metal container having a bright, polished surface, and to observe its temperature when the surface becomes clouded with condensed moisture. Suppose the dew point is observed in this way to be 10°C, when the air temperature is 20°C. We then know that the water vapor in the air is saturated at 10°C, hence its partial pressure is 8.94 mm, equal to the vapor pressure at 10°C. The pressure necessary for saturation at 20°C is 17.5 mm. The relative humidity is therefore

$$\frac{8.94}{17.5} \times 100 = 51\%.$$

A simpler but less accurate method of determining relative humidity employs a *wet-and-dry bulb thermometer*. Two thermometers are placed side by side, the bulb of one being kept moist by a wick dipping in water. The lower the relative humidity, the more rapid will be the evaporation from the wet bulb, and the lower will be its temperature below that of the dry bulb. The relative humidity corresponding to any pair of wet-and-dry bulb temperatures is read from tables.

The *hair hygrometer* makes use of the fact that human hair absorbs or gives up moisture from the air in an amount which varies with relative humidity, and changes its length slightly with moisture content. Several strands of hair are wrapped around a small pivoted shaft to which is attached a pointer. The hair is kept taut by a light spring, and changes in its length cause the shaft to rotate and move the pointer over a scale.

23–5 The Wilson cloud chamber. The Wilson cloud chamber is an extremely useful piece of apparatus for securing information about elementary particles such as electrons and α-particles. In principle (Fig. 23–9), it consists of a cylindrical enclosure having glass walls A and a glass top B and provided with a movable piston C. The enclosed space contains

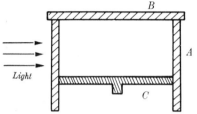

FIG. 23–9. The Wilson cloud chamber.

air, water vapor, and sufficient excess water so that the vapor is saturated. (Other liquids such as alcohol are often used in place of water.) When the piston is suddenly pulled down a short distance, the resulting adiabatic cooling lowers the temperature below the dew point. If the air is perfectly clean, the cooled vapor does not immediately condense. It is found, however, that any ions that may be present serve as very efficient nuclei upon which droplets of water will form. Hence if any ions were present just before the expansion, their presence is made evident by the appearance of a tiny droplet immediately after the expansion.

Electrons, protons, and α-particles are all capable of traveling several centimeters through air, but as they collide with or pass near the air molecules they may knock off one or more of their electrons and hence leave behind them a trail of ions. Therefore if such a particle passed through the chamber just before the expansion, a trail of droplets appears after the expansion indicating the path the particle followed. For photographing the tracks, an intense beam of light is projected transversely through the chamber and a camera mounted above it.

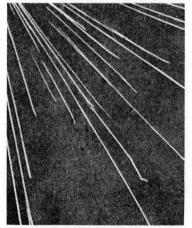

FIG. 23–10. Tracks of alpha particles in a Wilson cloud chamber.

Figure 23–10 is a photograph, made in this way, of the tracks produced by α-particles emanating from a bit of radioactive material in the chamber. (α-particles are doubly ionized helium nuclei.)

23–6 Thermodynamic surfaces. The equation of state of a substance is a relation between the three variables p, V, and T. If these quantities are laid off on three mutually perpendicular axes, the equation of state defines a *surface* in the p-V-T space. All possible states of the substance are represented by points on this surface, and all processes through which the substance may be carried are represented by lines in the surface. An isothermal process is a line in the surface at all points of which T is constant, or in other words, it is the intersection of the surface with a plane perpendicular to the temperature axis. Similarly, processes at constant pressure or volume are intersections of the surface with planes perpendicular to the pressure or volume axes. Such a surface is called a thermodynamic surface, although variables other than p, V, and T are often used.

Figure 23–11 illustrates the simplest of all p-V-T surfaces, that of an ideal gas. A few isotherms are indicated by the heavy lines. When this surface is viewed in a direction perpendicular to the p-V plane, the isotherms appear as in Fig. 22–1.

Figure 23–12 is a photograph of the p-V-T surface of a real substance, not to a uniform scale. When viewed perpendicular to the p-V plane, the isotherms appear as in Fig. 23–2. It will be seen that the region under the tongue-shaped curve in Fig. 23–2 is in reality a ruled surface, sloping back from the plane of that figure. When the model is viewed perpendicular to the p-T plane this ruled surface appears as a line, which is one of the boundary lines in Fig. 23–5 or 23–8. The triple "point" is actually not a point but a line, since, although the pressure and temperature are fixed at that point, the volume is not. That is, the volume depends on the relative masses of the substance which are in each phase. If, for example, the substance is mostly in the vapor phase, with only small amounts of solid and liquid present, the volume will be large.

The surface is not constructed to scale because of the large volume changes involved in changes of phase.

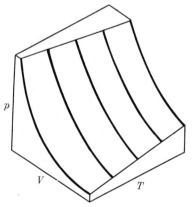

Fig. 23–11. p–V–T surface of an ideal gas.

23–7 The Van der Waals equation of state. Many attempts have been made to find an equation which represents a real gas as the relation $pv = RT$ does an ideal gas. We shall consider but one, first proposed by J. D. Van der Waals in 1873. Van der Waals reasoned that because of the small but finite volume occupied by the atoms of a gas, the volume available for them to move about in is less than the actual volume of the container—hence a term should be subtracted from v. Also, because of the forces of attraction between the molecules, the pressure term should be increased above the measured pressure. The correction to the pressure term should be greater at smaller volumes when the molecules are closer together, and, it turns out, should be inversely proportional to the square of the specific volume. Hence Van der Waals' equation is

$$\left(p + \frac{a}{v^2}\right)(v - b) = RT, \tag{23–2}$$

where a and b are constants which are to be suitably chosen for each particular gas. At very large specific volumes, a/v^2 becomes small and b is negligible compared to v. Hence at large specific volumes (low pressures) the equation goes over into the ideal gas equation, as it should. The Van der Waals equation does give better agreement with the observed properties of real gases than does the ideal gas equation, but it must be considered as only a "second approximation" to the true equation of state.

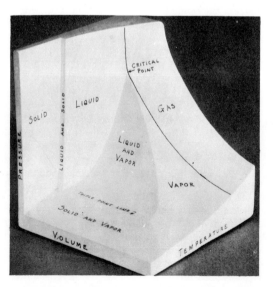

FIG. 23–12. p–V–T surface of a real substance. (Not to a uniform scale.)

PROBLEMS

23–1. Make two plots for a real gas, one showing pressure as a function of volume, and the other showing pressure as a function of temperature. Show on each graph the region in which the substance exists as (a) a gas or vapor, (b) a liquid, (c) a solid. Show also the triple point and the critical point.

23–2. A cylinder provided with a movable piston contains water vapor at 100°C and a pressure of 355 mm of mercury. The pressure is kept constant while heat is continuously removed from the cylinder and contents until the temperature falls to −10°C. Construct a graph showing the approximate relation between temperature and volume. Indicate on the graph the numerical values of the temperature at any discontinuities.

23–3. Use the data in Table XX to construct two graphs of the vapor pressure of water as a function of centigrade temperature. Let the first graph cover the temperature range from 0°C to 20°C, and the second the range from 0°C to 200°C. Let 1 inch horizontally equal 5 C° in the first graph and 50 C° in the second. Let 1 inch vertically equal 5 mm of mercury in the first, and 50 lb/in² in the second. Read from the graphs the following: (a) the boiling point of water under a pressure of 10 mm of mercury, (b) the pressure, in lb/in², at which the boiling point of water is 175°C.

23–4. From the data in Table XX, construct a graph of the vapor pressure of water in the temperature range from 60°C to 140°C, or use the second graph in problem 23–3 if it is available. Draw a tangent line to the graph at a temperature of 100°C and determine its slope. Compute the volume of 1 gm of water vapor at 100°C and 1 atm, assuming the vapor to behave like an ideal gas. Then with the help of the Clausius-Clapeyron equation compute the heat of vaporization of water at 100°C and compare with the measured value of 539 cal/gm. Be careful of units.

23–5. What pressure is necessary to lower the freezing point of water to −1°C? The density of ice is 0.92 gm/cm³. (Approximate finite changes by differentials.)

23–6. The triple point pressure of water is 4.5 mm of mercury. What is the triple point temperature? The melting point of ice at atmospheric pressure is 0°C and the density of ice is 0.92 gm/cm³.

23–7. At the top of a low mountain, water is observed to boil at a temperature of 97.0°C. What is atmospheric pressure at this point (in cm of mercury)? Use the data given in the example in Section 23–3, and approximate small changes by differentials.

23–8. Given the following data for a certain substance:

Abs. press (lb/in²)	Boiling point (°F)	Specific volume (ft³/lb)	
		Sat. liquid	Sat. vapor
99	317.27	.01773	4.474
100	318.00	.01774	4.432
101	318.71	.01775	4.391

Compute from these data the heat of vaporization of the substance under a pressure (absolute) of 100 lb/in². The measured value is 883 Btu/lb.

23–9. Show that if the specific volume of the liquid is neglected in comparison with that of the vapor, and if the vapor is assumed to behave like an ideal gas, the Clausius-Clapeyron equation may be written

$$\frac{dp}{dT} = \frac{pML}{RT^2},$$

where M is the molecular weight.

23–10. A barometer is made of a tube 90 cm long and of cross section 1.5 cm². Mercury stands in this tube to a height of 75 cm. The room temperature is 27°C. A small amount of nitrogen is introduced into the evacuated space above the mercury and the column drops to a height of 70 cm. How many grams of nitrogen were introduced?

23–11. The vapor pressure of sulfur dioxide (SO_2) at 27°C is 4.08 atm and the density of liquid SO_2 at this temperature and pressure is 0.733 gm/cm^3. A cylinder, provided with a tightly fitting piston, contains 1 mole of SO_2 at a pressure of 1 atm and a temperature of 27°C. The piston is slowly forced into the cylinder while the temperature is kept constant. Assume that the vapor behaves like an ideal gas. (a) What is the initial volume · of the system? (b) To what value must the volume be reduced before condensation begins? (c) When the volume has been reduced to 1000 cm^3, how many grams of SO_2 have condensed?

23–12. The small cylinders of CO_2 used for inflating Mae Wests have a volume of 10 cm^3 and contain 7.5 gm of CO_2. The vapor pressure of CO_2 at 65°F is 795.1 lb/in^2, the specific volume of saturated CO_2 vapor is 0.00558 m^3/kgm, and the density of liquid CO_2 is 49.14 lb/ft^3. (a) At a temperature of 65°F, what fraction of the volume of a cylinder is occupied by vapor and what fraction by liquid? (b) If the temperature is increased, will the volume of the liquid phase increase or decrease? (c)

When the cylinder is punctured, what volume of CO_2 at 65°F and atmospheric pressure is liberated?

23–13. (a) What is the relative humidity on a day when the temperature is 68°F and the dew point is 41°F? (b) What is the partial pressure of water vapor in the atmosphere? (c) What is the absolute humidity, in gm/m^3?

23–14. (a) What is the dew point temperature on a day when the air temperature is 20°C and the relative humidity is 60%? (b) What is the absolute humidity, in gm/m^3?

23–15. The temperature in a room is 40°C. A can is gradually cooled by adding cold water. At 10°C the surface of the can clouds over. What is the relative humidity in the room?

23–16. A pan of water is placed in a sealed room of volume 60 m^3 and at a temperature of 27°C. (a) What is the absolute humidity in gm/m^3 after equilibrium has been reached? (b) If the temperature of the room is then increased 1 C° how many more grams of water will evaporate?

23–17. An air-conditioning system is required to increase the relative humidity of 10 ft^3 of air per second from 30% to 65%. The air temperature is 68°F. How many pounds of water are needed per hour?

23–18. Use Table XX and Fig. 23–7 to find the heat of vaporization of water boiling under an absolute pressure of 225 lb/in^2.

CHAPTER 24

THE SECOND LAW OF THERMODYNAMICS

24–1 The second law of thermodynamics. The dominating feature of an industrial society is its ability to utilize, whether for wise or unwise ends, sources of energy other than the muscles of men or animals. Except for water power, where mechanical energy is directly available, most energy supplies are in the form of fuels such as coal or oil, where the energy is stored as chemical energy. The process of combustion releases the chemical energy and converts it to thermal energy. In this form the energy may be utilized for heating habitations, for cooking, or for maintaining a furnace at high temperature in order to carry out other chemical or physical processes. But to operate a machine, or propel a vehicle or a projectile, the thermal energy must be converted to mechanical energy, and one of the problems of the mechanical engineer is to carry out this conversion with the maximum possible efficiency.

There is only one type of process in which chemical energy can be converted directly to mechanical energy, and that is when the chemical substances can be combined in an electrolytic cell. All other methods involve the intermediate step of transforming chemical energy into thermal energy. The changes may be represented schematically by

$$\text{Chemical energy} \longrightarrow \text{Thermal energy} \longrightarrow \text{Mechanical energy}$$

The process represented by

$$\text{Chemical energy} \longrightarrow \text{Thermal energy}$$

presents few difficulties. The most common example is, of course, the combustion of coal, oil, or gas. The problem then reduces to

$$\text{Thermal energy} \longrightarrow \text{Mechanical energy}$$

It is evident in the first place that the transformation of thermal into

448

mechanical energy always requires the services of some sort of an *engine*, such as a steam engine, gasoline engine, or Diesel engine.

At first sight the problem does not seem difficult, since we know that 1 Btu = 778 ft·lb, and it appears that every Btu of thermal energy should provide us with 778 ft·lb of mechanical energy. A pound of coal, for instance, develops about 13,000 Btu when burned, and might be expected to provide 13,000 × 778 ft·lb of mechanical work. Actual steam engines, however, furnish only from about 5% to about 30% of this value. What becomes of the remaining 70% to 95%?

Stack and friction losses account for only a small part, by far the largest part appearing as heat rejected in the exhaust. No one has ever constructed a heat engine which does not throw away in its exhaust a relatively large fraction of the heat supplied to it, and it is safe to say that no one ever will. The impossibility of constructing an engine which with no other outstanding changes will convert a given amount of heat *completely* into mechanical work is a fundamental law of Nature, known as the second law of thermodynamics. The first law, it will be recalled, is a statement of the principle of the conservation of energy, and merely imposes the restriction that one can obtain *no more* than 778 ft·lb of mechanical work from every Btu of thermal energy. It does not in itself restrict the fraction of a given amount of thermal energy which an engine can convert into mechanical energy. The second law goes beyond the first, and states that 100% conversion is not possible by any form of engine. Of course, for that fraction of the thermal energy supplied to it which an engine *does* convert to mechanical form, the equivalence expressed by the first law must hold true.

A young French engineer, Sadi Carnot, was the first to approach the problem of the efficiency of a heat engine from a truly fundamental standpoint. Improvements in steam engines up to the time of Carnot's work in 1824 had either been along the lines of better mechanical design, or, if more basic improvements had been made, they had come about by chance or inspiration and had not been guided by any knowledge of basic principles. Carnot's contribution was a "theoretical" one, but it had more influence on the development of our industrial society in the 19th century than the work of any of the "practical" men who had preceded him in this field.

Briefly, what Carnot did was to disregard the details of operation of a heat engine and focus attention on its truly significant features. These are, first, the engine is supplied with energy, in the form of heat, at a relatively high temperature. Second, the engine performs mechanical work. Third, the engine rejects heat at a lower temperature. In Carnot's time the caloric theory of heat as an indestructible fluid was still believed to be true, and Carnot pictured the flow of heat through an engine, from a higher to a lower temperature, as analogous to the flow of water through a water wheel or turbine from a higher to a lower elevation. In any time interval, equal amounts of water enter the turbine and are discharged from it, but in the process some mechanical energy is abstracted from the water. Carnot believed that a similar process took place in a heat engine—some mechanical energy was abstracted from the heat but the amount of heat rejected by the engine in any time interval was equal to that delivered to the engine. We know now that this idea is incorrect, and that the heat rejected by the engine is less than the heat supplied to it by the amount that has been converted to mechanical work. In spite of his erroneous concept of the nature of heat, Carnot did in fact obtain the correct expression for the maximum efficiency of any heat engine operating between two given temperatures.

Since it is only heat and work that are of primary concern in a heat engine, we consider for simplicity an engine working in *closed cycles*. That is, the material that expands against a piston is periodically brought back to its initial condition so that in any one cycle the change in internal energy of this material, called the *working substance*, is zero. The condensing type of steam engine actually does operate in this way; the exhaust steam is condensed and forced back into the boiler so that the working substance (in this case, water) is used over and over again. The working substance then merely serves to transfer heat from one body to another and, by virtue of its changes in volume, to convert some of the heat to mechanical work.

The energy transformations in a heat engine are conveniently represented schematically by the *flow diagram* of Fig. 24–1. The engine itself is represented by the circle. The heat Q_2 supplied to the engine is proportional to the cross section of the incoming "pipeline" at the top of the diagram. The cross section of the outgoing pipeline at the bottom is proportional to that portion of the heat, Q_1, which is rejected as heat in the exhaust. The branch line to the right represents that portion of the heat supplied which the engine converts to mechanical work, W. Since the working substance is periodically returned to its initial state and the change in its internal energy in any number of complete cycles is zero, it follows from the first law of thermodynamics that

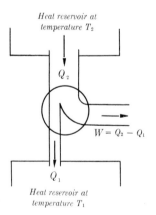

Fig. 24–1. Schematic flow diagram of a heat engine.

$$Q_2 = Q_1 + W,$$

$$W = Q_2 - Q_1.$$

That is, the mechanical work done equals the difference between the heat supplied and the heat rejected. (For convenience, W, Q_2, and Q_1 are all considered positive.)

The *efficiency* of the engine is the ratio of output to input. The output is the mechanical work W. The exhaust heat is not considered a part of the output. The input is the heat Q_2. Hence

$$\text{Efficiency} = \frac{\text{work output}}{\text{heat input}}$$

$$= \frac{W}{Q_2}$$

$$= \frac{Q_2 - Q_1}{Q_2}.$$

In terms of the flow diagram, the most efficient engine is the one for which the branch pipeline representing the work obtained is as large as possible, and the exhaust pipeline representing the heat rejected is as small as possible, for a given incoming pipeline or quantity of heat supplied.

We shall now consider, without going into the mechanical details of their construction, the internal-combustion engine, the Diesel engine, and the steam engine.

24–2 The internal-combustion engine. The common internal-combustion engine is of the four-cycle type, so called because four processes take place in each cycle. Starting with the piston at the top of its stroke, an explosive mixture of air and gasoline vapor is drawn into the cylinder on the downstroke, the inlet valve being open and the exhaust valve closed. This is the *intake* stroke. At the end of this stroke the inlet valve closes and the piston rises, performing an approximately adiabatic compression of the air-gasoline mixture. This is the *compression* stroke. At or near the top of this stroke a spark ignites the mixture of air and gasoline vapor, combustion taking place very rapidly. The pressure and temperature increase at nearly constant volume.

The piston is now forced down, the burned gases expanding approximately adiabatically. This is the *power stroke* or *working stroke*. At the end of the power stroke the exhaust valve opens. The pressure in the cylinder drops rapidly to atmospheric and the rising piston on the *exhaust*

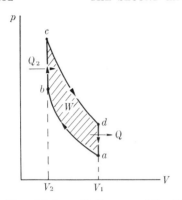

FIG. 24–2. p–V diagram of the Otto cycle.

stroke forces out most of the remaining gas. The exhaust valve now closes, the inlet valve opens, and the cycle is repeated.

For purposes of computation, the internal-combustion cycle is approximated by the *air standard* or *Otto* cycle illustrated in Fig. 24–2.

Starting at point a, air at atmospheric pressure is compressed adiabatically in a cylinder to point b, heated at constant volume to point c, allowed to expand adiabatically to point d, and cooled at constant volume to point a, after which the cycle is repeated. Line ab corresponds to the compression stroke, bc to the explosion, cd to the working stroke, and da to the exhaust of an internal-combustion engine. V_1 and V_2 in Fig. 24–2 are respectively the maximum and minimum volumes of the air in the cylinder. The ratio V_1/V_2 is called the *compression ratio*, and is about 7 for an internal-combustion engine.

The work output in Fig. 24–2 is the shaded area enclosed by the figure $abcd$. The heat *input* is the heat supplied at constant volume along the line bc. The exhaust heat is removed along da. No heat is supplied or removed in the adiabatic processes ab and cd.

The heat input and the work output can be computed in terms of the compression ratio, assuming air to behave like an ideal gas. The result is

$$ \text{Eff } (\%) = 100 \left(1 - \frac{1}{(V_1/V_2)^{\gamma-1}} \right). $$

For a compression ratio of 7 and a value of $\gamma = 1.4$, the efficiency is about 54%. It will be seen that the higher the compression ratio, the higher the efficiency. For an actual engine this ratio cannot be made much greater than 7 or pre-ignition and knocking will result.

Friction effects, turbulence, loss of heat to cylinder walls, etc., have been neglected. All these effects reduce the efficiency of an actual engine below the figure given above.

24–3 The Diesel engine. In the Diesel cycle, air is drawn into the cylinder on the intake stroke and compressed adiabatically on the compression stroke to a sufficiently high temperature so that fuel oil injected at the end of this stroke burns in the cylinder without requiring ignition by a spark. The combustion is not as rapid as in the gasoline engine, and the first part of the power stroke proceeds at essentially constant pressure. The remainder of the power stroke is an adiabatic expansion. This is followed by an exhaust stroke which completes the cycle.

The idealized air-Diesel cycle is shown in Fig. 24–3. Starting at point a, air is compressed adiabatically to point b, heated at constant pressure to point c, expanded adiabatically to point d, and cooled at constant volume to point a.

Since there is no fuel in the cylinder of a Diesel engine on the compression stroke, pre-ignition cannot occur and the compression ratio V_1/V_2 may be much higher than that of an internal-combustion engine. A value of 15 is typical. The *expansion* ratio V_1/V_3 may be about 5. Using these values, and taking $\gamma = 1.4$, the efficiency of the air-Diesel cycle is about 56%. Hence somewhat higher efficiencies are possible than for the Otto cycle. Again, the actual efficiency of a real Diesel must be smaller than the value given above.

24–4 The steam engine. The condensing type of steam engine performs the following sequence of operations. Water is converted to steam in the boiler, and the steam thus formed is superheated above the boiler temperature. Superheated steam is admitted to the cylinder, where it expands against a piston, connection being maintained to the boiler for the first part of the working stroke, which thus takes place at constant pressure. The inlet valve is then closed and the steam expands adiabatically for the rest of the working stroke. The adiabatic cooling causes some of the steam to condense. The mixture of water droplets and steam (known as "wet" steam) is forced out of the cylinder on the return stroke and into the condenser, where the remaining steam is condensed into water. This water is forced into the boiler by the feed pump, and the cycle is repeated.

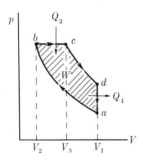

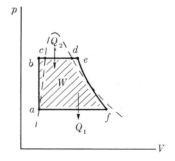

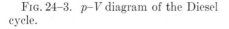

FIG. 24–3. $p–V$ diagram of the Diesel cycle.

FIG. 24–4. The Rankine cycle.

An idealized cycle (called the Rankine cycle) which approximates the actual steam cycle is shown in Fig. 24–4. Starting with liquid water at low pressure and temperature (point a), the water is compressed adiabatically to point b at boiler pressure. It is then heated at constant pressure to its boiling point (line bc), converted to steam (line cd), superheated (line de), expanded adiabatically (line ef), and cooled and condensed (along fa) to its initial condition.

The efficiency of such a cycle may be computed in the same way as was done in the previous examples, by finding the quantities of heat taken in and rejected along the lines be and fa. Assuming a boiler temperature of 417°F (corresponding to a pressure of 300 lb/in^2), a superheat of 63°F above this temperature (480°F), and a condenser temperature of 102°F, the efficiency of a Rankine cycle is about 32%. Efficiencies of actual steam engines are, of course, considerably lower.

24–5 The Carnot engine. Although their efficiencies differ from one another, none of the heat engines which have been described has an efficiency of 100%. The question still remains open as to what is the maximum attainable efficiency, given a supply of heat at one temperature and a reservoir at a lower temperature for cooling the exhaust. An idealized engine which can be shown to have the maximum efficiency under these conditions was invented by Carnot and is called a *Carnot engine*. The *Carnot cycle*, shown in Fig. 24–5, differs from the Otto and Diesel cycles in that it is bounded by two *isothermals* and two adiabatics. Thus all the heat input is supplied at a *single* high temperature and all the heat output is rejected at a *single* lower temperature. (Compare with Figs. 24–2 and 24–3, in which the temperature is different at all points of the lines bc and da.) If an ideal gas is carried through a Carnot cycle, it is not difficult to show that the efficiency is given by

$$\text{Eff } (\%) = 100 \,\frac{T_2 - T_1}{T_2}. \quad (24\text{–}1)$$

For example, a Carnot engine operated between the temperatures of 480°F (522°K) and 102°F (312°K) would have an efficiency of 41%, as compared with an efficiency of 32% for a Rankine cycle operated between the same two temperatures.

We shall show in Section 24–8 that no engine operating between these temperatures could be more

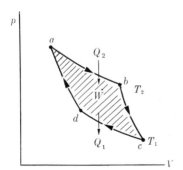

FIG. 24–5. The Carnot cycle.

efficient than a Carnot engine, and that all Carnot engines, whether using an ideal gas or not, have the same efficiency when operating between the same two temperatures.

Equation (24–1) points the way to the conditions which a real engine, such as a steam engine, must fulfill to approach as closely as possible the maximum attainable efficiency. These conditions are that the intake temperature T_2 must be made as high as possible and the exhaust temperature T_1 as low as possible.

The exhaust temperature cannot be lower than the lowest temperature available for cooling the exhaust. This is usually the temperature of the air, or perhaps of river water if this is available at the plant. The only recourse then is to raise the boiler temperature, T_2. Since, as shown in Fig. 23–3, the vapor pressure of all liquids increases rapidly with increasing temperature, a limit is set by the mechanical strength of the boiler. Another possibility is to use, instead of water, some liquid with a lower vapor pressure. Successful experiments in this direction have been made with mercury vapor replacing steam. At a boiler temperature of 200°C, at which the pressure in a steam boiler would be 225 lb/in^2, the pressure in a mercury boiler is only 0.35 lb/in^2.

The efficiency of a Carnot engine using an ideal gas is computed as follows. The work done in the four steps of the cycle is

$$W_{ab} = nRT_2 \ln \frac{V_b}{V_a},$$

$$W_{bc} = nC_v(T_2 - T_1),$$

$$W_{cd} = nRT_1 \ln \frac{V_d}{V_c},$$

$$W_{da} = nC_v(T_1 - T_2),$$

and the work done in the whole cycle is the sum of these terms. But this is the output of the engine. Hence

$$\text{Output} = W = nR\left(T_2 \ln \frac{V_b}{V_a} + T_1 \ln \frac{V_d}{V_c}\right)$$

$$= nR\left(T_2 \ln \frac{V_b}{V_a} - T_1 \ln \frac{V_c}{V_d}\right).$$

The input is the heat supplied along the path ab, or

$$\text{Input} = Q_2 = nRT_2 \ln \frac{V_b}{V_a}.$$

Now points a and d lie on the same adiabatic, as do points b and c. Hence from Eq. (22–26),

$$T_2 V_b^{\gamma-1} = T_1 V_c^{\gamma-1},$$

$$T_2 V_a^{\gamma-1} = T_1 V_d^{\gamma-1},$$

and

$$\frac{V_b}{V_a} = \frac{V_c}{V_d}.$$

The efficiency of the cycle is the ratio of the output to the input. Dividing and canceling, we find

$$\text{Efficiency} = \frac{T_2 - T_1}{T_2}.$$

24–6 The refrigerator. A refrigerator may be considered to be a heat engine operated in reverse. That is, a heat engine takes in heat from a *high* temperature source, converts a part of the heat into mechanical work output, and rejects the differences as heat in the exhaust at a *lower* temperature. A refrigerator takes in heat at a *low* temperature, the compressor supplies mechanical work *input*, and the sum is rejected as heat at a *higher* temperature.

The flow diagram of a refrigerator is given in Fig. 24–6. In terms of the processes in a household mechanical refrigerator, Q_1 represents the heat removed from the refrigerator by the cooling coils within it, W the work done by the motor, and Q_2 the heat delivered to the external cooling coils and removed by circulating air or water. It follows from the first law that

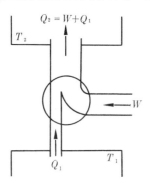

FIG. 24–6. Schematic flow diagram of a refrigerator.

$$Q_2 = Q_1 + W.$$

That is, the circulating air or water must absorb both the heat "pumped" out of the refrigerator and the heat equivalent of the work done by the motor.

From an economic point of view, the best refrigeration cycle is one that removes the greatest amount of heat Q_1 from the refrigerator for the least expenditure of mechanical work W. We therefore define the *coefficient of performance* (rather than the efficiency) of a refrigerator as the ratio Q_1/W, and since $W = Q_2 - Q_1$,

$$\text{Coefficient of performance} = \frac{Q_1}{Q_2 - Q_1}.$$

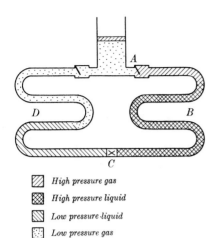

High pressure gas

High pressure liquid

Low pressure·liquid

Low pressure gas

FIG. 24–7. Principle of the mechanical refrigeration cycle.

The principles of the common refrigeration cycle are illustrated schematically in Fig. 24–7. Compressor A delivers gas (SO_2, NH_3, etc.) at high temperature and pressure to coils B. Heat is removed from the gas in B by water or air cooling, resulting in condensation of the gas to a liquid, still under high pressure. The liquid passes through the throttling valve or expansion valve C, emerging as a liquid at a lower pressure. In coils D, heat is supplied to the liquid, which evaporates, becomes a gas at low pressure, and enters compressor A to repeat the cycle. In a domestic refrigerator, coils D are placed in the ice compartment, where they cool the refrigerator directly. In a large refrigerating plant, these coils are usually immersed in a brine tank, cooling the brine, which is then pumped to the refrigerating rooms.

A simplified diagram of the so-called *gas refrigerator* is given in Fig. 24–8. In the generator a solution of ammonia in water is heated by a small gas flame. Ammonia is driven out of solution and ammonia vapor rises in the liquid lift tube, carrying with it some of the water in the same way that water is raised in the central tube of a coffee percolator. This water collects in the separator, from which point it flows back through the absorber, while the ammonia vapor rises to the condenser. Here the ammonia vapor is liquefied, its heat of condensation being removed by air circulating around the cooling vanes. The liquid ammonia then flows into the evaporator, located in the cooling unit of the refrigerator, where it evaporates and in so doing absorbs heat from its surroundings. The ammonia vapor continues on to the absorber, where it dissolves in the water returning from the separator. The ammonia-water solution then flows to the generator, completing the cycle.

The absorber and evaporator also contain hydrogen gas which is maintained in circulation by a convection process, brought about by the fact that the mixture of ammonia and hydrogen in the tube at the extreme left is denser than the pure hydrogen in the tube leading from the top of the absorber. This current of hydrogen, entering at the top of the evaporator, sweeps the ammonia vapor out of the evaporator and aids in rapid evaporation. Since ammonia is much more readily soluble in water than

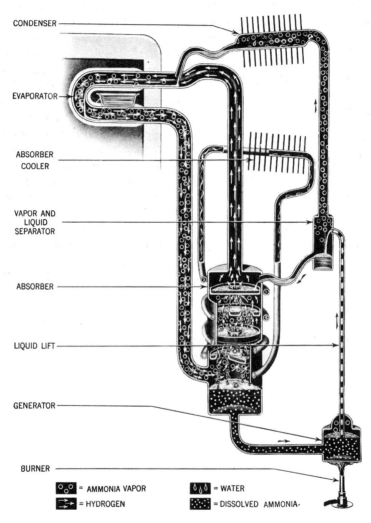

CONDENSER

EVAPORATOR

ABSORBER
COOLER

VAPOR AND
LIQUID
SEPARATOR

ABSORBER

LIQUID LIFT

GENERATOR

BURNER

○°○ = AMMONIA VAPOR   ◊◊◊ = WATER

➡ = HYDROGEN   ░░ = DISSOLVED AMMONIA.

FIG. 24–8. Simplified diagram of the gas refrigerator. (Courtesy of Servel-Electrolux)

is hydrogen, most of the ammonia is dissolved in the water trickling down through the absorber while the hydrogen passes upward through the absorber.

It is necessary that heat be removed from the absorber as well as from the condenser because heat is liberated when ammonia vapor dissolves in water.   This is accomplished by the auxiliary circuit made up of the cooling coils around the absorber, and the absorber cooler.

24–7 Entropy. There is no concept in the whole field of physics which is more difficult to understand than is the concept of entropy, nor is there one which is more fundamental.

The first law of thermodynamics is the law of energy, the second law of thermodynamics is the law of entropy, and every process that takes place in Nature, whether it be mechanical, electrical, chemical, or biological, must proceed in conformity with these two laws.

It has been said regarding some of the equations of thermodynamics, "Experience indicates that it is much less difficult to use [certain] formulae than to understand them." The same may be said of the entropy concept; it is much less difficult to use it than to understand it. A book of this nature is not the place for a thorough exposition of entropy and the second law of thermodynamics. We shall content ourselves with defining entropy, computing its changes in a few instances, and stating some of its properties.

A re-reading of Section 22–6 will recall to mind that when a system is carried from one state to another it is found by experiment that the difference between the heat added and the work done by the system, $(Q - W)$, has the same value for all paths. The fact that this difference does have the same value makes it possible to introduce the concept of internal energy, the change in internal energy being defined and measured by the quantity $Q - W$.

Entropy, or rather a change in entropy, may be defined in a similar way. Consider two states of a system and a number of quasi-static paths connecting them. (The restriction to quasi-static paths need not be made with regard to internal energy changes.)   While the heat added to the system is different along different paths, it is found by experiment that if the heat added at each point of the path is divided by the absolute temperature of the system at the point, and the resulting ratios summed for the entire path, this sum has the same value for all (quasi-static) paths between the same end points. In mathematical symbols,

$$\int_1^2 \frac{dQ}{T} = \begin{array}{l}\text{constant for all quasi-static} \\ \text{paths between states 1 and 2.}\end{array}$$

It is therefore *possible* (whether it is of any use or not one can only tell later) to introduce a function whose difference between two states, 1 and 2, is defined by the integral above. This function may be assigned any arbitrary value in some standard reference state, and its value in any other state will be a definite quantity. The function is called the entropy of the system, and we will denote it by S. We then have

$$S_2 - S_1 = \int_1^2 \frac{dQ}{T} \text{ (along any quasi-static path)}. \qquad (24\text{–}2)$$

If the change is infinitesimal,

$$dS = \frac{dQ}{T}.$$

From its definition, the units of entropy are calories per degree Kelvin, Btu/°F abs, or some similar unit.

Equation (24–2) may be considered the second law of thermodynamics, just as Eq. (22–15) may be considered the first law.

EXAMPLES. (1) 1 kgm of ice at 0°C is melted and converted to water at 0°C. Compute its change in entropy.

Since the temperature remains constant at 273°K, T may be taken outside the integral sign. Then

$$S_2 - S_1 = \frac{1}{T} \int dQ = \frac{Q}{T}.$$

But Q is simply the total heat which must be supplied to melt the ice, or 80,000 calories. Hence

$$S_2 - S_1 = \frac{80,000}{273} = 293 \text{ cal/°K},$$

and the increase in entropy of the system is 293 cal/°K. In any *isothermal* quasi-static process, the entropy change equals the heat added divided by the absolute temperature.

(2) 1 kgm of water at 0°C is heated to 100°C. Compute its change in entropy.

The temperature is not constant and dQ and T must be expressed in terms of a single variable in order to carry out the integration. This may readily be done, since

$$dQ = mc \, dT.$$

Hence

$$S_2 - S_1 = \int_{273}^{373} mc \frac{dT}{T} = mc \ln \frac{373}{273}$$

$$= 312 \text{ cal/°K}.$$

(3) A gas is allowed to expand adiabatically and quasi-statically. What is its change in entropy?

In an adiabatic process no heat is allowed to enter or leave the system. Hence $Q = 0$ and there is no change in entropy. It follows that every quasi-static adiabatic process is one of constant entropy, and may be described as *isentropic*.

(4) A real (not an ideal) gas is carried through a Carnot cycle. What is the efficiency of the cycle?

When a system is carried around a closed cycle, its entropy change is zero, since it returns to its initial state. Hence for any closed cycle $S_2 - S_1 = 0$, and therefore for the Carnot cycle $\int (dQ/T) = 0$. The Carnot cycle is bounded by two isotherms and two adiabatics. Let T_1 and T_2 be the temperature of the isotherms, Q_2 the heat added at the higher temperature T_2, and Q_1 the heat removed at the lower temperature T_1. From Example 3 the entropy change along the adiabatics is zero, and from Example 1 the entropy change along the isotherms is

$$\frac{Q_2}{T_2} - \frac{Q_1}{T_1}.$$

Since the entropy change in the whole cycle is zero,

$$\frac{Q_2}{T_2} - \frac{Q_1}{T_1} = 0,$$

or

$$\frac{Q_2 - Q_1}{Q_2} = \frac{T_2 - T_1}{T_2}.$$

But $(Q_2 - Q_1)/Q_2$ is the efficiency of the cycle, and the efficiency is the same as that of the ideal gas cycle derived on page 456. In fact, since no use whatever was made of any specific properties of a real gas, the result above would hold true no matter what substance (gas, liquid, or solid) were carried through the cycle. Hence by a very simple argument we have shown that *all* Carnot cycles have the same efficiency. The great power of the entropy concept begins to be evident.

24–8 The principle of the increase of entropy. One of the features which distinguishes entropy from such concepts as energy, momentum, and angular momentum is that *there is no principle of conservation of entropy*. In fact, the reverse is true. Entropy *can* be created at will and there is an increase in entropy in every natural process if all systems taking part in the process are considered. We shall illustrate with one example.

Consider the process of mixing a liter of water at 100°C with a liter of water at 0°C. Let us arbitrarily call the entropy of water zero when it is in the liquid state at 0°C. This is the reference state adopted in engineering work. Then from Example 2 in Section 24–7, the entropy of 1 liter ($= 1$ kgm) of water at 100°C is 312 cal/°K and the entropy of 1 liter at 0°C is zero. The entropy of the system, before mixing, is therefore 312 cal/°K.

After the hot and cold water have been mixed, we have 2 kgm of water at a temperature of 50°C or 323°K. From the results of Example 2, the entropy of the system is

$$mc \ln \frac{323}{273} = 2000 \times \ln \frac{373}{273} = 336 \text{ cal/°K.}$$

There has therefore been an increase in entropy of

$$336 - 312 = 24 \text{ cal/°K.}$$

Physical mixing of the hot and cold water is, of course, not essential in bringing about the final equilibrium state. We might simply have let heat flow by conduction, or be transferred by radiation, from the hot to the cold water. The same increase in entropy would have resulted.

This simple example of the mixing of substances at different temperatures, or the flow of heat from a higher to a lower temperature, is illustrative of all natural (i.e., nonquasi-static) processes. When all of the entropy changes in the process are summed up, the increases in entropy are always greater than the decreases. In the special case of a quasi-static process the increases and decreases are equal. Hence we can formulate the general principle, which is considered a part of the second law of thermodynamics, that in every process the entropy either increases or remains constant. In other words, *no process is possible in which the entropy decreases.*

This aspect of the second law can be used to show that no engine operating between two given temperatures can have an efficiency greater than that of a Carnot engine operated between the same two temperatures.

Let T_2 represent the higher and T_1 the lower temperature. The Carnot engine takes in heat Q_2 from the high-temperature reservoir, rejects heat Q_1 to the low-temperature reservoir, and converts the difference into mechanical work $W = Q_2 - Q_1$.

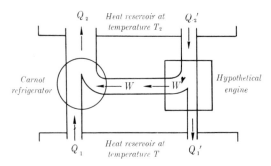

FIGURE 24–9

Now postulate a second engine operating between the same two temperatures but having a higher efficiency. Let Q_2' and Q_1' be the quantities of heat taken in and rejected by this engine, and let $W' = Q_2' - Q_1'$ be the mechanical work output. Without loss of generality, we can adjust the two cycles so that both this engine and the Carnot engine perform equal amounts of mechanical work per cycle. That is,

$$W = W' \quad \text{or} \quad Q_2 - Q_1 = Q_2' - Q_1'. \tag{24–3}$$

The second engine can then be connected to the Carnot engine *operated in reverse as a refrigerator*. The work output of the second engine just suffices to run the Carnot refrigerator and the complete system runs itself. The arrangement is illustrated schematically in Fig. 24–9, where the Carnot refrigerator (see Fig. 24–6) is represented by the circle and the second engine by the rectangle.

In one cycle of operations the second engine takes in heat Q_2' from the heat reservoir at temperature T_2 and gives up heat Q_1' to the reservoir at temperature T_1. The Carnot refrigerator takes in heat Q_1 from the reservoir at temperature T_1 and delivers heat Q_2 to the reservoir at temperature T_2. The net heat delivered to the latter reservoir is $Q_2 - Q_2'$ and the net heat removed from the former is $Q_1 - Q_1'$. We now show that these differences are equal and are not zero. That they are equal follows at once from Eq. (24–3),

or

$$Q_2 - Q_1 = Q_2' - Q_1',$$

$$Q_2 - Q_2' = Q_1 - Q_1'.$$

That they are not zero follows from a consideration of the efficiencies. The efficiency of the Carnot engine is W/Q_2, that of the second engine is W/Q_2', and by hypothesis

$$\frac{W}{Q_2'} > \frac{W}{Q_2}.$$

Hence

$$Q_2 > Q_2' \quad \text{and} \quad Q_2 - Q_2' > 0.$$

Hence in one cycle the increase in entropy of the high-temperature reservoir is

$$\Delta S_2 = \frac{Q_2 - Q_2'}{T_2}.$$

The *decrease* in entropy of the low-temperature reservoir is

$$\Delta S_1 = \frac{Q_1 - Q_1'}{T_1}.$$

The numerators are equal, but $T_1 < T_2$. Hence $\Delta S_1 > \Delta S_2$ and the decrease in entropy is greater than the increase. This violates the principle that the entropy must increase (or remain constant) in any process. Therefore the process is impossible and the second engine cannot have a greater efficiency than the Carnot engine, which was to be proved.

What is the significance of the increase of entropy that accompanies every natural process? The answer, or one answer, is that it represents the extent to which the Universe "runs down" in that process. Consider again the example of the mixing of hot and cold water. We *might* have used the hot and cold water as the high- and low-temperature reservoirs of a heat engine, and in the course of removing heat from the hot water and giving heat to the cold water we could have obtained some mechanical work. But once the hot and cold water have been mixed and have come to a uniform temperature, this opportunity of converting heat to mechanical work is lost, and, moreover, it is lost irretrievably. The lukewarm water will never *unmix* itself and separate into a hotter and a colder portion.* Of course, there is no decrease in energy when the hot and cold water are mixed, but there has been a decrease in the availability, or an *increase in the unavailability* of the energy, in the sense that a certain amount of energy is no longer available for conversion to mechanical work. Hence when entropy increases, energy becomes more unavailable, and we say that the Universe has "run down" to that extent.

The tendency of all natural processes such as heat flow, mixing, diffusion, etc., is to bring about a uniformity of temperature, pressure, composition, etc., at all points. One may visualize a distant future in which, as a consequence of these processes, the entire Universe has attained a state of absolute uniformity throughout. When and if such a state is reached, although there would have been no change in the energy of the Universe, all physical, chemical, and presumably biological processes, would have to cease. This goal toward which we appear headed has been described as the "heat death" of the Universe.

24–9 The Kelvin absolute-temperature scale. As previously stated, the efficiency of all Carnot engines operating between two given temperatures is the same, whatever the material carried through the cycle. Lord Kelvin proposed that this fact be used to define a temperature scale which would be independent of the properties of any particular substance, unlike the mercury scales and gas-thermometer scales described in Chapter 18.

The ratio of two temperatures on the Kelvin scale is *defined* as the ratio of the heat taken in at the higher temperature to the heat exhausted at the lower temperature by a Carnot engine operating between the two temperatures. For example, the ratio of the Kelvin temperature of the steam point to that of the ice point may be found by operating a Carnot engine between these temperatures, and measuring the heat taken in from the steam bath and the heat rejected to the ice bath. The same ratio would be found, whatever material were used in the engine.

---

*The branch of physics called "Statistical Mechanics" would modify this statement to read, "It is highly improbable that the water will separate spontaneously into a hotter and a colder portion, but it is not impossible."

The ratio alone does not completely fix the temperatures. If we arbitrarily set the *difference* between the temperatures at 100°, we obtain temperatures on the Kelvin centigrade scale. If the difference is set at 180°, we obtain temperatures on the Kelvin fahrenheit scale.

For example, any Carnot engine operating between the steam and ice points and taking in 1000 calories in each cycle from the steam bath would be found to give up (very nearly) 732 calories to the ice bath. Hence, if T_s and T_i are the Kelvin temperatures,

$$\frac{T_s}{T_i} = \frac{1000}{732}.$$

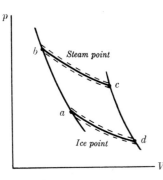

FIGURE 24–10

If we set $T_s - T_i = 100$, and solve simultaneously, we find

$$T_s = 373°, \qquad T_i = 273°.$$

If we set $T_s - T_i = 180°$, we find

$$T_s = 672°, \qquad T_i = 492°.$$

Temperatures between the ice and steam points may be defined as illustrated in Fig. 24–10. Isotherms at intervals of one degree on the Kelvin scale are those which divide the area *abcd* into areas 1/100 as great as that of *abcd*. Since the area of a cycle equals the work done in that cycle, we may consider the division to be made by using one hundred Carnot engines, each of which takes in the exhaust heat rejected by the engine above it, performs 1/100 as much work as the large Carnot engine, and rejects the remainder of the heat to the engine below it.

The same process may be extended above the steam point and below the ice point. It is evident, however, that we cannot continue indefinitely in the latter direction. As a quantitative example, consider the Carnot engine mentioned earlier, which rejected 732 calories at the ice point and performed $1000 - 732 = 268$ calories of work.

Each small engine must then perform $1/100 \times 268 = 2.68$ calories of work. The first engine below the ice point would take in 732 calories, perform 2.68 calories of work, and reject $732 - 2.68 = 729.32$ calories at a temperature of 272°K. Since each succeeding engine abstracts another 2.68 calories, the process must evidently cease after a number of cycles equal to the number of times 2.68 is contained in 732, or after

$$\frac{732}{2.68} = 273 \text{ cycles.}$$

That is, the process could not continue after a temperature 273° below the ice point had been reached. This is the true significance of the term absolute zero.

The University of Leiden, in Holland, was the center of low-temperature research before its work was interrupted by World War II. Temperatures as low as a few hundredths of a degree Kelvin have been obtained in the laboratory there.

PROBLEMS

24–1. What is the efficiency of an Otto cycle in which the compression ratio is 8 and $\gamma = 1.50$?

24–2. A Carnot engine whose high temperature reservoir is at 127°C takes in 100 cal of heat at this temperature in each cycle, and gives up 80 cal to the low temperature reservoir. Find the temperature of the latter reservoir.

24–3. A Carnot engine whose low temperature reservoir is at 7°C has an efficiency of 40%. It is desired to increase the efficiency to 50%. By how many degrees must the temperature of the high temperature reservoir be increased?

24–4. A Carnot engine is operated between two heat reservoirs at temperatures of 400°K and 300°K. (a) If in each cycle the engine receives 1200 cal of heat from the reservoir at 400°K, how many calories does it reject to the reservoir at 300°K? (b) If the engine is operated in reverse, as a refrigerator, and receives 1200 cal of heat from the reservoir at 300°K, how many calories does it deliver to the reservoir at 400°K? (c) How many calories would be produced if the mechanical work required to operate the refrigerator in part (b) were converted directly to heat?

24–5. (a) Show a graph of entropy as a function of temperature for a heat engine operating in a Carnot cycle. (b) Indicate graphically the heat absorbed and the heat rejected by the substance

during one cycle. (c) Use the principle of the increase of entropy to show that the efficiency is given by

$$\frac{T_2 - T_1}{T_2}.$$

24–6. A Carnot refrigerator takes heat from water at 0°C and discards it to the room at a temperature of 27°C. 100 kgm of water at 0°C are to be changed to ice at 0°C. (a) How many calories of heat are discarded to the room? (b) What is the required work in joules?

24–7. What is the efficiency of an engine which operates by taking an ideal monatomic gas through the following cycle? Let $C_v = 3$ cal/mole·C°.
(a) Start with n moles at p_0, V_0, T_0.
(b) Change to $2p_0$, V_0 at constant volume.
(c) Change to $2p_0$, $2V_0$ at constant pressure.
(d) Change to p_0, $2V_0$ at constant volume.
(e) Change to p_0, V_0 at constant pressure.

24–8. (a) A cylinder having an initial volume of 1 liter and containing 0.1 mole of H_2O at 10°C is isothermally expanded in a quasistatic manner to a volume of 10 liters. What is the change in the entropy of the system? (b) The system is now adiabatically compressed to its initial volume. What is the change in entropy during this second process?

24–9. In engineering work, the entropy of water is assumed zero when the water is in the liquid phase at 32°F and 1 atm. Entropy is expressed in Btu/°F abs. (a) What is the entropy of 100 lb of water in the liquid phase at 212°F? (b) What is the entropy of 100 lb of steam at 212°F? (c) What is the entropy of 100 lb of ice at 32°F and 1 atm? (Absolute zero = −459°F.)

24–10. Consider the entropy of water to be zero when in the liquid phase at 0°C and 1 atm. (a) What is the entropy of 500 gm of water at 80°C? (b) What increase in entropy results when 500 gm of water at 80°C are mixed with 500 gm at 0°C?

24–11. 10 gm of steam at 1 atm and 100°C are introduced into a mixture of 100 gm of ice and 200 gm of water at 0°C. (a) What is the final temperature? (b) What is the increase in entropy resulting from this mixing?

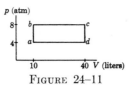

FIGURE 24–11

24–12. Two moles of an ideal gas occupy a volume of 10 liters under a pressure of 4 atm (see point a, Fig. 24–11). (a) If the specific heat of the gas at constant volume is 3 cal/mol·K°, compute the change in entropy along path abc, along path adc, along path $abcda$. (b) If the gas is carried through one complete cycle each second, what is the horsepower developed?

24–13. (a) One-tenth mole of an ideal gas at a temperature of 27°C expands isothermally and quasistatically from a volume of 1 liter to a volume of 4 liters. Compute the change in entropy of the gas, the surroundings, and the universe during the expansion. (b) One-tenth mole of an ideal gas at a temperature of 27°C makes a free expansion from a volume of 1 liter to a volume of 4 liters in a thermally insulated container. Compute the change in entropy of the gas, the surroundings, and the universe during the expansion.

24–14. Derive an equation for the change in entropy of an ideal gas in terms of pressure and temperature as follows. Consider an ideal gas in a state given by p_1 and T_1. Let the gas expand at constant pressure until the temperature is T_2. Then let the gas expand isothermally until the pressure drops to p_2. Find the change in entropy by adding the changes of entropy in the two steps. Express the answer in terms of p_1, p_2, T_1, T_2, and appropriate constants.

CHAPTER 25

KINETIC THEORY OF GASES

25–1 Derivation of ideal gas law. The kinetic theory of gases, which is sufficiently well founded so that it need no longer be considered a "theory," is one branch of the study of the molecular nature of matter in general. Kinetic theory proposes to explain the observed properties of gases on the basis of the laws of mechanics and a few simple assumptions regarding the nature of a gas. We assume the gas to be not a continuous fluid, but an enormous number of tiny particles which will be called atoms. The atoms are assumed to be separated by distances large in comparison with their own dimensions, to be in a continual state of random motion, and to exert no forces on one another except when they collide. Collisions with other atoms or with the walls of the containing vessel are assumed to be perfectly elastic.

To obtain an idea of the enormous number of atoms in a cubic centimeter of gas at ordinary conditions, it may be recalled first that one gram-molecular weight of any gas occupies a volume of 22,400 cm^3 at standard conditions, and second that the number of atoms in a mole (Avogadro's number) is 6.02×10^{23}. The number of atoms per cubic centimeter, at standard conditions, is therefore

$$\frac{6.02 \times 10^{23}}{22,400} = 2.68 \times 10^{19} \text{ atoms/cm}^3.$$

At a given temperature, the number of atoms per unit volume is directly proportional to the pressure. This may be shown as follows. Let N represent the total number of atoms in a sample of gas, and let A be Avogadro's number. The number of moles in the sample is

$$n = \frac{N}{A}.$$

Then from the gas law,

$$p = \frac{nRT}{V} = \frac{N}{V}\frac{RT}{A}.$$

The ratio N/V is the number of atoms per unit volume, and R and A are constants. Therefore at constant temperature N/V is proportional to p.

468

The best "vacuum" pumps obtainable are capable of lowering the pressure to about one ten-thousandth of a millimeter of mercury, or about 10^{-7} atm. At this pressure there are still $10^{-7} \times 2.68 \times 10^{19}$ or about 3,000,000,000,000 atoms in a cubic centimeter!

The mass of a single atom (or molecule) is the gram-atomic weight divided by Avogadro's number. For example, the mass of an atom of "atomic" hydrogen is

$$m_H = \frac{1}{6.02 \times 10^{23}} = 1.66 \times 10^{-24} \text{ gm.}$$

From this figure it follows at once that the mass of a hydrogen molecule is

$$m_{H_2} = 2 \times 1.66 \times 10^{-24} = 3.32 \times 10^{-24} \text{ gm,}$$

while the mass of an oxygen molecule is

$$m_{O_2} = 32 \times 1.66 \times 10^{-24} = 53.2 \times 10^{-24} \text{ gm,}$$

and so on.

In an actual gas not all atoms have the same speed; some travel more slowly and others more rapidly than the average. As a first approximation, however, we shall assume that all the atoms have the same speed, which we shall represent by c. Also, in an actual gas the directions of the velocities of the atoms are entirely at random. For simplicity we shall assume that one-third of the atoms move parallel to the X-axis, one-third parallel to the Y-axis, and one-third parallel to the Z-axis. Finally, we shall ignore any effects of collisions between atoms, which is equivalent to treating them as geometrical points. Each of the atoms, however, has the same mass, m.

Imagine the gas to be contained within a cubical box as in Fig. 25–1, with edges of length L parallel to the coordinate axes. Let N be the total number of atoms in the box, so that $N/3$ atoms are traveling back and forth parallel to the X-axis with a velocity of magnitude c. As each atom in turn collides with the end face $abcd$, its velocity reverses from $+c$ to $-c$. Hence its momentum

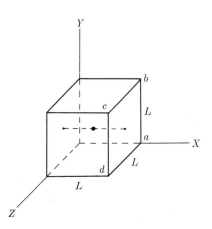

FIGURE 25–1

changes from $+mc$ to $-mc$, and the change of momentum is therefore $2mc$. This change of momentum is equal to the impulse of the force exerted by the atom on the face $abcd$. The observed pressure of the gas against this face arises from the combined effect of all of these tiny impulsive forces.

We have no way of knowing the length of time during which an atom is in contact with the wall and hence cannot find the force at any one collision. However, the time elapsing between successive collisions of any one atom with the face $abcd$ is the time required for an atom to travel to the opposite face and back, or over a distance $2L$. This time is

$$t = \frac{2L}{c}.$$

Hence a reversal of momentum of any one atom at one face occurs once in a time interval of duration $2L/c$. Therefore, since

Impulse = average force × time = change in momentum,

$$\text{Average force} \times \frac{2L}{c} = 2mc,$$

or

$$\text{Average force} = \frac{mc^2}{L} \text{ (one atom).}$$

Multiplying by $N/3$, the number of atoms which make collisions with face $abcd$, we get

$$\text{Average force} = \frac{N}{3} \frac{mc^2}{L} \text{ (all atoms colliding at face } abcd\text{).}$$

The average pressure exerted against this face is

$$\text{Average pressure} = \frac{\text{average force}}{\text{area}},$$

$$p = \frac{N}{3} \frac{mc^2}{L} \frac{1}{L^2}.$$

But since L^3 equals the volume of the container, V,

$$p = \frac{N}{3} \frac{mc^2}{V},$$

or

$$pV = \tfrac{1}{3}Nmc^2. \tag{25-1}$$

If, now, we set

$$\tfrac{1}{3}Nmc^2 = nRT, \qquad (25\text{–}2)$$

we obtain the ideal gas law,

$$pV = nRT.$$

This last step is a very significant one, as it provides us for the first time with a connecting link between the temperature concept and the concepts of mechanics. Let us rewrite Eq. (25–2) as follows. Since the number of moles, n, equals the number of atoms, N, divided by Avogadro's number, A,

$$\frac{1}{3}Nmc^2 = N\frac{R}{A}T. \qquad (25\text{–}3)$$

The ratio R/A occurs so frequently in kinetic theory that it is convenient to represent it by a single letter k. In cgs units,

$$k = \frac{R}{A} = \frac{8.31 \times 10^7}{6.02 \times 10^{23}} = 1.37 \times 10^{-16}\,\text{erg/atom·°K}.$$

Since it is a ratio of two universal constants k is also a universal constant and is called "the gas constant per atom" or the *Boltzmann constant*, after Ludwig Boltzmann (1844–1906). In terms of the Boltzmann constant, Eq. (25–3) becomes

$$kT = \tfrac{1}{3}mc^2 = \tfrac{2}{3}\tfrac{1}{2}mc^2. \qquad (25\text{–}4)$$

But $\tfrac{1}{2}mc^2$ is the translational kinetic energy of an atom, and hence *the Kelvin temperature of a gas is proportional to the translational kinetic energy of the atoms of which it consists.*

We can now compute the velocity with which the atoms of a gas are traveling. Equation (25–4) can be put in the following forms:

$$\tfrac{1}{2}mc^2 = \tfrac{3}{2}kT, \qquad (25\text{–}5)$$

$$c = \sqrt{\frac{3kT}{m}}. \qquad (25\text{–}6)$$

It follows from Eq. (25–5) that the kinetic energies of all species of molecules are the same at the same temperature and from Eq. (25–6) that the velocities of two different species, at the same temperature, are inversely proportional to the square roots of their masses.

At a temperature of 27°C, or 300°K, we find

$$c_{H_2} = 19.3 \times 10^4 \text{ cm/sec} = 4320 \text{ mi/hr,}$$

$$c_{He} = 13.1 \times 10^4 \text{ cm/sec} = 2930 \text{ mi/hr,}$$

$$c_{O_2} = 4.83 \times 10^4 \text{ cm/sec} = 1070 \text{ mi/hr,}$$

Traveling at this velocity, a hydrogen molecule could encircle the earth in about six hours.

25-2 Specific heats. We shall show next how the specific heat of an ideal gas can be computed from kinetic theory considerations. Note that while thermodynamic principles, as explained in Section 22-5, show that the difference between the specific heats of an ideal gas at constant pressure and at constant volume equals the gas constant R, thermodynamics alone can give no information regarding the absolute magnitude of either specific heat.

We have shown in Eq. (22-18) that the molal specific heat at constant volume is equal to the rate of increase of internal energy per mole, per unit increase in temperature.

$$C_v = \frac{1}{n} \frac{dU}{dT}.$$

It follows that if we can set up an expression for the internal energy of a gas in terms of its temperature, differentiation of this expression will give us the specific heat at constant volume.

The energy of a gas molecule consists of kinetic energy of translation of the molecule as a whole, together with kinetic energy of rotation and energy associated with the vibrations of its component atoms. These various forms of energy are associated with the so-called *degrees of freedom* a molecule may have. That is, in order to specify the translational motion of the center of mass of a molecule, the components of velocity along three mutually perpendicular axes must be given, and we accordingly say the molecule has *three translational degrees of freedom.* The rotational motion is specified by the components of the angular velocity vector along three axes, and hence *three rotational degrees of freedom* are possible. In addition, there may be vibrational degrees of freedom.

The total energy of the molecule is distributed in some way between these degrees of freedom, and the simplest assumption, to be made tentatively and discarded if it does not work, is that the energy is shared *equally* by the various degrees of freedom. This hypothesis is called *the principle of equipartition of energy.*

Now we have already derived an expression for the energy associated with the translational motion of the molecule. From Eq. (25–5), the translational kinetic energy is

$$\frac{1}{2}\,mc^2 = \frac{3}{2}\,kT$$

Since there are three translational degrees of freedom, the kinetic energy per translational degree of freedom is

$$\frac{1}{3}\left(\frac{1}{2}\,mc^2\right) = \frac{1}{2}\,kT.$$

Finally, if f represents the total number of degrees of freedom, and if all share equally in the energy, the total energy of the molecule is

$$\frac{f}{3}\left(\frac{1}{2}\,mc^2\right) = \frac{f}{2}\,kT.$$

The energy of N molecules, or the internal energy U, is therefore

$$U = \frac{f}{2}\,NkT.$$

Since $k = R/A$ and $N/A = n$, this can be put in the form

$$U = \frac{f}{2}\,nRT.$$

We have now expressed the internal energy in terms of the temperature and can therefore compute C_v.

$$C_v = \frac{1}{n}\frac{dU}{dT} = \frac{f}{2}\,R. \tag{25–7}$$

From the relation $C_p = C_v + R$ it follows that

$$C_p = \frac{f}{2}\,R + R = \frac{f+2}{2}\,R, \tag{25–8}$$

and finally

$$\gamma = \frac{C_p}{C_v} = \frac{(f+2/2)\,R}{(f/2)\,R} = \frac{f+2}{f}. \tag{25–9}$$

Hence with the help of the kinetic model of an ideal gas and the equipartition hypothesis, C_p, C_v, and γ can all be expressed in terms of the number of degrees of freedom.

How do the predictions above agree with experiment? Some data are given in Table XXI. The first two gases, argon and helium, are *monatomic*. They might therefore be expected to behave nearly like geometrical points with three translational degrees of freedom but no rotational or vibrational degrees of freedom. If we set $f = 3$ in Eqs. (25–7) and (25–9), we find

$$C_v = \frac{3}{2} R = \frac{3}{2} \times 1.98 = 2.97 \text{ cal/mole·C°,}$$

$$\gamma = \frac{3 + 2}{3} = \frac{5}{3} = 1.667,$$

and it will be seen that these figures are in good agreement with the observed values.

TABLE XXI

SPECIFIC HEATS OF GASES

Gas	c_v (cal/gm·C°)	C_v (cal/mole·C°)	γ
A	0.0752	3.03	1.668
He	0.753	3.02	1.660
O_2	0.155	4.99	1.401
N_2	0.177	4.95	1.404
Air	0.171	4.96	1.40
H_2	2.40	4.80	1.410
CO	0.169	4.96	1.404
CO_2	0.153	6.77	1.304
SO_2	0.117	7.72	1.29
$C_4H_{10}O$	0.412	30.9	1.08

The next four gases are diatomic and might be expected to have a total of six degrees of freedom, three of translation and three of rotation. We do not, however, get good agreement by setting $f = 6$, but if we try $f = 5$ we obtain

$$C_v = \frac{5}{2} R = 4.95 \text{ cal/mole·C°,}$$

$$\gamma = \frac{5 + 2}{5} = \frac{7}{5} = 1.400.$$

Apparently, then, these molecules behave as if their energy was shared equally among five degrees of freedom. The explanation is connected with the fact that the moment of inertia of a "dumbbell" molecule about an axis joining its atoms is much smaller than that about axes at right angles to this direction, and only two of the rotational degrees of freedom seem to share in the energy.

The more complex the molecule, the greater the number of degrees of freedom it can be expected to possess, and from Eqs. (25–7) and (25–9) the greater should be its specific heat, while the ratio of its specific heats should approach unity. Table XXI shows that this is indeed the general trend. However, while the examples cited show that there is at least a germ of truth in this simple theory of specific heats, it is, as a matter of fact, not very satisfactory, partly because a nonintegral number of degrees of freedom must be assumed to obtain agreement in many instances and partly because observed specific heats show a dependence on temperature which the theory does not predict. The equipartition principle is not the whole story, and only the concepts of the quantum theory afford a satisfactory explanation of the specific heats of gases.

25–3 Brownian motion. Atoms and molecules are much too small for any direct observation of their thermal motion. Indirect evidence of the existence of this motion can be obtained in many ways, but perhaps the most vivid is from a study of *Brownian motion*. In 1827 an English botanist, Brown, who was examining with a microscope a suspension of fine inanimate spores in water, observed that the spores were darting about in a constant state of random motion. He found that colloidal suspensions of inorganic substances showed the same type of movement, and it can also be observed in small particles (such as those of smoke) in air.

The dimensions of the particles in a colloidal suspension are of the order of 10^{-4} cm. They are thus about ten thousand times as large as an atom, but nevertheless small enough so that the numbers of atoms striking them on opposite sides do not always exactly balance. The result is that the particles are continually driven about in random directions and they may be considered as large molecules which share in the thermal energy of the surrounding atoms.

The French physical chemist Jean Perrin made a very thorough study of Brownian motion in colloidal suspensions. He was able to show that the colloidal particles obeyed the equipartition principle and that the number of particles per unit volume varied with height in exactly the same way as the number of gas particles per unit volume does in the earth's atmosphere. His measurements led to the first reasonably accurate determination of the Avogadro number, for which he found

$$A = 6.85 \times 10^{23} \text{ atoms per mole.}$$

25–4 Mean free path. Real atoms are not geometrical points and therefore make collisions with one another as they move about. The distance traveled by an atom between two consecutive collisions with other atoms is called a *free path*, and the average distance between collisions is the *mean free path*. A few free paths are indicated in Fig. 25–2, where one particular atom, shown by the black circle, is followed as it collides with other atoms shown by open circles. The mean free path of an atom may be computed as follows.

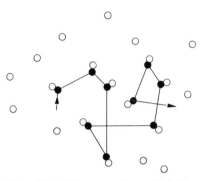

FIG. 25–2. The mean free path is the average distance traveled between collisions.

Let us assume the atoms to be spheres of diameter σ. A collision will then take place when the centers of two atoms approach within a distance σ of one another as in Fig. 25–3(a). The same number of collisions of any one atom would result if the diameter of that atom were increased to 2σ (or its radius to σ) and all the other atoms were reduced to geometrical points as in Fig. 25–3(b). As our single atom of diameter 2σ moves through the gas, it sweeps out in time t a cylinder whose cross section is $\pi\sigma^2$, and whose length is the distance traveled in time t, or ct. (The cylinder is not straight but is like a long, jointed stovepipe, with a joint at every collision.) In time t the atom under consideration makes a collision with every other atom whose center lies in this volume. Hence, if there are n atoms per unit volume, the number of other atoms in the cylinder or the number of collisions in time t is

$$\pi n\sigma^2 ct.$$

The number of collisions per unit time, or the collision frequency Z, is

$$Z = \pi n\sigma^2 c.$$

We shall show in the next section how molecular diameters can be "measured." These diameters are about the same for all gases, 2 or 3×10^{-8} cm. It was computed in Section 25–1 that at standard conditions

FIGURE 25–3

there are about 3×10^{19} atoms/cm^3 in a gas, and their velocities are about 10^5 cm/sec. Hence, in round numbers, the collision frequency in a gas at standard conditions is about

$$Z = 3.14 \times 3 \times 10^{19} \times (2 \times 10^{-8})^2 \times 10^5$$
$$= 4 \times 10^9$$

or 4,000,000,000 collisions per second!

The mean free path, L, is the average distance between collisions, or the total distance covered per unit time ($=c$) divided by the number of collisions per unit time. Hence

$$L = \frac{c}{Z} = \frac{c}{\pi n \sigma^2 c},$$

$$L = \frac{1}{\pi n \sigma^2}. \tag{25–10}$$

Introducing the numerical values used above, we find

$$L = \frac{1}{\pi \times 3 \times 10^{19} \times (2 \times 10^{-8})^2} = 3 \times 10^{-5} \text{ cm.}$$

This distance is slightly smaller than the wave length of light in the visible spectrum.

25–5 Viscosity of a gas. The kinetic theory affords a beautifully simple explanation of the viscosity of a gas. Figure 25–4, which corresponds to Fig. 17–11 on page 346, represents a moving upper plate, separated from a stationary lower plate by a layer of gas of thickness d, and being pulled to the right with velocity V by a force F. The gas atoms, in addition to their large thermal velocities, have a forward velocity component which equals V at the top plate and decreases uniformly to zero at the bottom plate. The forward velocity v at any height y above the lower plate can be found by proportion

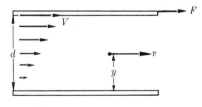

Fig. 25–4. Viscous flow between a stationary lower plate and a moving upper plate.

$$\frac{v}{y} = \frac{V}{d}, \quad \text{or} \quad v = \frac{y}{d}\, V.$$

Since the gas is viscous, a shearing force is exerted on the gas below any horizontal plane by the gas above it. The explanation of this force is that momentum is transmitted across any such plane by the atoms crossing it, because those coming from above, where the forward velocity is large, carry more forward momentum than those crossing from below, where the velocity is smaller.

On the average, each atom crossing a horizontal plane from above makes its last collision before crossing at a height of one mean free path above the plane. Its forward velocity when it crosses is the forward velocity of the gas at the height of this last collision. If y is the elevation of some plane in the gas, the forward velocity at a height L above it is

$$v = \frac{y + L}{d} V,$$

and each atom crossing from above carries across the plane a forward momentum of

$$mv = m \frac{y + L}{d} V.$$

Assume that $\frac{1}{3}$ of the atoms move along the Y-axis, $\frac{1}{6}$ up and $\frac{1}{6}$ down. Let c represent their thermal velocity. The number crossing an area A from above in time t is the number moving down and contained in a cylinder of base A and height ct. If the total number of atoms per unit volume is n, the number crossing from above in time t is

$$\tfrac{1}{6}n \times A \times ct,$$

and the momentum transported across the area from above is

$$\frac{1}{6} nActm \frac{y + L}{d} V. \tag{25–11}$$

Similarly, the momentum transported across the area by those moving up from below is

$$\frac{1}{6} nActm \frac{y - L}{d} V. \tag{25–12}$$

The increase of momentum of the gas below the area due to this cause is the difference between the expressions (25–11) and (25–12), or

$$\frac{1}{3} \frac{nActmLV}{d}.$$

The *rate* of increase of momentum, which we set equal to the average force, is

$$F = \frac{1}{3} \frac{nAcmLV}{d} = \frac{1}{3} nmcL \frac{AV}{d}.$$

But from the definition of the coefficient of viscosity, Eq. (17–10),

$$F = \eta \frac{AV}{d}.$$

Hence

$$\eta = \tfrac{1}{3} nmcL,$$

and when this is combined with Eq. (25–10) for the mean free path L we obtain

$$\eta = \frac{1}{3} \frac{nmc}{\pi n\sigma^2} = \frac{mc}{3\pi\sigma^2},$$

or

$$\sigma^2 = \frac{mc}{3\pi\eta}. \tag{25–13}$$

Equation (25–13) thus provides a relation from which atomic diameters can be computed, since m, c, and η can all be computed or measured.

25–6 The Maxwell-Boltzmann distribution of molecular speeds. As a consequence of collisions with one another, the atoms of a gas would not long retain any simple assignment of velocities, such as the one we have thus far assumed where one-third of the atoms move along each coordinate axis. The interatomic collisions would result in some atoms acquiring high speeds while others were brought nearly to rest. While no one atom would retain any one velocity for an appreciable time, we would expect to find the same *number* of atoms moving with any given speed at any time.

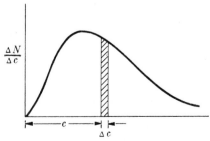

FIG. 25–5. The Maxwell-Boltzmann distribution of molecular speeds.

This is actually the case. The equation which gives the fractional number of atoms having any assigned speed was first derived by James Clerk Maxwell (1831–1879) in 1860 and was later put on a sounder theoretical basis by Boltzmann.

The Maxwell-Boltzmann distribution of atomic speeds is illustrated by the graph of Fig. 25–5, where the area of any small shaded strip under the curve, erected at an abscissa c, represents the number of atoms traveling with speed c. Strictly speaking, the area represents the number (ΔN) with speeds between c and $c + \Delta c$. The ordinate of the curve is $\Delta N/\Delta c$ and the width of the strip is Δc. Its area is therefore

$$\frac{\Delta N}{\Delta c} \times \Delta c = \Delta N.$$

It will be seen from the figure that very few atoms move with low speeds and very few with high speeds, most of them being grouped around the maximum of the curve, which is the *most probable speed*. More atoms have this speed than any other. Because of the nonsymmetrical shape of the curve, the *average speed* is slightly larger than the speed which is most probable. The square root of the average squared speed, or the *root-mean-square speed* is somewhat larger than either.

The equation of the Maxwell-Boltzmann distribution function is

$$\frac{\Delta N}{\Delta c} = A c^2 \epsilon^{-\frac{1}{2}mc^2/kT},$$

where A is a proportionality constant, ϵ is the base of natural logarithms, m is the atomic mass, k is Boltzmann's constant, and T the Kelvin temperature.

SUGGESTED BOOKS FOR COLLATERAL READING

AMERICAN INSTITUTE OF PHYSICS, *Temperature*. Reinhold.

KEENAN, J. H., *Thermodynamics*. Wiley.

LOEB, L. B., *Kinetic Theory of Gases*. McGraw-Hill.

TYNDALL, J., *Heat Considered as a Mode of Motion*. Appleton.

ZEMANSKY, M. W., *Heat and Thermodynamics*. McGraw-Hill.

PROBLEMS

25–1. What is the length, in cm, of the side of a cube which contains just one million molecules of a gas at standard conditions?

25–2. At what temperature is the velocity of an oxygen molecule equal to the velocity of a hydrogen molecule at a temperature of 27°C?

25–3. The velocity of sound in air at 27°C is about 1100 ft/sec. Compare this with the velocity of a nitrogen molecule at the same temperature.

25–4. (a) What is the translational kinetic energy, in ergs, of an oxygen molecule at 27°C? (b) If an oxygen molecule has five degrees of freedom, what is its total kinetic energy at this temperature? (c) What is the internal energy, in joules, of 1 mole of oxygen at this temperature?

25–5. (a) To what pressure must a flask containing oxygen be exhausted to increase the mean free path of the gas molecules within it to 20 cm? The temperature is 57°C. (b) What is the collision frequency? Assume a molecular diameter of 3×10^{-8} cm.

25–6. The viscosity of oxygen at a temperature of 15°C is 195 micropoises. Compute the effective diameter of an oxygen molecule.

25–7. Show that the kinetic theory of viscosity predicts that the viscosity of a gas should increase with increasing temperature. If the temperature is kept constant, how should the viscosity vary with pressure, according to the theory?

25–8. Starting with the expression for the average kinetic energy of a monatomic gas in terms of the temperature, determine the numerical value of C_p, the molal specific heat at constant pressure, of a monatomic gas.

25–9. (a) Compute the velocity of an air molecule at atmospheric pressure and 0°C. (b) Compute the number of molecular impacts per second on a square centimeter of area of the wall of the container under these conditions.

25–10. At what temperature is the average speed of hydrogen molecules equal to the escape velocity from the gravitational field of the earth (velocity with which a particle must be projected up from the surface of the earth to reach infinity)?

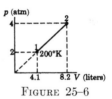

FIGURE 25–6

25–11. A monatomic ideal gas is caused to expand from point 1 to point 2 in Fig. 25–6 in such a way that the relation $p = kV$ holds at all stages of the process, where k is a constant. (a) Find the work done by the gas. (b) Find the Kelvin temperature at point 2. (c) Find the heat added to the gas in this process, expressed in calories. (Note that the process is not isothermal, isobaric, or adiabatic.)

25–12. (a) What would be the efficiency of the engine in Problem 24–7, if a diatomic gas were used instead of a monatomic gas? For maximum efficiency, should a diatomic or a monatomic gas be used as the work substance for (b) the Otto cycle, (c) the Carnot cycle?

SOUND

CHAPTER 26

WAVE MOTION

26–1 Introduction. Acoustics is that branch of physics and engineering which deals with the production and propagation of sound, with the nature of the hearing process, with instruments and apparatus for the measurement, recording, and reproduction of sound, and with the design of auditoriums for good hearing conditions. The subject of acoustics was first put on a firm theoretical and mathematical basis by Hermann Helmholtz (1821–1894). His work was later extended by Lord Rayleigh (1842–1919). Except for the development of new instruments and methods of measurement, little has been added to our basic knowledge of acoustics since the work of these two men.

The term *sound* is used in a subjective sense to designate the sensation in the consciousness of a human observer when the terminals of his auditory nerve are stimulated, and in an objective sense with reference to compressional waves in the air which are capable of stimulating the auditory nerve. Waves in solids and liquids are also called sound waves if they lie in the audible frequency range.

The source of sound waves in air is always to be found in the vibratory motion of some body in contact with the air, as, for example, the sounding board of a piano or the diaphragm of a drum or loudspeaker. In wind instruments the vibrating "body" is itself an air column. One type of vibratory motion has already been studied in some detail, namely, the harmonic motion of a pendulum or of a weight suspended from a spring. The vibrations of strings, air columns, and diaphragms are similar but more complex problems, and in themselves involve the concepts of wave motion as well as of vibration alone. We shall begin with an analysis of wave motion in a stretched string, because of its relative simplicity and because the string is a part of so many musical instruments.

26–2 Transverse waves in a string. Imagine a perfectly flexible string stretched with a tension T and being pulled from right to left with a velocity V through a piece of frictionless glass tubing bent into some arbitrary shape as in Fig. 26–1(a). A small portion of the string of length Δs is shown in part (b) of the figure. If the portion is sufficiently short, it may be considered the arc of a circle of radius R. Since the inner walls of the tube are presumed frictionless, the only forces on the element of string are the normal force N exerted by the tube and the tension T at

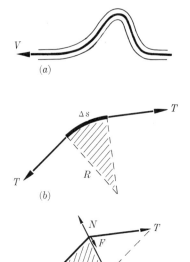

Fig. 26–1. Forces on an element of flexible string pulled through a frictionless curved tube.

each end of the element. We shall assume the distortion of the string to be sufficiently small so that the tension is not affected thereby.

The resultant force F on the element may be found as in part (c) of the figure, where f represents the resultant of the two tension forces. It will be seen that $F = f - N$.

The shaded sector in part (b) approximates a triangle, similar to the shaded triangle in (c). Hence

$$\frac{f}{T} = \frac{\Delta s}{R},$$

and therefore

$$F = T \frac{\Delta s}{R} - N.$$

The element of string is moving in a circle of radius R with a velocity V and a radial acceleration V^2/R. Let μ represent the mass per unit length of the string. The mass of the element is then $\mu \, \Delta s$. We now equate the resultant force on the element to the product of its mass and its radial acceleration.

$$T \frac{\Delta s}{R} - N = \mu \, \Delta s \, \frac{V^2}{R},$$

or

$$N = T \frac{\Delta s}{R} - \mu \, \Delta s \, \frac{V^2}{R}. \tag{26–1}$$

It will be seen from Eq. (26–1) that the force N is zero if the string is pulled through the tube with a velocity such that

$$T \frac{\Delta s}{R} = \mu \, \Delta s \, \frac{V^2}{R}, \qquad \text{or} \qquad V = \sqrt{T/\mu}. \tag{26–2}$$

Hence, if the string travels with a velocity equal to the square root of the tension divided by the mass per unit length, the tube exerts no force at all on the string. The tube could therefore be broken away and the displaced portion of the string would retain its form indefinitely as long as the string was kept in motion toward the left.

The preceding analysis would evidently hold equally well if the ends of the string had been fixed and the tube moved along it from left to right with the velocity $V = \sqrt{T/\mu}$. The tube could be broken away after the motion had been started and the *form* imparted by it to the string would continue to travel toward the right with the same velocity and without change in shape. The advancing disturbance is referred to as a *transverse pulse*, and the velocity V as the *velocity of propagation*.

The tension T in Eq. (26–2) may be expressed in pounds, newtons, or dynes, and μ in slugs per foot, kgm per meter, or grams per centimeter. Corresponding units of V are ft/sec, m/sec, or cm/sec.

Note carefully that when the ends of the string are kept fixed, it is not the particles of the string which travel but simply the form or shape of the disturbance. Any point of the string rises or falls as the disturbance passes it, and then returns to its equilibrium position.

The general mathematical expression for a transverse pulse traveling along the string from left to right may be deduced as follows. Construct a pair of rectangular axes with the X-axis along the equilibrium position of the string. At some instant of time, at which we may set $t = 0$, the shape of the string is represented by the equation

$$y = f(x) \qquad (t = 0).$$

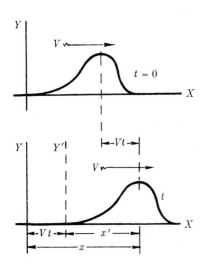

(See Fig. 26–2a.) At a later time t (Fig. 26–2b) the pulse has advanced, without changing its shape, a distance Vt. Construct a new axis Y' as in Fig. 26–2(b), also displaced a distance Vt to the right. Let x' be the coordinate of a point referred to the new origin. The equation of the string at time t, in terms of x', is evidently the same as its equation at time $t = 0$ in terms of x. That is,

$$y = f(x') \qquad (t = t).$$

But from the diagram we see that $x' = x - Vt$. Hence the equation at time t is

$$y = f(x - Vt). \qquad (26–3)$$

Equation (26–3) therefore represents a transverse pulse advancing toward the right with velocity V,

Fig. 26–2. In time t the pulse advances a distance Vt.

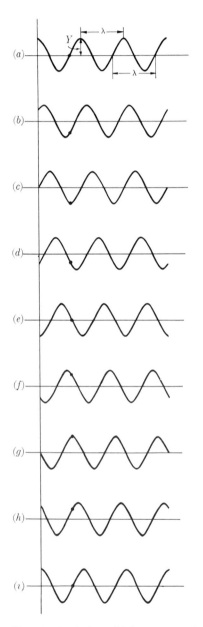

the shape of the pulse at time $t = 0$
being $y = f(x)$. If the pulse is
advancing toward the left its equation is

$$y = f(x + Vt).$$

The particular form of the function
$f(x)$ is, of course, determined by the
initial shape imparted to the string.
Specific examples will be discussed
shortly.

Suppose next that one end of a
stretched string is forced to vibrate
periodically in a transverse direction
with simple harmonic motion of
amplitude Y, frequency f, and
period $T = 1/f$. For the present we
shall assume the string to be long
enough so that any effects at the far
end need not be considered. A *continuous train* of transverse sinusoidal
waves then advances along the
string, rather than the single pulse
considered in the preceding discussion. The shape of a portion of
the string near the end, at intervals
of $\frac{1}{8}$ of a period, is shown in Fig. 26–3
for a total time of one period.
The string is presumed to have been
vibrating for a sufficiently long time
so that the shape of the string is
sinusoidal for an indefinite distance
from the driven end. It will be seen
from the figure that the wave form
advances steadily toward the right,
while any one point on the string
(see the black dot) oscillates about
its equilibrium position with *SHM*.

The distance between two successive maxima (or between any two
successive points in the same phase)
is the *wave length* of the wave and is
denoted by λ. Since the wave form,
traveling with constant velocity V,

Fig. 26–3. A sinusoidal wave traveling toward the right, shown at intervals
of $\frac{1}{8}$ period.

advances a distance of one wave length in a time interval of one period, it follows that

$$\lambda = VT, \qquad \lambda = V/f, \qquad V = f\lambda. \tag{26–4}$$

That is, the velocity of propagation equals the product of frequency and wave length.

We now wish to obtain the equation of this advancing sinusoidal wave train. If we set $t = 0$ at an instant when the string has the form of Fig. 26–3(c), then $f(x)$ at time $t = 0$ is

$$y = Y \sin \frac{2\pi}{\lambda} x,$$

and hence from Eq. (26–3) the equation of the traveling wave is

$$y = Y \sin \frac{2\pi}{\lambda} (x - Vt). \tag{26–5}$$

If we set $t = 0$ when the string has the form of Fig. 26–3(a), then $f(x)$ at time $t = 0$ is

$$y = Y \cos \frac{2\pi}{\lambda} x,$$

and the equation of the traveling wave is

$$y = Y \cos \frac{2\pi}{\lambda} (x - Vt). \tag{26–6}$$

If t is set equal to zero at some arbitrary time, the equation becomes

$$y = Y \cos \frac{2\pi}{\lambda} (x - Vt - x_0).$$

Since we are free to set $t = 0$ at any convenient time, let us for simplicity use Eq. (26–6) to represent the wave.

If a sinusoidal wave train is advancing toward the left, its equation is evidently

$$y = Y \cos \frac{2\pi}{\lambda} (x + Vt). \tag{26–7}$$

It will be left as an exercise to show that Eq. (26–6) is equivalent to

$$y = Y \cos 2\pi \left(\frac{x}{\lambda} - \frac{t}{T} \right) = Y \cos 2\pi f \left(t - \frac{x}{V} \right) = Y \cos \left(2\pi f t - \frac{2\pi x}{\lambda} \right). \tag{26–8}$$

Perhaps the best way of satisfying oneself that Eq. (26–5) or (26–6) actually represents a sine (or cosine) wave advancing toward the right is to construct a graph of the equation for a few values of t. See Problem 1 on page 502.

In any one of the preceding equations y represents the transverse displacement (from its equilibrium position) of a point on the string at a distance x from the origin, and at a time t. It will be recalled (see page 271) that the general expression for the displacement of a particle vibrating in the X-direction with SHM is

$$x = A \cos (2\pi ft + \theta_0),$$

where x is the displacement, A is the amplitude, and θ_0 is the initial phase angle or the epoch angle. For a vibration in the Y-direction, of amplitude Y, this equation becomes

$$y = Y \cos (2\pi ft + \theta_0).$$

Comparison with the last of Eqs. (26–8) shows that they are of exactly the same form. The term $-2\pi x/\lambda$ corresponds to the epoch angle θ_0. Hence all points on the string vibrate with SHM of the same frequency and amplitude, but as one proceeds along the string from the point $x = 0$, successive points get further and further out of phase with the point at the origin, since the angle θ_0 is proportional to x.

It is important to distinguish between the motion of the *wave form*, which moves with constant velocity V along the string, and the motion of the *particles of the string*, which is simple harmonic and transverse to the string. The transverse velocity of a particle of the string is the rate of change of the transverse displacement of the particle, or dy/dt. The transverse acceleration is dv/dt or d^2y/dt^2. Thus for a transverse traveling wave represented by the second of Eqs. (26–8), the transverse velocity v, at time t, of a particle whose coordinate is x, is

$$v = \frac{dy}{dt} = \frac{d}{dt}\left[Y \cos 2\pi f\left(t - \frac{x}{V}\right)\right]$$

$$= -2\pi f Y \sin 2\pi f\left(t - \frac{x}{V}\right).$$

The transverse acceleration, a, is

$$a = \frac{dv}{dt}$$

$$= -4\pi^2 f^2 Y \cos 2\pi f\left(t - \frac{x}{V}\right).$$

26–3 Fourier series. In general, if a traveling wave is set up by a disturbance of any sort, not necessarily sinusoidal, the equation of the wave is

$$y = f(x \pm Vt),$$

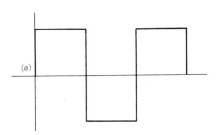

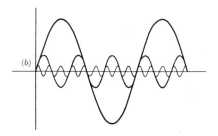

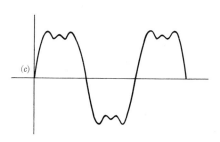

the plus sign applying to a wave traveling toward the left, and vice versa. The particular form of the function depends on the nature of the disturbance (i.e., on the shape of the wave pulses or the wave form). Eq. (26–6) is a special case where the function is a sine (or cosine) of a constant times $(x + Vt)$. The great utility of Eq. (26–6), however, is that *any periodic function* (that is, one which repeats itself at equal time intervals) *can be expressed as the sum of a number of sine or cosine functions.* This fact was discovered by the French mathematician Joseph Fourier in the year 1807, and is known as Fourier's theorem. We therefore need to work out the mathematics of a problem for sine waves only. If the equation of the arbitrary periodic function is known, the component sine waves out of which it can be built may be computed. If the equation is not known but a graph of the function is available, the component waves can be found by some one of a number of instruments known as *harmonic analyzers.* The sine or cosine terms which, when summed, are equal to the given function are called a Fourier series.

Fɪɢ. 26–4.   (a) A square-topped wave; (b) first three terms in its Fourier series; (c) sum of first three terms.

As an illustration of a Fourier series, consider the square-topped wave form shown in Fig. 26–4. The Fourier series of such a wave is

$$y = A \sin x + \tfrac{1}{3}A \sin 3x + \tfrac{1}{5}A \sin 5x + \cdots$$

It will be seen from the diagram that the sum of the first three terms of the series affords a reasonably good approximation to a square wave.

26–4 The wave equation. We shall next derive the differential equation of a traveling wave in a string. Let the X-axis in Fig. 26–5 be the equilibrium position of the string and the curved line its distorted shape. The actual displacement is assumed to be very small (as will be the case with the string of a musical instrument) but it is much exaggerated in the figure for clarity.

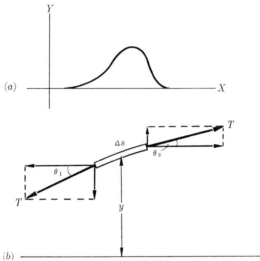

Fig. 26–5. Forces on an element of a string in which there is a transverse wave pulse.

Consider again an element of length Δs. Resolve the tension at each end into X- and Y-components. Then

$$\Sigma X = T \cos \theta_2 - T \cos \theta_1,$$
$$\Sigma Y = T \sin \theta_2 - T \sin \theta_1.$$

The series expansion of the cosine is

$$\cos \theta = 1 - \frac{\theta^2}{2!} + \frac{\theta^4}{4!} \cdots,$$

and since by hypothesis θ_1 and θ_2 are small angles, the terms in θ^2, θ^4, etc., are negligible. Hence

$$\cos \theta_2 - \cos \theta_1 = 1 - 1 = 0,$$

and ΣX is zero.

ΣY may be written

$$\Sigma Y = T \,(\sin \theta_2 - \sin \theta_1) = T\Delta \,(\sin \theta) = T\Delta \,(\tan \theta),$$

since for small angles $\sin \theta$ is nearly equal to $\tan \theta$.

Finally, since $\tan \theta = dy/dx$,

$$\Sigma Y = T\Delta \,(dy/dx),$$

where $\Delta \,(dy/dx)$ means the difference in slope between the ends of the element.

Let us next apply Newton's second law and equate the resultant Y-force acting on the element to the product of its mass and its Y-acceleration. Again let μ be the mass per unit length, and approximate the actual length Δs by its X-component Δx. Then

$$F = ma \,,$$

$$T \,\Delta \left(\frac{dy}{dx}\right) = \mu \,\Delta x \,\frac{d^2 y}{dt^2} \,,$$

$$T \,\frac{\Delta}{\Delta x} \left(\frac{dy}{dx}\right) = \mu \,\frac{d^2 y}{dt^2} \,.$$

In the limit, when $\Delta x \to 0$, the approximations become exact and

$$T \,\frac{d}{dx} \left(\frac{dy}{dx}\right) = \mu \,\frac{d^2 y}{dt^2} \,,$$

or finally

$$\frac{d^2 y}{dt^2} = \frac{T}{\mu} \,\frac{d^2 y}{dx^2} \,. \tag{26–9}$$

Equation (26–9) is the *differential equation of wave motion*, for the special case of "one-dimensional" waves. (Surface ripples on a liquid or waves in a stretched membrane are two-dimensional waves; sound waves or light waves which spread out in all directions are three-dimensional.) We shall now show that the differential equation is satisfied if y is any function of $(x \pm Vt)$, where $V = \sqrt{T/\mu}$. The proof is simply to let $y = f(x \pm Vt)$, compute $d^2 y/dt^2$ and $d^2 y/dx^2$ and substitute in the differential equation.

Note that the symbol f means "a function of" and is not to be confused with the frequency f. To say that "y is a function of $(x - Vt)$" means that the variables x and t occur only in the combination $x - Vt$. That is, $\cos \,(2\pi/\lambda) \,(x - Vt)$ and $\log \,(x - Vt)$ are functions of $(x - Vt)$, but $\sqrt{x^2 - Vt^2}$ is not.

For brevity, let $(x \pm Vt) = u$, and abbreviate $f(u)$ by f. Then*

$$\frac{dy}{dx} = \frac{df}{dx} = \frac{df}{du}\frac{du}{dx},$$

and since $du/dx = (d/dx)(x \pm Vt) = 1$,

$$\frac{dy}{dx} = \frac{df}{du}.$$

A second differentiation gives

$$\frac{d^2y}{dx^2} = \frac{d^2f}{du^2}\frac{du}{dx} = \frac{d^2f}{du^2}.$$

In the same way we find

$$\frac{dy}{dt} = \frac{df}{du}\frac{du}{dt} = \pm V\frac{df}{du},$$

and

$$\frac{d^2y}{dt^2} = V^2\frac{d^2f}{du^2}.$$

When these values are substituted in Eq. (26–9), we get

$$V^2\frac{d^2f}{du^2} = \frac{T}{\mu}\frac{d^2f}{du^2},$$

which is an identity if $V = \sqrt{T/\mu}$. This is the formal mathematical method of proving that the velocity of propagation of transverse waves in a stretched string equals the square root of the tension divided by the mass per unit length.

26–5 Sound waves in a gas. We consider next the question of sound waves in air. These differ from the *transverse* waves in a string in that the oscillations of the air particles are lengthwise, along the direction of propagation. The wave is said to be *longitudinal*. If unimpeded, the sound waves from a source will spread out in all directions and the problem is a three-dimensional one. We shall avoid the complexity of three dimensions by considering waves in a tube. Furthermore, the tube is the prototype of all wind instruments, as the stretched string is of all stringed instruments.

---

*Strictly speaking, all derivatives with respect to x and t should be written as partial derivatives.

Figure 26–6 represents one end of a long tube or pipe provided with a plunger. The vertical lines represent layers of air molecules, equally spaced in the top diagram when the air is at rest. (The thermal motion of the molecules is ignored here. See page 500.) If the plunger is suddenly pushed forward the layers of air in front of it are compressed. These, in turn, compress the layers beyond and a *compressional pulse* travels along the tube. If the plunger is pushed forward and quickly withdrawn, as in Fig. 26–6, a pulse of compression followed by one of *rarefaction* travels along the tube. These pulses are analogous in all respects to transverse pulses traveling along a string, except that the particle displacements are longitudinal rather than transverse. Evidently, if the plunger is caused to oscillate back and forth, a continuous train of condensations and rarefactions will travel along the tube. For the present the tube will be considered sufficiently long so that reflections at the far end need not be considered.

Let the X-axis be taken along the direction of the tube. Consider an elementary length of gas in the tube, bounded by planes whose coordinates are x and $x + \Delta x$ when the gas is in equilibrium (Fig. 26–7). Let the equilibrium pressure be p_0. (This will ordinarily be atmospheric pressure.) As the wave advances along the tube, the element oscillates about its equilibrium position (Fig. 26–6.) We shall use the letter y, as we did when discussing transverse waves in a string, to represent the displacement of any plane from its equilibrium position. In this case, of course, the displacement is to the right or left rather than up or down. In general, the boundary planes of an element will be displaced by different amounts. Hence the volume of the element changes and the pressure varies from point to point.

The displacement of the left face of the element is represented by y and that of the right face by $y + \Delta y$. Let the gauge pressure at the left face of the element be p and that at the right face, $p + \Delta p$. If the ele-

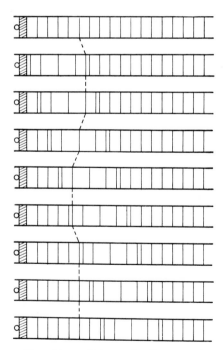

Fig. 26–6. Schematic diagram of a compressional pulse in a gas.

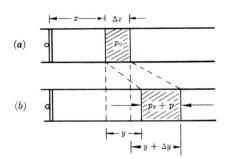

FIG. 26–7. Element of a gas in a tube in which there is a longitudinal wave; (a) equilibrium position; (b) displaced position.

ment is very short, Δp is small and p can be considered to be the gauge pressure to which the element as a whole is subjected. The *absolute* pressure on the element (gauge plus atmospheric) is $p_0 + p$, and the absolute pressures at its faces are $p_0 + p$ and $p_0 + p + \Delta p$.

If the cross section of the tube is A, the force on the right face of the element is $-(p_0 + p + \Delta p)A$ and that on the left face is $(p_0 + p)A$. The net restoring force is therefore $-\Delta pA$. Let ρ_0 be the density of the gas at its equilibrium pressure p_0. The mass of the element is then $\rho_0 A \, \Delta x$. Hence from Newton's second law, $F = ma$,

$$-\Delta p A = \rho_0 A \, \Delta x \, \frac{d^2 y}{dt^2},$$

$$\frac{d^2 y}{dt^2} = -\frac{1}{\rho_0} \frac{\Delta p}{\Delta x},$$

and in the limit when Δx is very small,

$$\frac{d^2 y}{dt^2} = -\frac{1}{\rho_0} \frac{dp}{dx}. \tag{26–10}$$

The volume of the element in its equilibrium position is $A \, \Delta x$. In its displaced position (see Fig. 26–7) the coordinate of its right face is $x + \Delta x + y + \Delta y$ and the coordinate of its left face is $x + y$. The length of the displaced element is $(x + \Delta x + y + \Delta y) - (x + y) = \Delta x + \Delta y$ and its volume is $A(\Delta x + \Delta y)$. The change in volume is therefore $A(\Delta x + \Delta y) - A \, \Delta x = A \, \Delta y$. From the general definition of compressibility k (see page 250),

$$k = -\frac{1}{\text{original volume}} \frac{\text{change in volume}}{\text{change in pressure}},$$

$$k = -\frac{1}{A \, \Delta x} \frac{A \, \Delta y}{(p_0 + p) - p_0} = -\frac{\Delta y}{p \, \Delta x}.$$

Hence

$$p = -\frac{1}{k}\frac{\Delta y}{\Delta x},$$

and in the limit

$$p = -\frac{1}{k}\frac{dy}{dx}.$$ (26–11)

Differentiating this with respect to x gives

$$\frac{dp}{dx} = -\frac{1}{k}\frac{d^2y}{dx^2}.$$

When this expression for dp/dx is substituted in Eq. (26–10), we get

$$\frac{d^2y}{dt^2} = \frac{1}{k\rho_0}\frac{d^2y}{dx^2}.$$ (26–12)

It will be seen that, except for the constant term, this equation is of precisely the same form as Eq. (26–9) on page 493 for transverse waves in a string. We can conclude, therefore, that compressional waves in a gas travel with a velocity of propagation

$$V = \sqrt{\frac{1}{k\rho_0}}.$$ (26–14)

Since the bulk modulus B is the reciprocal of the compressibility, Eq. (26–13) may be written

$$V = \sqrt{\frac{B}{\rho_0}}.$$ (26–13)

Although derived for waves in a gas, Eqs. (26–13) and (26–14) apply also to compressional waves in a liquid, but not to surface waves. A similar argument (which will not be given) shows that compressional waves in a rod or bar travel with a velocity

$$V = \sqrt{\frac{Y}{\rho_0}},$$ (26–15)

where Y is Young's modulus.

It is a familiar fact that compression of a gas causes a rise in its temperature (and vice versa) unless the heat of compression is withdrawn in some way. As a compressional wave advances through a gas, the regions which are compressed at any instant are slightly warmed while the rarefactions are slightly cooled. There will thus be a flow of heat from the compressions to the rarefactions. However, the distances between compressions and rarefactions are so large and the alternations of temperature occur so rapidly that, as a matter of fact, there is practically no interchange of heat between the warmer and cooler portions of the wave. The compressions are therefore *adiabatic* rather than isothermal and it is the adiabatic compressibility which must be used in the wave equation.

We have shown (see page 428) that the adiabatic compressibility of a gas is

$$k_{ad} = \frac{1}{\gamma p},$$

where p is the pressure and $\gamma = C_p/C_v$. Hence Eq. (26–13) may be written

$$V = \sqrt{\frac{\gamma p_0}{\rho_0}}, \tag{26–16}$$

where p_0 is the *absolute* equilibrium pressure and may be expressed in lb/ft^2, newtons/m^2, or dynes/cm^2. The density ρ_0 may be in slugs/ft^3, kgm/m^3, or gm/cm^3. (γ is a pure number.) Corresponding units of velocity are ft/sec, m/sec, and cm/sec.

From the ideal gas equation we have

$$\frac{p_0}{\rho_0} = \frac{RT}{M},$$

where M is the molecular weight. Another form of Eq. (26–16) is therefore

$$V = \sqrt{\frac{\gamma RT}{M}}, \tag{26–17}$$

and since for a given gas γ, R, and M are constants, we see that the velocity of propagation is proportional to the square root of the absolute temperature.

Let us use Eq. (26–17) to compute the velocity of sound waves in air. The mean molecular weight of air is 29, $\gamma = 1.40$, and $R = 8.3 \times 10^7$ ergs/mole. Let $T = 300°$K. Then

$$V = \sqrt{\frac{1.40 \times 8.3 \times 10^7 \times 300}{29}}$$

$$= 34{,}600 \text{ cm/sec} = 346 \text{ m/sec} = 1130 \text{ ft/sec.}$$

This is in excellent agreement with the measured velocity at this temperature.

The ear is sensitive to a range of sound frequencies from about 20 to about 20,000 cycles/sec. From the relation $V = f\lambda$, the corresponding wave-length range is from about 56 ft, corresponding to a 20-cycle note, to about 0.056 ft or $\frac{5}{8}$ in, corresponding to 20,000 cycles/sec.

26–6 Pressure variations in a sound wave. The acoustical engineer finds it more useful to deal with the pressure variations in a sound wave than with the actual displacements of the air particles. The relation between pressure and displacement can be obtained by differentiating Eq. (26–6) with respect to x and combining with Eq. (26–11). We have

$$y = Y \cos \frac{2\pi}{\lambda} (x - Vt), \tag{26–6}$$

$$\frac{dy}{dx} = -\frac{2\pi Y}{\lambda} \sin \frac{2\pi}{\lambda} (x - Vt).$$

Hence from Eq. (26–11)

$$p = \frac{2\pi Y}{k\lambda} \sin \frac{2\pi}{\lambda} (x - Vt).$$

Since $V = \sqrt{1/k\rho_0}$, this may be more conveniently written

$$p = \left[\frac{2\pi\rho_0 V^2 Y}{\lambda} \right] \sin \frac{2\pi}{\lambda} (x - Vt).$$

It will be recalled that p represents the gauge pressure, i.e., the excess or deficiency of pressure above or below atmospheric. The term in brackets evidently represents the maximum gauge pressure or the *pressure ampli-tude*. If this is denoted by P, then

$$p = P \sin \frac{2\pi}{\lambda} (x - Vt), \tag{26–18}$$

where

$$P = \frac{2\pi\rho_0 V^2}{\lambda} Y. \tag{26–19}$$

Equation (26–19) relates the pressure amplitude to the displacement amplitude, and the sound wave may be considered either as a displacement wave or as a pressure wave. If the former is written as a cosine function, the latter will be a sine function, and vice versa. Hence the displacement wave is 90° out of phase with the pressure wave. In other words, at a point where the displacement is a maximum or minimum, the excess pressure is zero; at a point where the displacement is zero, the excess or deficiency of pressure is a maximum.

Measurements of sound waves show that the maximum pressure variations, P, in the loudest sounds which the ear can tolerate, are of the order of magnitude of 280 dynes/cm^2 (above and below atmospheric pressure of about 1,000,000 dynes/cm^2). The corresponding maximum displacement, Y, may be computed from Eq. (26–19). For $\lambda = 35$ cm, corresponding to a frequency of about 1000 cycles/sec,

$$Y = \frac{\lambda P}{2\pi \rho_0 V^2}$$

$$= \frac{35 \times 280}{2\pi \times 0.00122 \times (3.46 \times 10^4)^2}$$

$$= 1.07 \times 10^{-3} \text{ cm,}$$

or about 10^{-3} cm. The displacement amplitudes, even in the loudest sounds, are therefore extremely small.

The maximum pressure variations in the *faintest* 1000-cycle sound which can be heard are only about 0.0002 dyne/cm^2. The corresponding displacement is about 10^{-9} cm, or 0.000000001 cm! By way of comparison, the wave length of yellow light is 5×10^{-5} cm, and the diameter of an atom about 10^{-8} cm. It will be appreciated that the ear is an extremely sensitive organ.

For simplicity, the molecular nature of a gas has been ignored in the preceding discussion and the gas has been treated as though it were a continuous fluid. Actually, we know that a gas is composed of molecules in a state of random motion, with spaces between the molecules which are large compared worth their diameters. The vibrations which constitute a sound wave are superposed on the random thermal motion. Hence in Fig. 26–7, for example, where a volume element of the gas is shown in its equilibrium and displaced positions, it must be realized that the individual molecules do not occupy the same positions in the element in both diagrams, and that during the displacement some molecules have crossed the boundaries of the element, their places being taken, of course, by others entering from neighboring elements.

Marlow

Since the molecules of a gas are not in actual contact, an impulse imparted to one molecule can be transmitted to another only after the first molecule has moved the intervening distance and collided with the second. One would therefore expect a close correlation between molecular velocities and the velocity of sound, and, in particular, would expect that the velocity of sound could not exceed the molecular velocity.

Simple kinetic theory gives for the average velocity of a molecule (see page 471), $c = \sqrt{3RT/M}$, while from Eq. (26–17) the velocity of a sound wave is $V = \sqrt{\gamma RT/M}$. Equation (26–17), although derived without any reference to the molecular picture, shows that the velocity of a sound wave and the mean molecular speed are closely related. Since γ is never larger than 1.66, the velocity of sound in a gas is always less than the molecular speed, although the two are of the same order of magnitude.

Another figure which is of interest is the mean free path of a gas molecule, which at atmospheric pressure is about 10^{-5} cm. The amplitude of a faint sound wave may be only one ten-thousandth of this amount. An element of gas through which a sound wave is traveling can be compared with a swarm of gnats where the swarm as a whole may be seen to oscillate slightly, while individual insects move about through the swarm apparently at random.

PROBLEMS

26–1. A traveling transverse wave on a stretched string is represented by the equation

$$y = Y \cos \frac{2\eta}{\lambda} (x - Vt).$$

Let $Y = 1$ inch, $\lambda = 2$ inch, and $V = \frac{1}{4}$ in/sec. (a) At time $t = 0$, compute the transverse displacement y at $\frac{1}{4}$-inch intervals of x (i.e., at $x = 0$, $x = \frac{1}{4}$ inch, $x = \frac{1}{2}$ inch, etc.) from $x = 0$ to $x = 4$ in. Show the results in a graph. This is the shape of the string at time $t = 0$. (b) Repeat the calculations, for the same values of x, at times $t = 1$ sec, $t = 2$ sec, $t = 3$ sec, and $t = 4$ sec. Show on the same graph the shape of the string at these instants. In which direction is the wave traveling?

26–2. Show that the three equations (26–8) are equivalent forms of Eq. (26–6).

26–3. The Fourier series expressing a "saw-tooth" wave is

$$y = A \sin x \frac{A}{2} \sin 2x$$

$$+ \frac{A}{3} \sin 3x - \frac{A}{4} \sin 4x \cdots$$

Plot the first four terms in the series and find their sum graphically. Let a distance of 2 inches on the X-axis equal π radians, and let $A = 2$ inches.

26–4. The equation of a sinusoidal traveling wave in a stretched string is

$$y = Y \cos \frac{2\pi}{\lambda} \left(x + \sqrt{\frac{T}{\mu}}\, t \right).$$

Prove that this is a solution of the differential equation (26–9), by evaluating d^2y/dt^2 and d^2y/dx^2 and substituting in Eq. (26–9).

26–5. The equation of a transverse traveling wave is

$$y = 2 \sin 2\pi \left(\frac{t}{.01} - \frac{x}{30} \right),$$

where x and y are in cm and t is in sec. What are the amplitude, wave length, frequency, and velocity of propagation of the wave?

26–6. The equation of a transverse traveling wave on a string is

$$y = 2 \cos [\pi(0.5x - 200t),$$

where x and y are in cm and t is in sec. (a) Find the amplitude, wave length, frequency, period, and velocity of propagation. (b) Sketch the shape of the string at the following values of t: 0, 0.0025, and 0.005 sec. (c) If the mass per unit length of the string is 5 gm/cm, find the tension.

26–7. The equation of a longitudinal traveling wave in a rod of density 15.2 slug/ft^3 is

$$y = 10^{-7} \sin 3400\pi \left(t - \frac{x}{17,000} \right),$$

where t is in sec, x and y are in feet. (a) Find the amplitude, wave length, frequency, and velocity of propagation. (b) Find an equation for the longitudinal strain dy/dx at any value of x and t. What is the maximum strain? What is the maximum stress? (c) Find an equation for the particle velocity dy/dt at any value of x and t. What is the maximum particle velocity?

26–8. (a) What is the maximum transverse velocity of a particle of the string in problem 26–1? (b) What is the maximum transverse acceleration of a particle of the string? (c) Compute the displacement, velocity, and acceleration of a particle of the string whose X-coordinate is $\frac{3}{4}$ inch, at time $t = 2$ sec. What is the significance of the negative signs?

26–9. A steel wire 6 m long has a mass of 60 gm and is stretched with a tension of 1000 newtons. What is the velocity of propagation of a transverse wave in the wire?

26–10. The velocity of sound waves in water is approximately 1450 m/sec at 20°C. Compute the adiabatic compressibility of water and compare with the isothermal compressibility listed in Table IV.

26–11. Provided the amplitude is sufficiently great, the human ear can respond to sound waves over a range of frequencies from about 20 cycles/sec to about 20,000 cycles/sec. Compute the wave lengths corresponding to these frequencies, (a) for sound waves in air, (b) for sound waves in water. (See Problem 26–10.)

26–12. The sound waves from a loud speaker spread out nearly uniformly in all directions when their wave length is large compared with the diameter of the speaker. When the wave length is small compared with the diameter of the speaker, much of the sound energy is concentrated in the forward direction. For a speaker of diameter 10 inches, compute the frequency for which the wave length of the sound waves, in air, is (a) 10 times the diameter of the speaker, (b) equal to the diameter of the speaker, (c) 1/10 the diameter of the speaker.

26–13. (a) By how many m/sec, at a temperature of 27°C, does the velocity of sound in air increase per centigrade degree rise in temperature? Hint: compute dV in terms of dT, and approximate finite changes by differentials. (b) By how many ft/sec does the velocity increase per fahrenheit degree? (c) Is the rate of change of velocity with temperature the same at all temperatures?

26–14. What must be the stress in a stretched wire of a material whose Young's modulus is Y, in order that the velocity of longitudinal waves shall equal 10 times the velocity of transverse waves?

26–15. At a temperature of 27°C, what is the velocity of sound waves in (a) argon, (b) hydrogen? Compare with the velocity in air at the same temperature.

26–16. Show that the relation between the pressure amplitude P and the displacement amplitude Y in a sound wave in a gas can be written

$$P = 2\pi\rho_0 f^2 \lambda Y = 2\pi\rho_0 f V Y.$$

26–17. (a) The pressure amplitude of the faintest sound wave of frequency 100 cycles/sec that can be heard by a person of good hearing is about 0.02 dyne/cm^2. What is the displacement amplitude of the wave, if in air? The pressure amplitude of the loudest tolerable sound wave of frequency 100 cycles/sec is about 200 dynes/cm^2. What is the displacement amplitude, (b) if the wave is in air, (c) if the wave is in water?

26–18. The velocity of sound in air at normal temperature and pressure is about 350 m/sec. The ratio of the specific heats is 1.4. (a) What is the wave length of a note of frequency 10 kilocycles/sec? (b) What is the magnitude of the temperature fluctuation at any point in the presence of a sound wave whose pressure amplitude is 10 dynes/cm^2? (c) What is the distance between points of maximum and minimum temperature for the wave in (a) above?

CHAPTER 27

VIBRATION OF STRINGS AND AIR COLUMNS

27-1 Boundary conditions. Let us now consider what will happen when a wave pulse or wave train advancing along a stretched string arrives at the end of the string. If fastened to a rigid support, the end must evidently remain at rest. The arriving pulse exerts a force on the support, and the reaction to this force "kicks back" on the string and sets up a *reflected* pulse traveling in the reversed direction. At the opposite extreme from a rigidly fixed end would be one which was perfectly free— a case of no great importance here (it may be realized by a string hanging vertically)—but which is of interest since its analogue does occur in other types of waves. At a free end the arriving pulse causes the string to "overshoot" and a reflected wave is also set up. The conditions which must be satisfied at the ends of the string (such as $y = 0$ at a fixed end) are called *boundary conditions*.

The multiflash photograph of Fig. 27–1 shows the reflection of a pulse at a fixed end of a string. (The camera was tipped vertically while the photographs were taken so that successive images lie one under the other. The "string" is a rubber tube and it sags somewhat.) It will be seen that the pulse is reflected with its displacement and its velocity both reversed. When reflection takes place at a free end, the direction of the velocity is reversed but the direction of the displacement is unchanged.

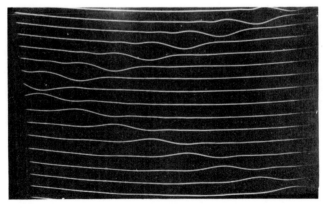

Fig. 27–1. A pulse starts in the upper right corner and is reflected from the fixed end of the string at the left.

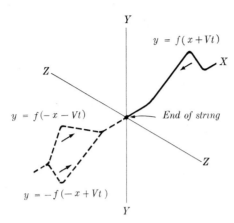

Fig. 27–2. A virtual pulse moves in from the left and combines with the original pulse to form the reflected pulse.

It is helpful to think of the process of reflection in the following way. Imagine the string to be extended indefinitely beyond its actual terminus. The actual pulse can be considered to continue on into the imaginary portion as though the support were not there, while at the same time a "virtual" pulse, which has been traveling in the imaginary portion, moves out into the real string and forms the reflected pulse. The nature of the reflected pulse depends on whether the end is fixed or free. The two cases are shown in Fig. 27–2. The upper virtual pulse (dotted), which corresponds to reflection at a free end, has the same form as would the optical image of the incident pulse in a plane mirror in the Y-Z plane. The lower virtual pulse, which corresponds to reflection at a fixed end, is the mirror image of the upper in the X-Z plane. Mathematically, the equation of the first type is obtained by changing x to $-x$ in the equation of the incident wave, while that of the second type is obtained by further changing f to $-f$. That is, if the equation of the wave in the string is $f(x + Vt)$ as in Fig. 27–2, the equation of a wave reflected at a free end, where the velocity only is reversed, is $f(-x + Vt)$, while the equation of a wave reflected at a fixed end is $y = -f(-x + Vt)$.

The displacement at a point where the actual and virtual pulses cross each other is the algebraic sum of the displacements in the individual pulses. Figures 27–3 and 27–4 show the shape of the end of the string for both types of reflected pulses. It will be seen that Fig. 27–3 corresponds to a free end and Fig. 27–4 to a fixed end. In the latter case, the incident and reflected pulses combine in such a way that the displacement of the end of the string is always zero.

27–2 Standing waves. When a continuous train of waves arrives at a fixed end of a string, a continuous train of reflected waves appears to originate at the end and travel in the opposite direction. Provided the elastic limit of the string is not exceeded and the displacements are sufficiently small for the approximations in Section 26–4 to hold, the actual displacement of any point of the string is the algebraic sum of the displacements of the individual waves, a fact which is called the *principle*

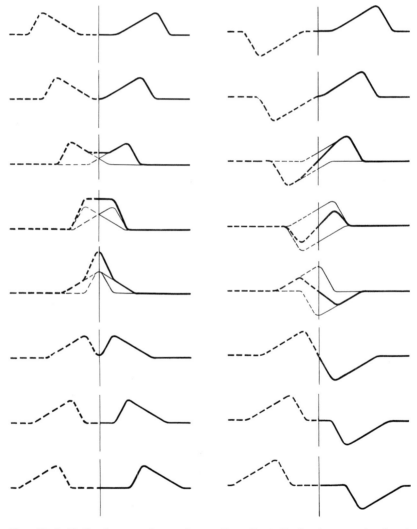

Fig. 27–3. Reflection at a free end.   Fig. 27–4. Reflection at a fixed end.

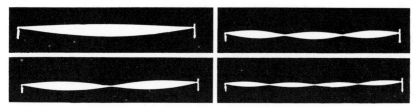

FIG. 27–5. (a) Standing waves in a stretched string (time exposure).

FIG. 27–5. (b) Multiflash photograph of a standing wave, with nodes at the center and at the ends.

of superposition.   This principle is extremely important in all types of wave motion and applies not only to waves in a string but to sound waves in air, to light waves, and, in fact, to wave motion of any sort. The general term *interference* is applied to the effect produced by two (or more) sets of wave trains which are simultaneously passing through a given region.

The appearance of the string under these circumstances gives no evidence that two waves are traversing it in opposite directions. If the frequency is sufficiently great so that the eye cannot follow the motion, the string appears subdivided into a number of segments, as in the time-exposure photograph of Fig. 27–5(a). A multiflash photograph of the same string, in Fig. 27–5(b), indicates a few of the instantaneous shapes of the string. At any instant (except those when the string is straight) its shape is a sine curve, but, whereas in a traveling wave the amplitude remains constant while the wave progresses, here the wave form remains fixed in position (longitudinally) while the amplitude fluctuates. Certain points, known as the *nodes*, remain always at rest. Midway between these points, at the *loops* or *antinodes*, the fluctuations are a maximum. The vibration as a whole is called a *standing wave.*

The analytic expression for a standing sinusoidal wave can be derived as follows. Suppose that a sinusoidal wave train given by

$$y_1 = Y \cos \frac{2\pi}{\lambda} (x - Vt)$$

is incident at a fixed end of a string. The equation of the reflected wave is obtained first by changing $(x - Vt)$ to $(-x - Vt)$, which accomplishes one of the reflections in Fig. 27–2, and second by changing Y to $-Y$, which accomplishes the second reflection. Hence the reflected wave train is given by

$$y_2 = -Y \cos \frac{2\pi}{\lambda} (-x - Vt),$$

or, since $\cos(-\theta) = \cos \theta$,

$$y_2 = -Y \cos \frac{2\pi}{\lambda} (x + Vt).$$

The resultant displacement, by the principle of superposition, is

$$y = y_1 + y_2 = Y \left[\cos \frac{2\pi}{\lambda} (x - Vt) - \cos \frac{2\pi}{\lambda} (x + Vt) \right].$$

Introducing the expressions for the cosine of the sum and difference of two angles, and combining terms, we obtain

$$y = 2Y \sin \frac{2\pi x}{\lambda} \sin 2\pi ft \qquad (27\text{–}1)$$

as the equation of a standing wave.

At any instant of time we may write

$$y = [2Y \sin 2\pi ft] \sin \frac{2\pi x}{\lambda}.$$

The shape of the string at each instant is, therefore, a sine curve whose amplitude (the expression in brackets) varies sinusoidally with time. At values of t such that $2\pi ft = 0, \pi, 2\pi$, etc., the amplitude is zero and the string is straight. When $2\pi ft = \pi/2, 3\pi/2, 5\pi/2$, etc., the amplitude is a maximum and equal to $2Y$, the sum of the amplitudes of the interfering waves.

On the other hand, if we fix our attention on some particular point of the string and write Eq. (27–1) as

$$y = \left[2Y \sin \frac{2\pi x}{\lambda} \right] \sin 2\pi ft,$$

we see that each point of the string performs simple harmonic motion of frequency f but with an amplitude (the expression in brackets) which depends on the position of the point. In contrast with a traveling wave (see page 490) all points between each pair of nodes vibrate *in phase* with one another.

For values of x such that

$$2\pi x/\lambda = 0, \pi, 2\pi, \text{etc.},$$

the amplitude is zero. In other words, these points always remain at rest and are the nodes. At points where

$$2\pi x/\lambda = \pi/2, 3\pi/2, \text{etc.},$$

the amplitude is a maximum and equals $2Y$. These points are the antinodes. The nodes are located at the points where

$$x = 0, \frac{\lambda}{2}, \frac{2\lambda}{2}, \frac{3\lambda}{2}, \text{etc.,}$$

and the antinodes where

$$\frac{\lambda}{4}, \frac{3\lambda}{4}, \frac{5\lambda}{4}, \text{etc.}$$

The nodes are therefore one-half a wave length apart, as are the anti-nodes.

27–3 String fixed at both ends. Thus far we have been discussing a long string fixed at one end and have considered the standing waves set up near that end by interference between the incident and reflected waves. Let us next consider the more usual case, that of a string fixed at both ends. A single pulse set up in the string travels back and forth from one end to the other as in Fig. 27–6 (friction being neglected). A continuous train of sine or cosine waves is reflected and re-reflected in the same way, and since the string is fixed at both ends, both ends must be nodes. Since the nodes are one-half a wave length apart, the length of the string may be $\lambda/2, 2(\lambda/2), 3(\lambda/2), \ldots$, or, in general, any integral number of half-wave lengths. Or, to put it differently, if one considers a particular string of length L, standing waves may be set up in the string by vibrations of a number of different frequencies, namely, those which give rise to waves of wave lengths $2L/1, 2L/2, 2L/3$, etc.

From the relation $f = V/\lambda$, and since V is the same for all frequencies, the possible frequencies are

$$\frac{V}{2L}, \quad 2\frac{V}{2L}, \quad 3\frac{V}{2L}, \quad \cdots$$

The lowest frequency, $V/2L$, is called the *fundamental* frequency f_0 and the others are the *overtones*. The frequencies of the latter are, therefore, $2f_0$, $3f_0$, $4f_0$, and so on. Overtones whose frequencies are integral multiples of the fundamental are said to form a *harmonic series*. The fundamental is the *first harmonic*. The frequency $2f_0$ is the *first overtone* or the *second harmonic*, the frequency $3f_0$ is the *second overtone* or the *third harmonic*, and so on.

We can now see an important difference between a spring-weight system and a vibrating string. The former has but one natural frequency, while the vibrating string has an infinite number of natural frequencies, the fundamental and all of the overtones. If a weight suspended from a spring is pulled down and released, only one frequency of vibration will ensue. If a string is initially distorted so that its shape is the same as

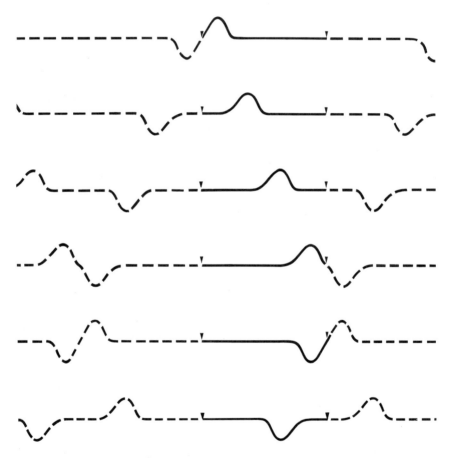

Fig. 27–6. Reflection of a single pulse at the fixed ends of a string.

any one of the possible harmonics, it will vibrate when released at the frequency of that particular harmonic. If a string is struck or plucked in some arbitrary way so that its shape is not sinusoidal, the original disturbance can be expressed by a Fourier series or a sum of sine waves of wave lengths equal to those of the possible harmonics of the string. The ensuing vibration is then a mixture of the standing waves, due to the Fourier components of the original disturbance. Thus, when a piano string is struck, not only the fundamental but many of the overtones are present in the sound which is emitted. Furthermore, while a spring-weight system resonates to only one frequency, a stretched string will resonate either to its fundamental frequency or any of its overtones.

The amplitudes of the individual overtones depend on the particular way in which the string is struck. The "pitch" of the musical sound emitted by the string is determined chiefly by the frequency of the fundamental, and the "quality" of the sound is dependent on the number of overtones present and on their amplitudes.

If the string were perfectly free to move transversely at either end, that end would be an *antinode* rather than a node. This condition cannot be realized experimentally with a string under tension but an analogous situation is shown in Fig. 27–7, which is a photograph of a vibrating rod.

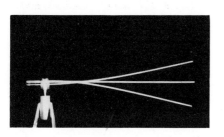

The fundamental frequency of the vibrating string is $f_0 = V/2L$, where $V = \sqrt{T/\mu}$. It follows that

$$f_0 = \frac{1}{2L} \sqrt{\frac{T}{\mu}}. \qquad (27\text{--}2)$$

FIG. 27–7. Multiflash photograph of transverse vibration of a rod.

Stringed instruments afford many examples of the implications of this equation. For example, all such instruments are "tuned" by varying the tension, an increase of tension increasing the frequency or pitch, and vice versa. The inverse dependence of frequency on length is illustrated by the long strings of the bass section of the piano or the bass viol compared with the shorter strings of the piano treble or the violin. One reason for winding the bass strings of a piano with wire is to increase the mass per unit length, so as to obtain the desired low frequency without resorting to a string which is inconveniently long.

27–4 Vibrations of membranes and plates. If a stretched flexible membrane, such as a drumhead, is struck a blow, a two-dimensional

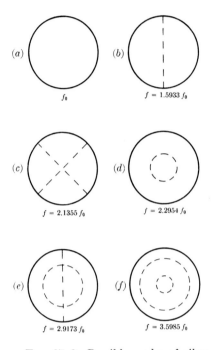

f_0

$f = 1.5933 f_0$

$f = 2.1355 f_0$

$f = 2.2954 f_0$

$f = 2.9173 f_0$

$f = 3.5985 f_0$

FIG. 27–8. Possible modes of vibration of a membrane, showing nodal lines. The frequency of each is given in terms of the fundamental frequency f_0.

pulse travels outward from the struck point and is reflected and re-reflected at the boundary of the membrane. If some point of the membrane is forced to vibrate periodically, continuous trains of waves travel along the membrane. Just as with the stretched string, standing waves can be set up in the membrane and each of these waves has a certain natural frequency. The lowest frequency is the fundamental and the others are overtones. In general, when the membrane is vibrating, a number of overtones are present.

The nodes of a vibrating membrane are lines (nodal lines) rather than points. The boundary of the membrane is evidently one such line. Some of the other possible nodal lines of a circular membrane are shown in Fig. 27–8, with the modes of vibration arranged in order of increasing frequency. The natural frequency of each mode is given in terms of the fundamental f_0. It will be noted that the frequencies of the overtones are *not* integral multiples of f_0. That is, they are not harmonics.

The restoring force in a vibrating flexible membrane arises from the tension with which it is stretched. A metal plate, if sufficiently thick, will vibrate in a similar way, the restoring force being produced by bending stresses in the plate. The study of vibrations of membranes and plates is of importance in connection with the design of loudspeaker diaphragms and the diaphragms of telephone receivers and microphones.

27–5 Standing waves in an air column. Compressional waves traveling along a tube of finite length are reflected at the ends of the tube in much the same way that transverse waves in a string are reflected at its ends. Interference between the waves traveling in opposite directions gives rise to standing waves and the tube, like the string, has an infinite number of natural frequencies.

If reflection takes place at a closed end, the displacement of the particles at that end must necessarily be always zero. Hence the end is a *displacement node*. However, since points of zero displacement are points of maximum pressure variation, the end is a *pressure antinode*.

If the end of the tube is open, the nature of the reflection is more complex and depends on whether the tube is wide or narrow compared with the wave length of the sound. If the tube is narrow compared with the sound wave length, which is the case with most musical instruments, the reflection is such as to make the open end a *displacement antinode* or a *pressure node*.* The latter seems reasonable, since if the end is open to the atmosphere, one would expect the pressure at the end to remain constant and equal to atmospheric pressure. Thus (displacement) waves in an air column are reflected at the closed and open ends of a tube in the same way that transverse waves in a string are reflected at fixed and free ends respectively.

The reflections at the opening where the instrument is blown are found to be such that a displacement antinode (pressure node) is located at or near the opening. The effective length of the air column of a wind instrument is thus less definite than the length of a string fixed at its ends.

The vibrations of the air columns of wind instruments are excited by some sort of vibration at one end of the column. These vibrations may be those of the player's lips (trumpet, trombone), or of a reed (clarinet, oboe), or may be set up by an air jet directed across an opening (flute) or against a sharp edge or lip (organ pipe). In any case, the air column picks out of the vibration those Fourier components having frequencies equal to its own fundamental and overtones and resonates to those frequencies.

Figure 27–9 shows the fundamental and first two overtones of an open and a closed organ pipe. The standing *pressure* waves in the pipe are shown; the displacement nodes are located at the pressure antinodes and vice versa. Remembering that the nodes of a standing wave pattern are separated by a half a wave length, it is evident from the diagram that the wave length of the fundamental vibration of an open pipe of length L is $2L$, while that of the fundamental of a closed pipe of the same length is $4L$. The corresponding fundamental frequencies (from $V = f\lambda$) are $V/2L$ for the open pipe and $V/4L$ for the closed pipe. The fundamental frequency of the open pipe is therefore twice that of the closed, and its pitch is an octave higher.

---

*The complete theory shows that the node is not precisely at the end of the tube, but that the effective tube length exceeds the actual length by about six-tenths of the radius.

It will be seen from Fig. 27–9 that the wave lengths of the overtones of the open pipe are $2L/2$, $2L/3$, $2L/4$, etc., and their frequencies are 2, 3, 4, etc., times the fundamental frequency. The overtones of a closed pipe, on the other hand, have wave lengths $4L/3$, $4L/5$, $4L/7$, etc., and therefore frequencies of 3, 5, 7, etc., times the fundamental frequency. It follows that *all* of the harmonics are present in the vibrations of an open pipe but only the *odd* harmonics are present in the vibrations of a closed pipe. The quality of the tone emitted by a closed pipe therefore differs from that of an open pipe even if the pitches or fundamental frequencies are made the same by proper choice of lengths.

27–6 Beats. Standing waves in an air column have been cited as one example of interference. They arise when two wave trains of the same amplitude and frequency are traveling through the same region in opposite directions. We now wish to consider another type of interference which results when two wave trains of equal amplitude but slightly different

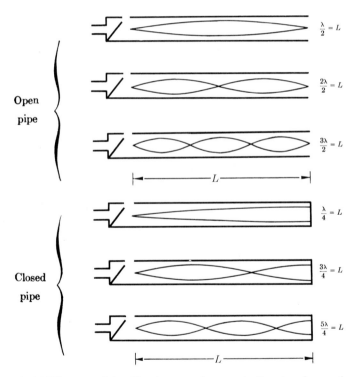

Fig. 27–9. Diagram of standing (pressure) waves in the air column of an organ pipe. A *displacement* antinode is located at each *pressure* node.

frequency travel through the same region. Such a condition exists when two tuning forks of slightly different frequency are sounded simultaneously or when two piano wires struck by the same key are slightly "out of tune."

If we consider some one point of space through which the waves are simultaneously passing, the displacements due to the two waves separately may be written

$$y_1 = Y \cos 2\pi f_1 t,$$

$$y_2 = Y \cos 2\pi f_2 t$$

(the amplitudes are assumed equal).

By the principle of superposition, the resultant displacement is

$$y = y_1 + y_2 = Y[\cos 2\pi f_1 t + \cos 2\pi f_2 t] ,$$

and since

$$\cos a + \cos b = 2 \cos \frac{a+b}{2} \cos \frac{a-b}{2} ,$$

this may be written

$$y = \left[2Y \cos 2\pi \left(\frac{f_1 - f_2}{2} \right) t \right] \cos 2\pi \frac{f_1 + f_2}{2} t. \qquad (27\text{–}3)$$

The resulting vibration can then be considered to be of frequency $(f_1 + f_2)/2$, or the average frequency of the two tones, and of amplitude given by the expression in brackets. The amplitude therefore varies with time at a frequency $(f_1 - f_2)/2$. If f_1 and f_2 are nearly equal this term is small and the amplitude fluctuates very slowly. When the amplitude is large the sound is loud and vice versa. The fluctuations in amplitude are called *beats*, and the number of maxima (or minima) per second is the number of beats per second. A beat, or a maximum of amplitude, will occur when $\cos 2\pi(f_1 - f_2/2)t$ equals 1 or -1. Since each of these values occurs once in each cycle, the number of beats per second is twice the frequency $(f_1 - f_2)/2$, or, *the number of beats per second equals the difference*

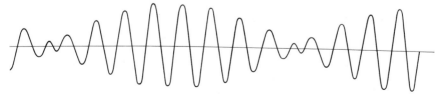

FIG. 27–10. Beats are fluctuations in amplitude (loudness) produced by two sound waves of slightly different frequency.

of the frequencies. Figure 27–10 is a graph illustrating the phenomenon of beats. Advantage is taken of this phenomenon in tuning two musical instruments to the same pitch or to "unison."

27–7 Combination tones. Beats between two tones can be detected by the ear up to a beat frequency of 6 or 7 per second. At higher frequencies, individual beats can no longer be distinguished and the sensation merges into one of *consonance* or *dissonance*, depending on the frequency ratio of the tones. A beat frequency, even though it lies within the frequency range of the ear, is not interpreted by the ear as a tone of that frequency. Nevertheless, a tone can be heard of frequency equal to the frequency difference between two others sounded simultaneously. Such

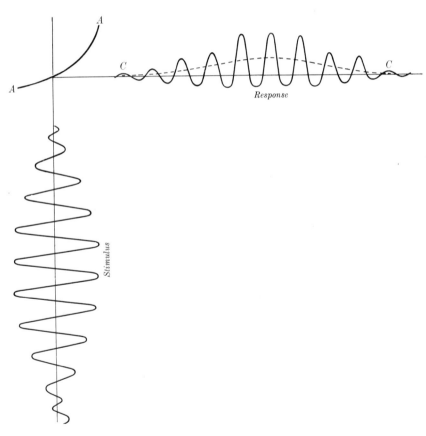

Fig. 27–11. Illustrates production of combination tones.

a tone is called a *difference* tone. Although not as easy to recognize, a frequency called a *summation* tone and equal to the sum of the frequencies of the two tones can also be heard. The general term applied to both difference tones and summation tones is *combination* tones.

Combination tones are similar to beats but are due to the nature of the hearing mechanism. The ear is one of a number of devices whose response is said to be *nonlinear*. A vacuum tube operated on the curved portion of its characteristic is another. Figure 27–11 illustrates the situation schematically. The response to a given stimulus is represented by the curved line AA. (In a vacuum tube the "response" would be the plate current and the "stimulus" the grid potential.) The curve extending vertically downward represents the displacement of the air at a point through which two waves of different frequency are passing simultaneously. The response to this stimulus is obtained by projecting upward from every point of this curve to the curve AA and then projecting across. It will be seen that because of the curvature (or nonlinearity) of AA, the response curve is not symmetrical but has an "average" upward sweep indicated by the dotted line. Hence, if the graph represents the motion of the eardrum, it is seen that the latter will vibrate at a frequency equal to the beat frequency or the difference in frequencies of the two tones. This vibration is superposed on the much higher frequency of the individual waves in the tone. Note carefully that if the response curve AA were *linear*, the curve CC would be symmetrical and no such motion of the eardrum would ensue.

PROBLEMS

27–1. Show how Eq. (27–1) follows from the equation immediately preceding it.

27–2. (a) Construct on the same X-axis, for values of x over a range from zero to 4 inches, graphs of the equations

$$y_1 = Y \cos \frac{2\pi}{\lambda} (x + Vt)$$

and

$$y_2 = Y \cos \frac{2\pi}{\lambda} (x - Vt)$$

at the instant when $t = 0$. Let $Y = 1$ inch, $\lambda = 2$ inches, and $V = \frac{1}{4}$ in/sec. Let these graphs represent an incident wave and a wave reflected at the point $x = 0$. Add the two graphs to obtain the resultant wave form at the instant $t = 0$.

(b) In a second diagram, construct graphs of y_1 and y_2 at the instant when $t = 1$ sec. Add the graphs to obtain the resultant wave form.

(c) Repeat, for the instant when $t = 2$ sec. Answer the following questions: In which direction is the first wave traveling? In which direction is the second wave traveling? At what values of x are the nodes located? The antinodes? What is the amplitude of vibration of a point whose X-coordinate is $\frac{1}{4}$ inch? $\frac{1}{2}$ inch?

27–3. The equation of a transverse wave in a stretched string is

$$y = 4 \sin 2\pi \left(\frac{t}{.02} - \frac{x}{400} \right),$$

where y and x are in centimeters and t is in seconds.

(a) Is the wave a traveling wave or a standing wave?

(b) What is the amplitude of the wave?

(c) What is its wave length?

(d) What is its velocity of propagation?

(e) What is its frequency?

(f) Find the transverse velocity of a point on the string whose X-coordinate is 800 cm, at time 0.08 sec.

FIGURE 27–12

27–4. A flexible rubber tube 10 ft long, of mass 0.4 slug, is stretched with a tension of 25 lb by a long light cord as shown in Fig. 27–12. Standing waves of frequency 3.125 cycles/sec are set up in the tube. Sketch the shape of the tube when the displacement at an antinode is a maximum, showing clearly the positions of the nodes and antinodes.

27–5. A steel wire of length $L = 100$ cm and density $\rho = 8$ gm/cm^3 is stretched tightly between two rigid supports. Vibrating in its fundamental mode, the frequency is $f = 200$ cycles per sec. (a) What is the velocity of transverse waves on this wire? (b) What is the longitudinal stress in the wire (in dynes/cm^2)? (c) If the maximum acceleration at the mid-point of the wire is 80,000 cm/sec^2, what is the displacement amplitude at the mid-point?

27–6. A stretched string is observed to vibrate with a frequency of 30 cycles per second in its fundamental mode when the supports are 60 cm apart. The amplitude at the antinode is 3 cm. The string has a mass of 30 gm. (a) What is the velocity of propagation of a transverse wave in the string? (b) Compute the tension in the string. (c) Write the equation representing this wave motion, using the constants given above and computed in (a).

27–7. A steel piano wire 50 cm long, of mass 5 gm, is stretched with a tension of 400 newtons. (a) What is the frequency of its fundamental mode of vibration? (b) What is the number of the highest overtone that could be heard by a person who can hear frequencies up to 10,000 cycles/sec?

27–8. Suppose the piano wire in Problem 27–7 is set vibrating at twice its fundamental frequency. (a) Sketch the shape of the wire at a few instants. (b) What is the wave length of transverse waves in the wire? (c) What is the wave length, in air, of the sound waves emitted by the wire?

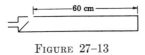

FIGURE 27–13

27–9. The closed organ pipe in Fig. 27–13 is 60 cm long and the standing waves within it have a wave length of 80 cm. (a) Show in a diagram the position of the pressure nodes and antinodes. Indicate clearly which is which. (b) Show in a second diagram the positions of the displacement nodes and antinodes. (c) If the displacement amplitude of the standing waves is 10^{-6} cm, what is the instantaneous displacement of the air particles at the center of the pipe, at an instant when the displacement at a pressure node is a maximum?

27–10. Find the frequencies of the fundamental and the first four overtones of an organ pipe 4 ft long (a) if the pipe is open, (b) if the pipe is closed. End corrections may be neglected.

27–11. Find the fundamental frequency and the first four overtones of a 6-inch pipe (a) if the pipe is open at both ends, (b) if the pipe is closed at one end. (c) How many overtones may be heard by a person having normal hearing for each of the above cases?

27–12. A long tube contains air at a pressure of 1 atm and temperature 77°C. The tube is open at one end and closed at the other by a movable piston. A tuning fork near the open end is vibrating with a frequency of 500 cycles/sec. Resonance is produced (standing waves are set up in the air column) when the piston is at distances 18.0, 55.5, and 93.0 cm from the open end. (a) From these measurements, what is the velocity of sound in air at 77°C? (b) From the above result, what is the ratio of the specific heats γ for air?

27–13. Compare the fundamental frequencies of an open organ pipe 1 meter long (a) when the pipe is filled with air, (b) when it is filled with hydrogen.

27–14. Two identical organ pipes are blown, one with hydrogen and one with helium. (a) If both gases are at the same temperature, which pipe has the higher fundamental frequency? Explain. (b) If the temperature of the hydrogen is 300°K, what should be the temperature of the helium so that both pipes emit a note of the same frequency?

27–15. An organ pipe A of length 2 ft, closed at one end, is vibrating in the first overtone. Another organ pipe B of length 1.35 ft, open at both ends, is vibrating in its fundamental mode. Take the velocity of sound in air as 1120 ft/sec. Neglect end corrections. (a) What is the frequency of the tone from A? (b) What is the frequency of the tone from B? (c) If both operate simultaneously, what is the beat frequency?

27–16. Two identical piano wires when stretched with the same tension have a fundamental frequency of 400 vibrations/sec. By what fractional amount must the tension in one wire be increased in order that 4 beats per sec shall occur when both wires vibrate simultaneously? (Approximate finite changes by differentials.)

27–17. A plate cut from a quartz crystal is often used to control the frequency of an oscillating electrical circuit. Longitudinal standing waves are set up in the plate with displacement antinodes at opposite faces. The fundamental frequency of vibration is given by the equation

$$f_0 = \frac{2.87 \times 10^5}{s},$$

where f_0 is in cycles/sec and s is the thickness of the plate in cm. (a) Compute Young's modulus for the quartz plate. (b) Compute the thickness of plate required for a frequency of 1200 kilocycles/sec (1 kilocycle = 1000 cycles). The density of quartz is 2.66 gm/cm^3.

27–18. An aluminum weight is hung from a steel wire. The fundamental frequency for transverse standing waves on the wire is 300 cycles/sec. The weight is then immersed in water so that one-half of its volume is submerged. What is the new fundamental frequency?

CHAPTER 28

SOUND WAVES. THE EAR AND HEARING

28-1 Intensity. From a purely geometrical point of view, that-which-is-propagated by a traveling wave is the *wave form*. From a physical viewpoint, however, something else is propagated by a wave, namely, *energy*. The most outstanding example, of course, is the energy supply of the earth, which reaches us from the sun via electromagnetic waves. The *intensity* I of a traveling wave is defined as *the time average rate at which energy is transported by the wave per unit area* across a surface perpendicular to the direction of propagation. More briefly, the intensity is the average power transported per unit area.

The energy in a medium through which a sound wave is traveling is in part potential, associated with the energy of compression of the medium, and in part kinetic, associated with the motion of the particles of the medium. Within any small volume the relative amounts of the two forms are continually changing, but the total remains constant. An analogous problem is that of a vibrating mass suspended from a spring (see Section 14–5) where, as we have shown, the energy alternates between potential and kinetic.

Let us, for brevity, call the energy associated with the compression of the medium, *pressure energy*. The work W done on a system in a compression process is

$$W = -\int p\, dv,$$

where v is the volume.

From the definition of compressibility, k,

$$dv = -kv_0\, dp.$$

Hence

$$W = -\int kv_0 p\, dp.$$

In any sound wave encountered in practice, the pressure variations are so small that k and v_0 can be considered constant. The work done in

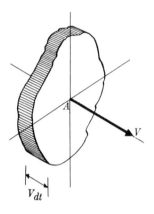

FIGURE 28–1

increasing the gauge pressure from 0 to p is then

$$W = kv_0 \int_0^P p \, dp = \tfrac{1}{2}kv_0 p^2.$$

The *pressure energy per unit volume* is

$$\frac{W}{v_0} = \frac{1}{2} kp^2.$$

At an instant when the gauge pressure p equals the maximum pressure P, the pressure energy is a maximum also. The kinetic energy is therefore zero at this instant, the energy is wholly pressure energy, and the pressure energy equals the total energy. But the total energy is the same at all instants, so the total energy per unit volume is

$$\frac{W}{v_0} = \frac{1}{2} kP^2. \tag{28–1}$$

Consider now the volume element in Fig. 28–1, whose ends of area A are perpendicular to the direction of propagation of the sound wave, and whose length is $V \, dt$, where V is the velocity of propagation. The sound energy in this volume is the product of the energy per unit volume and the volume of the element.

Energy in volume element $= \tfrac{1}{2}kP^2 \times AV \, dt$.

In time dt, all the energy in the volume element will cross the face of area A. Hence the energy crossing, per unit area and per unit time, or the intensity I, is

$$I = \frac{\tfrac{1}{2}kP^2 \times AV \, dt}{A \, dt}$$

$$= \tfrac{1}{2}kP^2V. \tag{28–2}$$

But

$$V = \sqrt{\frac{1}{k\rho_0}},$$

and

$$k = \frac{1}{\rho_0 V^2}.$$

When this expression for k is inserted in Eq. (28–2) we get

$$I = \frac{P^2}{2\rho_0 V} \cdot$$

(28–3)

It will be noted that the *intensity* is proportional to the *square of the amplitude*, a result which is true for any sort of wave motion.

The intensity of a sound wave of pressure amplitude $P = 280$ dynes/cm^2 (roughly, the loudest tolerable sound) is

$$I = \frac{(280)^2}{2 \times 0.00122 \times 3.46 \times 10^4}$$

$$= 940 \text{ ergs per second, per square centimeter}$$

$$= 94 \times 10^{-6} \text{ watt/cm}^2.^{(*)}$$

The pressure amplitude of the faintest sound wave which can be heard is about 0.0002 dyne/cm^2 and the corresponding intensity is about 10^{-16} watt/cm^2.

The total power carried across a surface by a sound wave equals the product of the intensity at the surface area, if the intensity over the surface is uniform. The average power developed as sound waves by a person speaking in an ordinary conversational tone is about 10^{-5} watt, while a loud shout corresponds to about 3×10^{-2} watt. The population of New York City is about six million persons, so that if everyone in New York City were to speak at the same time, the acoustical power developed would be about 60 watts, or enough to operate a moderate sized electric light. On the other hand, the power required to fill a large auditorium with loud sound is considerable. Suppose the intensity over the surface of a hemisphere 20 meters in radius is 10^{-4} watt per square centimeter. The area of the surface is about 25×10^6 cm^2. Hence the acoustic power output of a speaker at the center of the sphere would have to be

$$10^{-4} \times 25 \times 10^6 = 2500 \text{ watts}$$

---

*The "watt/cm^2" is a hybrid unit, neither cgs nor mks. We shall retain it to conform with general usage in acoustics.

or 2.5 kilowatts. The electrical power input to the speaker would need to be considerably larger, since the efficiency of such devices is not very high.

28–2 **Intensity level. The decibel.** Because of the large range of intensities over which the ear is sensitive, a logarithmic rather than an arithmetic intensity scale is more convenient. Accordingly the *intensity level* β of a sound wave is defined by the equation

$$\beta = 10 \log \frac{I}{I_0}, \qquad (28\text{–}4)$$

where I_0 is an arbitrary reference intensity which is taken as 10^{-16} watt/cm^2, corresponding roughly to the faintest sound which can be heard. Intensity levels are expressed in *decibels*, abbreviated db.*

If the intensity of a sound wave equals I_0 or 10^{-16} watt/cm^2, its intensity level is zero. The maximum intensity which the ear can tolerate,

TABLE XXII

NOISE LEVELS DUE TO VARIOUS SOURCES

(Representative values)

Source or description of noise	Noise level (db)
Threshold of pain	120
Riveter	95
Elevated train	90
Busy street traffic	70
Ordinary conversation	65
Quiet automobile	50
Quiet radio in home	40
Average whisper	20
Rustle of leaves	10
Threshold of hearing	0

---

*Originally, a scale of intensity levels in *bels* was defined by the relation

$$\text{Intensity level} = \log \frac{I}{I_0}.$$

This unit proved rather large and hence the decibel, one-tenth of a bel, has come into general use. The unit is named in honor of Alexander Graham Bell.

about 10^{-4} watt/cm^2, corresponds to an intensity level of 120 db. Table XXII gives the intensity levels in db of a number of familiar noises. It is taken from a survey made by the New York City Noise Abatement Commission.

The intensity of a sound wave is a purely objective or physical attribute of the wave, and can be measured by acoustical apparatus without making use of the hearing sense of a human observer. However, if we listen to a sound wave whose intensity is gradually increased, the sensation which we describe as *loudness* increases also. The term loudness is reserved to refer to this sensation, and since it is a sensation or a subjective attribute of a sound wave, loudness cannot be measured by physical apparatus. Nevertheless it is possible to establish a numerical scale of loudness sensation. For details, a more advanced text should be consulted. It is found that while an increase of intensity results in an increase in the loudness sensation, loudness is by no means proportional to intensity. That is, a sound of intensity 10^{-6} watt/cm^2 is not one hundred times as loud as one of intensity 10^{-8} watt/cm^2. Rather, the loudness sensation is more nearly, although not directly, proportional to the logarithm of the intensity or to the intensity level. In other words, the loudness sensation produced by a sound wave of intensity level 60 db exceeds the loudness sensation produced by a wave of intensity level 40 db by approximately the same amount as the sensation produced by the 40 db-wave exceeds that produced by a wave of intensity level 20 db.

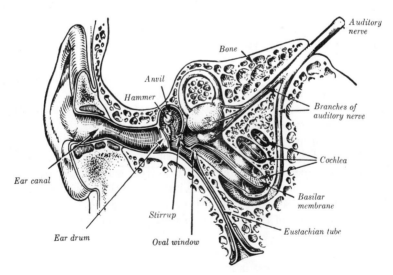

FIG. 28–2. Diagrammatic section of the right ear.

28–3 The ear and hearing. Figure 28–2 is a semidiagrammatic section of the right ear. The scale of the inner ear has been exaggerated in order to show details. Sound waves traveling down the ear canal strike the eardrum. A linkage of three small bones, the hammer, anvil, and stirrup, transmits the vibrations to the oval window. The oval window in turn transmits them to the inner ear, which is filled with fluid. The terminals of the auditory nerve, of which there are about 30,000 in each ear, are distributed along the basilar membrane which divides the spiral channel or cochlea into two canals. The 30,000 nerve terminals actually occupy an area only about 30 millimeters long and $\frac{1}{3}$ of a millimeter wide, a remarkable feat of engineering.

A great deal of work has been done in recent years, notably by Dr. Harvey Fletcher of the Bell Telephone Laboratories, on the processes by which the sound waves set up in the cochlea are picked up by the nerve endings. To represent the process graphically, the cochlea is drawn as a conventionalized spiral (Fig. 28–3). Each division along the spiral refers to a so-called "patch" of 1% of the nerve endings (about 300 terminals).

The width of the blackened strip in Fig. 28–4 shows the extent to which the corresponding nerve patches are stimulated by a 200-cycle tone at an intensity level of 90 db. Diagrams of this sort are called *auditory patterns*. Figure 28–5(a) shows how the extent of the stimulation varies with change in intensity level at a fixed frequency. Figure 28–5(b) shows how the location of the stimulated portions shifts with frequency, at a constant level of 90 db.

Although the response of the ear to a pure tone is not localized at any one point, it will be seen that notes of lower frequency stimulate chiefly those patches near the inner portion of the spiral and vice versa. The position of the region of maximum response to pure tones is shown in Fig. 28–6. When listening to a street noise or a symphony orchestra, all portions of the cochlea will be stimulated to a greater or lesser extent. Figure 28–7 is the auditory pattern of a steamboat whistle.

Defective hearing, or deafness, may be of two types, obstructive deafness and nerve deafness. Both types can, of course, occur in the same ear. In the former type the mechanism transmitting sound from the outside air to the inner ear is defective. In the latter type the nerve terminals are defective. The "bone conduction" type of hearing aid transmits sound directly to the bones of the head, by-passing the defective portions of the ear. It is effective in cases of obstructive deafness but not in nerve deafness.

A case of nerve deafness is represented in the auditory pattern of Fig. 28–8. The significance of the shaded portion is that an auditory pattern must extend above this portion in order to be heard. The blackened

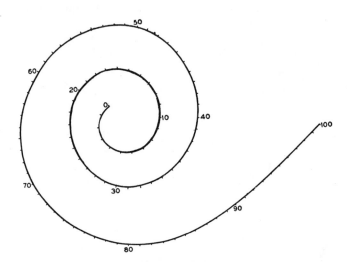

Fig. 28–3. Conventionalized diagram of the cochlea.
(Courtesy of Dr. Harvey Fletcher)

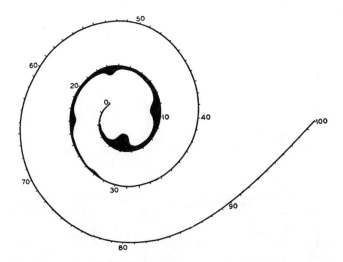

Fig. 28–4. Auditory pattern of a 200-cycle tone at an intensity level of 90 db.
(Courtesy of Dr. Harvey Fletcher)

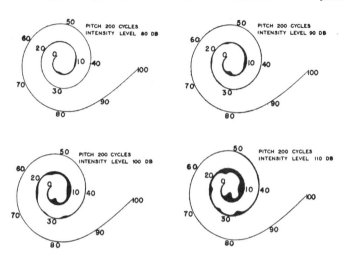

FIG. 28–5(a). Auditory patterns for a 200-cycle tone at various intensity levels
(*Courtesy of Dr. Harvey Fletcher*)

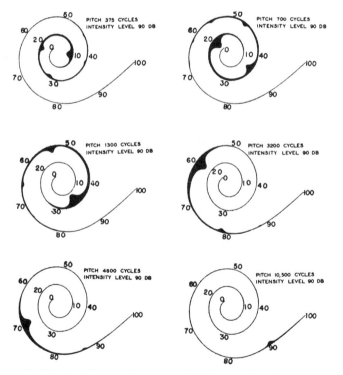

FIG. 28–5(b). Auditory patterns at 90 db for various frequencies.
(*Courtesy of Dr. Harvey Fletcher*)

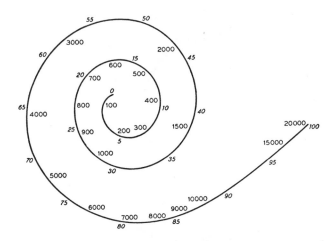

Fig. 28–6. Position in the cochlea for maximum response to pure tones.
(*Courtesy of Dr. Harvey Fletcher*)

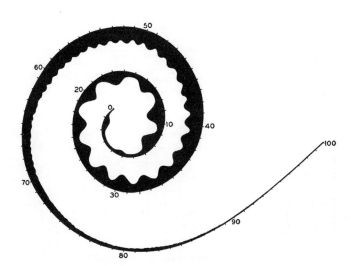

Fig. 28–7. Auditory pattern of a steamboat whistle.
(*Courtesy of Dr. Harvey Fletcher*)

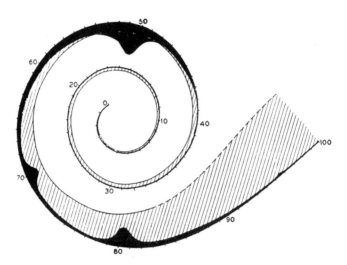

FIG. 28–8. Nerve deafness pattern.
(*Courtesy of Dr. Harvey Fletcher*)

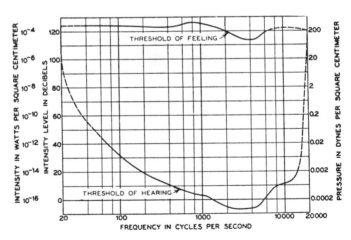

FIG. 28–9. Auditory area between threshold of hearing and threshold of feeling.
(*Courtesy of Dr. Harvey Fletcher*)

portions of the figure are the normal auditory patterns produced by tones of frequencies 2300 cycles and 6500 cycles. A portion of the former pattern projects above the shaded area and will therefore be heard, while the 6500-cycle note will not be heard, since it lies wholly within the shaded area.

The range of frequencies and intensities to which the ear is sensitive are conveniently represented by a diagram like that of Fig. 28–9, which is a graph of the *auditory area* of a person of good hearing. The height of the lower curve at any frequency represents the intensity level of the faintest pure tone of that frequency which can be heard. It will be seen from the diagram that the ear is most sensitive to frequencies between 2000 and 3000 cycles, where the *threshold of hearing*, as it is called, is about −5 db. The height of the upper curve at any frequency corresponds to the intensity level of the loudest pure tone of that frequency which can be tolerated. At intensitives above this curve, which is called the *threshold of feeling*, the sensation changes from one of hearing to discomfort or even pain. The height of the upper curve is approximately constant at a level of about 120 db for all frequencies. Every pure tone which can be heard may be represented by a point lying somewhere in the area between these two curves.

Only about 1% of the population has a threshold of hearing as low as the bottom curve in Fig. 28–9. 50% of the population can hear pure tones of a frequency of 2500 cycles when the intensity level is about 8 db, and 90% when the level is 20 db.

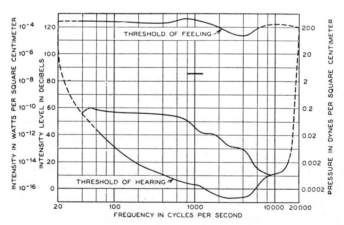

Fig. 28–10. Spectrogram of street noise.
(*Courtesy of Dr. Harvey Fletcher*)

It was stated earlier that the frequency range of the ear was from about 20 to about 20,000 cycles/sec and the intensity range from about 10^{-16} to 10^{-4} watt/cm^2, or zero to 120 db. We see now that these statements must be qualified. That is, for a loud tone of intensity level 80 db, the frequency range is from 20 to 20,000 cycles/sec, but at a level of 20 db it is only from about 200 to about 15,000 cycles/sec. At a frequency of 1000 cycles/sec the range of intensity level is from about 3 db to about 120 db, whereas at 100 cycles/sec it is only from 30 db to 120 db.

Sounds which are not pure tones are not associated with any one frequency and hence cannot be represented by a single point on the diagram. A sound such as that from a musical instrument, consisting of a mixture of a relatively few frequencies (the fundamental and overtones), can be represented by a set of points, each point giving the intensity and frequency of one particular overtone. A sound such as a street noise, while it cannot be considered as made up of a fundamental and overtones, can nevertheless be represented on the diagram in the following way. The sound is picked up by a microphone and sent through an electrical network which selects a narrow range of frequencies and measures the average intensity within this range. By repeating the process at a large number of frequencies throughout the audible range a series of points are obtained which can be plotted. A continuous curve drawn through them is called the *spectrogram* of the sound. A typical spectrogram of street noise is shown in Fig. 28–10.

The term "spectrogram" is borrowed from optics. The process just described is entirely analogous to the optical one of dispersing a beam of light waves into a spectrum by means of a prism and measuring the intensity at a number of points throughout the spectrum. The light emitted by a gas in an electrical discharge is a mixture of waves of a number of definite frequencies and corresponds to the sound emitted by a musical instrument. Most light beams, however, are a mixture of all frequencies and are therefore the optical analogue of noise.

The total intensity level of a noise can be found from its spectrogram by an integration process. There are also instruments known as noise meters which measure the level directly. The level of the street noise in Fig. 28–10 is about 85 db and is shown by the short heavy line.

28–4 The Doppler effect. When a source of sound, or a listener, or both, are in motion relative to the air, the observed pitch as heard by the listener is, in general, not the same as when source and listener are at rest. Perhaps the most common example is the sudden drop in pitch of an automobile horn which takes place just as one meets and passes an automobile proceeding in the opposite direction. The general term applied to this phenomenon is the *Doppler effect*.

FIGURE 28-11

In Fig. 28-11, S represents a source of sound moving toward the right with a velocity v_S and emitting sound waves of a frequency f_0. The listener, moving toward the right with a velocity v_L, is represented by L. For generality, let us assume that the medium (air) is also in motion toward the right with a velocity v_m. Velocities v_S, v_L, and v_m are all relative to the earth.

A sound wave emitted by the source S at time $t = 0$ advances *relative to the medium* with a velocity of propagation V. (The velocity of propagation of sound waves in or relative to a medium is a property of the medium only and is independent of the velocity of the source. The waves forget about the source as soon as they leave it.) Since the medium is moving to the right with velocity v_m, the velocity of the emitted waves toward the right, relative to the earth, is the sum of their velocity relative to the medium and the velocity of the medium relative to the earth, or it is $V + v_m$. Hence in time t, a wave advances a distance $(V + v_m)t$ toward the right. The source, in the same time, has advanced a distance $v_S t$ and has emitted $f_0 t$ waves. Hence $f_0 t$ waves occupy the distance between the source and the wave emitted at time $t = 0$, or a distance $(V + v_m)t - v_S t = (V + v_m - v_S)t$. The distance between any two consecutive waves, or the wave length, is therefore

$$\lambda = \frac{(V + v_m - v_S)t}{f_0 t} = \frac{V + v_m - v_S}{f_0}. \tag{28-5}$$

Consider next the listener. Sound waves traveling with a velocity $V + v_m$ are passing him, but his own velocity is v_L. Hence the velocity of the waves relative to the listener is $V + v_m - v_L$. The number of waves that pass the listener per unit time, or the apparent frequency f, is the ratio of the relative velocity to the wave length, or

$$f = \frac{V + v_m - v_L}{(V + v_m - v_S)/f_0},$$

$$\frac{f}{f_0} = \frac{V + v_m - v_L}{V + v_m - v_S}, \tag{28-6}$$

where f/f_0 is the ratio of apparent to true frequency.

Careful attention must be paid to the construction of the diagram and to algebraic signs when using this equation. The diagram must be drawn as in Fig. 28–11, with the source at the left of the listener, and all velocities shown by vectors. If, in a given case, any velocity is opposite to that in Fig. 28–11, its sign should be reversed in Eq. (28–6). The velocity of propagation, V, is considered positive always.

EXAMPLES. (1) A stationary source in still air emits a sound wave of frequency f_0. What is the apparent frequency heard by a listener approaching the source with a velocity of magnitude v_L?

See Fig. 28–12 (a). Since the medium and the source are at rest, $v_m = v_S = 0$, and since v_L is directed toward the left,

$$f = f_0 \frac{V + v_L}{V}.$$

(a)

The apparent frequency is higher than the true frequency, in agreement with common experience.

(2) A listener moves away from a stationary source in still air with a velocity of magnitude v_L. Find the ratio of apparent to true frequency.

(b)

FIGURE 28–12

See Fig. 28–12(b). Evidently

$$f = f_0 \frac{V - v_L}{V},$$

and the apparent frequency drops when the listener recedes from the source.

The preceding analysis is a special case in which v_S, v_L, and v_m are all parallel to the line joining the source and the listener. The interested reader can readily work out the more general case.

The Doppler effect is not confined to sound waves. The light from a star which is approaching the earth is of somewhat higher frequency or shorter wave length than it would be if the two were at relative rest. Much valuable information regarding stellar motions has been obtained in this way.

28–5 Reflection of sound waves. We have already mentioned the reflection of sound waves at the open or closed ends of an organ pipe. The familiar phenomenon of an echo also arises from the reflection of sound. The reflection of a three-dimensional sound wave from a surface which it strikes can be treated mathematically by an extension of the methods in Section 27–1 for analyzing the problem of reflection of transverse waves at the end of a string. We shall avoid the mathematics with the help of a useful graphical method called Huygens' principle. The same principle is extremely useful in dealing with light waves.

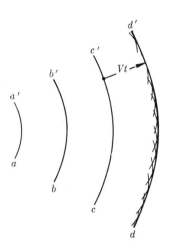

FIG. 28-13. Huygens' method for finding the shape of a wave.

In Fig. 28–13, the point S represents a small source of sound and the circular arcs aa', bb', etc., are traces of sound waves spreading out from the source with velocity V. Suppose we know that at some instant a wave has the shape cc' and we wish to find its shape after a time interval t. Huygens' principle states that every point of the wave cc' may be considered a "secondary" source from which there spread out spherical wavelets with the velocity V. In a time interval t each wavelet advances a distance Vt, and the new position of the wave is found by drawing the common tangent to the wavelets or, as it is called, their envelope. In Fig. 28–13, this is the wave dd'.

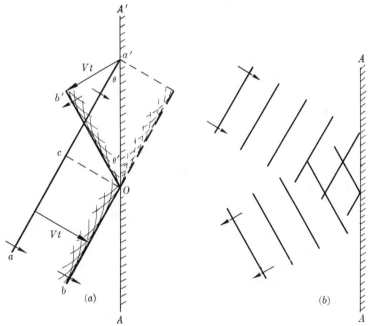

FIG. 28–14. (a) Two stages in the reflection of a plane wave from a plane surface. (b) Reflection of a train of plane waves from a plane surface.

We now apply Huygens' principle to the reflection of a sound wave. For simplicity, let a plane wave aa' in Fig. 28–14(a) strike a fixed plane surface AA'. Construct Huygens' wavelets of radius Vt from each point of aa'. The envelope of those from the lower part of aa' is the portion of the new wave bO. The wavelets from the upper part of aa', had the surface not been present, would have advanced to the positions of the dotted circles. Actually, since they cannot penetrate the surface, they spread out in the opposite direction and their envelope is the portion of the new wave Ob'.

The angle θ between the incident wave aa' and the surface is called the *angle of incidence;* the angle θ' between the reflected wave Ob' and the surface is the *angle of reflection.* It is easy to see that the right triangles $Ob'a'$ and Oca' are similar and hence $\theta' = \theta$. That is, a plane sound wave is reflected from a plane surface with the angle of reflection equal to the angle of incidence. It will be recognized that this is the same law governing the reflection of light waves from a plane mirror.

Figure 28–14(b) shows successive stages in the reflection of a plane wave from a plane surface, or it may be considered to represent a "snapshot" of a train of plane waves, incident on and reflected from the surface.

If the surface from which the sound wave is reflected is perfectly rigid, there is no loss of energy in the process of reflection. No actual surface fulfills this requirement, but yields to some extent under the pressure of the wave. Furthermore, if the surface is at all porous, the air in the pores or cavities is set into turbulent motion. Consequently, some heat is always developed when sound waves strike a surface and the reflected energy is less than the incident energy. The energy lost by the sound wave is said to be *absorbed* by the surface.

28–6 Acoustics of rooms.   Reverberation time.   The study of the proper design of an auditorium for best hearing conditions was first undertaken by Prof. W. C. Sabine of Harvard in 1895. Sabine's contributions to this part of the science of acoustics were, first, his discovery that the acoustic properties of a room are determined to a large extent by the rate at which sound energy is absorbed at its walls, floor, and ceiling, as well as by an audience which may be present, and second, that the rate of sound absorption is closely connected with the time required for a sound in the room to die out after the source of sound has been cut off.

When a source of sound is started in a room, a certain time is required for the system of sound waves to build up to a steady state. Although the source is continually supplying energy, the sound energy in the room does not increase indefinitely, because of absorption. If the source is suddenly cut off the sound does not cease immediately since a certain time is required for the sound energy in the room to reach the walls and be absorbed. The persistence of sound in a room after the source has been cut off is called *reverberation*. The *reverberation time* of a room is defined arbitrarily as the time required for the intensity to decrease to one one-millionth of its original value, or for the intensity level to decrease by 60 db. It is found that this time is nearly independent of the original sound level and of the quality of the sound. Instruments are now available which read the reverberation time directly.

If the sound absorption is large the reverberation time is small. When this is the case, the sound level which can be built up by a source of given acoustic power, such as a person speaking, is small, and the speaker will have difficulty in making himself heard throughout the room because of low intensity. The room is said to sound "dead." On the other hand, if the absorption is small and the reverberation time long, the speaker's words may be unintelligible because one syllable will still be heard with appreciable intensity after the next is spoken. It is found that for best hearing conditions the reverberation time should lie between one and two seconds.

The absorbing properties of a surface are defined quantitatively as follows. When a sound wave strikes a surface a certain fraction, say α, of its intensity is absorbed and the remainder, $(1 - \alpha)$, is reflected. The quantity α is called the *absorption coefficient* of the surface. Some typical values are listed in Table XXIII. If the intensity of the incident wave is I_0 (not to be confused with the reference level I_0 of 10^{-16} watt/cm^2 or 0 db) the intensity after one reflection is $I_0(1 - \alpha)$, after two reflections it is $I_0(1 - \alpha)^2$, and so on. To obtain the intensity after time t, we must compute the number of reflections in time t. This may be done by assuming some average distance between reflections, which is usually taken as

$$4 \times \frac{\text{Volume of room}}{\text{Area of room*}}.$$

(This is equivalent to a distance equal to $\frac{2}{3}$ of the length of one side if the room is a cube.)

TABLE XXIII

ABSORPTION COEFFICIENTS

(Representative values)

Material	Absorption coefficient (α)
Brick wall	0.03
Carpet	0.30
Celotex	0.35
Glass	0.02
Hair felt	0.50
Linoleum	0.02
Plaster	0.02

*Total area of floor, walls, and ceiling.

In time t, the wave travels a distance Vt, and the number of reflections in this time is the distance traveled divided by the average distance between reflections. Hence the intensity I at time t is

$$I = I_0(1 - \alpha)^{(Vt/4)\times(\text{Area}/\text{Volume})}.$$

The reverberation time is defined as the time when $I = 10^{-6} \times I_0$. If T represents this time,

$$10^{-6}I_0 = I_0(1 - \alpha)^{(VT/4)\times(\text{Area}/\text{Volume})},$$

or, taking natural logarithms of both sides,

$$2.3 \times (-6) = \frac{VT}{4} \times \frac{\text{Area}}{\text{Volume}} \ln(1 - \alpha).$$

Now $\ln(1 + \alpha) = \alpha - (\alpha^2/2) + (\alpha^3/3) \ldots$

It will be seen from Table XXIII that α is a small quantity for most surfaces. Hence we need retain only the first term and we obtain, approximately,

$$T = 0.16 \times \frac{\text{Volume}}{\text{Area} \times \alpha}. \tag{28-7}$$

(T in seconds, Volume in cubic meters, Area in square meters.)

In the preceding derivation, the absorption coefficient was assumed to be the same over the entire surface of the room. If it is not, the term Area $\times \alpha$ must be replaced by

$$A_1\alpha_1 + A_2\alpha_2 + \ldots = \Sigma A\alpha,$$

where A_1, A_2, etc., are those areas having absorption coefficients α_1, α_2, etc.

On the basis of Eq. (28-7), the acoustical engineer can compute in advance with sufficient precision what the reverberation time of any room will be.

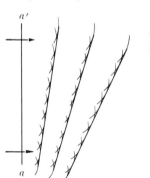

FIG. 28-15. Refraction of a sound wave.

28–7 Refraction of sound. We have shown that the velocity of propagation of a sound wave in a gas varies with the temperature. Consider a plane sound wave in air such as aa' in Fig. 28–15. The plane of the wave is vertical and the wave is advancing from left to right. Suppose the temperature of the air, and hence the velocity of propagation, increases with elevation. The Huygens' wavelets at the top of the wave then have a larger radius than those at the bottom, with the result that the wave continually alters direction as shown. This change in direction, caused by variations in velocity from point to point, is called *refraction*.

In general, the temperature of the air over the earth's surface is not the same at all points and it will be evident that refraction of sound is an important factor in the process of sound ranging, that is, the location of a source of sound such as a distant plane or an exploding shell or bomb.

28–8 Interference of sound waves. The standing waves in a stretched string or an air column have been cited as examples of interference, that is, the combined effect of two or more wave trains simultaneously passing through a given region. We consider next a similar problem in two dimensions. Let S and S' in Fig. 28–16 represent two sources of ripples on a liquid surface. The sources are in phase and emitting waves of the same

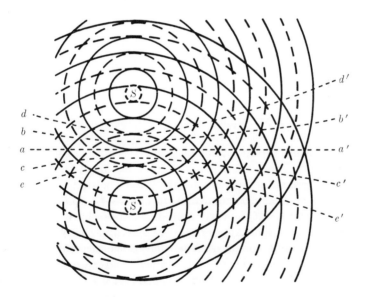

Fig. 28–16. Interference of two wave trains of the same frequency.

frequency and amplitude. The solid circles represent elevations and the dotted circles depressions in the liquid surface. At all points along the line aa' two elevations or two depressions arrive simultaneously from both sources. The resultant amplitude at points in this line is therefore twice that which would result from a single source. At all points along the lines bb' and cc' an elevation from one source arrives at the same instant as a depression from the other. Hence one wave cancels the other at all instants, and along these lines the surface remains at rest. The waves are said to interfere *constructively* along aa' and *destructively* along bb' and cc'. Along lines dd' and ee' we again have constructive interference, and so on.

From Fig. 28–16 the reader can mentally construct the corresponding diagram for three-dimensional sound waves by imagining a rotation of 180° about an axis through the two sources. The circles then trace out spheres in space, and the lines aa', etc., trace out surfaces of constructive or destructive interference.

If the ear of a listener is located anywhere on a surface of destructive interference, no sound is heard. On a surface of constructive interference, the amplitude of the sound is twice and its intensity four times that produced by either source alone. Thus a listener whose ear was on a surface such as bb' would hear a sound if either source alone were emitting waves, but would hear nothing if both sources were emitting simultaneously.

Interference effects of this sort between direct and reflected waves may sometimes be noted in auditoriums. Such effects are obviously undesirable.

28–9 Diffraction of sound. The reflecting surface in Fig. 28–14 was purposely taken considerably larger than the breadth of the train of sound waves. We may next inquire what would happen if a train of waves advancing as in Fig. 28–17 were to encounter an obstacle A having the relative dimensions shown. A complete analysis of the situation is beyond the scope of this book. It will suffice to state that if the dimensions of the obstacle are large compared with the wave length of the sound waves, Huygens' principle may be applied, as in Fig. 28–14, to find the shape of the reflected waves. If, however, the dimensions of the obstacle are of the same order of magnitude as the wave length of the sound waves, the process is much more complex. A portion of the incident wave "bends" or "flows" around the obstacle and continues to advance toward the right. Superposed on this wave is another so-called *scattered* wave which spreads out in all directions from the obstacle, but with an intensity which is different in different directions. The general term applied to the phenomenon is *diffraction*.

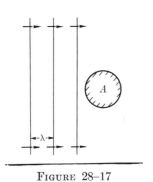

FIGURE 28–17

Diffraction effects are of importance in acoustical engineering, since Fig. 28–17 might represent the sound waves from a person speaking and A might be a microphone. The response of a microphone depends on the pressure variations at its surface and these in turn are determined by the precise nature of the diffraction effects. A text on the mathematical theory of sound should be consulted for further details.

Diffraction effects are also exhibited by light waves when they encounter an obstacle. The reason that diffraction of light is a less familiar phenomenon than, say, the regular reflection of light by a mirror, is that the wave lengths of visible light waves (about 5×10^{-5} cm) are so small that most optical instruments are large by comparison. On the other hand, the wave length in air of a 500-cycle sound wave is about 2 ft and evidently the dimensions of much acoustical apparatus, such as microphones and loudspeakers, are of the same order of magnitude. Hence diffraction effects are of relatively greater importance in acoustics than in optics.

The upper frequency limit of audible sound is about 15,000 cycles/sec, corresponding to a wave length in air of about one inch. Shorter wave lengths (higher frequencies) than this are called *supersonic* waves. Such waves are, of course, inaudible, but they can be produced and detected by the same type of mechanical or electrical instruments as are used in the audible region. The dimensions of apparatus are readily made much larger than the wave lengths of supersonic waves so that diffraction effects are of less importance and supersonic waves can readily be reflected, refracted, focused, etc., in the same way as light waves.

PROBLEMS

28–1. (a) If the pressure amplitude in a sound wave is tripled, by how many times is the intensity of the wave increased? (b) By how many times must the pressure amplitude of a sound wave be increased in order to increase the intensity by a factor of 16 times?

28–2. (a) Two sound waves, one in air and one in water, are equal in intensity. What is the ratio of the pressure amplitude of the wave in water to that of the wave in air? (b) If the pressure amplitudes of the waves are equal, what is the ratio of their intensities?

28–3. Relative to the arbitrary reference intensity of 10^{-16} watt/cm^2, what is the intensity level in db of a sound wave whose intensity is 10^{-10} watt/cm^2? What is the intensity level of a sound wave in air whose pressure amplitude is 2 dynes/cm^2?

28–4. (a) Show that if β_1 and β_2 are the intensity levels in db of sounds of intensities I_1 and I_2 respectively, the difference in intensity levels of the sounds is

$$\beta_2 - \beta_1 = 10 \log \frac{I_2}{I_1}.$$

(b) Show that if P_1 and P_2 are the pressure amplitudes of two sound waves, the difference in intensity levels of the waves is

$$\beta_2 - \beta_1 = 20 \log \frac{P_2}{P_1}.$$

(c) Show that if the reference level of intensity is $I_0 = 10^{-16}$ watt/cm^2, the intensity level of a sound of intensity I is

$$\beta = 160 + 10 \log I.$$

28–5. Two loud-speakers, A and B, radiate sound uniformly in all directions. The output of acoustic power from A is 8×10^{-4} watt, and from B it is 13.5×10^{-4} watt. Both loudspeakers are vibrating in phase at a frequency of 173 cycles/sec. (a) De-

termine the difference in phase of the two signals at a point C, 3 m from B and 4 m from A. (b) Determine the intensity at C from speaker A if speaker B is turned off, and the intensity at C from speaker B if speaker A is turned off. (c) With both speakers on, what is the intensity and intensity level at C?

28–6. The intensity due to a number of independent sound sources is the sum of the individual intensities. How many db greater was the intensity level when all five of the quintuplets cried simultaneously than when a single one cried? How many more crying babies would be required to produce a further increase in the intensity level of the same number of db?

28–7. A window whose area is 1 m^2 opens on a street where the street noises result in an intensity level, at the window, of 60 db. How much "acoustic power" enters the window via the sound waves?

28–8. (a) What are the upper and lower limits of intensity level of a person whose auditory area is represented by the graph of Fig. 28–9? (b) What are the highest and lowest frequencies he can hear when the intensity level is 40 db?

28–9. Two whistles, A and B, each have a frequency of 500 cycles/sec. A is stationary and B is moving toward the right (away from A) at a velocity of 200 ft/sec. An observer is between the two whistles, moving toward the right with a velocity of 100 ft/sec. Take the velocity of sound in air as 1100 ft/sec. (a) What is the frequency from A as heard by the observer? (b) What is the frequency from B as heard by the observer? (c) What is the beat frequency heard by the observer?

28–10. A railroad train is traveling at 100 ft/sec in still air. The frequency of the note emitted by the locomotive whistle is 500 cycles/sec. What is the wave length of the sound waves (a) in front of, (b) behind the locomotive? What would be the frequency of the sound heard by a stationary listener (c) in front of, (d) behind the locomotive? What frequency would be heard by a passenger on a train traveling at 50 ft/sec and (e) approaching the first, (f) receding from the first? (g) How is each of the preceding answers altered if a wind of velocity 30 ft/sec is blowing in the same direction as that in which the locomotive is traveling?

28–11. A source of sound waves S, emitting waves of frequency 1000 cycles/sec, is traveling toward the right in still air with a velocity of 100 ft/sec. At the right of the source is a large smooth reflecting surface moving toward the left with a velocity of 400 ft/sec. (a) How far does an emitted wave travel in 0.01 sec? (b) What is the wave length of the emitted waves in front of (i.e., at the right of) the source? (c) How many waves strike the reflecting surface in 0.01 sec? (d) What is the velocity of the reflected waves? (e) What is the wave length of the reflected waves?

28–12. The floor of a room measures 5 m by 10 m, and the room is 3 m high. (a) What is the reverberation time if the average absorption coefficient of all surfaces is 0.05? (b) To what value would the reverberation time be reduced if the ceiling were covered with Celotex and a carpet put on the floor?

28–13. Find the wave lengths, in air and in water, of supersonic waves of frequency (a) 50,000 cycles/sec, (b) 10^6 cycles/sec.

SUGGESTED BOOKS FOR COLLATERAL READING

CULVER, C. A., *Musical Acoustics.* Blakiston.

FLETCHER, H., *Speech and Hearing.* Van Nostrand.

MILLER, D. C., *The Science of Musical Sounds.* Macmillan.

MORSE, P. M., *Vibration and Sound.* McGraw-Hill.

COMMON LOGARITHMS

N	0	1	2	3	4	5	6	7	8	9
50	6990	6998	7007	7016	7024	7033	7042	7050	7059	7067
51	7076	7084	7093	7101	7110	7118	7126	7135	7143	7152
52	7160	7168	7177	7185	7193	7202	7210	7218	7226	7235
53	7243	7251	7259	7267	7275	7284	7292	7300	7308	7316
54	7324	7332	7340	7348	7356	7364	7372	7380	7388	7396
55	7404	7412	7419	7427	7435	7443	7451	7459	7466	7474
56	7482	7490	7497	7505	·7513	7520	7528	7536	7543	7551
57	7559	7566	7574	7582	7589	7597	7604	7612	7619	7627
58	7634	7642	7649	7657	7664	7672	7679	7686	7694	7701
59	7709	7716	7723	7731	7738	7745	7752	7760	7767	7774
60	7782	7789	7796	7803	7810	7818	7825	7832	7839	7846
61	7853	7860	7868	7875	7882	7889	7896	7903	7910	7917
62	7924	7931	7938	7945	7952	7959	7966	7973	7980	7987
63	7993	8000	8007	8014	8021	8028	8035	8041	8048	8055
64	8062	8069	8075	8082	8089	8096	8102	8109	8116	8122
65	8129	8136	8142	8149	8156	8162	8169	8176	8182	8189
66	8195	8202	8209	8215	8222	8228	8235	8241	8248	8254
67	8261	8267	8274	8280	8287	8293	8299	8306	8312	8319
68	8325	8331	8338	8344	8351	8357	8363	8370	8376	8382
69	8388	8395	8401	8407	8414	8420	8426	8432	8439	8445
70	8451	8457	8463	8470	8476	8482	8488	8494	8500	8506
71	8513	8519	8525	8531	8537	8543	8549	8555	8561	8567
72	8573	8579	8585	8591	8597	8603	8609	8615	8621	8627
73	8633	8639	8645	8651	8657	8663	8669	8675	8681	8686
74	8692	8698	8704	8710	8716	8722	8727	8733	8739	8745
75	8751	8756	8762	8768	8774	8779	8785	8791	8797	8802
76	8808	8814	8820	8825	8831	8837	8842	8848	8854	8859
77	8865	8871	8876	8882	8887	8893	8899	8904	8910	8915
78	8921	8927	8932	8938	8943	8949	8954	8960	8965	8971
79	8976	8982	8987	8993	8998	9004	9009	9015	9020	9025
80	9031	9036	9042	9047	9053	9058	9063	9069	9074	9079
81	9085	9090	9096	9101	9106	9112	9117	9122	9128	9133
82	9138	9143	9149	9154	9159	9165	9170	9175	9180	9186
83	9191	9196	9201	9206	9212	9217	9222	9227	9232	9238
84	9243	9248	9253	9258	9263	9269	9274	9279	9284	9289
85	9294	9299	9304	9309	9315	9320	9325	9330	9335	9340
86	9345	9350	9355	9360	9365	9370	9375	9380	9385	9390
87	9395	9400	9405	9410	9415	9420	9425	9430	9435	9440
88	9445	9450	9455	9460	9465	9469	9474	9479	9484	9489
89	9494	9499	9504	9509	9513	9518	9523	9528	9533	9538
90	9542	9547	9552	9557	9562	9566	9571	9576	9581	9586
91	9590	9595	9600	9605	9609	9614	9619	9624	9628	9633
92	9638	9643	9647	9652	9657	9661	9666	9671	9675	9680
93	9685	9689	9694	9699	9703	9708	9713	9717	9722	9727
94	9731	9736	9741	9745	9750	9754	9759	9763	9768	9773
95	9777	9782	9786	9791	9795	9800	9805	9809	9814	9818
96	9823	9827	9832	9836	9841	9845	9850	9854	9859	9863
97	9868	9872	9877	9881	9886	9890	9894	9899	9903	9908
98	9912	9917	9921	9926	9930	9934	9939	9943	9948	9952
99	9956	9961	9965	9969	9974	9978	9983	9987	9991	9996
100	0000	0004	0009	0013	0017	0022	0026	0030	0035	0039
N	0	1	2	3	4	5	6	7	8	9

COMMON LOGARITHMS

N	0	1	2	3	4	5	6	7	8	9
0	....	0000	3010	4771	6021	6990	7782	8451	9031	9542
1	0000	0414	0792	1139	1461	1761	2041	2304	2553	2788
2	3010	3222	3424	3617	3802	3979	4150	4314	4472	4624
3	4771	4914	5051	5185	5315	5441	5563	5682	5798	5911
4	6021	6128	6232	6335	6435	6532	6628	6721	6812	6902
5	6990	7076	7160	7243	7324	7404	7482	7559	7634	7709
6	7782	7853	7924	7993	8062	8129	8195	8261	8325	8388
7	8451	8513	8573	8633	8692	8751	8808	8865	8921	8976
8	9031	9085	9138	9191	9243	9294	9345	9395	9445	9494
9	9542	9590	9638	9685	9731	9777	9823	9868	9912	9956
10	0000	0043	0086	0128	0170	0212	0253	0294	0334	0374
11	0414	0453	0492	0531	0569	0607	0645	0682	0719	0755
12	0792	0828	0864	0899	0934	0969	1004	1038	1072	1106
13	1139	1173	1206	1239	1271	1303	1335	1367	1399	1430
14	1461	1492	1523	1553	1584	1614	1644	1673	1703	1732
15	1761	1790	1818	1847	1875	1903	1931	1959	1987	2014
16	2041	2068	2095	2122	2148	2175	2201	2227	2253	2279
17	2304	2330	2355	2380	2405	2430	2455	2480	2504	2529
18	2553	2577	2601	2625	2648	2672	2695	2718	2742	2765
19	2788	2810	2833	2856	2878	2900	2923	2945	2967	2989
20	3010	3032	3054	3075	3096	3118	3139	3160	3181	3201
21	3222	3243	3263	3284	3304	3324	3345	3365	3385	3404
22	3424	3444	3464	3483	3502	3522	3541	3560	3579	3598
23	3617	3636	3655	3674	3692	3711	3729	3747	3766	3784
24	3802	3820	3838	3856	3874	3892	3909	3927	3945	3962
25	3979	3997	4014	4031	4048	4065	4082	4099	4116	4133
26	4150	4166	4183	4200	4216	4232	4249	4265	4281	4298
27	4314	4330	4346	4362	4378	4393	4409	4425	4440	4456
28	4472	4487	4502	4518	4533	4548	4564	4579	4594	4609
29	4624	4639	4654	4669	4683	4698	4713	4728	4742	4757
30	4771	4786	4800	4814	4829	4843	4857	4871	4886	4900
31	4914	4928	4942	4955	4969	4983	4997	5011	5024	5038
32	5051	5065	5079	5092	5105	5119	5132	5145	5159	5172
33	5185	5198	5211	5224	5237	5250	5263	5276	5289	5302
34	5315	5328	5340	5353	5366	5378	5391	5403	5416	5428
35	5441	5453	5465	5478	5490	5502	5514	5527	5539	5551
36	5563	5575	5587	5599	5611	5623	5635	5647	5658	5670
37	5682	5694	5705	5717	5729	5740	5752	5763	5775	5786
38	5798	5809	5821	5832	5843	5855	5866	5877	5888	5899
39	5911	5922	5933	5944	5955	5966	5977	5988	5999	6010
40	6021	6031	6042	6053	6064	6075	6085	6096	6107	6117
41	6128	6138	6149	6160	6170	6180	6191	6201	6212	6222
42	6232	6243	6253	6263	6274	6284	6294	6304	6314	6325
43	6335	6345	6355	6365	6375	6385	6395	6405	6415	6425
44	6435	6444	6454	6464	6474	6484	6493	6503	6513	6522
45	6532	6542	6551	6561	6571	6580	6590	6599	6609	6618
46	6628	6637	6646	6656	6665	6675	6684	6693	6702	6712
47	6721	6730	6739	6749	6758	6767	6776	6785	6794	6803
48	6812	6821	6830	6839	6848	6857	6866	6875	6884	6893
49	6902	6911	6920	6928	6937	6946	6955	6964	6972	6981
50	6990	6998	7007	7016	7024	7033	7042	7050	7059	7067
N	0	1	2	3	4	5	6	7	8	9

NATURAL TRIGONOMETRIC FUNCTIONS

Angle	Sine	Cosine	Tangent	Angle	Sine	Cosine	Tangent
0°	0.000	1.000	0.000				
1°	.018	1.000	.018	46°	.719	.695	1.036
2°	.035	0.999	.035	47°	.731	.682	1.072
3°	.052	.999	.052	48°	.743	.669	1.111
4°	.070	.998	.070	49°	.755	.656	1.150
5°	.087	.996	.088	50°	.766	.643	1.192
6°	.105	.995	.105	51°	.777	.629	1.235
7°	.122	.993	.123	52°	.788	.616	1.280
8°	.139	.990	.141	53°	.799	.602	1.327
9°	.156	.988	.158	54°	.809	.588	1.376
10°	.174	.985	.176	55°	.819	.574	1.428
11°	.191	.982	.194	56°	.829	.559	1.483
12°	.208	.978	.213	57°	.839	.545	1.540
13°	.225	.974	.231	58°	.848	.530	1.600
14°	.242	.970	.249	59°	.857	.515	1.664
15°	.259	.966	.268	60°	.866	.500	1.732
16°	.276	.961	.287	61°	.875	.485	1.804
17°	.292	.956	.306	62°	.883	.470	1.881
18°	.309	.951	.325	63°	.891	.454	1.963
19°	.326	.946	.344	64°	.899	.438	2.050
20°	.342	.940	.364	65°	.906	.423	2.145
21°	.358	.934	.384	66°	.914	.407	2.246
22°	.375	.927	.404	67°	.921	.391	2.356
23°	.391	.921	.425	68°	.927	.375	2.475
24°	.407	.914	.445	69°	.934	.358	2.605
25°	.423	.906	.466	70°	.940	.342	2.747
26°	.438	.899	.488	71°	.946	.326	2.904
27°	.454	.891	.510	72°	.951	.309	3.078
28°	.470	.883	.532	73°	.956	.292	3.271
29°	.485	.875	.554	74°	.961	.276	3.487
30°	.500	.866	.577	75°	.966	.259	3.732
31°	.515	.857	.601	76°	.970	.242	4.011
32°	.530	.848	.625	77°	.974	.225	4.331
33°	.545	.839	.649	78°	.978	.208	4.705
34°	.559	.829	.675	79°	.982	.191	5.145
35°	.574	.819	.700	80°	.985	.174	5.671
36°	.588	.809	.727	81°	.988	.156	6.314
37°	.602	.799	.754	82°	.990	.139	7.115
38°	.616	.788	.781	83°	.993	.122	8.144
39°	.629	.777	.810	84°	.995	.105	9.514
40°	.643	.766	.839	85°	.996	.087	11.43
41°	.656	.755	.869	86°	.998	.070	14.30
42°	.669	.743	.900	87°	.999	.052	19.08
43°	.682	.731	.933	88°	.999	.035	28.64
44°	.695	.719	.966	89°	1.000	.018	57.29
45°	.707	.707	1.000	90°	1.000	.000	∞

ANSWERS TO ODD-NUMBERED PROBLEMS

NOTE: The data given in the problems should be assumed correct to three significant figures (i.e., 2 cm implies 2.00 cm) and, in general, your answers should be carried out to three significant figures or "slide rule accuracy." Only two significant figures are given in the answers to most of the problems. The purpose of the answers is not to give you a figure which should be checked exactly, but to let you know if you are on the right track.

CHAPTER 1

1–3. (a) 19 lb, (b) 9.2 lb.

1–5. 47 lb in the negative Y-direction.

1–7. (a) 7 lb, 2.9 lb, (b) 7.6 lb, (d) 11 lb.

1–9. $R = 0$.

CHAPTER 2

2–5. (a) 150 lb in A, 180 lb in B, 200 lb in C.
(b) 200 lb in A, 280 lb in B, 200 lb in C.
(c) 550 lb in A, 670 lb in B, 200 lb in C.
(d) 170 lb in A, 58 lb in B, 130 lb in C.

2–7. (a) Parts (b) and (c) can be solved.
(b) In part (a) another side or angle is necessary.

2–9. (a) 7.5 lb, (b) 94 lb, (c) 310 lb.

2–11. 1400 lb.

2–13. (a) Zero, (b) 5 lb, (c) 8 lb, (d) 4 lb, (e) 4 lb.

2–15. $\mu w/(\cos \phi + \mu \sin \phi)$.

2–17. Sliding down, 53°; sliding up, 37°.

2–19. (b) 10 lb, (c) 30 lb.

2–21. $w\,\dfrac{\sin \beta + \mu \cos \beta}{\cos \alpha + \mu \sin \alpha}$

2–23. (a) $F = w/(2 \sin \theta)$,
(b) $T = (w \cot \theta)/2$.

CHAPTER 3

3–1. The resultant of a couple is zero, but zero force does not produce the same effect as a couple.

3–3. (a) $F_x = -6$ lb, $F_y = 10$ lb.
(b) 5/3.
(c) 12 lb.
(d) 2 ft from right end of bar.

3–5. (a) $F_D = 64$ lb, $F_E = 36$ lb.
(b) $V_A = 70$ lb, $V_B = 30$ lb.
(c) At point A, 90 lb to right, 70 lb down.

At point C, 90 lb to left, 34 lb up.
At point E, 36 lb up.

3–7. 1200 lb.

3–9. (a) 16 lb, (b) 52 lb, 28 lb.

3–11. Each guide exerts 200 lb.

3–13. $T_1 = 6$ tons, $T_2 = 7.4$ tons.
$F_1 = 6$ tons, $F_2 = 8.2$ tons.

3–15. 663 lb.

547

3–17. $R = 0$.

3–21. 7.3 ft from small end.

3–23. (a) 1, (b) 45°.

3–25. (b) 86 lb at 21° with vertical.

3–27. (a) 15 lb.

(b) 5 lb on each front leg, 20 lb on each rear leg.
(c) 8.8 lb on each front leg, 16 lb on each rear leg.
(d) 3.3 ft above floor.

3–29. 24,000 lb total on front wheels, 27,000 lb total on rear wheels.

CHAPTER 4

4–1. (a) 0.0037 mi/sec.
(b) 13 mi/hr.
(c) 600 cm/sec.
(d) 20 ft/sec.

4–3. (a) 5 cm/sec, −4 cm/sec.
(b) $8 - 6t - 3\,\Delta t$ cm/sec.
(c) $8 - 6t$ cm/sec.
(d) $t = 4.3$ sec.
(e) 8 cm/sec².

4–5. (a) $+1.5$ ft/sec² right.
(b) -1.5 ft/sec² left.
(c) -1.5 ft/sec² left.
(d) $+1.5$ ft/sec² right.
(e) -4 ft/sec² left.
(f) 4 ft/sec² right.
(g) only in (b) and (d).

4–7. 80 ft/sec.

4–9. (a) 91 ft, (b) 300 ft.

4–11. (a) 24 ft/sec.   (c) 0.38 sec.
(b) 0.75 sec.   (d) 6.8 ft.

4–13. (a) 94 ft/sec.   (b) 120 ft.
(c) 53 ft/sec.   (d) 150 ft/sec².
(e) 2 sec.   (f) 93 ft/sec.

4–15. (a) $v_0 = 48$ ft/sec.
(b) 36 ft.
(c) 80 ft/sec.

4–17. (a) 4 m/sec².
(b) 6 m/sec.
(c) 4.5 m.

4–19. $v = v_0 + 8t$
$x = x_0 + v_0 t + 4t^2$
Initial velocity and displacement.

4–21. (a) 8 ft/sec.
(b) At $t = \pm 2$ sec.
(c) -32 ft/sec².

4–23. (a) $x_1 = 10 + 20t + 8t^2 - t^3$.
(b) $x_2 = 10t + 8t^2 - t^3$.
(c) $v_{12} = 10 + 16t - 3t^2$.
$a_{12} = 16 - 6t$.

4–25. (a) 4 ft.
(b) -8 ft/sec².
(c) $t = 2$ sec.
(d) Moves in positive direction with decreasing speed until $t = 2$ sec, moves in negative direction with increasing speed thereafter.

4–27. $v = \sqrt{4x + 6x^2 + 100}$.

4–29. Man in rowboat, 40 min; man walking, 30 min.

4–31. (a) 0, 10 mph westward.
(b) 17 mph down; 20 mph at 30° W of vertical.

CHAPTER 5

5–1. 980 lb.

5–3. (a) 2.5 cm, 1 cm/sec.
(b) 20 cm, 2 cm/sec.

5–5. (a) 1.6×10^{-10} dyne.
(b) 3.3×10^{-9} sec.
(c) 1.8×10^{17} cm/sec².

5–7. 33° W of N, 1.1 meters/sec².

5–9. (a) 16 ft/sec² downward.
(b) $a = 0$.
(c) 16 ft/sec² upward.

5–11. (a) 59 newtons.
(b) velocity = constant.
(c) 6.3 m/sec.

5–13. (b) $t = 40$ sec.
    (c) 50 cm.
    (d) 250 cm.

5–15. (a) 7.2 newtons.
    (b) 82 meters.
    (c) 41 m/sec.

5–17. 240 ft.

5–19. (a) 630 ft, (b) 1000 lb.

5–21. (c) Friction force $= 15$ lb.

5–23. 0.44.

5–25. $a = g(\sin \theta - \tan \alpha \cos \theta)$.

5–27. (a) 37°.
    (b) 6.4 ft/sec^2.
    (c) 2.5 sec.

5–29. (a) 1 lb, (b) 1.03 lb.

5–31. 127.5 lb.

5–33. (a) 0.83 sec, (b) 59 newtons.

5–35. (a) 200 cm/sec^2.
    (b) 310,000 dynes.

5–37. (a) to left, (b) 2.1 ft/sec^2,
    (c) 57 lb.

5–39. $2m_2g/(4m_1 + m_2)$;
    $m_2g/(4m_1 + m_2)$.

5–41. (a) 16 ft/sec^2.
    (b) 27 lb.
    (c) 21 lb.

5–43.

	a_1	a_2
(a)	0	0
(b)	0	0
(c)	0	16 ft/sec^2.
(d)	4 ft/sec^2.	28 ft/sec^2.
(e)	16 ft/sec^2.	48 ft/sec^2.

CHAPTER 6

6–1. 180 ft/sec.

6–3. (a) 0.44 ft, (b) 1.8 ft, (c) 4 ft,
    (d) 16 ft.

6–5. $h = 5000$ ft, $t = 18$ sec, $R = 5200$
    ft, $v = 640$ ft/sec, $\theta = 63°$.

6–7. $\tan \theta = 2v^2/gh$.

6–9. (a) 40 sec, (b) 55°.

6–11. 17 ft/sec.

6–13. (a) 200 ft, (b) 170 ft/sec.

6–15. (a) 90 cm, (b) horizontally.

6–17. (a) 9.5°, 81°.
    (b) 38 ft, 3.1 sec, 1400 ft, 19 sec.

6–19. (a) 7 sec.
    (b) $x = 670$ ft, $y = 110$ ft.

6–21. (a) 670 ft/sec.
    (b) 2700 ft.
    (c) horizontal component $= 530$
    ft/sec.
    vertical component $= 560$
    ft/sec.

CHAPTER 7

7–1. No. For example, see Problem
    7–5.

7–3. If masses are arranged in nu-
    merical order in the clockwise di-
    rection with the 10-gm mass at
    the origin, $\bar{x} = 14$ cm, $\bar{y} = 10$ cm.

7–5. 44 cm from end nearest 20-gm
    weight.

7–7. 3000 mi from earth's center.

7–9. $\dfrac{L}{3}\left(\dfrac{3\rho_0 + 2aL}{2\rho_0 + aL}\right)$

7–11. 2 ft (plank moves 10 ft).

7–13. (a) 40 ft, (b) 64 ft from bldg.

7–15. Throw out some baseballs. If no
    external force is acting, the cen-
    ter of mass continues to move in
    a straight line.

7–17. $\frac{3}{4}$ of the longest dimension from
    the force F.

7–19. (a) 14 ft/sec^2.
    (b) On each front leg, 26 lb.
    On each rear leg, 5.7 lb.

7–21. 73 ft, 1.7.

7–23. 1800 lb tension, 450 lb.

7–25. (a) On axis, 4 cm from base.
(b) 98,000 dynes.

7–27. (a) $T = m_1 m_2 g/(m_1 + m_2)$.
(b) $y = l(m_1 + m_2)/2m_2$.

7–29. (a) 64 lb.
(b) 96 ft/sec^2.
(c) 32 ft/sec^2.

CHAPTER 8

8–1. 63×10^6 ft·lb.

8–3. 470 ft·lb.

8–5. 350 joules.

8–7. $+6$ lb against force, $+48$ lb against force, 23 ft·lb.

8–9. 3×10^4 cm/sec.

8–11. 4.5×10^{-10} erg, 4.5×10^{-17} joule.

8–13. 130 ft·lb, 33 ft·lb, 2.1 ft·lb.

8–15. 160 ft·lb.

8–17. (a) 3300 ft·lb.
(b) 1300 ft·lb.
(c) 1500 ft·lb.
(d) 500 ft·lb. Goes into heat.
(e) $b + c + d = a$.

8–19. 80,000 lb.

8–21. 0.25.

8–23. 10 cm.

8–25. (a) 5000 ft, 10,000 ft.
(c) 2500 ft.

8–27. (a) 16 ft/sec.
(b) 0.25 ft.
(c) 50 ft·lb.

8–29. 6%.

8–31. (a) 8 ft/sec, (b) 28 ft/sec^2.

8–33. 1100 hp.

8–35. (a) 10^3 ft/sec.
(b) 10^6 ft·lb.
(c) 1.06×10^6 ft·lb.

8–37. 150 hp.

8–39. Three dollars.

8–41. (a) 550,000 ft·lb.
(b) 130 ft/sec.

8–43. 73 ft/sec.

8–45. 67 lb tension in A.

CHAPTER 9

9–1. (a) 28,000 slug·ft/sec.
(b) 60 mi/hr.
(c) 43 mi/hr.

9–3. (a) 8×10^5 m/sec^2.
(b) 4×10^4 newtons.
(c) 5×10^{-4} sec.
(d) 20 newton·sec.

9–5. (a) 28,000 slug·ft/sec.
(b) 15 mph.
(c) 300,000 ft·lb.

9–7. (a) Zero.
(b) 4 m/sec to the left, 1.5 m/sec to the left.
(c) 33 joules.

9–9. 4.2 lb.

9–11. 0.65 mi/hr.

9–13. 280 meters/sec. Note that initial KE of bullet is *not* equal to final PE of pendulum.

9–15. (a) 0.16.
(b) 240 joules.
(c) 0.32 joules.

9–17. (a) 5×10^6 ergs.
(b) 100 cm/sec.
(c) 10^4 cm/sec.

9–19. (a) $5mv^2/11$.
(b) 0.33.

9–21. (a) 10 ft/sec.
(b) 11 ft/sec.
(c) 8 ft/sec.

9–23. (a) 10 sec.
(b) 8-lb fragment: 100 ft/sec; 2-lb fragment: 400 ft/sec.

(c) 25 lb·sec.
(d) 25,000 lb.
(e) 335 ft/sec.

9–25. (a) 0.7
(b) 2 ft·lb.

9–27. (a) 41 cm, $h_n = 100 (0.64)^n$ cm,
10.
(b) 0.72 sec, $t_n = (0.9)(0.8)^n$.
(c) 0.8

9–29. (a) $v_1 = \sqrt{2gL}$.

(b) $v_2 = \sqrt{2gL \cos \theta}$.

(c) $V = \dfrac{m}{M} (\sqrt{2gL})$
$\times (1 - \sqrt{\cos \theta})$.

9–31. 3×10^{-4} gm.

CHAPTER 10

10–1. (a) 1.5 radians.
(b) 1.57 radians, 90°.
(c) 120 cm, 120 ft.

10–3. (a) 20 ft/sec.
(b) 230 rpm.

10–5. 5 rad/sec², 1000 rad.

10–7. 20 rad/sec².

10–9. (b) $a_T = 0$, $a_R = \omega^2 R$.
(c) Yes, $a_T = R\alpha$; Yes, $a_R = R\omega^2$.

10–11. (a) 40π rad/sec.
(b) 130π rad.
(c) 600π in/sec.
(d) 14,000 in/sec².

10–13. (a) 0, 24 rad/sec².
(b) 20 rad.

10–15. (a) 13.2 ft/sec² at 54° to original direction.
(b) 13.9 ft/sec² at 72° to original direction.
(c) 14 ft/sec² at 81° to original direction.
(d) 14.1 ft/sec² at 90° to original direction.

10–17. 160,000 rpm.

10–19. (a) 135° from top.
(b) 2.4 ft/sec², 11 ft/sec².

10–21. (a) 28 ft/sec².
(b) 19 ft/sec².

10–23. $\dfrac{mg + \mu Mg}{M\omega^2}$.

10–25. (a) 130 ft/sec².
(b) 480 lb.
(c) 480 lb.
(d) 16 ft/sec.

10–29. $v = \sqrt{gbR/2h}$.

10–31. (a) $\theta = 11°$.
(b) 6.2 × (mass of plane in slugs).
(c) 32.8m (using $g = 32.2$ ft/sec²).

10–33. (a) Friction force upward to equal mg. Normal force inward.
(b) 40 ft/sec.
(c) 320 lb, 800 lb.

10–35. (a) 10 tons downward.
(b) 2 tons upward.
(c) Safety factor, 2.

10–37. (a) 15 lb, (b) 5 lb.

10–39. 13 oz.

10–41. (a) 4400 lb·ft.
(b) 5900 lb.
(c) 190 ft/sec.

10–43. (a) 16 hp.
(b) 280 lb·ft, 840 lb·ft.
(c) 240 lb, 720 lb.

CHAPTER 11

11–1. (a) $2L/3$ (b) $3mL^2/2$
(c) $mL^2/6$ (d) $1.2L$.

11–3. $7ML^2/48$.

11–5. 0.29 ft.

11–7. (a) Smallest: (A); Largest: (D). (More mass is further from axis.)

(b) $4MR^2/3$.

11–9. $13MR^2/24$.

11–11. $MR^2/4$.

11–13. (a) 4 rad/sec².
  (b) $\omega = 14$ rad/sec.
  (c) 32,000 ft·lb.

11–15. 4.2 lb.

11–17. (a) 17 kgm·m².
  (b) -1.8 newton·m.
  (c) 92 rev.

11–19. (a) 9.8×10^5 gm·cm².
  (b) 1.1 rad/sec.

11–21. (a) 99 sec.
  (b) 4100 rev.

11–23. (a) 21 lb tension.
  (b) 52 ft/sec.
  (c) 2.5 sec.

11–25. (a) 5.7 ft/sec.
  (b) 16 lb·ft.
  (c) Remains constant.

11–27. (a) 240 cm/sec, 320 cm/sec.
  (b) 5×10^5 gram·cm².
  (c) 1600 cm/sec².
  (d) 780 cm/sec².
  (e) 51,000 dynes tension.

11–29. (a) $\omega = 2\sqrt{g/3R}$.
  (b) $4gM/3$.

11–31. (a) $2g/3$.
  (b) $\omega = 2\sqrt{g/3R}$.

11–33. $I = I_0 + Ma^2$ is a minimum when $a = 0$.

CHAPTER 12

12–1. (a) 60 ft/sec.
  (b) Velocity is zero.
  (c) 120 ft/sec in same direction.
  (d) 85 ft/sec at 45° below horizontal.
  (e) 6.7 ft/sec opposite to direction of locomotive.
  (f) Points 3 ft from instantaneous axis, direction of velocity is 60° to horizontal.

12–3. (a) $Mv^2(1 + k_0^2/R^2)/2$.
  (b) $Mv^2(k/R)^2/2$.

12–5. (a) 50 lb.
  (b) 60 lb.

12–7. (a) 16 ft/sec².
  (b) 16 rad/sec².
  (c) 2π ft.

12–9. (a) 0.25.
  (b) 100 cm.

12–11. (a) 1.2.
  (b) 1 (including rotational kinetic energy).

12–13. (a) Force of gravity and normal force, both perpendicular to surface.
  (b) 2 sec.
  (c) 160 cm.

12–15. 15 min.

12–17. $a_R = 20g/7$, $a_T = 5g/7$.

12–19. (a) 330 cm/sec.
  (b) $K_T = 5.6 \times 10^6$ ergs.
    $K_R = 2.2 \times 10^6$ ergs.
  (c) 79 cm.

12–21. (a) $v = \sqrt{10gh/7}$.
  (b) $\sqrt{10gh/7}/r$.
  (c) $h = 7(R + r)/17$.

12–23. (a) 88.4 cm/sec², 44.2 rad/sec².
  (b) 9.8×10^6 ergs.
  (c) 2.22 cm/sec.

12–25. 2 ft/sec², 2.5 lb.

12–27. (a) 24 slug·ft²/sec.
  (b) 6 rad/sec.
  (c) 24 ft·lb, 72 ft·lb. The difference comes from the work done in pulling in the dumbbells.

12–29. (a) 6 rpm, (b) 6 rpm.

12–31. -0.04 rad/sec, 60°, 72°.

12–33. (a) Zero, (b) 2 rad/sec counterclockwise.

12–35. (a) -1000 dyne·sec, $1.5 + 10^4$ ergs.
  (b) 2×10^4 ergs.

(c) In the first case the pivot exerts an impulse, so the *total* impulse is *not* the same in the two cases.

12–37. (a) 40,000 dyne·sec.
  (b) 3 rad/sec.

12–39. (a) 180,000 dynes.
  (b) 4300 rpm.

CHAPTER 13

13–1. 25×10^6 lb/in^2.

13–3. 1.4×10^7 lb/in^2, 1.6×10^4 lb/in^2.

13–5. (a) 1.8 ft.
  (b) 30,000 lb/in^2 in steel, 12,000 lb/in^2 in copper.
  (c) 0.001 in steel, 0.0006 in copper.

13–7. 0.028 inch.

13–9. $S/2$.

13–11. 2000 lb.

13–13. 64.0071 lb/ft^3.

13–15. -6×10^{-4} (decrease), 7×10^{-4} (increase).

13–17. (a) 0.0033 edges parallel to F.

0.00093 edge perpendicular to F.
  (b) 0.0015.

13–19. (a) Along Z-axis, $F(1 + \sigma)/YL^2$.
  Along Y-axis, $-F(1 + \sigma)/YL^2$.
  Along X-axis, zero.
  (b) Zero.

13–21. (a) Shearing stress in A is 2400 lb/in^2; in B, zero.
  (b) 900-lb tension in A.
  (c) 900-lb compression in B.

13–23. 53 newton-meters.

13–25. 0.005.

13–27. (a) 18 lb·ft.
  (b) $0.037°$ or 6.7×10^{-4} radian.

CHAPTER 14

14–3. (a) $x = 10 + 5 \sin \pi t$.
  (b) $F = -75\pi^2 \sin \pi t$.
  (c) 10 cm.
  (d) 5, 15 cm.

14–5. (a) 2.5 cm.
  (b) 1.3 cm.
  (c) 4.3 cm below equilibrium position of platform and 0.24 sec after it leaves platform.

14–7. (a) 3.5 sec^{-1}.
  (b) 160,000 ergs.

14–9. $A_1 > A_2$.

14–11. (a) 35 cm, 25 cm.
  (b) 1 sec.

14–13. 979.8 cm/sec^2.

14–15. (a) 40 rad/sec.
  (b) 34 rad/sec.
  (c) 120 rad/sec^2 clockwise.

14–17. (a) 9 vib/sec.
  (b) 20×10^6 lb/in^2.

14–19. (a) 29 cm.
  (b) 1.5 sec.

14–21. (a) 6.9 in.
  (b) 70 rad/sec.

14–23. Torsion pendulum, 60 vibrations; other pendulums, 40 vibrations.

14–25. 99 cm below pivot.

14–27. (a) $(8\pi \sqrt{L/g})/3$.
  (b) $(2\pi \sqrt{17L/g})/3$.

CHAPTER 15

15–1. (a) 1.9×10^{16} tons.
(b) 450 miles.

15–3. (a) 5×10^{-6} cm/sec^2 along perpendicular bisector of line joining A and B.
(b) 6.5×10^{-3} cm/sec.

15–5. $G_A = 13\gamma$ dynes/gm at a 29° angle with the line joining A with the solid sphere.
$G_B = 2.3\gamma$ dynes/gm toward solid sphere.
$V_A = -620\gamma$ erg/gm.
$V_B = -5100\gamma$ erg/gm.

15–7. Pull due to sun is 2.4 times as great as pull due to earth.

15–9. (a) 1.7×10^3 m/sec.
(b) 0.87 m.

(c) 22 ft.
(d) 2.4 sec.

15–11. $v = \sqrt{2ghR/(R+h)}$, reducing to $\sqrt{2gh}$ when $h << R$.

15–13. $x = 1.4R$.

15–15. (a) 25γ dynes.
(b) 1.1γ at 27° with line AD.
(c) -15γ erg.
(d) The same as it was at point B, since the potential at D is the same as at B.

15–17. 1.8×10^{27} kgm.

15–19. $G_P = \gamma M/(L+A)A$, reducing to $\gamma M/A^2$ when $A >> L$.

15–21. (a) $F = \gamma mm'x/R^3$ if $x << R$.
(b) $f = \sqrt{\gamma m/R^3}/2\pi$.

CHAPTER 16

16–1. 21 lb/in^2.

16–3. 120 lb/in^2, 18,000 lb/ft^2.

16–5. 270,000 ft^3, 9.3 tons lift using helium.

16–7. (a) 4500 lb.
(b) 10^4 lb.
(c) 230 lb.

16–9. (a) 100 lb/ft^3.
(b) E will read 5 lb, D will read 15 lb.

16–11. 0.78 gm/cm^3.

16–13. (b) 4 lb.
(c) 1 ft^3.

16–15. 62,000 lb·ft.

16–17. (a) 42×10^6 lb·ft.
(b) $\tau_g > \tau$ and the dam is stable.

16–19. 14,000 dynes/cm^2.

16–21. (a) 70.7 cm of mercury.
(b) 71.2 cm of mercury.
(c) 11 cm.

16–23. 4.3 cm.

16–25. 3.8 cm.

16–27. 40 dynes/cm^2.

16–31. $2\alpha/R$.

CHAPTER 17

17–1. 36 ft/sec, 0.2 ft^3/sec.

17–3. (a) 0.056 ft^3/sec.
(b) 3 ft below bottom.

17–5. (a) 10 cm.
(b) 11 sec.

17–7. (a) 0.2 ft^3/sec.
(b) 4.7 ft.
(c) 3 ft.

17–9. (a) 95 ft from a point directly below the hole.
(b) 21 lb.
(c) 120 lb.

17–11. (a) 2.5 sec.
(b) 6300 lb/ft^2 = 43 lb/in^2.
(c) 4700 lb/in^2.

17–13. $v_1 = \sqrt{2ghA_2^2/(A_2^2 - A_1^2)}$

17–15. 10 lb/in^2.

17–17. (a) 5 ft/sec in wide part, 20 ft/sec in narrow part.
(b) 2.5 lb/in^2.
(c) 0.47 ft.

17–19. A has largest and B has smallest.

17–21. (a) 0.083 ft^3/sec.
(b) −11 lb/in^2.

17–23. 1.6×10^6 cm·dynes.

17–25. 25.0 centipoises. SAE number not given.

17–27. (b) $2\rho g w a^3/3\eta$.

17–29. 0.03 cm.

17–31. (a) 0.33 cm/sec.
(b) 55 cm/sec.

CHAPTER 18

18–1. (a) 1800°F, (b) 7.8×10^{-6} per F°, (c) 42×10^{-6} per C°, (d) −40°F = −40°C.

18–3. 1.6 ft.

18–5. 0.2506 inch.

18–7. 2×10^{-5} per centigrade degree.

18–9. 0.14.

18–11. (a) 1.2×10^{-4}.
(b) 5.2 sec per day gained.

18–13. 0.63 cm^3.

18–15. 180,000 lb/in^2 or 12×10^9 dynes/cm^2.

18–17. Tension in steel, 72,000 lb/in^2 (assuming the area of the brass bar is so large its length is relatively unaffected by the tensile stress).

18–19. 480 atm.

18–21. 70°C.

18–23. 9.8×10^8 dynes/cm^2.

CHAPTER 19

19–1. 74 ft^3.

19–3. 400 Btu.

19–5. 88 watts.

19–7. 20 C°.

19–9. 1.0, 0.83, 0.35.

19–11. 700°F.

19–13. (a) $m(at_1 + bt_1^3/3)$.
(b) $(a + bt_1^2/3)$.
(c) $(a + bt_1^2/4)$.

19–15. (a) 1.1×10^{-2} ft^3.
(b) 650 ft·lb.
(c) 1800 Btu.
(d) 1.4×10^6 ft·lb (the work done by the oil is negligible).

CHAPTER 20

20–1. 86,000 cal/day.

20–3. 1.1 cal/sec, 4% through steel, 96% through copper.

20–5. (a) 320 cal/sec, (b) 310 cal/sec.

20–7. (a) 3.1% higher in hot arm.
(b) 3.0×10^4 dynes/cm^2.

20–9. (a) 40°C, (b) 2.4 cal/sec.

20–11. (a) $dH = x\,dx$ (b) 200 cal/sec.

20–13. (a)
$$H = \frac{K_0 A[(t_2 - t_1) + a(t_2^2 - t_1^2)/2]}{L}$$
(b) 62°C.

CHAPTER 21

21–3. 0°C with 0.2 gm ice left.

21–5. 24°C.

21–7. 40°C.

21–9. 1.8 kgm.

21–11. 110°C.

21–13. 0.04 lb.

21–15. (a) 57,000 ft·lb.
(b) 900 Btu.

21–17. 360 meters/sec.

CHAPTER 22

22–1. 3.6 lb/in^2.

22–3. (a) 0.75 atm, (b) 2 gm.

22–5. (a) 7 ft, (b) 100 lb/in^2.

22–7. (a) 0.26 cm, (b) 0.29 cm.

22–9. (a) Zero, (b) 4200 cal,
(c) 18,000 joules, (d) 230
liters.

22–11. 270°C.

22–13. (a) 150°K, 0.62 atm, 4.9 liter·
atm, zero heat absorbed,
−4.9 liter·atm.
(b) 300°K, 1.2 atm, zero, zero,
zero.

22–15. 11 ft^3 (substance is not an ideal
gas).

22–17.

Process	W
ab	33 lit·am
bc	−46 lit·atm
ca	0
	Q
ab	2000 cal
bc	−1100 cal
ca	−1200 cal
	ΔU
ab	1200 cal
bc	0
ca	−1200 cal

22–19. (a) ab is adiabatic (steeper
slope).
(b) $P_b = P_c = 2$ atm.
(c) $T_b = 300°K$, $T_c = 600°K$.
(d) $V_c = 8$ liters.

22–21. 0.026 atm^{-1}

22–23. (a) 50 ft/sec, (b) 36 ft/sec.

CHAPTER 23

23–3. (a) 11°C, (b) 130 lb/in^2.

23–5. 140 atm.

23–7. 68 cm of mercury.

23–11. (a) 25 liters.
(b) 6 liters.
(c) 54 gm SO_2 condensed.

23–13. (a) 37%.
(b) 6.5 mm of mercury.
(c) 6.4 gm/m^3.

23–15. 16%.

23–17. 13 lb/hr.

CHAPTER 24

24–1. 65%.

24–3. 93 C°.

24–7. 15%.

24–9. (a) 32 Btu/F° (b) 180 Btu/F°
(c) −29 Btu/F°.

24–11. (a) 0°C.
(b) 5.9 cal/°K.

24–13. (a) 0.28 cal/°K, −0.28 cal/°K,
zero.
(b) zero, zero, zero.

CHAPTER 25

25–1. 3.3×10^{-5} cm.

25–3. 1700 ft/sec.

25–5. (a) 7.8×10^{-7} atm.
(b) 2500 collisions/sec.

25–7. η varies at $\sqrt{T}$ but is independent of pressure.

25–9. (a) 4.9×10^{4} cm/sec.
(b) 2.2×10^{23} impacts/cm^2, per second.

25–11. (a) 12 liter·atm, (b) 800°K,
(c) 1200 calories.

CHAPTER 26

26–5. 2 cm, 30 cm, 100 vib/sec, 3000 cm/sec.

26–7. (a) Amplitude, 10^{-7} ft; wave length, 10 ft; frequency, $1700\,\mathrm{sec}^{-1}$; velocity, 17,000 ft/sec.
(b) $-2\pi \times 10^{-8} \cos 3400\,\pi\,(t - x/17{,}000)$; $2\pi \times 10^{-8}$; 280 lb/ft^2 (Y for this bar must be 4.4×10^{9} lb/ft^2).
(c) $3.4\pi \times 10^{-4} \times \cos 3400\pi (t - x/17{,}000)$; $3.4\pi \times 10^{-4}$ ft/sec.

26–9. 320 meters/sec.

26–11. (a) 17 m, 0.017 m.
(b) 73 m, 0.073 m.

26–13. (a) 0.57 m/sec·C°.
(b) 1 ft/sec·F°.
(c) dV/dT varies as $1/\sqrt{T}$.

26–15. (a) 320 m/sec, (b) 1300 m/sec, 350 m/sec.

26–17. (a) 7.1×10^{-7} cm.
(b) 7.1×10^{-3} cm.
(c) 2.2×10^{-6} cm.

CHAPTER 27

27–3. (a) Traveling, (b) 4 cm,
(c) 600 cm,
(d) 20,000 cm/sec, (e) 33 cycles/sec, (f) 840 cm/sec.

27–5. (a) 4×10^{4} cm/sec, (b) 1.3×10^{10} dynes/cm^2.
(c) 0.05 cm.

27–7. (a) 200 cycles/sec, (b) 49th overtone.

27–9. (c) $y = 0.71 \times 10^{-6}$ cm.

27–11. (a) 1100, 2300, 3400, 4500, 5700 cycles/sec.
(b) 570, 1700, 2800, 4000, 5100 cycles/sec.
(c) 16, 17.

27–13. (a) $170\,\mathrm{sec}^{-1}$, (b) $660\,\mathrm{sec}^{-1}$.

27–15. (a) 420 cycles/sec, (b) 415 cycles/sec, (c) 5 cycles/sec.

27–17. (a) 8.8×10^{11} dynes/cm^2.
(b) 0.24 cm.

CHAPTER 28

28–1. (a) 9 times, (b) 4 times.

28–3. (a) 60 db, (b) 77 db.

28–5. (a) π radians, (b) due to A: 4×10^{-10} watt/cm; due to B: 12×10^{-10} watt/cm.
(c) 2.1×10^{-10} watt/cm.

28–7. 10^{-6} watt.

28–9. (a) 454 cycles/sec, (b) 462 cycles/sec,
(c) 8 cycles/sec.

28–11. (a) 11 ft, (b) 1 ft, (c) almost 15 waves,
(d) 1100 ft/sec, (e) 0.49 ft.

28–13. (a) 2.2×10^{-2} ft, 9.5×10^{-2} ft.
(b) 1.1×10^{-3} ft, 4.7×10^{-3} ft.

INDEX

564 INDEX